国际材料前沿丛书
International Materials Frontier Series

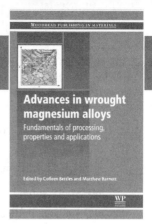

Colleen Bettles,
Matthew Barnett 编著

变形镁合金研究进展：加工原理、性能和应用

Advances in Wrought Magnesium Alloys:
Fundamentals of Processing,
Properties and Applications

影印版

中南大学出版社
www.csupress.com.cn

·长沙·

国字：18－2017－167

Advances in Wrought Magnesium Alloys: Fundamentals of Processing, Properties and Applications

Colleen Bettles, Matthew Barnett

ISBN：9781845699680

Copyright © 2012 by Elsevier Ltd. All rights reserved.

Authorized English language reprint edition published by the Proprietor.

Copyright © 2017 by Elsevier (Singapore) Pte Ltd. All rights reserved.

Elsevier (Singapore) Pte Ltd.
3 Killiney Road
#08－01 Winsland House I
Singapore 239519
Tel：(65) 6349－0200
Fax：(65) 6733－1817

First Published ＜2017＞

＜2017＞年初版

Printed in China by Central South University Press under special arrangement with Elsevier (Singapore) Pte Ltd. This edition is authorized for sale in China only, excluding Hong Kong SAR, Macao SAR and Taiwan. Unauthorized export of this edition is a violation of the Copyright Act. Violation of this Law is subject to Civil and Criminal Penalties.

本书英文影印版由Elsevier (Singapore) Pte Ltd.授权中南大学出版社在中国境内独家发行。本版仅限在中国境内(不包括香港、澳门以及台湾)出版及标价销售。未经许可之出口，视为违反著作权法，将受民事及刑事法律之制裁。

本书封底贴有Elsevier防伪标签，无标签者不得销售。

内容简介

本书介绍了变形镁合金的类型与性能、加工工艺以及应用。总结了镁合金的加工过程对其性能的影响，讨论了可用于开发新一代高性能应用合金的方法，介绍了铸造、挤压、轧制和锻造技术的主要变形行为以及锻造镁合金在汽车和生物医学工程中的应用。该书作者 Colleen Bettles 教授为澳大利亚莫纳什大学轻金属设计优化 ARC 中心副主任，Matthew Barnett 为墨尔本迪肯大学教授。该书为变形镁合金领域的必备参考书，适用于镁合金材料技术、研究人员以及相关本科生、研究生使用。

序

"最早的商用合金含有铝、锌、锰、硅和铈,这些元素仍然是当今绝大多数商用二元和三元合金的主要成分。"1940年W. Buchmann这样写道,当时全世界镁的产量正呈上升趋势。70多年后,这些话大体上还是没有错的。

70多年来,镁的产量波动明显,但最近镁及镁合金在越来越多的主流应用方面已经复苏。传统上,铸造一直是镁合金占主导地位的加工路线,但在过去十年中变形镁合金已在许多结构应用中变得重要。

市场对镁合金的最终性能要求越来越高,而希望价格越来越低。变形镁合金产品的未来将高度依赖巧妙的合金设计(结合化学和镁合金基本变形行为的理解)和优化的二次加工阶段,例如轧制、锻造和挤压。

本书提供了一个独特的机会,让来自众多不同学科的专家同心协力,推动变形镁合金成功实现在非尖端领域的广泛应用。这本书分为3个部分,首先介绍最新发展趋势,我们可能最后会脱离W. Buchmann的观点,接下来探讨变形行为的最重要的基本原理。这些章节是优化加工步骤的基础,且第2部分着重阐述大量不同加工工艺。结论章节讨论未来变形镁合金产品的应用。

我们要感谢本书的所有撰稿人,他们为本书付出了辛勤劳动,作出了专业贡献。我们还要感谢伍德海德出版社的编辑人员,他们提出出版变形镁合金著作的建议并将相关内容汇集,使本书出版成为可能。

目 录

作者联系方式 ·················· ix

绪论 ···························· xiii

第一部分　镁合金的类型与性能

1　变形镁合金的研究和开发 ········ 3
 1.1　变形镁合金发展概况 ········ 3
 1.2　变形镁合金发展现状 ········ 6
 1.3　变形镁合金研究进展 ······ 31
 1.4　变形镁合金发展趋势 ······ 55
 1.5　参考文献 ·················· 59

2　镁合金的变形机制 ············ 63
 2.1　引言 ······················ 63
 2.2　基面滑移 ·················· 65
 2.3　〈a〉型位错非基面滑移 ··· 72
 2.4　〈c+a〉型位错非基面滑移 ···
  ························ 78
 2.5　形变孪生 ·················· 83
 2.6　动态形变 ·················· 90
 2.7　结论 ······················ 92
 2.8　致谢 ······················ 93
 2.9　参考文献 ·················· 93

3　孪生及其在变形镁合金中的作用
 ························ 105
 3.1　引言 ···················· 105
 3.2　孪生的基本原理 ·········· 107
 3.3　孪生对力学响应的影响
  ······················ 118
 3.4　结构和工艺对孪生的影响
  ······················ 125
 3.5　结论 ···················· 136
 3.6　致谢 ···················· 136
 3.7　参考文献 ················ 136

4　大塑性变形镁合金的超塑性
 ························ 144
 4.1　引言 ···················· 144
 4.2　热机械加工过程中的微观
  组织演变 ················ 147
 4.3　超塑性行为 ·············· 157
 4.4　超塑性变形中的断裂 ······ 168
 4.5　形变机制与模型 ·········· 175
 4.6　结论 ···················· 179
 4.7　参考文献 ················ 179

5　镁合金的动态再结晶 ·········· 186
 5.1　引言 ···················· 186
 5.2　镁合金的动态再结晶机制
  ······················ 187

1

5.3 原始组织对动态再结晶的影响 …… 207
5.4 不同镁合金中的动态再结晶 …… 211
5.5 大塑性变形过程中的动态再结晶 …… 215
5.6 结论 …… 218
5.7 致谢 …… 219
5.8 参考文献 …… 219

第二部分 镁合金的加工

6 镁合金挤压坯与轧制厚板的直接激冷铸造 …… 229
  6.1 引言 …… 229
  6.2 热流与流体流 …… 235
  6.3 镁合金直接激冷铸造技术与工程学 …… 245
  6.4 凝固组织与缺陷形成 …… 249
  6.5 结论 …… 261
  6.6 参考文献 …… 262

7 镁合金的双辊铸轧 …… 272
  7.1 引言 …… 272
  7.2 工业前景 …… 274
  7.3 双辊铸轧工艺 …… 276
  7.4 凝固与带状组织 …… 282
  7.5 热力学计算 …… 284
  7.6 过程模拟与仿真 …… 292
  7.7 结论 …… 300
  7.8 参考文献 …… 300

8 提高变形镁合金的可挤压性 …… 304
  8.1 引言 …… 304
  8.2 变形镁合金的可挤压性 …… 305
  8.3 挤压过程中的微观组织变化 …… 309
  8.4 镁合金挤压研究进展 …… 319
  8.5 结论 …… 321
  8.6 参考文献 …… 321

9 镁合金的静液挤压 …… 323
  9.1 引言 …… 323
  9.2 工艺基础 …… 324
  9.3 研究与开发 …… 333
  9.4 发展趋势 …… 342
  9.5 延伸阅读 …… 343
  9.6 结论 …… 343
  9.7 参考文献 …… 344

10 镁合金的轧制 …… 346
  10.1 引言 …… 346
  10.2 镁薄板的发展前景 …… 347
  10.3 薄板轧制工艺及其对性能的影响 …… 348
  10.4 合金化对薄板性能的影响 …… 354
  10.5 镁薄板的可成形性 …… 362
  10.6 使能技术与发展趋势 …… 365
  10.7 延伸阅读 …… 368
  10.8 参考文献 …… 370

11 镁合金的锻造技术 …… 376
  11.1 引言 …… 376

11.2 锻造技术 ………… 376
11.3 镁合金的锻造 ………… 378
11.4 镁合金的近净成形 …… 384
11.5 结论 …………………… 388
11.6 致谢 …………………… 389
11.7 参考文献 ……………… 389

第三部分 镁合金的应用

12 镁合金在汽车工程中的应用
………………………… 393

12.1 引言 …………………… 393
12.2 材料性能 ……………… 394
12.3 合金开发 ……………… 398
12.4 制造技术开发 ………… 408
12.5 镁合金在汽车中的应用
………………………… 413

12.6 发展趋势 ……………… 417
12.7 结论 …………………… 421
12.8 致谢 …………………… 421
12.9 参考文献 ……………… 422

13 镁合金在生物医学中的应用
………………………… 427

13.1 引言 …………………… 427
13.2 镁合金植入物的功能 … 429
13.3 心血管植入物 ………… 440
13.4 骨科和其他植入物 …… 445
13.5 发展趋势 ……………… 449
13.6 延伸阅读 ……………… 451
13.7 结论 …………………… 452
13.8 参考文献 ……………… 453

索引 ……………………… 455

Contents

Contributor contact details		*ix*
Introduction		*xiii*

Part I Types and properties of magnesium alloys — 1

1 Current developments in wrought magnesium alloys — 3
M. O. PEKGULERYUZ, McGill University, Canada

1.1	Introduction: overview of wrought magnesium alloy development	3
1.2	Current developments in magnesium wrought alloys	6
1.3	Progress in wrought magnesium alloys	31
1.4	Future trends in wrought magnesium alloy development	55
1.5	References	59

2 Deformation mechanisms of magnesium alloys — 63
S. R. AGNEW, University of Virginia, USA

2.1	Introduction	63
2.2	Basal slip	65
2.3	Non-basal slip of <a> type dislocations	72
2.4	Non-basal slip of <c+a> type dislocations	78
2.5	Deformation twinning	83
2.6	Dynamic deformation	90
2.7	Conclusions	92
2.8	Acknowledgements	93
2.9	References	93

3 Twinning and its role in wrought magnesium alloys — 105
M. R. BARNETT, Deakin University, Australia

3.1	Introduction	105
3.2	Fundamentals of twinning	107
3.3	The impact of twinning on mechanical response	118
3.4	The impact of structure and processing on twinning	125

3.5	Conclusion	136
3.6	Acknowledgements	136
3.7	References	136

4 Superplasticity in magnesium alloys by severe plasticity deformation 144
R. Lapovok and Y. Estrin, Monash University, Australia

4.1	Introduction	144
4.2	Microstructure evolution during thermomechanical processing	147
4.3	Superplastic behaviour	157
4.4	Fracture during superplastic deformation	168
4.5	Mechanisms and models	175
4.6	Conclusions	179
4.7	References	179

5 Dynamic recrystallization in magnesium alloys 186
R. Kaibyshev, Belgorod State University, Russia

5.1	Introduction	186
5.2	Dynamic recrystallization (DRX) mechanisms operating in magnesium alloys	187
5.3	Effect of initial structure on DRX	207
5.4	DRX in different magnesium alloys	211
5.5	DRX during severe plastic deformation	215
5.6	Conclusions	218
5.7	Acknowledgements	219
5.8	References	219

Part II Processing of magnesium alloys 227

6 Direct chill casting of magnesium extrusion billet and rolling slab 229
J. F. Grandfield, Grandfield Technology Pty Ltd, Australia

6.1	Introduction	229
6.2	Heat and fluid flow	235
6.3	Magnesium direct chill (DC) casting technology and engineering	245
6.4	Solidified structures and defect formation	249
6.5	Conclusions	261
6.6	References	262

7 Twin roll casting of magnesium 272
E. Essadiqi, CANMET Materials Technology Laboratory, Canada, I.-H. Jung, McGill University, Canada and M. A. Wells, University of Waterloo, Canada

7.1	Introduction	272
7.2	Industrial perspective	274

7.3	Twin roll casting (TRC) process	276
7.4	Solidification and strip microstructure	282
7.5	Thermodynamic calculations	284
7.6	Process modeling and simulation	292
7.7	Conclusions	300
7.8	References	300

8	**Enhancing the extrudability of wrought magnesium alloys**	304
	A. G. Beer, Deakin University, Australia	
8.1	Introduction	304
8.2	Extrudability of magnesium alloys	305
8.3	Microstructural development during extrusion	309
8.4	Recent extrusion alloy development	319
8.5	Conclusions	321
8.6	References	321

9	**Hydrostatic extrusion of magnesium alloys**	323
	W. H. Sillekens, TNO, The Netherlands and J. Bohlen, Helmholtz-Zentrum Geesthacht, Germany	
9.1	Introduction	323
9.2	Process basics	324
9.3	Research and development issues	333
9.4	Future trends	342
9.5	Sources of further information	343
9.6	Conclusions	343
9.7	References	344

10	**Rolling of magnesium alloys**	346
	J. Bohlen, G. Kurz, S. Yi and D. Letzig, Helmholtz-Zentrum Geesthacht, Germany	
10.1	Introduction	346
10.2	Potential of magnesium sheets	347
10.3	The sheet rolling process and its influence on sheet properties	348
10.4	Alloying effects on sheet properties	354
10.5	Formability of magnesium sheets	362
10.6	Enabling technologies and future trends	365
10.7	Sources of further information	368
10.8	References	370

11	**Forging technology for magnesium alloys**	376
	B.-A. Behrens, I. Pfeiffer and J. Knigge, Leibniz Universität Hannover, Germany	
11.1	Introduction	376
11.2	Forging technology	376

11.3	Forging of magnesium alloys		378
11.4	Near-net-shape forming of magnesium alloys		384
11.5	Conclusions		388
11.6	Acknowledgements		389
11.7	References		389

Part III Applications of magnesium alloys — 391

12	Applications of magnesium alloys in automotive engineering	393
	A. A. Luo and A. K. Sachdev, General Motors Global Research and Development, USA	
12.1	Introduction	393
12.2	Materials properties	394
12.3	Alloy development	398
12.4	Manufacturing process development	408
12.5	Automotive applications of magnesium alloys	413
12.6	Future trends	417
12.7	Conclusions	421
12.8	Acknowledgements	421
12.9	References	422

13	Biomedical applications of magnesium alloys	427
	W. H. Sillekens, TNO, The Netherlands and D. Bormann, Leibniz Universität Hannover, Germany	
13.1	Introduction	427
13.2	Functionality of magnesium implants	429
13.3	Cardiovascular implants	440
13.4	Orthopaedic and other implants	445
13.5	Future trends	449
13.6	Sources of further information	451
13.7	Conclusions	452
13.8	References	453

Index — *455*

Contributor contact details

(* = main contact)

Editors

C. J. Bettles*
ARC Centre of Excellence for
 Design in Light Metals
Department of Materials
 Engineering
Building 27, Clayton Campus
Monash University
Victoria 3800
Australia
Email: colleen.bettles@monash.edu

M. R. Barnett
ARC Centre of Excellence for
 Design in Light Metals
Institute for Frontier Materials
Deakin University
Geelong
Victoria 3217
Australia
Email: matthew.barnett@deakin.edu.au

Chapter 1

M. O. Pekguleryuz
Department of Mining and Materials
 Engineering
Wong Building, Rm2m090
McGill University
3610 University Street
Montreal
Quebec H3A 2B2
Canada
Email: mihriban.pekguleryuz@mcgill.ca

Chapter 2

S. R. Agnew
Department of Materials Science
 and Engineering
University of Virginia
Charlottesville
Virginia 22904-4745
USA
Email: agnew@virginia.edu

Chapter 3

M. R. Barnett
ARC Centre of Excellence for
 Design in Light Metals
Institute for Frontier Materials
Deakin University
Geelong
Victoria 3217
Australia
Email: matthew.barnett@deakin.edu.au

Chapter 4

R. Lapovok* and Y. Estrin
Centre for Advanced Hybrid
  Materials
Department of Materials
  Engineering
Monash University
Clayton
Victoria 3800
Australia
Email: Rimma.Lapovok@.monash.edu

Chapter 5

R. Kaibyshev
Belgorod State University
Pobeda 85
Belgorod 308015
Russia
Email: rustam_kaibyshev@bsu.edu.ru

Chapter 6

J. F. Grandfield
Grandfield Technology Pty Ltd
37 Mattingley Crescent
Brunswick West
Victoria 3055
Australia
Email: grandfieldtechnology@gmail.
  com

Chapter 7

E. Essadiqi*
CANMET Materials Technology
  Laboratory
183 Longwood Road South
Hamilton
Ontario L8P 0A1
Canada
Email: essadiqi@NRCan.gc.ca

I.-H. Pung
Mc Gill University, Canada

M. A. Wells
University of Waterloo, Canada

Chapter 8

A. G. Beer
CAST Cooperative Research Centre
Institute for Frontier Materials
Deakin University
Locked Bag 20000
Geelong
Victoria 3220
Australia
Email: aiden.beer@deakin.edu.au

Chapter 9

W. H. Sillekens*
TNO
P.O. Box 6235
5600 HE Eindhoven
The Netherlands
Email: wim.sillekens@tno.nl

J. Bohlen
Magnesium Innovation Centre
  MagIC
Helmholtz-Zentrum Geesthacht
Max-Planck-Str. 1
21502 Geesthacht
Germany
Email: jan.bohlen@hzg.de

Chapter 10

J. Bohlen*, G. Kurz, S. Yi and
  D. Letzig
Magnesium Innovation Centre
  MagIC

Helmholtz-Zentrum Geesthacht
Max-Planck-Str. 1
21502 Geesthacht
Germany
Email: jan.bohlen@hzg.de

Chapter 11

B.-A. Behrens, I. Pfeiffer*
and J. Knigge
Institute of Forming Technology and Machines
Leibniz Universität Hannover
An der Universität 2
30823 Garbsen
Germany
Email: pfeiffer@ifum.uni-hannover.de

Chapter 12

A. A. Luo* and A. K. Sachdev
Light Metals for Powertrain and Structural Subsystems Group
Chemical Sciences & Materials Systems Lab
General Motors Global Research and Development
30500 Mound Road
Mail Code 480-106-224
Warren
Michigan 48090-9055
USA
Email: alan.luo@gm.com

Chapter 13

W. H. Sillekens*
TNO
P.O. Box 6235
5600 HE Eindhoven
The Netherlands
Email: wim.sillekens@tno.nl

D. Bormann
Institute of Materials Science
Leibniz Universität Hannover
An der Universität 2
30823 Garbsen
Germany

Introduction

C. BETTLES, Monash University, Australia and
M. BARNETT, Deakin University, Australia

'Even the earliest of these commercial alloys contained aluminium, zinc, manganese, silicon, as well as cerium, all of which remain today the principal constituents of the majority of commercial binary and ternary alloys'. So wrote W. Buchmann[1] in 1940, when the world-wide production of magnesium was on an upwards trajectory. Those words are still true today, by and large, some 70 years later.

Production tonnages have fluctuated markedly over this time but recently we have seen a resurgence in the use of magnesium and its alloys in an increasing number of mainstream applications. Traditionally, casting has been the dominant processing route, but in the last decade wrought products have found a place in many structural applications.

The final property requirements have been increasing, seemingly in an inverse fashion to pricing requirements. The future of wrought magnesium products will be highly dependent on clever alloy design (combining chemistry and an understanding of the fundamental deformation behaviours of magnesium alloys) and optimisation of the secondary processing stages such as rolling, forging and extrusion.

This book provided a unique opportunity to bring together experts from the many and varied disciplines that are necessary to successfully achieve the widespread adoption of wrought magnesium in non-niche applications. The book is divided broadly into three parts, beginning with an update on alloy trends, showing that we may at last be breaking away from the comment by Buchmann, and following this with chapters discussing the most important fundamental aspects of deformation behaviour. These chapters are the building blocks from which the optimisation of the processing steps can be constructed, and the second part of the book looks at a number of different processing routes. The concluding chapters are used to discuss the applications that will be available to wrought magnesium products in the future.

We would like to thank all the contributors to this book for their hard work and dedication to their particular fields of endeavour. We would also like to thank the editorial staff at Woodhead Publishing for firstly suggesting that wrought magnesium deserved a book of its own, and secondly for all their efforts in bringing the contents together and making the book possible.

Reference

1. Buchmann, W. (1940). In *The Technology of Magnesium and its Alloys*. Ed: Beck, A. London, FA Hughes.

Part I
Types and properties of magnesium alloys

1
Current developments in wrought magnesium alloys

M. O. PEKGULERYUZ, McGill University, Canada

Abstract: Recent activities in magnesium wrought alloy development aim at improving formability by weakening basal textures, reducing defects (e.g. edge cracking during sheet rolling), or increasing extrudability, where a combination of second phases, solutes, grain refining and reduced axial ratio (c/a) are seen to be important. The early projects focused on alloy modification via trace elements (Ce, Y, Sr, Ca). Mg-rare earth (RE) alloys were developed to produce weaker or alternate textures. Low levels of Li additions lead to reduced axial ratio, which modifies the twinning behavior and reduces edge cracking during rolling. Mg-Mn and Mg-Zn alloys with Ce and Sr lead to weaker textures.

Key words: magnesium, wrought alloys, rare earths, Ce, Sr, Li, Mn, Zn, AZ31, texture, axial ratio, particle-stimulated nucleation, rolling, extrusion, twinning-related brittleness, edge cracking.

1.1 Introduction: overview of wrought magnesium alloy development

While aluminum is best suited for wrought applications due to its face-centered-cubic (fcc) crystal structure, magnesium, with its hexagonal close-packed (hcp) structure, has a limited amount of slip systems and limited ductility, especially at room temperature. Mg wrought applications have been considerably hampered by the poor ductility and formability of magnesium at room temperature. Until the late 1990s, the work on wrought magnesium has concentrated largely on extrusions and forgings. Extrusion alloys AZ61 (Mg-6Al-1Zn), ZK60 (Mg-6Zn-0.6Zr), and forging alloys AZ80 (Mg-8Al-0.5Zn), ZK60 (Mg-6Zn-0.6Zr) have been developed and used for ambient temperature applications (all compositions in this chapter are in wt%). Sheet alloys developed were the commercial AZ31 (Mg-3Al-1Zn), AZ61 (Mg-6Al-1Zn), ZK60 (Mg-6Zn-0.3Zr), ZK31 (Mg-3Zn-0.6Zr), and experimental alloys containing Li, Zr and/or Mn.

Recent alloy development activities in wrought Mg have started due to the interest in using lightweight magnesium as automotive sheet and extrusions for vehicle body applications. As Bettles et al.[1] discuss, the focus on Mg alloys has been revived at the turn of the millennium, the common alloy considered today, AZ31, is now 50 years old. It is also being acknowledged that this alloy may not be suitable or indeed able to meet the critical property and production requirements imposed by high-volume, low-margin consumer applications. The challenges in using wrought Mg for automotive applications lie in cost as well as in properties. Major shortcomings of the common Mg alloys in wrought condition are: poor formability (r-value, n-exponent), basal texture, yield asymmetry associated with twinning, edge cracking in rolling related to compression banding and twinning, low tensile strength at high extrusion speeds, low ductility at high strength.

Some of the research and development requirements for successful use of light metals for body applications are given below:

- Wrought magnesium alloys with improved strength and toughness.
- Wrought magnesium alloys with improved formability and corrosion resistance.
- Processes that enhance the formability of magnesium.

Extrusion alloy requirements have been summarized by Bettles et al.[1] Sheet alloy requirements are also similar (Table 1.1).

1.1.1 Yield symmetry and weakened texture

Combating the strong basal texture in rolled Mg alloys (c-axis perpendicular to rolling plane) and the strong fiber texture (c-axis in the radial direction) are the basis of alloy design in Mg wrought alloy development. Activating non-basal slip and promoting recrystallization micro-mechanisms that lead to texture randomization are important components of the alloy design strategy. Here, it should be remembered that particle-stimulated nucleation (PSN) and twin-induced nucleation (TIN) lead to the recrystallization of new grains with orientations that are different from the parent grains, while grain boundary bulging (GBB) leads to the recrystallization of grains with orientations closely associated with the original grains. Non-basal slip can theoretically be activated via changes in the lattice parameters of Mg even at moderate temperatures. A small grain size enhances non-basal slip at grain boundaries via plastic compatibility stresses. Suppression of twinning is another way to weaken textures by inducing alternative deformation mechanisms, such as grain boundary or bulk non-basal slip, that lead to texture randomization.

Table 1.1 Magnesium wrought alloy requirements

Extrusion alloys	Sheet alloys
Extrudability (higher extrusion speed combined with high tensile yield strength)	Yield symmetry (weakened textures) that can be achieved via fine grain size and prevention of twinning
Yield symmetry (weakened textures) that can be achieved via fine grain size and prevention of twinning	Minimum edge cracking

1.1.2 Minimum edge cracking

The phenomenon of edge cracking during rolling is not clearly understood, however, it is known to be related to the twinning modes in hcp metals. It is known that the twinning shear is related to the axial ratio in hcp crystals[4–6] as presented in Fig. 1.1 where the twinning systems with positive slope indicate compression (contraction) twins and those with negative slopes indicate tension (extension) twins along the c-axis. Figure 1.1 also presents the axial ratios of the hcp metals and the most common twinning modes activated in them. Within the range of the axial ratios of the hcp metals (1.568–1.886), the $\{10\bar{1}2\}$-<1011> shear direction reverses at an axial ratio of √3, where this twinning mode becomes compressive for Zn and Cd, but is tensile for all other hcp metals (Be, Ti, Zr, Re, Mg). HCP metals that exhibit high ductility (Re, Zr, Ti) twin profusely in both tensile and compressive modes and those that are very brittle (Be, Zn) twin only by the most common type $\{10\bar{1}2\}$ tension twin.[4] In Mg, the $\{10\bar{1}2\}$ tension twins are followed by $\{10\bar{1}1\}$ compression twins. Generally, the lower the value of twinning shear of a given twin mode, the higher is the frequency of its occurrence. This correlation has been generally observed as indicated by Fig. 1.1 and also by those experiments in which the axial ratio was changed by alloying;[4] the additional factors are the ease with which atomic shuffling can occur to nucleate the twin and the ease of gliding parameters. The absence of $(10\bar{1}1)$ $[10\bar{1}2]$ compression twins in Be despite the low twinning shear is related to the complex atom shuffling, while the occurrence of $\{11\bar{2}1\}$ 1/3 $[\bar{1}\bar{1}26]$ twinning in Ti, Zr, and Re in spite of

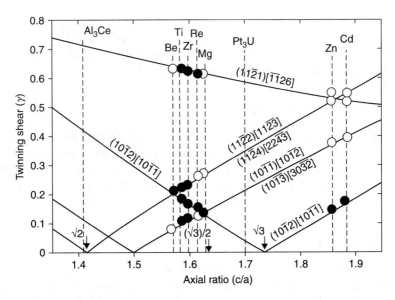

1.1 Twinning shear versus axial ratio. For the seven HCP metals, the filled symbols indicate that the twin mode is active.[4, 5]

its relatively large twinning shear is attributed to the relatively simple atomic shuffling.[4] It has been observed that subsequent slip-twin interactions largely determine the ductility of the hcp metals that undergo twinning.[4] If the twin boundaries become sinks for gliding dislocations the metal is ductile; if, however, the dislocations are repelled by the twin boundary they pile-up at the interface leading to eventual crack nucleation. In Mg, the basal dislocations are repelled by $\{10\bar{1}2\}$ tension twins[4] leading to edge cracking. It is known that the twinning modes in hcp metals has been altered via alloying,[4] which changes the axial ratio; e.g. twinning has been stopped in Mg-Cd alloy with c/a = $\sqrt{3}$, leading to extreme brittleness. It is therefore possible to influence the twinning behavior of Mg alloys via changes in the axial ratio.

1.1.3 Extrudability

Extrudability is defined as the maximum extrusion rate without producing visible surface defects.[7] Usually low ductility, low melting intermetallics and brittle inclusions (oxides) lower the extrudability of a material. Acceptable mechanical properties need to accompany extrudability. Higher speeds and the resultant temperature increase can influence dynamic recrystallization and the recrystallized grain size.

1.2 Current developments in magnesium wrought alloys

Current alloy development activities on wrought Mg started in around 2000 and initially focused on the modification of the Mg-Al alloys. Rare earth (RE) additions to Mg as well as Sn additions combined with REs or with Ca were investigated and Mg-Li compositions were studied.

1.2.1 Modified Mg-Al alloys

Mg-Al based alloys are the most widely used sheet (AZ31) and extrusion (AZ61) alloys. Modifications of AZ31 and AZ61 were effected via trace additions of RE. Other elements (Ca, Sr, Sb) were also used to optimize the AZ31 alloy. A different route considered the elimination of Zn from AZ31 for extrusion to lead to the development of the AM30 alloy.

Modified AZ31 alloy

AZ31 (Mg-3 wt%Al-1 wt%Zn, which is the most common Mg sheet alloy, suffers from edge-cracking during rolling, a strong basal texture and a non-homogeneous recrystallized grain-size or partial recrystallization leading to limited room temperature formability. Recently, it has been shown that twin-roll-cast (TRC)

AZ31 sheet can develop weaker textures upon annealing due to the metastable cast/deformed structure of TRC.[8,9]

Alloy modification has also been explored for improving the performance of the AZ31 alloy. The initial alloy strategies emphasized the grain refining effect of alloy additions as well as the role of second phases in work-softening of the deformed alloys. Rare and alkaline earths that are surface active in Mg and which can refine the grain structure and the microstructure[9–11] have been used for microlevel alloying of AZ31.

Laser et al.[10] have investigated the additions of Ce mischmetal (0.3–0.5 wt%) and Ca (0.8 wt%) to AZ31 and studied the effects on extrusion at 250–350 °C and mechanical properties. A reduction in final grain size was observed after extrusion, the smallest grain size of $D_{mean} = 8$ μm occurring in AZ31-0.4 wt%Ce-0.8 wt%Ca after 250 °C extrusion. The extruded alloy rods had $\{11\bar{2}0\}+\{10\bar{1}0\}$ fiber texture (c-axis in the radial direction) and showed a high degree of yield asymmetry (Fig. 1.2). Small differences in the yield asymmetry were observed amongst the alloys, which were attributed to small differences in texture. At 350 °C the Ca and Ce alloyed AZ31 showed lower yield asymmetry than the other alloys.

Smaller grain size and weaker textures were observed with Ca alloying. The weaker texture was explained by the PSN of recrystallized grains at Ca intermetallics and the reduced mobility of the grain boundaries due to the second phases. As is known, PSN and TIN can weaken the 'basal' deformation texture of Mg alloys during hot deformation while GBB would not.

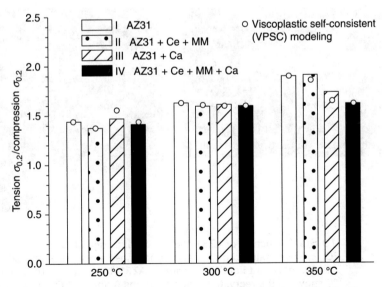

1.2 Yield stress asymmetry of AZ31 and AZ31 with Ca and/or Ce.[10]

Sadeghi et al.[11, 12] conducted an in-depth study of the recrystallization behavior, microstructure and texture evolution of AZ31 alloyed with 0.4 and 0.8 wt% Sr during extrusion at 200–350 °C. Extrusion at 250 °C resulted in a microstructure consisting of fine recrystallized and large elongated grains along with Al_4Sr stringer precipitates (Fig. 1.3a). Extrusion at 350 °C yielded a larger and more uniform grain size (Fig. 1.3b). Grain refinement and nucleation of new grains were associated with sub-grain formation in the elongated grains, grain boundary bulging and nucleation at the particle interfaces (Fig. 1.3d).

The texture that was described as a distribution of prism planes along the extrusion direction (Fig. 1.4) showed an increase in the basal ring fiber-texture from the undeformed zone toward the die opening in all alloys. Compared to the AZ31 alloy without Sr, it was seen that the texture intensity increases with increasing Sr at 250 °C, but decreases at 350 °C.

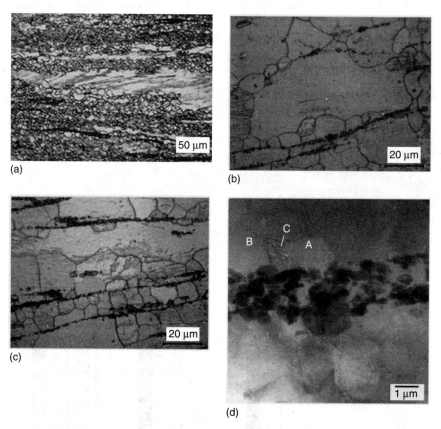

1.3 Grain structure of AZ31+0.8 wt%Sr extruded at (a) 250 °C, (b) 350 °C. (c) Recrystallized grains at second phase stringers. Alloy extruded at 350 °C. (d) Bright field TEM image of AZ31+0.8 wt%Sr extruded at 350 °C showing the recrystallization of grains A and B at the Al-Sr stringer and consuming the C deformation zone.[11]

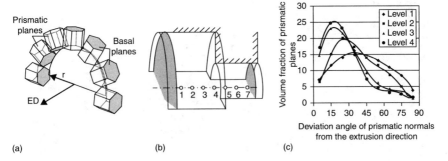

1.4 (a) Basal and prism planes in the fiber texture of extrusion. (b) Texture measurements locations. (c) The distribution of prism planes along the extrusion direction (ED) in AZ31+0.8 wt%Sr extruded at 250 °C.[11]

The differences in the effect of Sr on texture strength with temperature were explained by the different micro-mechanisms operating at these temperatures.[11] The Al solubility in the α-Mg matrix decreases with increasing Sr and therefore the Al solute atmosphere around the dislocations weakens. Due to the reduction of the solute drag effect, dislocation movement is facilitated and dynamic recrystallization becomes more intensive, resulting in a strengthening of the texture during extrusion at 250 °C. On the other hand, the decrease in texture intensity at 350 °C extrusion was attributed to the activation of PSN that leads to the nucleation of new grains with high orientation mismatch to the parent grains. Figure 1.5 presents the electron backscattered diffraction (EBSD) image of recrystallized grains at a stringer in AZ31+0.8 wt%Sr. In Fig. 1.5b the parent grain in dark gray and the recrystallized grains in light gray, and the inverse pole figure (Fig. 1.5c), confirms that the parent grain was in basal orientation and the recrystallized grains have different orientations.[11]

Sadeghi et al.[12] have further studied the hot-deformation behavior of Sr alloyed AZ31 via extrusion and compression tests in the 200–350 °C range and saw that at different extrusion temperatures and levels of Sr, different dynamic recrystallization (DRX) mechanisms become dominant, which affects the final texture (Fig. 1.6). At lower temperatures and at low levels of Sr, the bulging of grain boundaries was activated, resulting in a necklace grain structure. Furthermore, a strong deformation texture of <10.0> parallel to extrusion direction develops as a result of GBB. Twinning that was activated in the early stages of deformation acted as nucleation sites for DRX. In contrast, at high temperatures and high levels of Sr, PSN becomes significant, weakening the overall texture with the recrystallization of new grains of random orientation. In order to prevent the surface cracking in the extruded sample, a limit for Sr concentration and deformation temperature was determined. A shoulder region was found between the bulging and PSN-dominant areas where both mechanisms are active. The texture measured as the volume fraction of prismatic planes less than 20° deviated

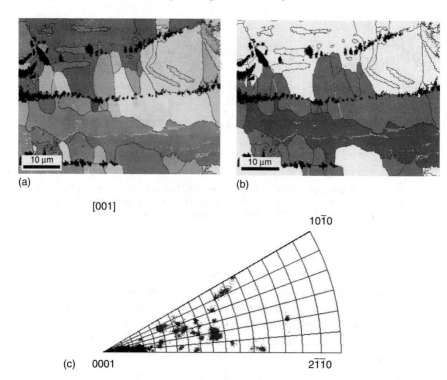

1.5 (a) EBSD image of recrystallized grains at a stringer in AZ31+0.8 wt%Sr. (b) Its color-coded map (parent grain in dark gray, the recrystallized grains in light gray). (c) The inverse pole figure.[11]

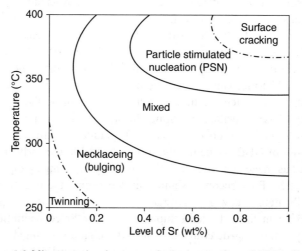

1.6 Micro-mechanism map during extrusion of AZ31 alloy containing up to 1wt%Sr.[12]

from the extrusion direction (F_v) was also mapped in the range of 0–0.8 wt%Sr addition to AZ31 and in the 250–400 °C temperature interval. The results are summarized as follows:

1. At low temperatures or low levels of Sr, the main micro-mechanism of DRX is grain boundary bulging (and the formation of a necklace structure until full recrystallization); the prismatic planes of the recrystallized grains are aligned perpendicular to the extrusion direction resulting in a strong <10.0> fiber texture.
2. At high temperatures and Sr concentrations, the contribution of PSN is considerable and the overall texture is weakened with the nucleation of new grains with random orientations at the precipitate boundaries.
3. At low temperatures, ε_c (critical strain for recrystallization) and F_v drop with increasing Sr. This is attributed to reduced Al solubility, which decreases the solute drag effect and enhances recrystallization.
4. At high levels of Sr, ε_c and F_v decrease with increasing temperature, explained by the fact that, at high temperatures, dislocation accumulation behind the boundaries is decreased and the contribution of the particles in blocking dislocations becomes more significant. Consequently, the effect of PSN becomes considerable and the final texture is weakened.
5. At very high temperatures and levels of Sr, cracks initiate at the extruded bar surface. At low temperatures and low levels of Sr, twinning is active and the twin boundaries act as nucleation sites for DRX.

An effect of Ca and REs was also seen in ECAPed AZ31. Masoudpanah et al.[13] have made combined additions of Ca and RE (0.6 wt% and 0.3 wt% of each) to AZ31 and investigated the effect on extruded and ECAPed AZ31 alloy. They found that Ca and RE formed second phases that resulted in the suppression of dynamic recrystallization during hot-deformation and led to grain refinement. The second phases caused dispersion strengthening and the texture modification improved the tensile ductility.

Modified AZ61 alloy

AZ61 is a wrought alloy containing twice the amount of aluminum compared to AZ31. Fang et al.[14] studied the microstructures and mechanical properties of rolled AZ61 alloys containing different levels of Y (0%, 0.5%, 0.9%, 1.4% Y). Y refines the microstructures of rolled alloys. The finest average grain size is obtained in the alloy containing 0.9%Y. Y forms the Al_2Y second phase, which is finely broken during rolling. The researchers suggest that these fine precipitates may suppress grain growth. However, the role of Al_2Y on DRX and texture has not been verified.

AM30 and AM50+Ce

AM30 (Mg-3%Al-0.4%Mn) alloy was developed by removing Zn from the AZ31 alloy. The alloy design was inspired by the development of high-ductility AM casting alloys by excluding Zn from the Mg-Al compositions. The alloy had similar strength to AZ31 but 50% higher ductility at temperatures below 200 °C and 20–50% higher maximum extrusion speed (Fig. 1.7).[15] The AM30 had slightly better formability than AZ31 at room and moderate temperatures due to higher strain hardening rate and exponent. Twinning and slip are the major deformation mechanisms for these alloys below 150 °C and DRX is described above.

In a modified version of AM30, the effect of 1.5 wt% Ce on the rolling behavior AM30 was studied.[16] The alloy was homogenized at 415 °C and rolled at 400 °C to a reduction of 64%, with preheating and intermittent heating at 400 °C for 10 minutes. Twins are clearly observed and the acicular $Al_{11}Ce_3$ phase present in the as-cast alloy was seen to break into many small sections during hot rolling.

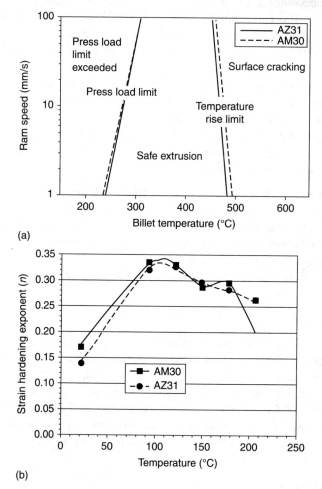

1.7 (a) Extrusion limit diagram; (b) strain hardening exponent vs temperature; (c) strain hardening rate at room temperature, for AM30 and AZ31.[15]

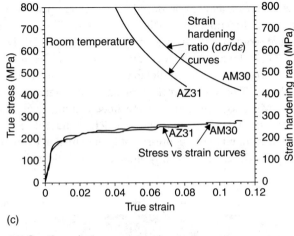

(c)

1.7 Continued.

The structure was dynamically recrystallized with an average grain size of about 15 μm. The role of $Al_{11}Ce_3$ phase has not been clearly determined and no texture analysis was conducted.

Luo et al.[7] have investigated the effects of Ce (0.2 and 0.5%) on the tube extrusion characteristics of the A3 (Mg-3%Al) alloy. 0.2%Ce addition improved the extrudability and mechanical properties of the extrusion (i.e. lowered surface defects) as seen in Table 1.2, which was attributed to grain refinement and dispersion strengthening provided by the $Mg_{12}Ce$ particles and the altered texture (with c-axes of the grains less than 90° to the extrusion direction) that was obtained.

1.2.2 Mg-rare-earth alloys

Mg wrought alloys containing various RE have recently seen renewed interest, mainly for the improvement of strength in wrought magnesium. An important finding was the appearance of a different texture component in Mg-RE alloys called the 'rare earth texture' during extrusion. Stanford et al.[17] have investigated texture development during the extrusion of several binary

Table 1.2 Tension-compression yield ratios of extruded tubes[7]

Alloy	Extrusion speed (mm/s)		
	10	30	90
Pure Mg		3.13	
AM30	2.00	2.03	2.35
AM30 + 0.2% Ce	1.89	2.02	2.05
AM30 + 0.5% Ce	2.10	1.95	1.92

alloys and have seen that La and Gd produced a new texture peak with <11$\bar{2}$1> parallel to the extrusion direction. Other alloying elements Al, Sn or Ca did not produce this texture component. It was proposed through EBSD analysis (Fig. 1.8) that the <11$\bar{2}$1> texture component arises from the different orientation of recrystallized grains (with <11$\bar{2}$1> parallel to the extrusion direction) nucleated at shear bands in Mg-Gd. The bands of recrystallized grains originating within the deformed grains in Mg alloys give rise to the fiber texture, while the grains originating from the shear band regions have an RE texture. The 'rare earth texture' component was found to be suppressed at high extrusion temperatures. A number of Mg-RE alloys have been recently studied. These alloys do not contain

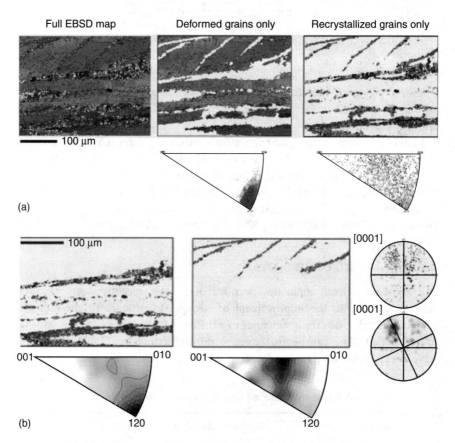

1.8 (a) EBSD map of alloy Mg-Gd extruded at 415 °C. Region shown is from the toe of the extrusion that first exits the extrusion die. Extrusion direction is left to right, with the higher strain region on the left. Inverse pole figures refer to the extrusion direction. (b) EBSD map of recrystallized regions. Inverse pole figures refer to the extrusion direction. {0001} pole figures are shown in the same orientation as the corresponding map, i.e. with the extrusion direction horizontal.[17]

aluminum or zinc. A distinct group that contains RE, as well as Sn, has also been developed.

Mg alloys containing various RE elements

Recently wrought magnesium alloys containing RE elements Gd and/or Y, Nd, Dy have been developed. The following alloys have good strength at both room and elevated temperatures and good creep resistance.[18-23]

- Mg-Gd-Y-Mn
- Mg-Y-Nd
- Mg-Gd-Y
- Mg-Gd-Y-Zr
- Mg-Gd-Y-Zr(Mn)
- Mg-Gd-Y-Nd
- Mg-Gd-Nd
- Mg-Gd-Nd-Zr
- Mg-Gd-Dy
- Mg-Gd-Dy-Zr
- Mg-Gd-Dy-Nd-Zr
- Mg-Y-Sm-Zr
- Mg-Gd-Er-Zr
- Mg-Gd-Ho-Zr
- Mg-Yb-Zr.

Table 1.3 presents some of the mechanical properties of these alloys.[24] It is noted that Gd-containing alloys have high mechanical properties in the extruded and extruded and heat-treated conditions.

It is expected that Mg alloys containing REs develop weaker fiber textures or alternate textures. Random or weak textures can promote formability, and improved mechanical isotropy in Mg alloys. Li et al.[25] have investigated the Mg-12Gd-3Y-0.6Zr alloy (GW123), extruded conventionally and hydrostatically. The average grain sizes in the two extruded conditions were 35 μm and 4 μm, respectively. The size of precipitates varied between the two extrusion microstructures and lay between 20–80 nm in length. Selected area diffraction (SAD) patterns taken with the incident beam parallel to [0001]α showed that the peak-aged microstructure predominantly consists of the α matrix and the metastable β' phase known to form in Mg-Gd alloys. The best combination of mechanical properties (strength and ductility) was achieved by RT hydrostatic extrusion following conventional extrusion at 430 °C, with the ultimate tensile strength (UTS), tensile yield strength (TYS) and elongation being 485 MPa, 413 MPa and 5.2% at room temperature.

Figure 1.9 shows inverse pole figures (IPF) of the as-extruded GW123 alloy. The texture results indicate that the c-axis of most grains was aligned preferentially

Table 1.3 Mechanical properties of recent Mg-RE wrought alloys[24]

Experimental alloy	State	Mechanical properties at RT		
		UTS (MPa)	YS (MPa)	E (%)
Mg-10Gd-3Y-0.5Zr	As extruded	290	192	13
	Extruded T5 (200 °C, 100 h)	341	228	11
	Extruded T5 (250 °C, 10 h)	397	228	5
Mg-5Y-4Nd	Extruded T5 (200 °C, 72 h)	351	274	7
Mg-5Yb-0.5Zr	As cast	189	106	7.5
	As extruded	304	268	12.9
Mg-9Gd-4Y-0.6Zr	Extruded-T6 (480 °C, 2 h, 150 °C, 100 h)	336	310	11.2
	Extruded-T5 (225 °C, 24 h)	360	320	5
Mg-7Gd-5Y-1.3Nd-0.5Zr	As cast	210	172	3.5
	Extruded T5 (250 °C, 16 h)	411	322	10.3
Mg-4Y-3Nd-0.5Zr	As cast	190	132	6.5
	As extruded	260	200	12
	Extruded T5 (250 °C, 16 h)	310	265	7.5

perpendicular to the extrusion direction, forming a typical extrusion Mg fiber texture. Texture development during hot conventional extrusion is highly affected by DRX during extrusion or static recrystallization (SRX) during cooling down. It was suggested that RX is responsible for the texture weakening and that hydrostatic extrusion at room temperature strengthens the fiber texture due to lack of recrystallization, and the texture strength is nearly two times higher than that of the hot extruded material. This was in agreement with the higher elongation to fracture of the hot extruded alloy due to the weaker extrusion texture. Given that considerable texture weakening has been observed when dynamic recrystallization occurs, it is likely that the micromechanism of DRX in the alloy is partly PSN or TIN.

Mg-RE-Sn alloys

It has been reported that Sn addition into Mg-RE alloys improves the rollability of the alloys.[26] Lim et al.[26] have investigated the properties of rolled Mg-RE-Sn alloys where REs were added in the form of mischmetal (MM). The specimens were fabricated by hot-rolling. The tensile tests were performed at room temperature and 200 °C. The mechanical properties improved with the addition of Al to Mg-2MM-2Sn alloy (ET22 AND ETA221 alloys). The yield strength and elongation-to-fracture of ET22 alloy were 100 MPa and 6%, respectively. The strength and elongation-to-fracture increased with small Al addition to 140 MPa and 20%, respectively. The SADP of that the new phases in ETA221 alloy (Fig. 1.10) showed that the Al_2MM phase observed in the alloys had an fcc

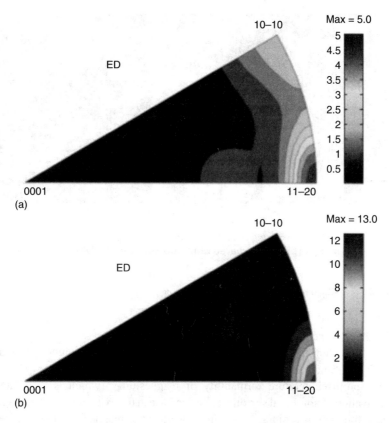

1.9 IPF of GW123. (a) Extrusion, 430 °C; (b) Conventional extrusion + RT hydrostatic extrusion.[25]

structure (S.G.: Fd-$3m$, a = 0.812 nm) similar to Al_2Ce (S.G.: Fd-$3m$, a = 0.806 nm) and Al_2La (S.G.: Fd-$3m$, a = 0.816 nm,) phases, but with slightly different lattice parameters due to the presence of a variety of RE elements in the MM. Small amounts of Al were found in the α-Mg matrix and in a small rod-shaped phase. Due to the formation of the Al_2MM phase, the amount of RE elements, i.e. Ce and La, was seen to decrease slightly in the matrix of the Al-containing ETA221.

The results of a conical cup test at 200 °C with punch speed of 50 mm/min show that the addition of Al also improves the formability of Mg-2MM-2Sn alloy. The CCV (conical cup value) was obtained by dividing the diameter of the base of the conical cup formed after test by the diameter of the original specimen (lower CCV equals better formability). The CCV of ET22 decreased from 0.9 to 0.8 with the Al addition in ETA221. The (0 0 0 2) poles of ETA221 alloy are spread more widely than those of ET22 alloy confirming the improved formability of the ETA221.

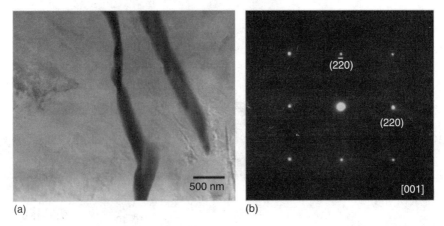

1.10 TEM images of Al2MM phase in ETA221 alloy. (a) Bright field image; (b) Selected area diffraction pattern (SADP).[26]

1.2.3 Recent lithium-containing alloys

Conventional Mg-Li alloys have usually depended on large Li additions that change the hcp structure of Mg to BCC.[27] In the Mg-Li binary system, when the Li content is greater than 5.7 wt%, β phase (Li solid solution) forms in the alloys.[3,4] Small amounts of lithium alloying additions are also known to improve the room temperature formability of magnesium,[27] which is associated with activation of non-basal slip due to the decrease in the axial ratio (c/a) of magnesium as a function of lithium.[28] As lithium is added to magnesium, the c value of the hcp crystal falls faster than the a value, which leads to a decrease in the c/a ratio,[29] and results in the Peierls stress for basal slip increasing relative to that for prismatic slip, and hence increases activation of non-basal prismatic planes. Texture modeling and TEM studies indicate that lithium additions also increase the glide of <c+a> dislocations on {11$\bar{2}$2} pyramidal planes.[30,31] There is evidence that <c+a> dislocations dissociate into partial dislocations,[32] and lithium may lower the energy of the stacking fault, and thereby increase the stability of the glissile dislocation configuration.[31] Lithium has been observed to alter the crystallographic texture by altering the balance of deformation mechanisms, which in turn influences the texture.[30]

Mg-Li alloys with Al and/or Zn

Mackenzie *et al.*[33] have recently studied the combined effects of Li additions and process parameters on rolling microstructures and textures of Mg-(1-3)Li and Mg-3Al-1Zn-(1-3)Li alloys. The alloys were cast into a plate mold and annealed at 380 °C. Test specimens were machined from the annealed material and rolled in seven passes to a true strain of 0.5 at 150 °C and 350 °C. Large differences were observed between the specimens following rolling at the lower temperature of 150 °C. Mg-3Li had significantly lower levels of cracking than Mg-1Li, and AZ31-3Li exhibited slightly lower levels of cracking than AZ31, indicating that Li additions improve the rollability of pure Mg and AZ31. The improvement in the binary alloys could be related to c/a reduction associated with Li in solid solution. Since edge cracking is related to c/a and to the twinning behavior in hcp alloys, the change in c/a would have influenced the twinning modes and the twinning-related cracking.

In both alloys, the basal planes are aligned with the sheet surface, but the basal poles are characteristically rotated approximately 15° towards the rolling direction (RD), which is associated with the glide of <c+ a> dislocations on the $\{11\bar{2}2\}$ planes. Li additions increase the rotation and splitting of the poles in the transverse direction (TD).[30] According to a study by Styczynski et al.,[34] increased activity of prismatic planes is related to increased spread of basal poles in the TD.

While AZ31 was partially recrystallized and the Mg-1Li alloy was fully recrystallized, no recrystallization was observed in AZ31-3Li alloy (Fig. 1.11).

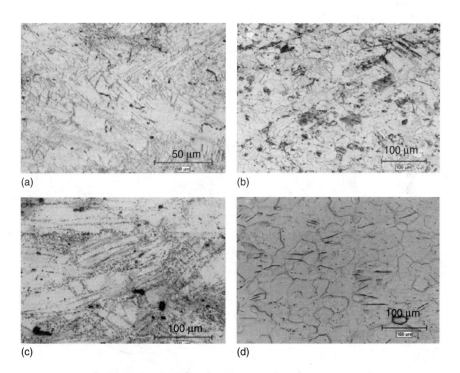

1.11 Optical micrographs of alloys rolled to a true strain of 0.5 at 350 °C. (a) AZ31 recrystallized grains at deformation features and grain boundaries. (b) Fine precipitates in AZ31–3Li (c) Recrystallized Mg–1Li. (d) Recrystallized Mg–1Li rolled to a true strain of 0.5 at 350 °C and annealed for 10 min. at 400 °C.[33]

Recrystallization in AZ31+3Li seems to be retarded by the fine precipitates (Fig. 1.11b). It is noted that the Mg-1Li alloy that is fully recrystallized (Fig. 1.11c) exhibits grain growth upon post annealing at 400 °C (Fig. 1.11d). Pole figures of post-annealed alloys shown in Fig. 1.12 indicate that the double peaks in the basal poles in the rolled material were replaced by a single peak in the basal poles following annealing at 400 °C. Microstructural changes during annealing suggest that grain growth, rather than primary recrystallization, was responsible for this evolution.

Mg-Li-Al alloys with higher Li content (8%) have also been studied.[35] Mg-8Li-1Al, Mg-8Li-1Zn and Mg-8Li-1Al-1Zn alloys were homogenized at 280 °C for 24 h, extruded with an extrusion speed of 10 mm/s at 280 °C and rolled at 200 °C. Because of the 8%Li, the alloys displayed a dual α-Mg+β-Li structure in the as-cast state and they were recrystallized during extrusion. The microstructure of the Zn-containing alloys has, in addition to the α and β phases, ZnO and Li_2O_2 as determined via XRD. The Mg-8Li-1Al alloy in the extruded/rolled condition had the highest tensile strength (314 MPa). The extruded Mg-8Li-1Zn alloy has the maximum elongation (44%). The extruded/rolled Mg-8Li-1Al-1Zn alloy has both high tensile strength and elongation (233 MPa and 9%, respectively).

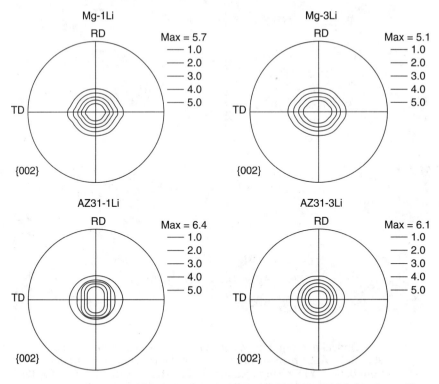

1.12 Textures of Mg-1Li, Mg-3Li, AZ31-1Li and AZ31-3Li following rolling to a true strain of 0.5 at 350 °C and annealing for 10 min. at 400 °C.[33]

The effect of low levels of Li on the c/a ratio and grain size of Mg was investigated by Becerra et al.[36,37] in Mg-Zn-Li and Mg-Zn-In-Li solid solution alloys. The mono-valent Li decreased the axial ratio (c/a) of magnesium from 1.624 to 1.6068 within the 0–16at%Li range (Fig. 1.13). This was attributed to the decrease in e/a causing electron overlap from the second Brillouin to the first Brillouin zone, which leads to a contraction of the c-spacing in real space and a decrease in c/a. The di-valent zinc showed no effect on c/a in the 0.2–0.7 at% range since, as explained by Vegard's law, the atom size caused a similar change in both a- and c-parameters. The tri-valent In increased the c/a of magnesium to 1.6261 as In increased towards 3.3 at%. An important effect of c/a was seen in the twinning behavior of the alloys and, in turn, on edge cracking during 150 °C rolling.[40]

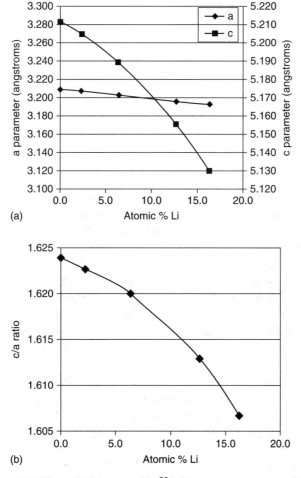

1.13 Effect of lithium on Mg.[36] (a) Lattice parameters; (b) axial ratio.

The changes in axial ratio and lattice parameters influence slip systems and twinning modes in hcp metals. The change in the deformation mechanism with reduced axial ratio can be attributed to the change in interplanar spacing, d, since the shear stress required to move dislocations is given by the Peierls stress,

$$\tau_\pi = P \cdot e^{\{-2s\, d/[b(1-\eta)]\}} \quad [1.1]$$

where P is a factor depending on the shear modulus and Poisson's ratio η, b is the magnitude of Burgers vector of the dislocation, and d is the interplanar spacing. The critical resolved shear stress, τ_{CRSS}, is related to the Peierls stresses τ_p. In a recent study, Uesugi et al.[38] have calculated τ_p from first principles as

$$\tau_p = \frac{Kb}{a'} \exp\left(-\frac{Kb}{2a'\tau_{max}}\right) \quad [1.2]$$

where K is energy factor (depends on elastic constants and the type of dislocation); a' is interplanar spacing in the direction of dislocation sliding on slip plane ($a' = a/2$ for basal slip where a is the lattice parameter and $a' = a$ for prismatic plane); b = Burgers vector; $b = (a/3)^{1/2}$ for Shockley partial dislocations of the basal plane and $b = a$ for edge dislocations of the prismatic plane. τ_{max} is the maximum restoring force (= maximum slope of the generalized stacking fault energy in the analysis). Equation 1.2 interestingly indicates that changing the a-spacing would not influence Peierls stresses and CRSS for prismatic slip (because $b = 2a$ for edge dislocations), but it would alter the basal slip of partial dislocations because their $b = (a/3)^{1/2}$.

The other important effect of c/a is on the twinning behavior of Mg as explained in Section 1.1 and Fig. 1.1. In the study by Pekguleryuz et al.,[39] the differences in the twinning modes of AZ31 and Mg-2Li-1Zn and Mg-6Li-0.4Zn-0.2In, which governed the edge cracking index (Ic) during 150 °C rolling, have been related to the axial ratio (Fig. 1.14). AZ31 significantly edge-cracked during rolling (Fig. 1.15), twinned in the tensile mode (Fig. 1.16.a.), and had a c/a of 1.6247. At the c/a of 1.6247, the shears of $\{10\bar{1}2\}$ tension twins and $\{10\bar{1}1\}$ compression twins start to differ, making the atomic reshuffling more difficult (see Section 1.1) impeding double twinning in AZ31 and leaving a higher propensity of tension twins that are potent sites for crack nucleation. The Li-containing alloys (Mg-2Li-1Zn and Mg-6Li-0.4Zn-0.2In) had lower c/a (Table 1.4), double-twinned (Fig. 1.16b) during rolling, exhibited a lower degree of twin-related brittleness and did not edge crack during rolling (Fig. 1.15).

Interesting effects of cast grain size were seen on the texture evolution of Mg alloys.[39] Rolling at 150 °C produced deformation structures and texture, and static recrystallization occurred during post annealing (Fig. 1.17). Maximum texture intensity during 150 °C rolling and subsequent annealing correlated strongly with the cast grain size (Table 1.4, Fig. 1.18). The texture weakened with decreasing grain size, which was attributed to the activation of non-basal slip by plastic compatibility stresses at grain boundaries. The a lattice parameter exerted a

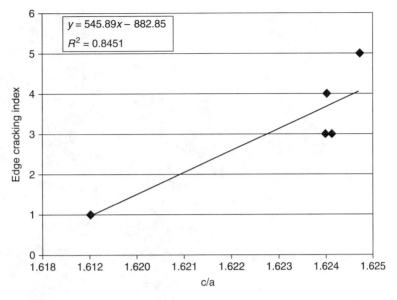

1.14 Edge cracking index vs c/a.[39]

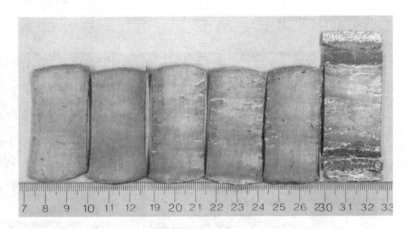

1.15 Alloys rolled in 15 passes at 150 °C (left to right: Mg-Li-Zn, Mg-Li-Zn-In, Mg-Zn-In, Mg-Zn, Mg and AZ31).[39]

weaker influence on texture, possibly by facilitating the cross slip of partial dislocations at grain boundaries.

Mg-Li alloys with RE additions

Mg-Li alloys with RE additions have been investigated to address the strength issues in Mg-Li alloys. Wu et al.[40] have studied RE additions to Mg-5Li-3Al-2Zn alloy. The microstructure and mechanical properties of extruded (at 280 °C with an

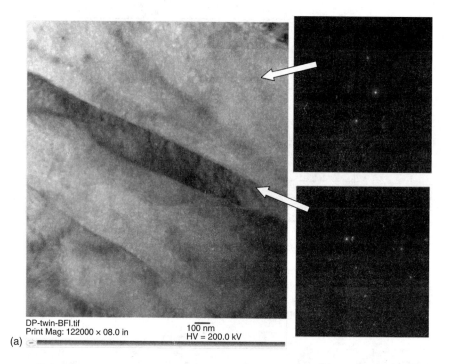

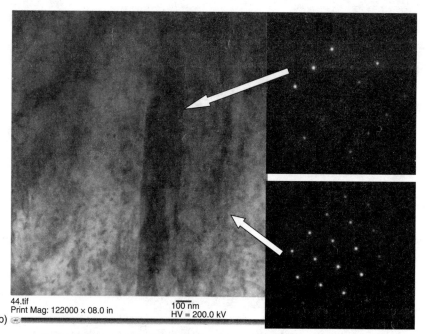

1.16 TEM image of (a) a tension twin in AZ31 and (b) a double twin in Mg-2Li-1Zn matrix, both following 15 passes of rolling at 150 °C.[39]

Current developments in wrought magnesium alloys

Table 1.4 Axial ratio, grain structure and size, edge cracking index and texture intensity[39]

Alloy (wt%)	a	c/a	Effective cast grain size D (mm)	Recrystallized grain size, d (µm)	Edge cracking index	Maximum intensity	
						Roll	Roll and anneal
Mg	3.2088	1.6240	6.1	46	High/4	12.2	17.9
Mg-1.8Li-1Zn	3.2009	1.6190	1.8	48	Minimum/1	4.2	2.6
Mg-6Li-0.4Zn-0.2In	3.1991	1.6190	1.0	49	Minimum/1	6.2	3.0
AZ31	3.1993	1.6247	0.15	51	Severe/5	3.4	2.4

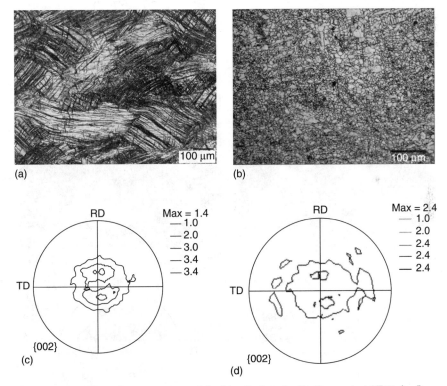

1.17 The microstructures of: (a, b) rolled and rolled/annealed AZ31, (e, f) rolled and rolled/annealed Mg-2Li-1Zn, and (i, j) rolled and rolled/annealed Mg-6Li-0.4Zn-0.2In. Rolling and annealing textures of: (c, d) AZ31, (g, h) Mg-2Li-1Zn and (k, l) Mg-6Li-0.4Zn-0.2In.[39]

(Continued)

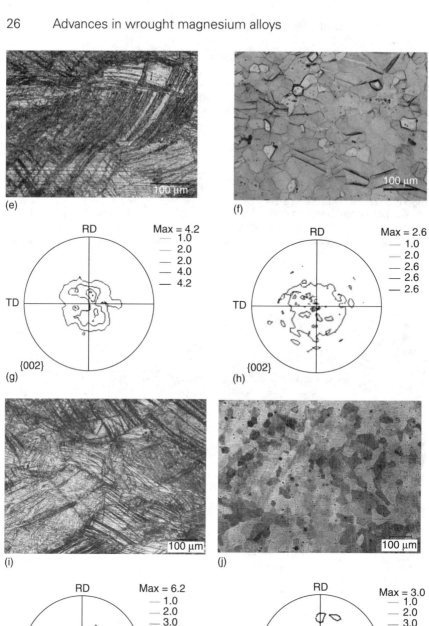

1.17 Continued.

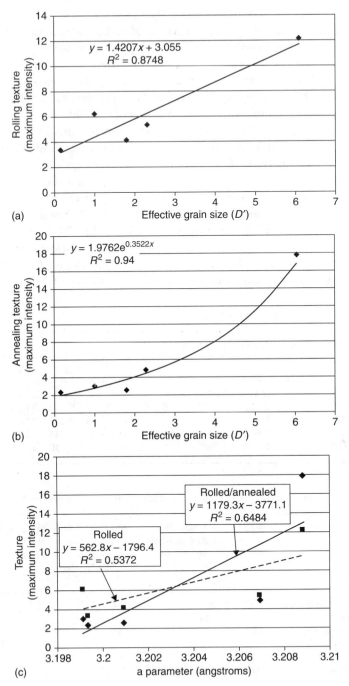

1.18 The effect of grain size on (a) rolling texture and (b) annealing texture, and (c) the effect of lattice parameter a on both rolled and annealed textures.[39]

extrusion ratio of 15) specimens were studied (Table 1.5). RE elements caused microstructural refinement and the formation of Al$_3$La (and decreased the AlLi phase present in the Mg-Al-Li alloys), which resulted in the improvement of mechanical properties. The optimal RE content for Mg-5Li-3Al-2Zn alloy was at 2 wt%, which yielded the finest microstructure and the highest mechanical properties. The further increase of RE coarsened the microstructure and changed the morphology of Al$_3$La from particle-like to rod-like or pearlite-like eutectic shape, leading to poor mechanical properties. It was noted that the extruded LAZ532-2RE and LAZ532-3RE alloys recrystallized during extrusion, but LAZ532-1RE and LAZ532-6RE did not. The LAZ532-3RE alloy showed some grain growth compared to the LAZ532-2RE and this was attributed to the finer precipitates in the LAZ532-2RE pinning grain boundaries and retarding grain growth. The properties of LAZ532-2Li were compared to data for AZ31 (Table 1.6). The researchers[41] further investigated the microstructure and properties of Mg-5.6Li-3.37Al-1.68Zn-1.14Ce alloy extruded at 250 °C (Table 1.6). The alloy recrystallized at a lower temperature. On the other hand, a dual-phase alloy (Mg-8Li-1Al-1Ce with no Zn) extruded at 220 °C showed no recrystallization and lower strength but higher ductility than the

Table 1.5 Compositions of Mg-Li-Al-Zn-RE alloys[40]

Nominal composition	Measured composition (wt%)						
	Li	Al	Zn	La	Pr	Ce	RE (La+Pr+Ce)
LAZ532	5.615	2.885	1.656				
LAZ532-1RE	5.622	2.890	1.726	0.495	0.197	0.136	0.828
LAZ532-2RE	5.628	2.876	1.537	0.967	0.358	0.287	1.621
LAZ532-3RE	5.626	2.884	1.981	1.559	0.676	0.341	2.576
LAZ532-6RE	6.506	2.758	1.922	2.771	0.816	0.636	4.223

Table 1.6 Mechanical properties and density of Mg-Li-Al-Zn-RE compared to AZ31[40-42]

Property	AZ31	LAZ532-2RE (Mg-5.6Li-2.9Al-1.59Zn-1.6RE) extruded at 280 °C	Mg-5.6Li-3.4Al-1.7Zn-1.14Ce extruded at 250 °C	Mg-8Li-1Al-1Ce extruded at 220 °C
Ultimate strength (MPa)	280	290	286	187
Elongation (%)	14	25	10	33
Specific strength (MPa/(g/cm^3))	157	181	188	117
Density (g/cm^3)	1.78	1.6*	1.6*	1.6*
Recrystallized	Partially	Fully	Fully	No

* Mg-Li alloys have density lower than 1.6. The density of new alloy compositions is assumed to be 1.6.

Zn-containing compositions.[42] It is noted that slight changes in the Li:Al:Zn:RE ratio affects the recrystallization behavior of the alloys.

Yttrium (Y) has been added to Mg-Li-Zn alloys to explore the possibility of strengthening via the formation of the icosahedral quasicrystalline I-phase (Mg_3Zn_6Y). Xu et al.[43] have investigated extruded Mg-Li-Zn-Y alloys (Table 1.7) comprising α-Mg, β-Li, LiMgZn and I-phases. As Zn/Y ratio increased (Alloy III) the W phase also formed in the matrix (Fig. 1.19). Due to solidification segregation I-phase formation was locally suppressed in favor of W and the LiMgZn precipitates. Tensile properties improved with increasing I-phase (Table 1.7).

A Mg-Li-Zn alloy that contains both Y and Ce (Mg-7.8%Li-4.6%Zn-0.96%Ce-0.85%Y-0.30%Zr) has been studied by Chen et al.[44] The microstructure (Fig. 1.20) consists of α-Mg, β-Li and Mg-Zn-Zr-RE compounds that form in the β phase (Fig. 1.20a). The hot-deformation behavior of the extruded alloy was studied via compression tests in the temperature range of 250–450 °C and a strain rate range of 0.001–10 s^{-1}. The flow stress data obtained from the tests were used to develop a processing map (Fig. 1.20b) – a contour plot of the iso-efficiency values on the temperature-strain rate field. In the plot, power dissipation through microstructural changes are expressed in terms of an efficiency of power dissipation given by $\eta = 2m/(m + 1)$, where m is the strain rate sensitivity of flow stress. The dissipation characteristics vary for the different microstructural mechanisms, and each domain on the map corresponds with a single dominant mechanism operating under those conditions of the domain.

The different efficiency domains corresponding to various microstructural characteristics identified in the study for the alloy were:

1 Domain I – temperature range of 250–275 °C and strain rate range of 1–10 s^{-1}, with a peak efficiency of about 50% at 250 °C/10 s^{-1}. Incomplete DRX in β phase and practically no DRX in the α phase.
2 Domain II – temperature range of 250–275 °C and strain rate range of 0.001–0.003 s^{-1}, with a peak efficiency of about 42% at 250 °C/0.001 s^{-1}. Incomplete DRX in β and α.
3 Domain III – temperature range of 400–450 °C and strain rate range of 1–10 s^{-1}, with a peak efficiency of about 42% at 450 °C/10 s^{-1}. Complete DRX in β and α.

Table 1.7 Compositions and tensile properties of Mg-Li-Zn-Y alloys[43]

Normal alloys	Chemical composition (wt%)				Zn/Y ratio	Mechanical properties		
	Mg	Zn	Y	Li		$\sigma_{0.2}$ (MPa)	UTS (MPa)	Elongation (%)
Alloy I	Balance	3.12	0.61	8.04	5.11	148	222	30.7
Alloy II	Balance	6.47	1.26	7.86	5.13	159	239	20.4
Alloy III	Balance	9.25	1.79	7.67	5.17	166	247	17.1

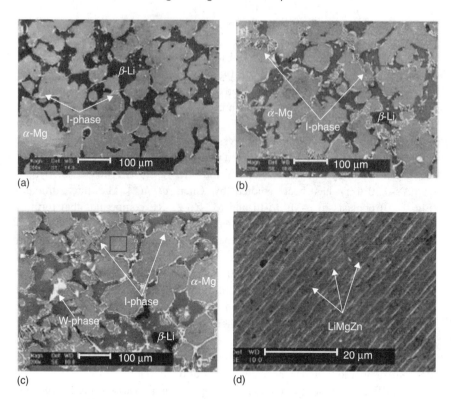

1.19 The microstructure of the as-cast Mg-Li-Zn-Y alloys: (a) Alloy I; (b) Alloy II; (c) Alloy III; (d) high-magnification observation of the location outlined by a square in image (c).[43]

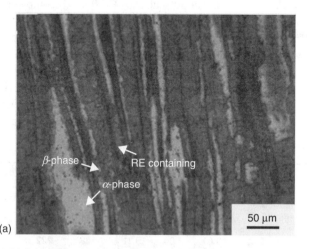

1.20 (a) Microstructure of the extruded alloy; bright zones are the α-phase and the dark zones are the β-phase; (b) hot-deformation map of the alloy.[44]

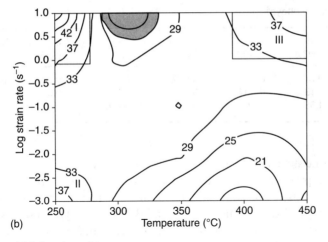

1.20 Continued.

The optimum parameters for hot working of the alloy were 250 °C/10 s^{-1} and 250 °C/0.001 s^{-1}, where a fine dynamic recrystallization microstructure is obtained. It was found that the RE-containing intermetallics and the α phase accelerate the DRX in the β phase. The softer β phase reduces the driving force for and retards the DRX in the α phase.

1.3 Progress in wrought magnesium alloys

The requirements for weakened or randomized textures for improved formability in wrought Mg alloys, combined with the need for adequate strength, are the driving forces for investigating new alloy development strategies. Two alloy systems studied recently are the Mg-Mn based alloys and novel Mg-Zn based alloys.

1.3.1 Mg-Mn based alloys

Mg-Mn alloys have been used as sheet material due to the good rolling behavior, weldability and corrosion resistance combined with a medium strength.[45] These alloys usually have a coarse grain structure and insoluble Mn particles.[45] Mg-Mn sheet alloys were developed with a relatively high manganese content (1.0–2.0 wt%).[28,46] These compositions exhibit a peritectic reaction resulting in the formation of Mg solid solution (α-Mg) and α-Mn.

Recently, the influence of coarse particles on the recrystallization behavior of three Mg-Mn alloys (Table 1.8) has been studied[47] during plane strain compression

Table 1.8 Mg-Mn alloys[47]

Alloy	Mn (wt%)	Zn (wt%)	Al (wt%)	Fe (wt%)	Si (wt%)	Ni (wt%)
M03	0.34	0.004	0.002	0.013	0.03	0.0007
M12	1.19	0.004	0.003	0.007	0.01	0.0009
M16	1.55	0.004	0.002	0.009	0.01	0.001

testing to a true strain of 0.4 and 1.2 at 300 °C. The alloys were heat-treated (32 h at 615 °C) to produce a distribution of large manganese-rich particles that also contain high levels of iron and silicon impurities. The particles were seen to lead to PSN of recrystallization as determined via EBSD (Fig. 1.21). These new grains had a random orientation (Fig. 1.22) as opposed to other grains that had recrystallized through the grain boundary bulging mechanism. Since only a few particle clusters took part in PSN, the randomization effect of PSN grains was small.

RE additions to Mg-Mn alloys have been explored in order to increase the strength and alter the texture. Masoumi et al.[48] have investigated the microstructure and texture development in rolled and rolled/annealed Mg-1wt%Mn-based alloys containing different levels of Ce (from 0.05–1.0 wt%) (Table 1.9). The alloys were cast into plates, homogenized for 8 h at 450 °C, then preheated at 400 °C for 15 min. before the first rolling pass and for 5 min. before each subsequent pass. The rolled sheet samples were subsequently annealed at 450 °C for 15 min. after the last rolling pass. The Ce addition refined the as-cast and rolled/annealed grain structure of the Mg-1wt%Mn alloy. Moreover, the overall texture intensity of the basal pole was weakened for rolled as well as rolled/annealed Mg-Mn-Ce alloys compared to the Mg-1wt%Mn (M1) alloy. It was also noted that, with increasing Ce, the angular distribution of basal poles was broadened in both the RD and TD. The texture was characterized by plotting as the percentage of basal planes as a function of the angle of deviation from the sheet normal direction ND where the effect of Ce on basal pole spread can be observed. It was also noted that the effect of Ce on texture and grain size becomes less significant above 0.1% Ce. The texture weakening was attributed to the solid solubility of Ce in Mg rather than PSN or c/a ratio alteration, since the maximum effect was seen at the maximum solubility of Ce in Mg at the rolling temperature.

Single additions of the RE elements cerium, yttrium or neodymium have been made to magnesium-manganese alloys (Table 1.10) by Bohlen and co-workers[49] in order to investigate their influence on the microstructure and texture formed during indirect extrusion at an extrusion ratio of 1:30 and two extrusion rates (1 and 10 mm/min with temperature profiles of 340 °C and 390 °C, respectively) and the resulting mechanical properties. The processes resulted in different degrees of

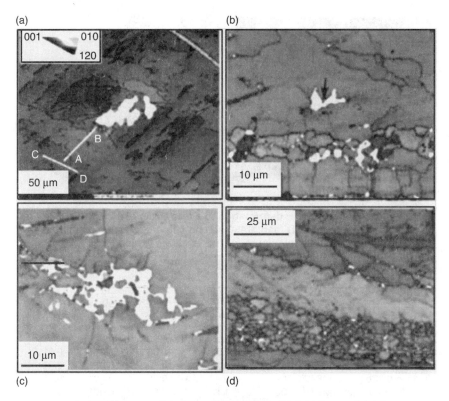

1.21 EBSD maps of deformed alloy M16. (a) A large particle cluster surrounded with new grains exhibiting lattice rotation after deformation to a true strain of 1.2. (b) A particle cluster at which a new grain is forming after deformation to a true strain of 0.53. (c) A number of new grains associated with a cluster of coarse particles after deformation to a true strain of 1.2. (d) Extensive dynamic recrystallization at a prior grain boundary after deformation to a true strain of 1.2.[47]

recrystallization in the alloys (Fig. 1.23). The binary Mg-Mn alloy M1 exhibited the typical <10.0> or <10.0>–<11.0> fiber texture depending on the extrusion rate, the RE-containing alloys produced weaker recrystallization textures and the formation of a new texture component (Fig. 1.24), which is similar to the RE component found in Mg-RE alloys.

Nd was found to modify the texture to a greater extent than Ce or Y in Mg-Mn alloys. Alloy MN11 exhibited a weak and different texture. The grain structures after both slow and fast extrusion are associated with complete recrystallization in this alloy. The texture was mainly characterized by orientations along the arc of the <10.0> and the <11.0> poles but with an additional tilt of the basal planes away from the extrusion direction. This new texture component dominated and weakened the texture. The texture weakening in the alloys was explained by the

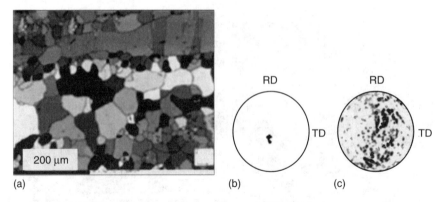

1.22 Alloy M03 after deformation to a true strain of 1.2 and 1 h annealing: (a) EBSD map; (b) (0001) pole figure for unrecrystallized grains; (c) (0001) pole figure for recrystallized grains.[47]

Table 1.9 Mg-Mn-Ce alloy compositions[48]

Alloy	Alloy symbol	Mn (wt%)	Ce (wt%)	Zn (wt%)	Al (wt%)
Mg-1Mn	M1	1.08	<0.001	0.001	0.010
Mg-1Mn-0.05Ce	ME10(0.05Ce)	1.05	0.06	0.01	0.01
Mg-1Mn-0.1Ce	ME10(0.1Ce)	1.15	0.10	0.01	0.01
Mg-1Mn-0.2Ce	ME10(0.2Ce)	0.97	0.20	0.001	0.010
Mg-1Mn-0.5Ce	ME11(0.5Ce)	1.11	0.50	0.001	0.010
Mg-1Mn-1.0Ce	ME11(1.0Ce)	1.04	0.94	0.001	0.010

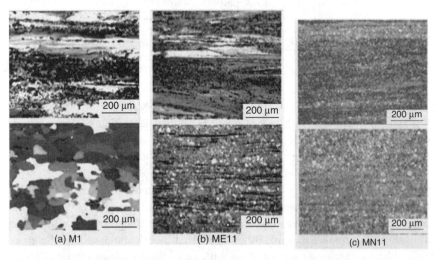

1.23 Longitudinal sections of extruded round bars. (a) M1; (b) ME11; (c) MN11. Upper micrographs: extrusion at 1 mm/min. Lower micrographs: extrusion at 10 mm/min.[49]

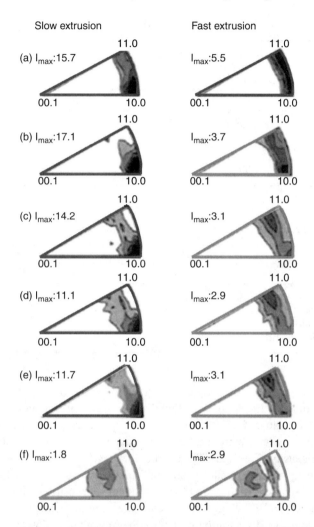

1.24 Inverse pole figures from extruded bars. (a) M1; (b) ME11; (c) MWO1; (d) MW11; (e) MW21; (f) MN11. Left-hand column: slow extrusion. Right-hand column: fast extrusion. Contour intensity levels: 1, 1.5, 2, 3, 5, 7.[49]

prevention of the growth of those grains having <11.0> parallel to the extrusion direction with different orientations.

Sr additions to Mg-Mn alloy were investigated in a recent study by Masoumi et al.[50] The microstructure and texture of as-rolled and annealed Mg-1Mn-xSr (MJ) alloys (Table 1.11) showed that the Sr addition refines the grain structure of the rolled/annealed Mg-1Mn (M1) alloy and, compared to the Mg-1Mn alloy, the texture intensity of basal poles weakens for rolled as well as rolled/annealed MJ alloys containing 0.5 and 1.0%Sr. The M1 alloy exhibits a strong basal texture,

Table 1.10 Mg-Mn-Re alloy compositions[49]

Alloy	Composition* (wt%)			
	Mn	Ce	Y	Nd
M1	0.99			
ME11	0.95	0.86		
MW01	0.20		1.32	
MW11	0.96		1.14	
MW21	1.74		1.08	
MN11	0.95			0.94

*Balance Mg.

Table 1.11 MJ alloy compositions[50]

Alloy	Alloy symbol	Mn (wt%)	Sr (wt%)
Mg-1%Mn	M1	1.08	<0.001
Mg-1%Mn-0.2%Sr	MJ10(0.2Sr)	0.94	0.17
Mg-1%Mn-0.5%Sr	MJ1(0.5Sr)	0.92	0.48
Mg-1%Mn-1.0%Sr	MJ11(1.0Sr)	0.94	1.02

having an intensity of more than eight times random in the basal pole figure. A small amount of Sr (0.2%) to the M1 alloy does not have any significant effect on the texture of M1 alloy. MJ11 alloys still have a basal texture, but the addition of Sr in excess of 0.5% has a significant texture weakening effect, as indicated by the lower peak intensity (Fig. 1.25). The addition of 0.5%Sr and 1.0%Sr reduces the texture intensity to approximately one-half and one-third of the M1 alloy (Fig. 1.25c, d). Moreover, Sr alloying broadens the angular distribution of basal poles in both the RD and TD (Fig. 1.25). The texture contour broadens from 33° in M1 alloy towards 42° by addition of 1 wt% Sr. Annealing at 450 °C for 15 minutes has a minor effect on the texture of the alloys; the maximum intensity decreases but the general features of pole figure do not alter (Fig. 1.25e–h).

Sr produces second phases that are thermally stable at the rolling temperature (400 °C) and their amount increases with increasing Sr (which increases the potential sites for PSN). Particle-stimulated recrystallization (PSN) was in fact determined to be the likely mechanism underlying texture weakening with increasing Sr. It can be observed that there is a close association between some of the recrystallized grains and the stringers ($Mg_{17}Sr_2$, Sr_5Si_3) (as marked by white frames in Fig. 1.26a, b). In those areas, the recrystallized grains grow till they impinge upon another row of stringers or newly recrystallized grains. An EBSD analysis performed on MJ11(0.5Sr) annealed for 4 min, i.e. a duration that is not sufficiently long to attain a fully recrystallized structure (Fig. 1.26a, b) showed the

Current developments in wrought magnesium alloys

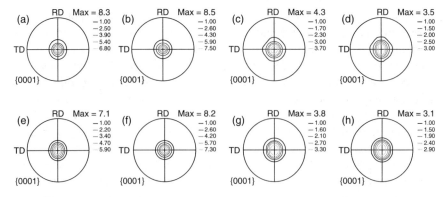

1.25 Surface texture pole figures of alloys, rolled at 400 °C. (a) M1; (b) MJ10(0.2Sr); (c) MJ1(0.5Sr); (d) MJ11(1.0Sr), annealed at 450 °C for 15 min; (e) M1; (f) MJ10(0.2Sr); (j) MJ1(0.5Sr); (h) MJ11(1.0Sr).[50]

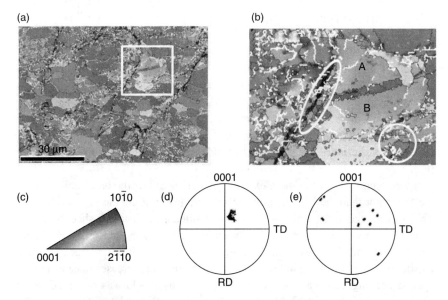

1.26 MJ11(0.5Sr) after 4 min. annealing at 450 °C. (a) EBSD map on the RD-TD plane. (b) Magnified view of the area in the white square. (c) Inverse pole figure. (d) Texture pole figure of grain A and B. (e) Texture pole figure of fine grain close to a stringer.[50]

effect of the PSN grains on texture. Figure 1.26b is the magnified view of the area inside the white square in Fig. 1.26a. In these two figures, the high angle boundaries (>15°) and low angle boundaries (>3°) are delineated with black and white lines, respectively. Some small recrystallized grains, with little or no deformation substructure typical of strain-free recrystallized grains, associated with stringers can be observed in Fig. 1.26b. These recrystallized grains that are associated with the precipitates exhibit a random texture. Figure 26c and d show the pole figure of parent grains (grains A and B) and the recrystallized grains next to a stringer. As seen in Fig. 1.26c, the parent grains show a near basal texture, whereas the recrystallized grains (Fig. 1.26d) show a much wider spread of orientations. This confirms the new recrystallized grains assisted by PSN have different orientations that contribute to the texture weakening. It was observed that some of the recrystallized grains are adjacent to high angle boundaries and are likely to have nucleated through the grain boundary bulging mechanism. These grains have orientations very similar to those of the parent grains (shown in Fig. 1.26b with white circle). The nucleation of new grains in the MJ alloys was associated with both grain boundary bulging and second-phase particles, and that the latter leads to texture weakening. Here, Sr seems to lead to considerable texture weakening, most likely due to the formation of thermally stable second-phase stringers during rolling.

1.3.2 Mg-Zn based alloys

Mg-Zn-RE alloys

Bohlen et al.[51] have observed that the rolling textures of Mg-Zn-RE alloys containing different levels of zinc and RE (e.g. MM or Y) additions (ZK10, ZE10, its modification with zirconium ZEK100, an alloy with higher zinc content ZEK410, a similar alloy containing yttrium ZW41) exhibit lower basal pole intensity aligned with the sheet ND (Fig. 1.27) than for conventional alloys such as AZ31. Furthermore, the strong basal pole intensity is tilted 20° towards the RD (Fig. 1.28).

It was seen that[51] when there exists a greater spread of the basal poles toward the TD rather than the RD, the flow stresses are lower along the TD than the RD. This is attributed to the grains with the c-axis tilted away from the sheet ND being more favorably oriented for 'soft' deformation mechanisms such as basal slip and tensile twinning. During tensile testing along TD, the alloy undergoes profuse tensile twinning, resulting in linear work hardening that increases plastic stability against necking. Also, a reduced planar anisotropy (r ~1) was observed, when compared to conventional Mg alloys. This was also related to the weaker textures with a larger volume fraction of grains oriented favorably to accommodate in-plane tensile deformation by basal slip and twinning facilitating sheet thinning.

The rolled and annealed Mg-1Zn and Mg-1Zn-xCe were characterized by Mackenzie et al.[52] The textures of the alloys (Table 1.12) rolled and annealed at 400 °C or rolled at 150 °C and annealed at 400 °C were studied. Mg-1Zn exhibited basal textures: the basal poles aligned with the ND. With the addition of Ce (as MM), the texture was basal when recrystallization was limited; during recrystallization the basal texture component weakened, to be replaced by a component with basal poles rotated 45° towards the TD.

Current developments in wrought magnesium alloys 39

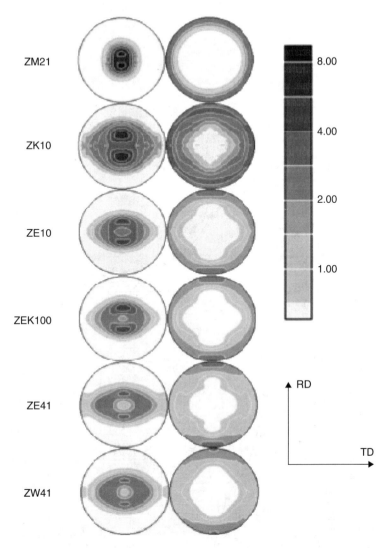

1.27 Texture measurements on rolled samples.[51]

During annealing, a slight texture evolution is seen in Z1 but a basal texture is retained, typical of other conventional Mg alloys. However, Ce was found to have a significant effect on the annealing texture. A non-basal texture developed over a range of alloying additions (0.3–1.0 wt% Ce), and irrespective of the rolling temperature (150 °C or 400 °C). This texture remained stable during relatively long periods of annealing (4 h). The non-basal texture was associated with grains that recrystallize in shear bands (Fig. 1.29), the so-called shear-band nucleation (SBN). No evidence of PSN or TIN was found and the micro-mechanism of

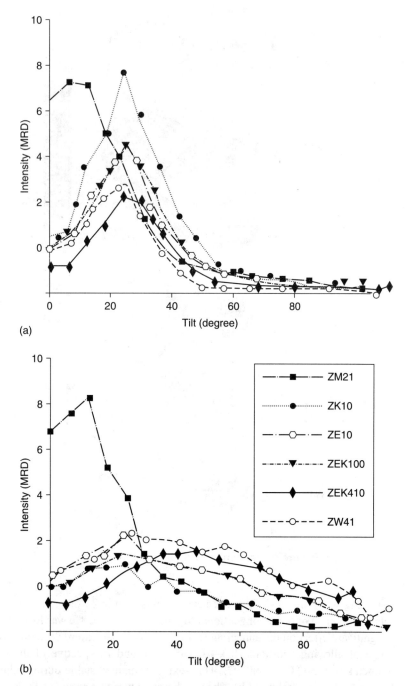

1.28 Intensity of basal pole figures versus tilt from the sheet normal direction (ND) toward: (a) rolling direction (RD); (b) transverse direction (TD).[51]

Table 1.12 Alloy compositions* and approximate grain size[52]

Alloy	Grain size (μm)	Zn (wt%)	Ce (wt%)	Fe (wt%)	Si (wt%)	Ni (wt%)
Mg-1Zn	800	0.93	<0.001	0.003	0.002	0.009
Mg-1Zn-0.3Ce	250	0.90	0.34	0.004	0.002	0.005
Mg-1Zn-0.6Ce	250	1.194	0.648	0.007	0.002	0.004
Mg-1Zn-01.0Ce	250	1.189	1.030	0.006	0.002	0.003

*Balance Mg.

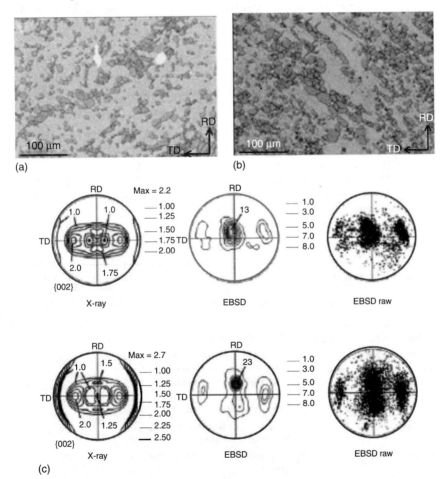

1.29 ZE10 rolled at 150 °C and annealed at 400 °C. (a) 60 s: EBSD orientation micrograph highlighting non-basal (dark)/basal (light), X-ray bulk texture and local EBSD orientation data. (b) 180 s: EBSD orientation micrograph highlighting non-basal (dark)/basal (light), X-ray bulk texture and local EBSD orientation data. (c) Pole figure contour lines are marked as appropriate.[52]

texture randomization was attributed to growth selection of randomly oriented grains due to particle pinning or solute drag that can alter grain boundary mobility.

Usually, the RE additions to Mg alloys are made in the form of MM, which contains a combination of various RE (Ce, La, Nd, Pr). Al-Samman and Li[53] have investigated Mg-Zn-Zr (Table 1.13) alloys containing small additions of single RE elements (ZEK100 alloys). MM alloy was also used as a control. The alloys were cast, homogenized at 450 °C for 12 h, quenched and warm rolled at 400 °C at a reduction per pass of 15–20% with 10 min. intermediate anneals at 400 °C between passes. Recrystallization annealing was carried out at 400 °C for 1 h.

The microstructures and macrostructures differed for the different alloys (Fig. 1.30, Table 1.14). Rolling microstructures (Fig. 1.31) showed equiaxed

Table 1.13 Chemical compositions* of ZEK100 alloys[53]

Alloy	Zn (wt%)	Zr (wt%)	Ce (wt%)	La (wt%)	Nd (wt%)	Gd (wt%)	Pr (ppm)	Fe (ppm)
Ce-series	0.68	0.18	0.78					27
La-series	0.89	0.22		0.69				9
Nd-series	0.61	0.27			0.58			26
Gd-series	0.73	0.24				0.73		9
MM-series	0.84	0.33	0.59	0.21	0.057		232	39

*Balance Mg.

Table 1.14 Micro and macro features of alloys[53]

Alloys	As-cast structure		Average grain size (µm)	
	Average grain size (µm)	Second phases	As-rolled	Annealed
Ce series	33 (uniform)	Mg-Zn-Ce (GB and TJ)	18	20
La series	24 and 80 (bi-modal)	GB and TJ precipitates; Mg-Zr colonies; high amount of randomly distributed fine (<1 µm) precipitates and 6 µm precipitates in the GI	36	36
Nd series	88	Very fine (<1 µm) Mg-Zn-Nd in the GI (low amount)	17	28
Gd series	96	Fine Mg-Zn-Gd in the GI (low amount)	29	42
MM series		Interdendritic plate-like; Mg-Zr colonies; very high amount of randomly distributed fine (<1 µm) precipitates and 6 µm precipitates in the GI	26	31

GB: grain boundary; TJ: triple junction; GI: grain interior.

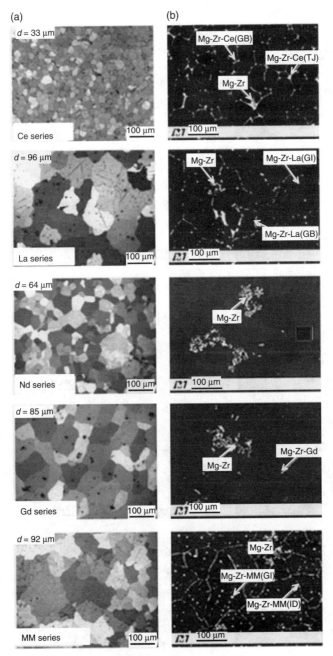

1.30 Cast ZEK100 alloys.[53] (a) Optical micrographs. (b) SEM micrographs.

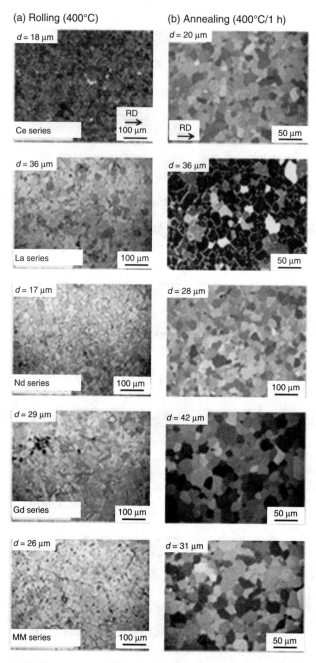

1.31 Microstructures (in the RD–TD plane of the sheet) after (a) warm rolling; (b) annealing.[53]

grains with second-phase particles at the grain boundaries and grains as well as some deformation twins. The average grain sizes after rolling ranged from 18 μm (finest for Ce-series) to 36 μm (coarsest for La-series).

The texture pole figures of the alloys with selective RE additions in the rolled and rolled/annealed conditions are given in Fig. 1.32. Rolling textures show a widening of basal pole distribution toward the TD as well as weakened maximum intensity. The prismatic $\{10\bar{1}0\}$ poles are aligned parallel to the RD. Gd additions result in the most pronounced rolling texture modification, in comparison to the other RE elements investigated. The Gd texture shows weak basal pole intensity located at about 40° from ND toward TD. Annealing textures are weaker than the rolling textures in these alloys, irrespective of the added RE element. The location of the maximum intensity of basal poles remains unchanged, upon annealing, except for the Nd and Gd-alloys. The annealed Nd-containing sheet shows two texture components: maximum intensity of basal poles located close to ND, as well as one tilted by 35° from ND toward TD. In the annealed Gd sheet, the peak intensity of basal poles is no longer in the ND-TD plane of the basal pole figure

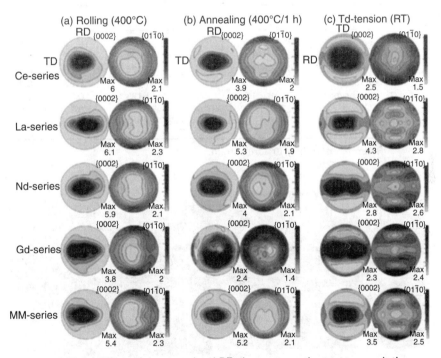

1.32 (a) Effect of the examined RE elements on the texture evolution during warm rolling. (b) Subsequent annealing at 400 °C for 1 h. (c) Tension to failure at room temperature in the transverse direction of the rolled sheets. Textures are represented by means of (0002)- and {1010}- recalculated pole figures.[53]

and is, instead, tilted away from ND by 42° toward the circumference. Figure 1.31 shows that annealing results in minor grain growth.

As observed, Nd and Gd-alloys seem to be more prone to grain growth compared with the other precipitate-containing alloys (Ce, La and MM), wherein grain boundary motion would be impeded by the second-phase particles. Indications of PSN, associated with the large Mg-Zr particles (Fig. 1.33) and not with the RE-containing precipitates, were found in the annealed rolled sheets, but this was not the most important micro-mechanism of texture alteration in the alloys. The high solubility of RE elements such as Gd and Nd were found to play a more significant role, exerting solute-related effects on, for example, grain boundary motion, which could alter recrystallization and grain growth. It is also found that the RE additions and the resultant texture lead to improved isotropy as seen from r-values (Lankford coefficients) as summarized in Table 1.15.

Mg-Zn-Gd alloys were further investigated by Wu et al.[54] in a composition containing higher amounts of Gd (Mg-2%Gd-1%Zn and Mg-3%Gd-1%Zn). The

Table 1.15 Sheet anisotropy from r-values obtained via tension tests[53]

Alloys	r_{RD}	R_{45RD}	r_{TD}	r	Δr	Δr$_2$
Ce-series	1.2	1.3	1.7	1.375	0.15	0.5
La-series	1.0	1.1	0.9	1.025	−0.15	0.2
Nd-series	1.0	1.1	0.9	1.025	−0.05	0.2
Gd-series	1.0	1.0	1.1	1.025	0.05	0.1
MM-series	1.2	1.1	1.1	1.125	0.05	0.1
AZ31	2.2		4	3.05	0.1	1.8
AZ31				2.27	0.3	

Room temperature, 10% true strain.

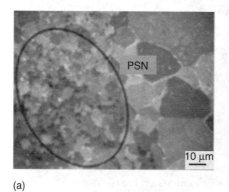
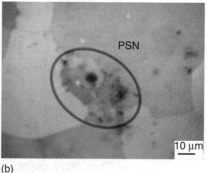

1.33 Highly magnified optical microstructures containing sheets showing indications of particle-stimulated nucleation (PSN) of recrystallization after annealing. (a) Ce. (b) Gd.[53]

rolled sheets recrystallized and had a large amount of homogeneously distributed tiny particles in the matrix. The basal plane texture intensity was quite low and the location of the maximum intensity of basal poles is 30° from ND toward both RD and TD. The sheets exhibit 50% elongation at fracture, uniform elongation >30%, and the Erichsen values of ~8 at room temperature. The two Mg-Gd-Zn alloy sheets have very good linear hardening. The excellent stretch formability at room temperature could be attributed to the non-basal texture and low texture intensity, which led to a lower 0.2% proof stress, a larger uniform elongation, a smaller Lankford value and a larger strain hardening exponent than AZ31 (Table 1.16).

The additions of REs (0.5–3%) to the conventional extrusion alloy ZK60 (Mg-6Zn-0.4Zr) alloys were studied by Ma et al.[55] The alloys were extruded at 300 °C, 340 °C, 380 °C and 420 °C with an extrusion ratio of 10:1. The RE additions decreased the extruded grain size from 20 μm down to 2–3 μm at 3% RE. The grain refinement was associated with as-cast grain-size refinement as well as with growth restriction by Mg-Zn-RE intermetallics during extrusion. The alloys had a deformation structure after extrusion at 300 °C and 340 °C, were partially recrstallized at 380 °C and almost fully recrystallized at 420 °C.

He et al.[56] added 1.3%Gd to ZK60 and studied the extruded-T6 properties. Gd addition refined the extruded structure both at the T5 and T6 heat-treated conditions (Fig. 1.34). The grain refining effect of Gd was attributed to the grain-boundary pinning effect of Mg-Zn-Gd phase particles. The grain refinement increased the ductility but resulted in some loss of age hardening strength due to the reduction of Zn solubility in the Mg matrix. The loss of strength was partly compensated by the refinement in grain size.

The effect of yttrium additions on Mg-Zn alloys has also been studied. Mg-Zn-Y-Zr alloys with different Y content (0–3 wt%), extruded at 390 °C with an extrusion ratio of 10:1 were evaluated for microstructure and texture.[57] The icosahedral quasicrystal phase (I-phase, Mg_3Zn_6Y) that forms in the alloys was

Table 1.16 Mechanical response of Mg-Gd-Zn sheets[54]

Alloy	r_{RD}	$R_{45°}$	r_{TD}	r	Δr	RD	45°	TD
GZ21	0.8	0.6	0.8	0.75	0.2			
GZ31	0.8	1	1	0.8	0.3			
AZ31[64]	2.2	3	4	3.1				
Strain-hardening exponent of the Mg-Gd alloys along different directions								
GZ21						0.24	0.27	0.29
GZ31						0.24	0.28	0.29
Low carbon steel[65]						0.20–0.25		
Al alloy[65]						0.20–0.30		
AM30[15]						0.17		
Rolled AZ31[15]						0.14		

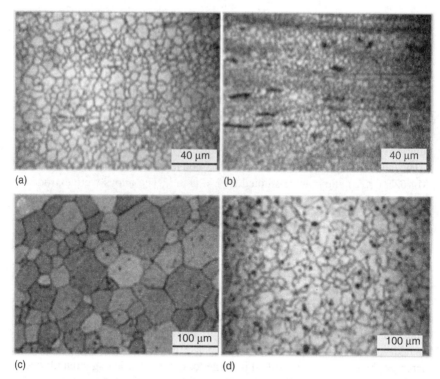

1.34 Optical micrographs of the alloys extruded at 350 °C. (a) ZK60,T5; (b) ZK60 + 1.3Gd,T5; (c) ZK60,T6; (d) ZK60 + 1.3Gd,T6.[56]

found to be more effective in grain refining than the cubic W-phase ($Mg_3Zn_3Y_2$). Yttrium had some effect on extruded texture: with Y content increasing, the orientation of the basal plane tends to TD but the intensity in (0 0 0 2) pole figures do not change much. Extruded samples tensile tested parallel to the extrusion and transverse directions indicated that the I-phase could effectively improve the strength (yield strength and ultimate tensile strength) of the alloys while the W-phase was not a significant strengthener and at high amounts decreased the strength. With the increase of Mg-Zn-Y phases, the anisotropy of the ultimate tensile strength (UTS) between the longitudinal and transverse directions also increases, which was attributed to the role of the distribution and quantity of the Mg-Zn-Y precipitates and especially the W phase.

Kim et al.[58] investigated the effect Al microalloying on the elevated temperature deformation behavior of Mg-Zn-Y system. Al leads to the formation of Al_2Y (Fig. 1.35), which occupies interdendritic regions together with the I-phase. Grain refinement and decreased Y solid solubility are the other effects of Al alloying to Mg-Zn-Y. Mg-4Zn-0.8Y-0.2Al and Mg-4Zn-0.8Y were rolled 400 °C, annealed at 300 °C for 20 minutes and subjected to uniaxial tensile and strain rate change

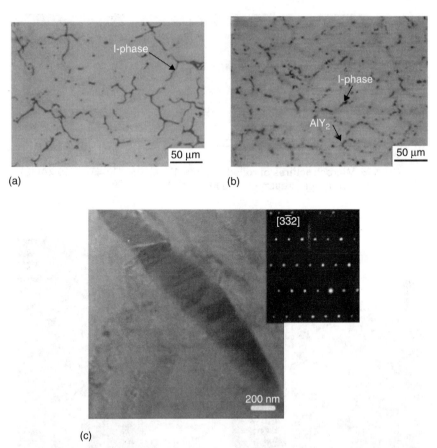

1.35 (a) As-cast Mg-4Zn-1Y. (b) As-cast Mg-4Zn-1Y-0.2Al. (c) Bright field TEM image of AlY$_2$ in Mg-4Zn-1Y–0.2Al and corresponding selected area diffraction pattern (SADP).[58]

(SRC) tests in the temperature interval of 300–400 °C and strain rate range of 1×10^{-4} s^{-1} to 1×10^{-2} s^{-1}. Both alloys were recrystallized with average grain size of 10–11 μm (Fig. 1.36). The alloys exhibited superplastic behavior at 1×10^{-4} s^{-1}; the elongation increased from 260% to 370% with Al microalloying and the mechanism of deformation changed from solute drag to grain boundary sliding (Fig. 1.37). The Al affected the change by decreasing Y solubility and solute drag and activating grain boundary sliding.

Mg-Zn-alkaline earth alloys

Alkaline earth additions exert effects similar to RE in Mg alloys. The one advantage of alkaline earths over RE is their thermal stability, usually related to

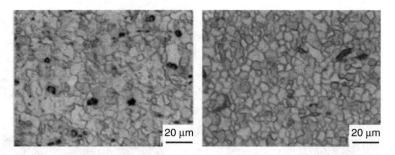

1.36 Microstructures of rolled Mg-4Zn-1Y (a) and Mg-4Zn-1Y-0.2Al (b) alloys after heat treatment at 300 °C for 20 min.[58]

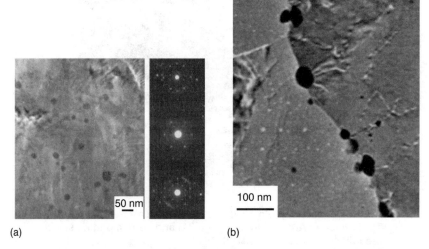

1.37 TEM images of Mg-4Zn-1Zr-0.2Al. (a) Rolled/annealed and (b) SADP showing 2-, 3- and 5-fold rotational symmetry zones of the I-phase. Grain boundary pinning by I-phase. Deformed at 300 °C with a strain rate of 1×10^{-4} s^{-1}.[58]

the fact that individual alkaline earths (Sr, Ca) can easily be added, whereas RE are usually added in the form of MM or a group of RE. Ca additions to Mg-Zn have been investigated in relation to ductility and fracture toughness,[59] which could be improved via the control of grain size and precipitate dispersion.[59] Extruded Mg-1.8Zn-0.3Ca alloy yields fine grain size (1 μm), thermally induced very fine dispersoids ($Mg_6Zn_3Ca_2$) and coarser, 2–3 μm, precipitates (Fig. 1.38) that are crack initiators.

Mg-6Zn alloy was microalloyed with Ca, Ag and/or Zr to investigate the ageing behavior and properties.[60] Mg-6Zn, Mg-6Zn-0.6Zr, Mg-6Zn-0.2Ca, Mg-6Zn-

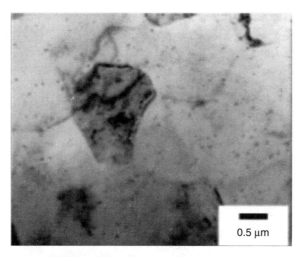

1.38 TEM image of the extruded Mg-Zn-Ca alloy showing fine and larger precipitates.[59]

0.4Ag, Mg-6Zn-0.4Ag-0.2Ca and Mg-6Zn-0.4Ag-0.2Ca-0.6Zr alloys were hot extruded and aged. Zr refined the grain size but did not affect the ageing response, whereas Ca and Ag additions refined the β' precipitate size (Fig. 1.39), and increased the number density, thereby improving ageing response and peak hardness (Fig. 1.40). The combined addition of Ag+Ca worked better than the addition of the individual elements. From 3D-atom probe data, the role of Ca was understood to be the formation of co-clusters in the pre-precipitation stage with

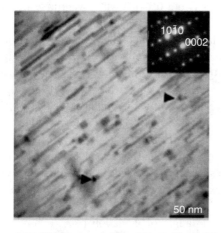

1.39 TEM image of Mg-6Zn-0.4Ag-0.2Ca extruded and T6 treated.[60]

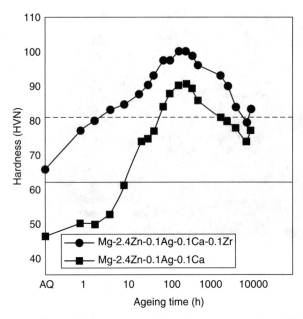

1.40 Hardening response of the alloys extruded at 350 °C, aged at 160 °C. Horizontal lines: as-extruded (350 °C) hardness (dashed: with Zr; solid: without Zr).[60] The values in the key are in atomic percent.

Zn to provide nucleation sites, while that of Ag in the early stages of ageing could not be clearly determined.

In the study by Masoumi et al.,[61] the microstructure and texture of rolled (400 °C) and rolled/annealed (400 °C/400 °C, 5 min.) alloys based on Mg-1wt%Zn and containing various levels of Sr (from 0.1–1.0 wt%) were investigated. The Sr addition refined the as-cast and rolled/annealed grain structure of the Mg-1wt%Zn alloy (Fig. 1.41, 1.42). Moreover, the overall texture intensity of basal pole was weakened for rolled as well as rolled/annealed Mg-Zn-Sr (ZJ) alloys compared to the Mg-1wt%Zn (Z1) alloy. The peak intensity of the basal poles was inclined from the sheet ND towards the RD forming a two-pole basal texture. The texture weakening effect is minor in the ZJ10(0.1) alloy, but the addition of higher amount of Sr further decreases the maximum intensity of the ZJ alloys, and at 1 wt%Sr, the texture intensity is about two-thirds of Z1 alloy. The (0001) texture contour of rolled ZJ alloys expands in the TD more than the RD. Sr has no significant effect on the angular distribution of basal poles in the RD, but the texture contour significantly broadens in the TD, from 30° in Z1 alloy towards 57° at ZJ11(1wt%Sr). The percentage of basal planes plotted as a function of the angle of deviation ($\Delta\theta_{ND}$) from the sheet

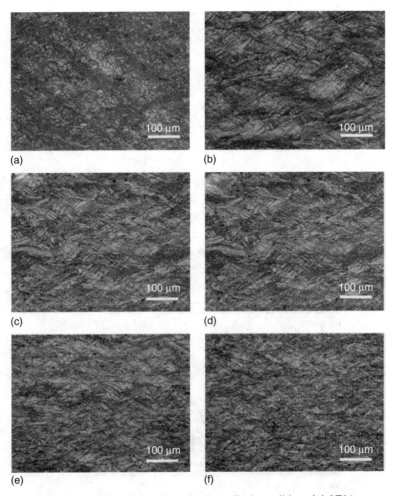

1.41 Microstructures of the alloys in the rolled condition. (a) AZ31; (b) Z1; (c) ZJ10(0.1Sr); (d) ZJ10(0.3Sr); (e) ZJ11(0.5Sr); (f) ZJ11(1.0Sr).[61]

ND shows that, with Sr > 0.3 wt%, the $\Delta\theta_{ND}$ of basal-pole peak intensity in ZJ alloys moves from ±15° seen in Z1, ZJ10(0.1Sr) and AZ31 to ±25° (Fig. 1.43).

Texture weakening in the Mg-Zn-Sr system has been attributed to PSN of recrystallization.[61] EBSD analysis performed on partially recrystallized ZJ11(1.0Sr) annealed at 400 °C for 3 min. Figure 1.44 shows small recrystallized grains, associated with stringers (here the high angle boundaries >15° are delineated with black lines and low angle boundaries >3° with white lines). The pole figure of one parent grain (grain A in Fig. 1.44b) and the recrystallized grains

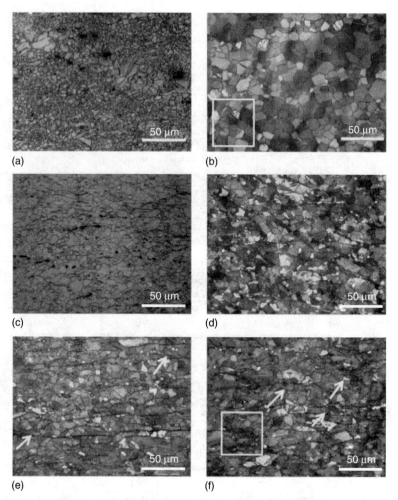

1.42 Microstructures of the alloys in the rolled/annealed condition. (a) AZ31; (b) Z1; (c) ZJ10(0.1Sr); (d) ZJ10(0.3Sr); (e) ZJ11(0.5Sr); (f) ZJ11(1.0Sr).[61]

next to stringer (area inside the white oval in Fig. 1.44b) are given in Fig. 1.44c, d. The parent grains show a near basal texture, while the recrystallized grains shave a much wider spread of orientations. As the Sr level in the alloy increases, the PSN becomes more abundant and the texture is increasingly weakened. There are also recrystallized grains adjacent to high angle boundaries that have orientations similar to the parent grain. These are likely to have nucleated through the grain boundary bulging mechanism.

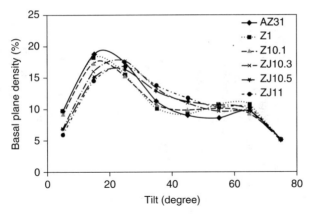

1.43 Density of basal planes versus deviation angle from sheet normal direction ($\Delta\theta_{ND}$) in as-rolled condition.[61]

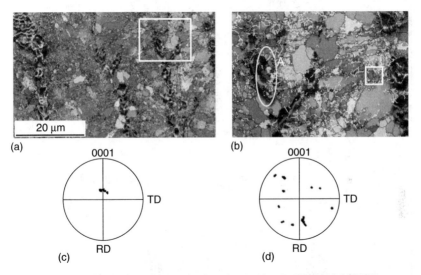

1.44 ZJ11(1.0Sr) after annealing for 3 min. at 400 °C. (a) EBSD map on the RD–TD plane. (b) Magnification of area in the white square in panel (a); small white square here shows recrystallized grains adjacent to high angle boundaries with orientations similar to the parent grain. (c) Texture pole figure of grain A. (d) Texture pole figure of fine grain close to stringer.[61]

1.4 Future trends in wrought magnesium alloy development

Recent research and development activities on Mg wrought alloys have begun to shed light on alloy composition–microstructure–texture–mechanical behaviour relationships. Table 1.17 gives a summary of the key trends in these relationships.

Table 1.17 Current development and progress in Mg wrought alloys

Alloying addition(s)	Wrought grain structure	Second phases	Texture/mechanisms	Yield asymmetry	Ductility (%e)/strength formability
Modified Mg-Al based alloys					
AZ31+Ca (hot extruded)[10]	Refined grains	Ca intermetallics	Weaker fiber texture/PSN	1.3	
AZ31+Sr (hot extruded)[11,12]	Mixed mode* (250 °C), fine uniform (350 °C)	Al$_4$Sr intermetallics	Strong texture at 250 °C and low solute (Al) drag. Weak texture at 350 °C/PSN	1.4–1.3 (low T and Sr and low T/high Sr)	High %e at 0.8%Sr High strength at 0.05%Sr
AM30+Ce (hot rolled)[7]	Refined grains	Mg$_{12}$Ce dispersion	Weaker texture/possibly PSN	1.9–2.0	Improved extrudability
Mg-RE alloys Mg-Gd-Y-T6 (hot extruded)[24,25]	Refined grains	β' phase	Weak texture		High %e
Mg-RE-Sn (Al) (hot rolled)[26]		Al$_2$MM	Weaker texture		High %e, and formability Low strength
Mg-Li-based alloys					
AZ31+3Li (hot rolled)[33]	Deformed grain structure (no DRX)		Basal poles split in TD; single peak when annealed due to growth selection		Improved rollability
Mg-Li-Al-Zn (hot extruded and rolled)[35]		α–β dual structure			Good combination of strength and ductility
Mg-Zn-Li, Mg-Zn-Li-In (150 °C roll 400 °C anneal)[39]	Refined grain size	Solid solution alloys	Weaker texture due to refined grain size and α-lattice parameter		Reduced edge cracking (double twins rather than tensile due to c/a change

Alloy	Processing	Phases/Precipitates	Texture	r
Mg-Li-Zn-Al-RE (extruded)[40]	DRX at 2 and 3%RE; deformed at 1 and 6%RE†	Al$_3$La		Depends on extrusion T, Li:Al:Zn:RE ratio and Zn content[40–41]
Mg-Li-Zn-Y-Ce[44]	DRX depends on extr. T and strain rate (best: 250 °C/10 s^{-1} and 250 °C/0.001 s^{-1})	α-Mg, β-Li, and Mg-Zn-Zr-RE in the β. Mg-Zn-Zr-RE accelerates DRX, β retards DRX,		
Mg-Mn based alloys				
Mg-Mn-Ce (400 °C roll, 400 °C anneal)[48]	Refined grains (both rolled and post annealed)	Mg-Ce and α-Mn precipitates	Basal poles spread in RD and TD. Weaker texture due to solute drag	
Mg-Mn-Nd (extruded)[49]		Fine precipitates	Weaker textures than Y or Ce additions; basal poles tilted away from ED‡	
Mg-Mn-Sr (400 °C roll, 400 °C anneal)[50]	Refined grains	Mg-Sr intermetallics	Weak texture, basal pole distribution widens in the RD and TD/PSN	
Mg-Zn-based alloys				
Mg-Zn-Zr-RE (rolled)[51]			Weaker textures, basal poles tilted 20° from RD towards TD	r ~ 1
Mg-Zn-Ce (150 °C roll 400°C anneal)[52]			Weak rolling texture: basal poles 45° to TD; non-basal annealing texture (growth selection of randomly oriented grains in shear bands)	

(Continued)

Table 1.17 Continued

Alloying addition(s)	Wrought grain structure	Second phases	Texture/mechanisms	Yield asymmetry	Ductility (%e)/strength formability
Mg-Zn-REs (rolled and post-annealed)[53] (1) Mg-Zn-Gd (2) Mg-Zn-Nd	18–36 μm	(1) Fine Mg-Zn-Gd (2) Very fine (<1 μm) Mg-Zn-Nd	Weak texture tilted 40° to TD. Weaker annealing textures: PSN on Mg-Zr precipitates (1) Basal poles tilted 42° away from ND. (2) Two texture components (poles close to ND and tilted 35° from ND towards TD). Solute effects	r ~ 1	
Mg-Zn-Gd (rolled)[54]		Tiny particles	Weak texture, basal poles tilted 30° from ND both to RD and TD.		
Mg-Zn-Zr-Y (extruded)[57]		I-phase, (Mg$_3$Zn$_6$Y), W-phase (Mg$_3$Zn$_3$Y$_2$)	Basal pole rotation towards TD	Anisotropy increases especially with the W-phase	High ductility, good formability (cup test), high strain hardening, r value close to 1 I-phase increases strength, W-phase decreases strength
Mg-Zn-Sr (400 °C roll, 400 °C anneal)[61]	Refined grains		Weak texture with two components: poles at ND and at RD. Poles spread in TD ~57° at 1wt% Sr/PSN		

%e, percentage elongation.
*Mixture of fine recrystallized, elongated deformed grains; †Li:Al:Zn:RE ratio affects recrystallization; ‡extrusion direction.

In most cases, the effort has concentrated on understanding the changes in microstructure and texture and the micro-mechanisms of texture alteration. Texture weakening in these alloys has been related to PSN,[7,10–12,50,61–63] solute effects,[11,12,48,53] refined grain size[39] or growth selection of randomly oriented new grains.[33,52] Characterization of most alloys is incomplete to date, especially in relating texture to anisotropy, formability and mechanical properties.

It is noted that weak textures are obtained by adding REs and alkaline earths to binary or ternary Mg alloy compositions. The underlying mechanisms are the generation of second-phase stringers that nucleate new recrystallized grains with orientations different from parent grains (PSN). Change in solute solubility at the hot working temperature is also important. Reduced solute strengthens basal texture by preventing solute drag (facilitating basal slip) and increased solute can weaken texture via solute drag. Growth selection of randomly oriented grains can further weaken texture during annealing. Alkaline earths (especially Sr) act similarly to REs[53,61] producing a two-component texture in Mg-Zn alloys. Mg-Zn based alloys seem to offer good opportunity for further development. Zn refines grains, improves both strength and ductility and, with further alloying with RE or alkaline earths, produces alloy compositions that offer weak and more random textures and improved formability.

1.5   References

1. Bettles, C. J. and Gibson, M. (2005) 'Current wrought magnesium alloys: Strengths and weaknesses', *JOM*, **57**, 5, 46–49.
2. Industry Canada (1999) *Use of Magnesium in Energy Applications*, Industry Canada, Ottawa, ON.
3. Diamond, S., Sklad, P., Novoa, L., Tolani, N. and Atchison, R. (ORNL Report HSWR-001, Sep 2002) *Research and Development Plan for High-Strength/Weight Reduction Materials*, Oakridge National Library, Oakridge, TN.
4. Yoo, M. H. (1981) 'Slip, twinning, and fracture in hexagonal close-packed metals', *Metall. Trans. A*, **12A**, 409–418.
5. Yoo, M. H. and Lee, J. K. (1991) 'Deformation twinning in h.c.p. metals and alloys', *Phil. Mag. A*, **63**, 5, 987–1000.
6. Wang, J., Hirth, J. P. and Tome, C. N. (2009) 'Twinning nucleation mechanisms in hexagonal-close-packed crystals', *Acta Mater.*, **57**, 5521–5530.
7. Luo, A., Wu, W., Mishra, R. K., Jin, L., Sachdev, A. K. *et al.* (2010) 'Microstructure and mechanical properties of extruded magnesium-aluminum-cerium alloy tubes' *Metall. Mater. Trans. A*, **41A**, 2662–2674.
8. Masoumi, M., Zarandi, F. and Pekguleryuz, M. (2010) 'Alleviation of basal texture in twin roll cast AZ31 magnesium alloy', *Scripta Mater.*, **62**, 11, 823–826.
9. Masoumi, M., Zarandi, F. and Pekguleryuz, M. (2011) 'Microstructure and texture studies on twin-roll cast AZ31 alloy and the effect of thermomechanical processing', *Mater. Sci. Eng. A*, **528**, 1268–1279.
10. Laser, T., Hartig, Ch., Nurnberg, M. R., Leitzig, D. and Bormann, R. (2008) 'The influence of calcium and cerium mischmetall on the microstructural evolution of

Mg-3Al-1Zn during extrusion and resulting mechanical properties', *Acta Mater.*, **56**, 2791–2798.
11. Sadeghi, A. and Pekguleryuz, M. (2011) 'Recrystallization and texture evolution of Mg–3%Al–1%Zn–(0.4–0.8)%Sr alloys during extrusion', *Mater. Sci. Eng. A*, **528**, 1678–1685.
12. Sadeghi, A., Hoseini, M. and Pekguleryuz, M. (2011) 'Effect of Sr addition on texture evolution of Mg–3Al–1Zn (AZ31) alloy during extrusion', *Mater. Sci. Eng. A*, **528**, 3096–3104.
13. Masoudpanah, S. M. and Mahmudi, R. (2009) 'Effects of rare earth elements and Ca on the microstructure and mechanical properties of AZ31 magnesium alloy processed by ECAP', *Mater. Sci. Eng. A*, **526**, 22–30.
14. Fang, X–Y., Yi, D-Q., Wang, B., Luo, W-H and Gu, W. (2006) 'Effect of yttrium on microstructures and mechanical properties of hot rolled AZ61 wrought magnesium alloy', *Trans. Nonferrous Met. Soc. China*, **16**, 5, 1053–1058.
15. Luo, A. and Sachdev, A. (2007) 'Development of a new wrought magnesium-aluminum-manganese alloy AM30', *Metall. Mater. Trans. A*, **38**, 1184–1192.
16. Wang, J., Liao, R., Wang, L., Wu, Y., Cao, Z. et al. (2009) 'Investigations of the properties of Mg-5Al-0.3Mn-xCe (x = 0–3, wt.%) alloys', *J. Alloys Compd.*, **477**, 341–345.
17. Stanford, N. and Barnett, M. R. (2008) 'The origin of "rare earth" texture development in extruded Mg-based alloys and its effect on tensile ductility', *Mater. Sci. Eng. A*, **496**, 399–408.
18. Yang, Z., Li, J. P., Zhang, J. X., Lorimer, G. W. and Robson, J. (2008) 'Review on research and development of magnesium alloys', *Acta Metall. Sin. (Engl. Lett.)*, **21**, 5, 313–328.
19. He, S. M., Zeng, X. Q., Peng, L. M., Gao, X., Nie, J. F. et al. (2007) 'Microstructure and strengthening mechanism of high strength Mg–10Gd–2Y–0.5Zr alloy', *J. Alloys Compd.*, **427**, 316–323.
20. Yang, Z., Li, J. P., Guo, Y. C., Liu, T., Xia, F. et al. (2007) 'Precipitation process and effect on mechanical properties of Mg-9Gd-3Y-0.6Zn-0.5Zr alloy' *Mater. Sci. Eng. A*, **454–455**, 274–280.
21. Li, J. P., Yang, Z. and Liu, T. (2007) 'Microstructures of extruded Mg-12Gd-1Zn-0.5Zr and Mg-12Gd-4Y-1Zn-0.5Zr alloys', *Scripta Mater.*, **56**, 137–140.
22. Zhang, P., Ding, W., Lindemann, J. and Leyens, C. (2009) 'Mechanical properties of the hot-rolled Mg-12gd-3Y magnesium alloy', *Mater. Chem. Phys.*, **118**, 2/3, 457–458.
23. Liu, X., Chen, R-Sh. and Han, E-H. (2008) 'High temperature deformations of Mg-Y-Nd alloys fabricated by different routes', *Mater. Sci. Eng. A*, **497**, 326–332.
24. Sheng, F-Sh., Zhang, J., Wang, J-F., Yang, M-B., Han, E-H. et al. (2010) 'Key RD activities for new types of wrought Mg alloys in China', *Trans. Nonferrous Met. Soc. China*, **20**, 1249–1258.
25. Li, X., Liu, C., Al-Samman, T., (2011) 'Microstructure and mechanical properties of Mg –12Gd–3Y–0.6Zr alloy upon conventional and hydrostatic extrusion', *Mater. Lett.*, **65**, 1726–1729.
26. Lim, H. K., Kim, D. H., Lee, J. Y., Kim, W. T. and Kim D. H. (2009) 'Effects of alloying elements on microstructures and mechanical properties of wrought Mg–MM–Sn alloy', *J. Alloys Compd.*, **468**, 308–314.
27. Raynor, G. V. (1959) *The Physical Metallurgy of Magnesium and its Alloys*, Pergamon Press, London.
28. Hauser, F. E., Landon P. R. and Dorn J. E. (1958) 'Deformation and fracture of alpha solid solutions of lithium in magnesium', *Trans. ASM*, **50**, 856–883.

29. Busk, R. S. (1950) 'Lattice parameters of magnesium alloys', *Trans. Met. Soc. AIME*., **188** 1460–1464.
30. Agnew, S. R., Yoo, M. H. and Tome, C. N. (2001) 'Application of texture simulation to understanding mechanical behavior of Mg and solid solution alloys containing Li or Y', *Acta Mater.*, **49**, 4277–4289.
31. Agnew, S. R., Horton, J. A. and Yoo, M. H. (2002) 'TEM investigation of dislocation structures in Mg and Mg-Li α-solid solution alloys', *Metall. Mater. Trans.*, **33A**, 851–858.
32. Stohr, J. F and Poirier, J. P. (1972) 'Etude en microscopie electronique du glissement pyramidal (1122) 1123) dans le magnesium', *Philos. Mag.*, **25**, 1313–1329.
33. Mackenzie, L. W. F. and Pekguleryuz, M. (2008) 'The influences of alloying additions and rolling parameters on the rolling microstructures and textures of Mg alloys', *Mater. Sci. Eng. A*, **480**, 189–197.
34. Styczynski, A., Hartig, Ch., Bohlen, J. and Letzig, D. (2004) 'Cold rolling textures in AZ31 wrought magnesium alloy', *Scripta Mater.*, **50**, 943–947.
35. Meng, X., Wu, R., Zhang, M., Wu, L. and Cui, Ch. (2009) 'Microstructures and properties of superlight Mg–Li–Al–Zn wrought alloy', *J. Alloys Compd.*, **486**, 722–725.
36. Becerra, A. and Pekguleryuz, M. (2008) 'Effects of lithium, indium and zinc on the lattice parameters of magnesium', *J. Mater. Res.*, **23**, 12, 3379–3386.
37. Becerra, A. and Pekguleryuz, M. (2009) 'Effects of lithium, indium and zinc on the grain size of magnesium', *J. Mater. Res.*, **24**, 5, 1722–1729.
38. Uesugi, T., Kohyama, M., Kohzu, M. and Higasi, K. (2003) 'Generalized stacking fault energy and dislocation properties for various slip systems in magnesium: A first principles study', *Mater. Sci. Forum, Trans. Tech.*, **419–422**, 225–230.
39. Pekguleryuz, M., Celikin, M., Hoseini, M., Becerra, A. and Mackenzie, L. (2011) 'Study on edge cracking and texture evolution during 150°C rolling of magnesium alloys: The effects of axial ratio and grain size', *J. Alloys Compd.*, doi:10.1016/j.jallcom.2011.08.093.
40. Wu, R. Z., Deng, Y. S. and Zhang, M. L. (2009) 'Microstructure and mechanical properties of Mg–5Li–3Al–2Zn–xRE alloys', *J. Mater. Sci.*, **44**, 4132–4139.
41. Wang, T., Zhang, M-L. and Wu, R-Z. (2007) 'Microstructure and mechanical properties of Mg-5.6Li-3.37Al-1.68Zn-1.14Ce alloy', *Trans. Nonferrous Met. Soc. China*, **17**, 444–447.
42. Wang, T., Zhang, M-L. and Wu, R-Z. (2008) 'Microstructure and properties of Mg–8Li–1Al–1Ce alloy', *Mater. Lett.*, **62**, 1846–1848.
43. Xu, D. K., Liu, L., Xua, Y. B. and Han, E. H. (2007) 'The strengthening effect of icosahedral phase on as-extruded Mg–Li alloys', *Scripta Mater.*, **57**, 285–288.
44. Chen, Z., Li, Z. and Yu, C. (2011) 'Hot deformation behavior of an extruded Mg–Li–Zn–RE alloy', *Mater. Sci. Eng. A*, **528**, 961–966.
45. Emley, E. F. (1966) *Principles of Magnesium Technology*, Pergamon Press, London.
46. Friedrich, H. E. and Mordike, B. L. (2006) *Magnesium Technology: Metallurgy, Design Data*, Springer, Berlin, New York.
47. Robson, J. D., Henry, D. T. and Davis, B. (2009) 'Particle effects on recrystallization in magnesium–manganese alloys: Particle-stimulated nucleation', *Acta Mater.*, **57**, 2739–2747.
48. Masoumi, M., Hoseini, M. and Pekguleryuz, M. (2011) 'The influence of Ce on the microstructure and rolling texture of Mg-1%Mn alloy', *Mater. Sci. Eng. A*, **528**, 7–8, 3122–3129.

49. Bohlen, J., Yi, S., Letzig, D. and Kainer, K. U. (2010) 'Effect of rare earth elements on the microstructure and texture development in magnesium–manganese alloys during extrusion', *Mater. Sci. Eng. A*, **527**, 26, 7092–7098.
50. Masoumi M. and Pekguleryuz, M. (2011) 'Effect of Sr on the Microstructure and texture of rolled Mg-Mn-based alloys', M. Masoumi, in *Microstructure And Texture Studies On Magnesium Sheet Alloys*, PhD Thesis, McGill University, Montreal, Canada.
51. Bohlen, J., Nurmberg, M. R., Senn, J. W., Letzig, D. and Agnew, S. R. (2007) 'The texture and anisotropy of magnesium–zinc–rare earth alloy sheets', *Acta Mater.*, **55**, 2101–2112.
52. Mackenzie, L. W. F. and Pekguleryuz, M. O. (2008) 'The recrystallization texture of magnesium-zinc-cerium alloys', *Scripta Mater.*, **59**, 6, 665–668.
53. Al-Samman, T. and Li, X. (2011) 'Sheet texture modification in magnesium-based alloys by selective rare earth alloying', *Mater. Sci. Eng. A*, **528**, 3809–3822.
54. Wu, D., Chen, R. S. and Han, E. H. (2011) 'Excellent room-temperature ductility and formability of rolled Mg–Gd–Zn alloy sheets', *J. Alloys Compd.*, **509**, 2856–2863.
55. Ma, C., Liu, M., Wu, G., Ding, W., and Zhu, Y. (2003) 'Tensile properties of extruded ZK60/RE alloys', *Mater. Sci. Eng. A*, **349**, 207–212.
56. He, S. M., Peng, L. M., Zeng, X. Q., Ding, W. J. and Zhu, Y. P. (2006) 'Comparison of the microstructure and mechanical properties of a ZK60 alloy with and without 1.3 wt.% gadolinium addition', *Mater. Sci. Eng. A*, **433**, 175–181.
57. Xu, D. K., Liu, L., Xu, Y. B. and Han, E. H. (2007) 'Effect of microstructure and texture on the mechanical properties of the as-extruded Mg–Zn–Y–Zr alloys', *Mater. Sci. Eng. A*, **443**, 248–256.
58. Kim, D. H., Lee, J. Y., Lim, H. K., Kim, W. T. and Kim, D. H. (2008) 'Effect of Al addition on the elevated temperature deformation behavior of Mg–Zn–Y alloy', *Mater. Sci. Eng. A*, **487**, 481–487.
59. Somekawa, H. and Mukai, T. (2007) 'High strength and fracture toughness balance on the extruded Mg–Ca–Zn alloy', *Mater. Sci. Eng. A*, **459**, 366–370.
60. Mendis, C. L., Oh-ishi, K., Kawamura, Y., Honma, T., Kamado, S. *et al.* (2009) 'Precipitation-hardenable Mg–2.4Zn–0.1Ag–0.1Ca–0.16Zr (at.%) wrought magnesium alloy', *Acta Mater.*, **57**, 749–760.
61. Masoumi, M. and Pekguleryuz, M. (2011) 'The influence of Sr on the microstructure and texture evolution of rolled Mg–1%Zn alloy', *Mater. Sci. Eng. A*, doi:10.1016/j.msea.2011.09.019.
62. Sadeghi, A. and Pekguleryuz, M. (2011) 'Microstructural investigation and thermodynamic calculations on the precipitation of Mg–Al–Zn–Sr alloys', *J. Mater. Res.*, **26**, 7, 1–8.
63. Sadeghi, A. and Pekguleryuz, M. (2011) 'Microstructure, mechanical properties and texture evolution of AZ31 alloy containing trace levels of strontium', *Mater. Charact.*, **62**, 742–750.
64. Chino, Y. and Mabuchi, M. (2009) 'Enhanced stretch formability of Mg-Al-Zn alloy sheets rolled at high temperature', *Scripta Mater.*, **60**, 447–450.
65. Duygulu, O. and Agnew, S. R. (2003)' The effect of temperature and strain rate on the tensile properties of textured magnesium alloy AZ31B sheet', *Magnes. Technol.*, 237–242.

2
Deformation mechanisms of magnesium alloys

S. R. AGNEW, University of Virginia, USA

Abstract: This chapter discusses the deformation mechanisms responsible for the complex mechanical behaviors exhibited by Mg and its alloys. The use of new experimental (e.g. electron backscattered diffraction (EBSD) and *in situ* neutron diffraction) and computational (e.g. crystal plasticity, molecular dynamics, and electron density functional theory) approaches has recently provided a wealth of new understanding. These findings are enumerated for each of the basal and non-basal dislocation slip mechanisms as well as deformation twinning and kink banding. The impacts of alloy and microstructure design strategies to improve strength, ductility, formability, and tolerance for dynamic loading are discussed for the individual mechanisms.

Key words: solute strengthening, precipitation strengthening, long period stacking order (LPSO), texture.

2.1 Introduction

Perhaps no topic of Mg metallurgy has been more actively studied than the deformation mechanisms that are responsible for the complex mechanical behaviors exhibited by Mg and its alloys. Historically, single crystal studies provided the bulk of information on the subject. Recently, the use of new experimental and computational approaches has brought significant, new understanding of polycrystal deformation, which is of the greatest relevance to engineering application. Distinctions between the prior single-crystal-based understanding and the new polycrystal-based understanding have been determined to arise due to the combined effects of alloying and microstructure effects that were not possible to probe using single-crystal methods. This new understanding is helping to provide guidance to current alloy and microstructure design strategies to improve strength, ductility and toughness. There are significant issues of controversy pertaining to each of the deformation mechanisms discussed below. Until these issues are resolved, uncertainty will remain as to the absolute best alloy and microstructure design strategies to employ for particular materials property goals. There are two major classes of properties that are of interest.

Firstly, many researchers are interested in developing Mg-based materials with higher strength. Indeed, the engineering potential of a material with the density of Mg and the strength of heavier materials such steel or titanium is obvious. Given a more complete understanding of the deformation mechanisms that accommodate plasticity, one may hypothesize more effective schemes for impeding these mechanisms, thus, strengthening the material. Secondly, even though Mg alloys typically exhibit comparable ductility to other light alloys, the low temperature formability is severely limited, and a number of researchers are seeking to develop alloys and thermomechanical processing schemes to improve the forming performance. Again, there has been some controversy over the most effective strategies to employ for improving the formability. For example, the effects of grain size, solute additions, and crystallographic texture have been debated in the literature. Developing a clearer understanding of the interaction between the strain-carrying defects (dislocations and twins) with each of these (and other) microstructural features represents a critical step forward in the effort to develop more formable alloys.

There are already a number of book chapters with extensive reviews of the deformation mechanisms in Mg and its alloys.[1,2] These reviews largely focus on the single crystal behavior of pure Mg, with more qualitative treatment of alloying effects on polycrystalline materials. It is not the author's intention to supplant those earlier reviews but to highlight areas where the field has advanced significantly in the recent decades and to highlight remaining gaps in our understanding. Of particular interest are issues of polycrystalline deformation, since Mg alloys are invariably utilized in the polycrystalline condition. Secondly, issues of solid solution alloying and precipitation strengthening, and their effects on individual deformation mechanisms will be examined. The new characterization techniques of electron backscattered diffraction (EBSD) and *in situ* X-ray and neutron diffraction have been used to develop a clearer understanding of deformation mechanism in Mg. The traditional tool of diffraction contrast imaging of dislocations and other crystal defects within the transmission electron microscope (TEM) has been extensively applied to understand the deformation behavior of Mg and its alloys since the aforementioned reviews were written in the late 1950s[1] and early 1960s.[2] Additionally, new diffraction line broadening techniques have provided insight. Finally, atomistic simulation is rapidly advancing our understanding of the details of dislocation core structure and its response to applied stresses, as well as the details of twinning (both nucleation and growth mechanisms). Again, this chapter is by no means an exhaustive review. To make such a claim for such a short work, would be preposterous, given the amount of research that has been performed in the last 15 years alone. Instead, a subset of topics that have experienced considerable advances will be discussed and some opportunities for further work will be highlighted. Attention is restricted to low temperature deformation (e.g. <0.5 T_m) and sufficiently high strain rates that mechanisms controlled by diffusional flow may be ignored.

This chapter is organized in the following way. The dislocation slip mechanisms will be discussed in turn, first basal slip of <a> type dislocations (the easiest slip mechanism in Mg and its alloys), followed by non-basal slip of <a> type dislocations, and finishing with a discussion of the slip of <c+a> dislocations. Because the [c] type dislocations are only highly stressed under conditions where the easy basal slip mechanism is strongly preferred, this final dislocation type is considered to be sessile and is not discussed as an important deformation mechanism, though such dislocations have been observed in Mg alloys. Next, a short discussion of deformation twinning will be provided, though a more detailed discussion of twinning is provided in Chapter 3. Finally, a brief discussion of high strain rate deformation mechanisms is provided. For each of the major deformation mechanisms, the distinct effects of alloying and microstructure will be considered. The goals of this review are to help readers place current work within the context of the many excellent historical studies of Mg alloy deformation, highlight some of the most noteworthy accomplishments within the past decade or so, and point out areas where significant questions remain open for future research.

2.2 Basal slip

Basal slip is the softest mechanism of plastic deformation available to magnesium alloys, and it is often responsible for accommodating a significant fraction of the total plastic strain during ambient temperature deformation, particularly for near-randomly textured polycrystals, such as many castings. Single crystal-based characterization of basal slip in Mg actually served as the basis for the Schmid 'law' of crystal plasticity.[3] Though there are notable exceptions to this law among materials of various crystal systems, it is the frame of reference upon which all discussions of crystal plasticity begin. Basal slip in Mg and its alloys clearly abides by the Schmid law; when the stress on the slip plane and along the slip direction reaches a critical value (the critical resolved shear stress, or CRSS) slip ensues. Estimates of the CRSS of basal slip at room temperature (all ~0.5 MPa), based upon single-crystal experiments, suggest that the various strength levels observed by different authors may be correlated with impurity levels.[2]

Although basal slip is the softest mechanism available to magnesium, it is incapable of accommodating arbitrary strains. There are three slip systems with 1/3<11.0> (or <a>) type Burgers vector dislocations gliding on basal (00.1) planes. Obviously, there are only two independent co-planar slip systems of this type. This falls far short of the five independent easy slip systems required for uniform arbitrary strain of polycrystalline aggregates (the so-called Von Mises[4] or Taylor[5] criterion for uniform polycrystalline ductility.) This is the primary reason why this chapter contains detailed discussions of other deformation mechanisms. Without these additional mechanisms, the finite elongation to failure exhibited by Mg and its alloys could not be explained. Beryllium and some intermetallic compounds, like NiAl, are examples of materials in which the soft slip mechanisms provide only two or three independent slip systems, respectively. The hard deformation mechanisms in Be and NiAl are exceedingly difficult to activate at room temperature and, therefore, they exhibit almost no polycrystalline ductility at all. Magnesium and its alloys, on the other hand, tend to exhibit tensile elongations to failure of 5%, 10%, 20% and even greater than 30% depending upon composition, grain size, temper, and texture.

2.2.1 Stacking fault energy: core structure of basal dislocations

Perfect basal dislocations in magnesium are dissociated into Shockley partial dislocations of type 1/6<10.0>. This dissociation reaction is analogous with that observed on the octahedral planes in face-centered cubic (fcc) metals of low stacking fault energy (SFE). The SFE of metals and alloys with fcc crystal structures has been widely associated with their deformation behavior. Those with low SFE are termed planar slip metals because dislocation glide is primarily restricted to a single slip plane and planar slip traces are observed on the surfaces of single crystals (or individual grains.) Those with high stacking fault energy are termed wavy slip metals because dislocations can readily cross-slip from one octahedral plane to another, leaving behind wavy slip traces on sample surfaces. It is noted that even intrinsically wavy slip metals can exhibit planar slip behavior if they contain shearable GP (Guinier–Preston) zones, precipitates or radiation damage. The distinctions between wavy and planar slip fcc metals are quite striking in their fatigue behavior (e.g. cyclic hardening and softening[6,7]) and cracking behavior[8]) and their deformation texture evolution (developing the so-called copper or brass textures, respectively, during cold-rolling of fcc metal plates or sheets).

There are various estimates of the stacking fault energy of basal stacking faults in Mg available in the literature, ranging from 10–280 mJ/m^2. These estimates are based upon direct measurements, i.e. by TEM assessment of partial dislocation separation or by various atomistic simulation methods. (See the following papers for reviews of this literature.[9,10]) While these estimates suggest a great deal of uncertainty, ranging from values that would suggest wavy slip behavior (see experimental estimates) to strongly planar slip behavior (e.g. the predictions of molecular dynamics approaches). Recent *ab initio* modeling results[9] favor the intermediate values $\gamma = 34$ mJ/m^2, while some of the atomistic studies based upon empirical pair potentials predict very low values. From a practical standpoint, Mg is a planar slip metal in which glide is strongly preferred on the basal plane. Unlike fcc metals in which there is a multiplicity of {111} octahedral glide planes, cross-glide of basal dislocations requires transitioning to an alternative glide plane type, such as the {10.0} prismatic or {10.1} pyramidal planes. Such cross-glide is considered a distinct deformation mechanism and will be discussed below.

2.2.2 Thermal activation of basal slip

Early single crystal measurements[11] demonstrated that basal slip is a thermally activated deformation mechanism. However, like many fcc and hexagonal close-packed (hcp) metals, room temperature is close to or above the critical temperature above which basal slip is athermal. The further decrease in CRSS, above the critical temperature, is primarily associated with the decrease in the shear modulus with increasing temperature. Perhaps it is due to this fact that thermal activation of basal slip has not been an active area of research, save its role in the discussion of solute strengthening studies below.

2.2.3 Solute strengthening

Basal slip can be significantly strengthened by solid solution alloying, and the study by Akhtar and Teghtsoonian[12] stands out as an exceptional example of experimental diligence. They surveyed a large number of alloying additions and measured the CRSS as a function of temperature for a number of compositions. Further analysis revealed that the critical stress is best described by Labusch sampling statistics.[13] In other words, basal dislocations find 'wiggly' low-energy valleys in which to reside between solute atoms, while the stress rises to a sufficiently high level to push the dislocation past and pull it away from the weak pinning points represented by the solute atoms. Thus, the concentration dependence expected for most solid solution additions is $c^{2/3}$, rather than the $c^{1/2}$ dependence predicted by Friedel sampling statistics, which are appropriate for strong, sparse pinning points. Recent experimental work has emphasized the anomalous solid solution strengthening potential of rare earth elements, including polycrystalline Mg-Y at creep tested high temperatures,[14] as well as Mg-Y and Mg-Dy single crystal experiments performed over a wide range of temperatures.[15] The latter study concluded that the atomic size mismatch was insufficient to explain the observed level of strengthening. They suggest that there may be a secondary effect due a higher level of forest dislocations in the alloy single crystals.

There is obviously more experimental work to be done, as there are numerous alloying additions that have been examined in only a cursory way, including the rather exhaustive 1940s study of MacDonald reviewed previously.[1,2] However, there are new computational tools that promise predictive power, and the work of Yasi et al. is exemplary.[16] They developed an electron density functional theory (DFT) approach to prediction of basal slip strengthening by solute additions and examined the solute-dislocation core interactions of 29 different solute types. Included in the calculation are both the traditional 'size effect' as well as chemical effects, such as the so-called Suzuki effect (solute-stacking fault interaction). Based upon the contours in their final map of solid-solution strengthening potency, the atomic misfit does account for a great deal of the potency.[16] However, the chemical or stacking fault interactions exhibit a significant modification to that first-order approximation, for some alloying additions.

Finally, there is a great deal of interest in rare earth alloying elements, from the perspective of solid solution strengthening, development of new strengthening phases (including the long period stacking order (LPSO) phases discussed below), as well as their replacement by less costly and more broadly available alloying elements. As concerns the 'anomalous' strengthening effect of rare earth additions,

Gao et al.[17] have used DFT to demonstrate that rare earth additions not only affect the Mg-RE bonding, but also the Mg-Mg bonding itself. While the energetics of dislocation solute interaction may be explored with DFT modeling, the critical configurations associated with thermally activated slip must rely upon molecular dynamics modeling with empirical bond potentials or higher-scale models,[16] such as continuum elasticity theory, which underpins dislocation dynamics approaches.

2.2.4 Order strengthening

Ordered Mg alloys have not historically been an active area of research within the structural materials community. Outside this mechanical metallurgy community, the MgB_2 intermetallic compound has received a great deal of attention as the metallic material with the highest superconducting critical temperature, $T_c = 39$ K, e.g.[18] However, the recent discovery of LPSO alloys with exceptional strength ($\sigma_y > 600$ MPa) has generated great interest in the potential of order strengthening.[19] Considerable study of the details of the stacking order have been performed by the original group that discovered the materials,[20, 21] as well as others, e.g.[22,23] In summary, the structures have been variously reported as 8R, 14H, 18R, etc., where R stands for rhombohedral and H for hexagonal and the preceding number indicates the number of stacked planes in a repeated unit cell. Disordered magnesium solid solutions are obviously of the 2H structure, and alloys frequently exhibit LPSO grains alongside grains of Mg solid solution.

Since the original discovery of high-strength alloys containing the LPSO phase, there has been significant effort expended seeking to determine the role of the LPSO structure in enhancing the strength. Based upon selected area electron diffraction (SAED), high-angle annular dark field (HAADF), Scanning transmission electron microscopy (STEM) and 3D atom probe (3DAP) characterization, Nie et al.[24] emphasize that there is in-plane ordering, as well as the ordering along the stacking direction. Only such in-plane ordering would lead to significant strengthening against basal slip, since basal slip would not disrupt the out-of-plane order. Indeed, a large increase in hardness is observed upon adding a small amount of Zn to the Mg-RE alloy. It is shown that this is not due to the formation of any precipitate clusters; however, 3DAP data suggests local ordering of the Zn and Gd solute (termed co-segregation and speculated to form dimers[24]). Such in-plane short range ordering could give rise to a strengthening effect on basal slip. However, estimates of the CRSS of basal slip, based upon examinations of directionally solidified samples, are not that impressive (10–30 MPa).[25] Although the strength is much larger than that of pure Mg or Mg-Zn solid solution alloys, it is not much greater than Mg-1at% Y. Thus, the in-plane ordering energy of the LPSO alloys examined must not be very high. A more significant effect has been observed on extension twinning (discussed below).

It may be that the effect of LPSO on secondary deformation mechanisms is responsible for the exceptional strength of these alloys. In other words, it appears that the LPSO structure is even more plastically anisotropic than Mg and disordered solid solutions.[26] Hagihara et al.[26] report that even very weakly textured polycrystals containing grains with the LPSO exhibit macroscopic plastic anisotropy. Additionally, it is found that alloys that contain the LPSO phase undergo more rapid grain refinement during thermomechanical processing than traditional alloys. Finally, while such alloys are shown to obey the Hall–Petch relation for grain size strengthening, the alloys containing the LPSO phase exhibit an additional strengthening element attributed to the hard phase reinforcement. Apparently, even though some directions would have a similar strength to the matrix, the very strong anisotropy of the LPSO grains can have a strengthening effect on the surrounding matrix.

2.2.5 Precipitate strengthening

Nie[27] recently referred to a question that had been considered previously, but had apparently lain dormant for some decades, 'Why is the aging response of Mg alloys so poor, in comparison with Al alloys?' The fact that Al alloys may be so potently strengthened by precipitates is one of the major reasons why they have proven such potent competition to other light metals (Mg and Ti), especially for aerospace applications. Many have wondered why precipitation strengthening efforts in Mg have been less fruitful. One reason is that there are relatively few elements with significant high temperature solubility, a prerequisite for precipitating a significant volume fracture of precipitates.

The major reason highlighted by Nie[27] is the geometric relationship between the precipitate shapes and basal slip. Presuming that basal slip is the mechanism that is responsible for yielding, it is demonstrated that plate-shaped precipitates located on the basal plane are the least effective possible precipitate geometry (per unit volume fraction of precipitate) for impeding slip on the basal planes. Ironically, this precipitate geometry is the one observed in the most commonly employed Mg alloy system, Mg-Al. The $Mg_{17}Al_{12}$ strengthening phase precipitates as plates on basal planes. Recent work on a weakly-textured Mg-Gd-Zn-Zr alloy with basal plate precipitates confirms this hypothesis.[28]

The second least effective precipitate geometry identified by Nie is the c-axis rod-shaped precipitate observed in the 'high strength' Mg-Zn alloying system. Although there is some dispute over the detailed crystallography of the strengthening phase (i.e. $MgZn_2$ vs Mg_4Zn_7[29,30]), there does not appear to be any controversy regarding the preferred c-axis rod geometry of these precipitates.[31] Although the strengths of high Zn alloys such as ZK60 are good, they do not nearly rise to the level of high strength aluminum alloys of the 2000 and 7000 series.

Nie suggests that plates on prismatic planes are the most effective geometry for impeding basal slip. Interestingly, this is exactly the precipitate geometry that is observed in the Mg-Y-rare earth (RE) alloys, such as WE43 and WE54. While these alloys are not as strong as the Al alloys mentioned above, their high temperature strength is exceptional. This is presumably due to the combination of optimal precipitate geometry, slow coarsening kinetics and low diffusivity solute. Notably, altering the precipitate geometry in Al alloys has been used to advantage. For example, Al-Cu-Mg-Li alloys precipitate so-called T1 plate-shaped precipitates on {111} planes, which impart incredible strengthening effect.[32] Further, microalloying of the Al-Cu-Mg system with Ag demonstrates the potential to reorient the dominant precipitate phase θ from the {100} in Al-Cu-Mg, to the Ω phase, which precipitates as plates on the {111} plane in Al-Cu-Mg-Ag.[33] One wonders if there is a microalloying addition to Mg that is yet to be discovered, which could revolutionize the precipitate strengthening behavior of a current (or new) Mg alloy system.

Another reason for the mild age-hardening response of Mg alloys, which has been highlighted by a number, is the relatively low number densities of precipitates. Microalloying has been used to great advantage in other systems to promote higher number densities of fine precipitates. Such work has only just begun in Mg alloys. For example, a strong effect has been observed due to Ca microalloying additions to Mg-Si alloys.[34] Recent work by Mendis et al.[35,36] has revealed the strong impact that Na microalloying can have on the age-hardening response of Mg-Sn-Zn alloys. The incredible increases in age hardening response that have been achieved in Al alloys by minor alloying additions to a small number of base alloy systems provides hope that Mg alloys could undergo a similar revolution. One also hopes that these improvements could be achieved over a significantly shorter period of time, given the increase in knowledge and computational power, which exists now, as compared to that which existed when alloy developers first started making systematic quaternary and quintenary additions to Al alloys.

2.2.6 Grain size effects

Grain boundary strengthening is an important strengthening mechanism in Mg alloys. Indeed, die casting and wrought processing both induce fine grain sizes in many magnesium alloys, so the non-trivial contribution of boundary strengthening cannot be ignored. Although empirical Hall–Petch type relationships have been successfully developed for Mg and its alloys,[37-40] the author cautions against using such relations blindly. These relationships are highly sensitive to the material history (especially as pertains to texture). Even for castings, which typically have random or very weak textures, the different grain size dependencies shown to exist for different deformation mechanisms (e.g. slip vs twinning[40]) suggest that a single, simple Hall–Petch relation cannot adequately capture the grain size dependence of strength. If a Mg alloy polycrystal has a weak (near random) texture, the microyield behavior will certainly be controlled by basal slip. However, even for weak texture, other mechanisms, such as extension twinning and prismatic slip are observed to play important roles in determining the flow stress at low strains.[41]

It is suggested that a full crystal plasticity-based approach is best, rather than the Taylor-factor-based approach that has been successfully employed for modeling fcc and body-centered cubic (bcc) metals and alloys and has been attempted for Mg.[39,42] If a crystal plasticity approach is used, the model will naturally account for the presence of different textures and correctly predict the effect of grain size. It is admitted that it is difficult to determine the effect of grain size reduction on individual deformation mechanisms. An attempt was made by Jain et al.[43] Their approach was employed more recently by Raeisinia et al.,[41,44] though there is a competing view, which will be discussed in some greater length below, in connection with the balance of basal and non-basal slip mechanisms.[45]

Regarding the effect of grain size on ductility and practical formability, there has been significant debate. There are many who have proposed that grain size reduction could give rise to significant increases in formability, based on the early results of Chapman and Wilson.[46] One of the reasons for the debate is that changes in grain size and texture are frequently difficult to separate, which was even recognized by Wilson and Chapman[47] themselves. In addition, there are recent studies that show the overall ductility of coarse-grained samples can be inferior to fine-grained,[48] and this is largely associated with suppressing failure mechanisms associated with deformation twinning.[49] However, recent results of Kang et al.[50] call the practical benefit of a grain refinement approach into question. They found that the strain hardening rate of larger-grained samples was higher than that of fine-grained samples, and therefore the uniform elongation and stretch formability, as assessed by a limiting dome height test, was actually improved by an increase in grain size. Obviously, there are still unanswered questions concerning the impact of grain size on ductility.

2.2.7 Texture effects

Crystallographic texture has an impact on the activities and roles of all the slip systems. In many cases, it is sufficient to consider the initial orientation of the c-axes, particularly if one simply wants to determine the relative importance of deformation mechanisms since c-axis orientation strongly influences the ease of basal slip and deformation twinning mechanisms. For many castings, which have nearly random initial textures, basal slip and {10.2} tension twinning dominate the initial yielding response.[42] For strongly textured polycrystals, typical of wrought Mg alloys, basal slip cannot possibly accommodate all imposed strains and non-basal slip mechanisms are significantly more important.

One area that has developed considerable interest is that of controlling texture via unique thermomechanical treatments, such as equal channel angular extrusion[51–53] or shear rolling,[54–56] since these have been observed to improve the ductility and formability. In some cases, effects have been mistakenly attributed to texture randomization,[51] when particular texturing was responsible.[53] One thing is clear; by producing textures that permit the material to deform by basal slip at low stresses and by promoting strain hardening via twinning[57] or otherwise, the uniform elongation can be improved. Textures that result in high flow stresses, particularly those that promote certain compression or double-twinning mechanisms[58,59] negatively affect ductility. Ball and Prangnell[60] revealed another strategy for texture alteration: alloying with rare-earth elements. This subject has received considerable attention, due to the possibility that such alloys with weaker textures would have improved formability. While there has been considerable research to determine the origin of the so-called rare earth texture effect, the question is still considered open. The interested reader is referred to the following works, which stress the effects of rare earth additions in promoting shear banding,[61,62] shear band nucleation during dynamic recrystallization,[63–65] solute drag[66,67] and most recently, Zener pinning due to second-phase particles.[68] Finally, it is worthwhile to document that this alloy-based strategy to alter the texture has been demonstrated to have a positive impact on formability by numerous research groups.[69,70,71]

2.3 Non-basal slip of <a> type dislocations

In past years, the role of non-basal slip was de-emphasized. Early experiments on pure Mg single crystals had illustrated that the CRSS of prismatic slip of dislocations with 1/3<11.0> (or <a>) type Burgers vectors is 50–100 times greater than that of basal slip at room temperature.[72,73] Because of observations such as this, and high temperature experiments that show that the strength differential between the two mechanisms remains high at mildly elevated temperatures, it has often been quoted that the temperature must exceed 225 °C before significant prismatic (or pyramidal) slip is observed.[2,73] The fact is, prismatic slip was observed via single surface slip trace analysis of polycrystalline Mg samples deformed at room temperature, albeit restricted to the grain boundary region, at an early date by Hauser et al.[74]

2.3.1 Role in polycrystal deformation

When one recalls that basal slip only provides two independent slip systems, it becomes apparent that Mg must possess other hard slip mechanisms to alleviate the compatibility stresses that arise between grains of distinct orientations, since Mg polycrystals generally exhibit finite strains to failure (not the infinitesimal levels of ductility exhibited by other slip system deprived metals like Be or NiAl). Prismatic slip provides two additional independent slip modes, bringing the total to four. Pyramidal slip of dislocations with <a> Burgers vectors provides four independent slip modes by itself, although the sorts of strain that can be accommodated are the same as prismatic and basal slip combined. (It does not provide additional independent slip modes.) The remaining type of strain (tension or compression along the c-axis) cannot be provided by any dislocation with an <a> Burgers vector, regardless of the slip plane.

Polycrystal modeling has revealed that non-basal slip of <a> dislocations is much more active in Mg and its alloys than previously suspected, based upon single crystal CRSS values. Such modeling work has emphasized that incorporating a significant amount of non-basal slip of <a> dislocations is necessary to explain the observed plastic anisotropy (e.g. r-values) of textured magnesium alloy, AZ31, sheets.[75,76] Indeed, the same studies showed that the significant activity of prismatic <a> dislocations is required to simulate the observed anisotropy also predicts a texture evolution that agrees with experimental observations. Styczynski *et al.*[77] performed a rather systematic study aimed at predicting the texture evolution of Mg alloy, AZ31, during cold-rolling. They considered the slip of <a> dislocations on basal, prismatic, and pyramidal planes, <c+a> slip on pyramidal planes, and {10.2} extension twinning. They achieved the best agreement between simulated and experimentally observed texture for the case involving a specific set of critical resolved shear strength ratios between basal, prismatic, pyramidal <c+a> slip modes. Ebeling *et al.*[78] have even developed an automated scheme for parameter determination for polycrystal modeling of magnesium alloy deformation. Invariably, the results of this approach show that the activity of non-basal slip in polycrystalline Mg alloys is much more than one would expect, based upon the ratios of CRSS values of the different slip modes measured for pure Mg single crystals.

The experimental EBSD-based work of Keshavarz *et al.*[79] and TEM-based observations of Koike[80] were also important in establishing the very significant, and previously unexpected, role of non-basal slip of <a> type dislocations, in the room temperature deformation of Mg alloy, AZ31. *In situ* neutron diffraction has provided additional evidence for extensive non-basal slip of <a> dislocations, by correlating the measured internal strains developed within different grain orientation types with those predicted by crystal plasticity codes. Initial efforts by the present author and collaborators showed that prismatic slip of <a> dislocations and pyramidal slip of <c+a> dislocations was required to model the observed internal stress developments.[81] However, they later examined the possibility of pyramidal <a> dislocations,[82] and concluded that prismatic <a> slip provided better agreement with the experimental data. The latest study of this type, by Muránsky *et al.*,[83] suggests that there must either be some pyramidal <a> slip or some cross-slip between basal and prismatic planes, which would appear similarly in the experimental results. Although the experimental method has improved with time, the present author does not think that this is a controversy that will be resolved without complementary experimental and simulation-based techniques. It is suggested that techniques such as careful transmission electron microscopy, dislocation dynamics simulation, and atomistic modeling efforts (described below) should be applied in these investigations.

There have been a number of proposals as to why there appears to be a discrepancy between early single crystal measurements, of the ratio between the CRSS for non-basal slip and that of basal slip, and the more recent polycrystal measurements. One possible explanation relates to the effect of solid solution alloying,[44] since solutes are known to potently harden basal slip,[12,13] while they can in some instances soften prismatic slip,[84] as discussed in more detail below. Another possibility is that the presence of grain boundaries alters the balance of slip system strength. A number of studies have put forward the notion that fine-grain alloys exhibit more non-basal slip than coarse-grained.[45,81] Along these lines, the possibility exists that the Hall–Petch strengthening effect of grain boundaries is equal for basal and non-basal slip of <a> dislocations. If this is the case, then the relative strength of the two mechanisms would be closer and closer, the finer the grain size.

Based upon analyses of hardening and softening, some authors continue to insist that 'cross-slip', which presumably is the main source mechanism for non-basal slip, is only significantly active at temperatures greater than about 150 °C in commercial alloys such as AM60.[85] Notably, this is in direct contradiction with the findings of numerous other authors that non-basal slip is surprisingly active, even at room temperature. One possible reason for this apparent contradiction is that these authors[85] examined a die casting, which is expected to have a nearly random texture, while many of the aforementioned authors examined strongly textured samples. It is known that stronger textures, which place fewer grains in orientations that are favorable for basal <a> slip, tend to promote non-basal slip.

2.3.2 Thermal activation

While some crystal plasticity modellers have preferred to discriminate between prismatic and pyramidal slip of <a> dislocations[84,86] the present author considers both to be variants of a single theme, that is, cross-slip of basal <a> dislocations. In the foregoing discussion of basal stacking fault energy, it was stated that Mg behaves as a low stacking fault energy metal. The <a> dislocations tend to be dissociated by a finite distance on the basal plane (more on this topic below). In order for a screw dislocation to cross-slip onto the prismatic (or pyramidal) plane, the Shockley partial dislocations must constrict. Püschl has recently presented a rather exhaustive review of cross-slip mechanisms in a variety of crystal structures.[87] He emphasizes that, for Mg, the evidence[88,89] favors a so-called jog-pair mechanism where the critical configuration consists of a faulted screw dislocation having cross-slipped two atomic planes and bound by two edge-type jogs, as was originally proposed by Yoshinaga,[90] rather than the single bowed-constricted dislocation originally proposed by Friedel.[91] Couret and Caillard[89] noted that they could not discriminate between prismatic and pyramidal slip of <a> dislocations during their *in situ* TEM observations. They imagined that it was possible that dislocations cross-slipped onto prism planes and then double-cross-slipped back onto the basal plane. Such a mechanism may not be discriminated from pyramidal slip when a more macroscopic perspective (crystal plasticity modeling) is adopted. As discussed below, if one uses atomistic modeling to examine the core structure of the dislocations directly, prismatic slip appears to be the preferred mode of cross-slip in Mg.

2.3.3 Core structure

Groh *et al.*[92] performed a molecular dynamics study of the stacking fault energies, critical resolved shear stresses required to induce <a> slip on the prism and pyramidal planes, and viscous mobilities of the same. They found that the stacking fault energy on the basal plane is 44.6 or 54.9 mJ/m^2 with a dissociation separation of 1.9 or 1.4 nm, using the interatomic potential of Liu *et al.*[93] or Sun *et al.*,[94] respectively. They do not observe any dissociation of edge dislocations on either the prismatic or pyramidal planes, though there was some spreading of the core in the prismatic plane. Screw dislocations show similar results: dissociation within the basal plane (provided the simulation began with initially dissociated partial dislocations), no dissociation in the prism plane, and non-coplanar dissociation (into the basal plane) for prismatic dislocations. In other words, <a> type screw dislocations were not even stable within the prismatic plane. Simulations performed to reveal the critical resolved shear stress (the Peierls stress) showed that pyramidal slip may be approximately equivalent in strength to prismatic (using the interatomic potential of Liu *et al.*[93]) or much greater for pyramidal (using the potential of Sun *et al.*[94]). First-principles atomistic simulations have also been used to examine the core structure of dislocations with <a> Burgers vectors, and Yasi *et al.*[95] suggest that the potential due to Sun *et al.* provides much better agreement with the *ab initio* modeling results.

2.3.4 Solute strengthening and softening

In general, solute additions lead to strengthening of dislocation slip accommodated plasticity and, as discussed in the section on basal <a> slip above, magnesium alloys also abide by this rule. However, there are exceptions to this rule, and prismatic slip of <a> dislocations in Mg appear to be an archetypal example of the phenomenon of solute softening. An extreme example is provided by the Mg–Li system. The Dorn group studied this system extensively in the 1950s because it was found that hcp Mg–Li alloys could exhibit enhanced ductility.[96] They associated this improvement in ductility with increased activity of prismatic slip. Single surface slip trace analysis showed that prismatic slip in polycrystalline Mg–Li alloys was homogeneous, whereas it only appeared to occur near grain boundaries in pure Mg. The explanation for this observation was levied in terms of a reduced c/a ratio, since the Peierls–Nabarro model of dislocation glide resistances suggests that slip will occur on the most widely spaced planes in the lattice that contain the Burgers vector. However, it is pointed out that the change in c/a ratio by Li additions is modest.[97] Further, Be has the lowest c/a ratio of any elemental metal and it slips almost exclusively on basal planes. Thus, c/a ratio must be considered a secondary concern, relative to the aforementioned issues of stacking fault energy and core configuration discussed above. These are, in turn, sensitive to the electronic structure of the metal in question. Urakami et al.[98,99] later examined the Mg-Li system in detail and reported findings that are consistent with the theory of solid solution softening later proposed by Sato and Meshii.[100] At about the same time, Akhtar and Teghtsoonian[84] performed their landmark study of a series of Mg single crystal alloys, including the two systems that are most important from a commercial perspective, Mg-Al and Mg-Zn. They observed solute softening of prismatic slip at low temperatures and solute strengthening at high temperatures.[84] Recent results obtained in the Mg-Y system suggest that non-basal slip may also be promoted by such rare earth alloying additions as well.[14]

Sato and Meshii[100] were the first to publish an exhaustive examination of the phenomenon of solute softening and provide a model to explain it. The focus of their work was bcc metals, and it is now established that only metals with thermally activated slip undergo this phenomenon. While there has been some dispute of their proposed mechanism,[101] there now appears to be some consensus in the literature that solute softening is a real 'intrinsic' phenomenon[102,103] and not only due to substitutional solute/interstitial solute complexing.[101] Only a brief, qualitative explanation is provided here and the interested reader is referred to the original publications. For thermally activated slip mechanisms (such as slip of ½<111> type screw dislocations in bcc metals and non-basal slip of <a> type dislocations in Mg), the critical configuration is a double-kink (or -jog) segment, which may readily glide to the sides, advancing the entire dislocation one step forward. Since screw dislocations may adopt a wiggly configuration due to the torque exerted by solute atom stress fields, one may consider that the dislocation is partially over the intrinsic activation barrier for slip (or cross-slip) even before an external stress is applied to the material. This is the basis of solute softening. It must be acknowledged that the solute still represents a barrier to the sideways motion of the kinks (or jogs) and, therefore, the effect of solute softening will be overwhelmed by normal hardening effects, once the temperature or solute concentration exceed critical levels.[104] As suggested earlier in this chapter, solute softening of non-basal slip is very likely one reason that commercial alloys have been observed to undergo much more non-basal slip than would be expected based on single crystal measurements of pure Mg.[45] However, even the most sophisticated atomistic studies[104] do not provide an explanation as to why the degree of solute softening appears to be so strong, for the particular systems of interest, e.g. Mg-Li, Mg-Al and Mg-Zn.

2.3.5 Precipitation strengthening

To the present author's knowledge, very little attention has been paid to the potential for impeding non-basal slip of <a> dislocations by precipitation. Most alloy developers seem to imagine that basal slip is the major strain accommodation mechanism and, therefore, impeding prismatic slip is of a lesser concern. If the alloy in question is to be used in a cast condition (with very little crystallographic texture) this logic is probably correct. On the other hand, if the alloy in question is wrought (extruded or rolled), the possibility of a strong crystallographic texture cannot be ignored. Since it has been shown that strongly textured Mg alloys can actually accommodate the majority of the applied strain via prismatic slip[82] it would seem wise to develop a precipitate microstructure design paradigm that included considerations of this secondary slip mechanism.

2.3.6 Grain size effects

There is a complementary view to the one based upon solute softening expressed above, which suggests that fine grain sizes suppress basal slip more than they do non-basal slip. Two papers written in support of this concept were published in the Proceedings of the Platform Science and Technology for Advancement Magnesium Alloys (PSTAM 2003).[105,106] The view of these authors and their collaborators is amplified in a subsequent paper,[107] which highlights the role of grain compatibility on the activation of non-basal slip. It is suggested that the entire grain is constrained if it is small, while the grain interior is only affected to a limited degree in coarse-grained material. Supporting evidence is provided by a comparison between material of 7 μm and 50 μm grain size, respectively.[105] While this argument is compelling, a systematic study of the anisotropy induced by crystallographic texture and slip system activity by Jain et al.[43] suggests that there is a similar level of basal vs non-basal slip system activity in all of the samples they examined, with grain sizes between 13–89 μm. As suggested in the earlier section on grain size effects (on the basal <a> slip mode), controversy on this subject continues to the present time. This is clearly a topic in need of resolution. One complicating factor that has not been actively incorporated in previous considerations is the competing role of mechanical twinning in strain accommodation. It is now well-established that grain size can have a strong impact on the stress required for mechanical twinning (discussed in detail in Chapter 3). Thus, any consideration of the relative importance of non-basal slip modes must include consideration of twinning as well. This is particularly critical, now that it has been suggested that the ease of non-basal slip may in some way control the activation of twinning.[108]

2.4 Non-basal slip of <c+a> type dislocations

The most controversial slip mechanism in Mg alloys is that of dislocations with <c+a> Burgers vectors. Early single crystal work did not even reveal the activity of <c+a> dislocations.[109] Rather, the crystals that were well-aligned for <c+a> slip, i.e. compression along the c-axis, yielded by a mechanical twinning mechanism that only narrowly preceded fracture.[109] Later, once the TEM was applied to the problem, it was revealed that <c+a> dislocations were active in aligned single crystals[110] as well as in polycrystals,[111,112] despite the fact that these large Burgers vector (|<c+a>| ~ 1.9·|<a>|) defects must have a very large self-energy. While TEM analysis may be criticized as only incorporating very small volumes of material, which may include exceptional characteristics, X-ray diffraction-based techniques generally poll a large volume of material. A recent X-ray diffraction line-broadening study revealed that <c+a> dislocations were present in all deformed polycrystalline samples investigated, with increasing density as the temperature is increased from room temperature up to 300 °C.[113] Finally, recent work involving polycrystal plasticity modeling has invariably incorporated <c+a> slip as a required mechanism to explain a number of observed effects: texture evolution,[77,78,97] anisotropy of textured polycrystals,[75,76] and internal stress development.[81–83]

The reason why <c+a> slip was not observed in the earliest single crystal experiments, yet is so frequently invoked in modern explanations of polycrystal deformation, is similar to the question raised about non-basal slip of <a> dislocations earlier in this chapter. It is best framed by the age-old question of whether such a hard mechanism is source or mobility limited. Yoo et al.[114] speculated that the issue was one of source limitation, based upon the observation of a high density of <c+a> dislocations in polycrystalline samples of Mg-Li alloys, relative to pure Mg. A very recent molecular dynamics study by Kim et al.[115] provides some support for the notion that <c+a> slip is a source-limited mechanism. They find that grain boundaries are potent sources of <c+a> dislocations in their simulations of nanocrystalline Mg. Perhaps the paucity of <c+a> dislocation source mechanisms in pure Mg single crystals explains why they were not observed in some of the earliest single crystal studies.

2.4.1 Role in polycrystal deformation

Only dislocations with a <c+a> type Burgers vector are capable of accommodating strains along the c-axis of magnesium crystals (twinning mechanisms notwithstanding). There is, however, outstanding debate regarding the role of <c+a> slip. Some seem to view it as a helpful mechanism, which promotes homogeneity in plastic strain accommodation and overall ductility[76] while others seem to view it as a mechanism of plastic instability and failure,[116] based upon single crystal observations. Even the first reported observation of <c+a>

dislocations in Mg provides a possible explanation; upon yielding by <c+a> slip, the material was observed to rapidly strain harden, then undergo mechanical twinning, and then fracture after only a small amount of plastic strain.[110] The stress to activate mechanical twinning has been shown to be even more strongly grain size-dependent than slip,[40] particularly the compression twinning mechanisms that would be relevant to this case.[60] Perhaps fine-grained polycrystals can accommodate significant strain by <c+a> slip prior to the onset of compression twinning, which has been repeatedly associated with the initiation of a ductile fracture mechanism,[58,59] in a way that is unobservable in unconstrained single crystal tests. As discussed below, some very recent atomistic modeling results provide a possible intrinsic explanation for plastic instability associated with <c+a> slip. Although the use of crystal plasticity modeling and modern characterization tools (TEM, EBSD, *in situ* diffraction) have provided much greater clarity regarding the role of <c+a> slip in the deformation of Mg and its alloys, there are clearly a number of outstanding issues to be resolved.

Another controversy that persists in the literature regards whether <c+a> dislocations slip on first-order {10.1} or second-order {11.2} pyramidal planes. This controversy has existed from some of the earliest single crystal experiments performed to explore the possibility of <c+a> slip in Mg. However, all of the early experimental research concluded that, similar to observations in Cd and Zn, and contrary to observations in Ti and Zr, <c+a> dislocations slip on {11.2} planes, rather than {10.1}. This appears to be another example of where intuition based upon c/a ratio would fail. Zn and Cd have a large c/a ratio > $\sqrt{3}$, whereas Mg, Ti, and Zr all have c/a ratios less than the ideal packing ratio, $\sqrt{(8/3)}$. As mentioned in the discussion of non-basal <a> dislocations above, the salient features are stacking fault energy and core configuration, which depend on electronic structure, rather than the c/a ratio referred to by previous generations of metallurgists, who were focused more on atomic structure. Initial *in situ* neutron diffraction work, combined with crystal plasticity modeling, was unable to resolve this controversy.[82] More recent analysis, of more accurate neutron diffraction measurements, has shown rather conclusively, that slip must be on {11.2} planes,[83] in agreement with early single crystal work. It is interesting to consider this evidence, in light of the new core structure information being provided by atomistic simulations.

2.4.2 Core structure

'The cores of the <c+a> dislocations are quite complex, spatially extended and subject to reconfiguration under the action of applied loads . . .'[115] This is not surprising, given the large Burgers vector of this dislocation. The large elastic strain energy demands that the material seek a low energy configuration, such as a dissociation into partial dislocations or spreading of the core. Indeed molecular dynamics (MD) simulation has been used to reveal a number of new insights for hcp metals in general,[117,118] More recently, the approach has been repeatedly applied to Mg, in particular, e.g.[115,119–121] All of these MD studies have reported complex non-coplanar dissociation and core spreading configurations, frequently involving faults on the basal plane as well as twin-nuclei on various pyramidal planes, such as the {11.1}.[121] Interestingly, {11.1} twins are an active deformation mechanism in the their own right, in transition metals Ti and Zr, but not in Mg.

While there is qualitative similarity in the results predicted using the Liu et al.[93] and Sun et al.[94] embedded atom method potentials, the former promotes more <c+a> slip than the latter.[116,121] This distinction is mainly associated with the ease of <c+a> slip relative to <a> slip. The Sun et al. potential predicts a much lower CRSS for basal slip, which is in closer agreement with single crystal experiments. As mentioned previously, the Sun et al. potential results in stacking fault energies that are in closer agreement with first principles electron DFT calculations.[95] Curiously, all of the most recent modeling results suggest that <c+a> slip would be favored on first-order {10.1} pyramidal planes, rather than second-order {11.2} planes suggested by experiments. The gamma surface calculations using DFT agree with the predictions of both MD interatomic potentials, i.e. dissociations are much lower energy in the first-order {10.1} pyramidal planes, rather than second-order {11.2} planes. However, unlike the case for <a> type dislocations discussed above, the large size of the <c+a> dislocation core precludes the use of electron DFT approaches to compute the complete core configuration, given current computational limits.[95] This is an issue that would benefit from high-resolution TEM imaging of <c+a> dislocation cores (an approach that has been applied to other systems[122,123]).

Concerning defect energies, the various atomistic simulation methods all agree that the <c+a> dislocation is a high-energy defect, and thus, it may be source-limited, as suggested previously, by Yoo et al.[114] However, some of the core configurations were shown to be much more glissile than one would expect, based upon early single crystal measurements that revealed CRSS values in the range of 40 MPa at 300 K.[110] Specifically, Nogaret et al.[121] predicted a Peierls resistance of 5 MPa or less, depending upon stressing direction and dislocation character at 300 K. The combination of these observations (glissile, high-energy dislocations) could explain why some researchers view <c+a> slip as a mechanism of plastic shear instability, rather than a mechanism of homogeneous plasticity that would enhance ductility. In short, if it is very difficult to nucleate these dislocations (i.e. it would require very high stresses) and then they can glide at relatively low stresses, it would indeed represent a potential mechanism of plastic instability, particularly in single crystals where there are few other obstacles for dislocation motion.

2.4.3 Thermal activation

The slip of <c+a> dislocations is, without a doubt, a thermally activated mechanism. All of the available data, both experimental and simulated show a strong downward trend in the critical resolved shear stress with increasing temperature.[110,121] However, some of the single crystal data also suggest a strength anomaly. The work of Ando and Tonda[124] reveals that this is a general feature of <c+a> slip in a number of hcp metals. Given the complexity of the core structure highlighted above, this is not surprising. Recall that the strength anomaly in intermetallic compounds, like Ni_3Al, has been related to thermally activated non-coplanar dissociation or core spreading out of the glide plane.[125] Because of the important role the <c+a> slip can have in strain accommodation within polycrystalline Mg alloys, this is an issue that merits further simulation and experimental investigation.

2.4.4 Solute effects

The study of the effect of solute on <c+a> dislocation motion is in its infancy. Although many published studies on Mg alloys refer to <c+a> slip in a cursory way, there are only a few studies that touch on this question with any level of sophistication. The work of Suzuki et al.[14] showed that Y additions promote <c+a> slip under high temperature creep conditions. The single crystal tension testing of Ando and Tonda[126,127] showed that <c+a> slip is easier in Mg-Li alloys, relative to pure Mg. They also show that <c+a> slip exhibits the yield strength anomaly, in Mg-Li, that they have first observed in pure Mg.[124] The work of Agnew et al. showed that Li additions promote <c+a> slip during room temperature compression of polycrystalline samples as well.[112] As mentioned previously, Yoo et al.[114] speculated that the reason for the enhancement in <c+a> slip activity in Mg-Li was secondary, i.e. the direct impact was on the ease of prismatic <a> slip (discussed in detail, earlier in this chapter). Although they were not able to test their hypothesis, the authors of a recent paper on MD simulation of <c+a> dislocations suggested the same source mechanism, junction formation between a sessile c-dislocation and a glissile prismatic <a> dislocation and subsequent bowing out into a pyramidal plane under the influence of an applied stress.[121] As discussed for basal dislocations, there could be a direct effect of solute on the mobility of <c+a> dislocations. In general, solute would be expected to have a strengthening effect. However, it could stabilize a particularly glissile core configuration or it could lower the thermal activation barrier for <c+a> glide by inducing a torque on the dislocation line (as they have been shown to do for prismatic <a> dislocations[98,99]).

2.4.5 Precipitation and composite effects

There does not yet appear to be much published work that relates particles to the activity of <c+a> dislocations. One exception is a recent work on the effect of nanocomposite reinforcements on the ductility and fatigue performance of Mg. The ductility of carbon nanotube (CNT)-reinforced Mg composites was significantly higher than pure Mg,[128] and it was suggested that this is due to the higher activity of <c+a> dislocations in the composite, relative to the pure metal. 'More dislocations are generated in the ... nanocomposite as compared to monolithic Mg. This can be explained by the thermal expansion and elastic modulus mismatch between the CNT and Mg matrix.'[129] The same group's previous work on Y_2O_3 nanoparticles-reinforced Mg composites supports the notion that if <c+a> slip is enhanced by the composite microstructure, the ductility will be enhanced.[129] The Y_2O_3 nanoparticles-reinforced Mg did not show significantly increased <c+a> dislocation activity and no significant improvement in ductility was observed either. Although there appears to be a positive impact upon tensile ductility, the impact upon fatigue resistance is negative. Fatigue experiments show that the Mg-CNT composite cyclically hardens more rapidly than pure Mg, and <c+a> dislocations are observed in the fatigued composite but not the pure metal.[129] Again, the Mg-Y_2O_3 nanocomposite exhibited very similar fatigue performance to pure Mg, and little enhancement in <c+a> dislocation activity was observed.[130] Thus, it may be concluded that certain reinforcements will enhance the activity of <c+a> dislocation activity more than others and that such enhancement can lead to an improvement in tensile ductility. However, there may be a concomitant decrease in fatigue resistance, presumably due to a lowered resistance to crack formation. It should be remembered that a dislocation is tantamount to a mode two crack (and pile-ups of dislocations even more so). The larger Burgers vector of the <c+a> dislocation would render it a more potent incipient crack than the <a> type dislocations. Thus, it would seem worthwhile for researchers to investigate role of <c+a> dislocations in crack formation.

2.4.6 Latent hardening

There are some topics related to the deformation of Mg and its alloys that may be considered as nascent. One such topic is latent hardening (i.e. the hardening of a slip/twinning system by other active systems). This subject has been actively researched for cubic metals, where the role of latent hardening in the activation of secondary slip systems and in texture evolution is well recognized. Unlike the case of fcc metals, where latent hardening theory and experiments[131–134] are well-established, there is very little detailed work in this area of hcp metals, like Mg. There is inadequate knowledge of how basal slip systems are hardened by non-basal, etc. Latent hardening is a subject that is relevant to every slip and twinning mode. It is raised in the section on <c+a> slip for the simple reason that the only significant experimental study in hcp magnesium, of which the author is aware, was performed to explore the role of <c+a> dislocations.[135] It is urged that similar sets of experiments need to be performed to explore other effects, such as the latent hardening of basal <a> slip by the presence of non-basal <a> dislocations, etc. As mentioned earlier in this chapter, it has been suggested that latent hardening effects may be responsible for the large solute strengthening effect that additions of Y and Dy can have.[15] Since Y has been noted to enhance the density of non-basal dislocations,[14] it is not unreasonable to suggest there may be a secondary hardening of basal dislocations.

The dislocation dynamics approach has promise for answering some of these questions. Indeed, it has already been used to examine interactions between prismatic slip systems in zirconium.[136] Capolungo has used the dislocation dynamics approach to demonstrate that junction formation is an issue that must be considered to develop a physically-based model for latent hardening.[137] More recently Capolungo et al.[138] have examined the effect of elastic anisotropy, which is not expected to have a strong impact on Mg, but is demonstrated to strongly influence the hardening of Hf. As was the case for the experimental studies of latent hardening, the <c+a> slip mechanism figures prominently in the theoretical discussions above. Again, it is urged that more general cases must be considered. Finally, as will be highlighted below, twinning appears to have a strong latent hardening effect on other slip systems.[139] The origins of the apparent latent hardening effects of twinning are debated and it remains an active topic of research.

2.5 Deformation twinning

Deformation twinning is an important mechanism of plastic strain accommodation in Mg and its alloys. A good explanation for this is the case is provided by Kocks and Westlake,[140] who suggested that Mg has 4.5 independent 'slip' modes if one considers basal and non-basal slip of <a> dislocations plus the role of the main {10.2} twinning mode (discussed below.) Their description helps one to understand why Mg is not brittle, since it has (according to this construct) nearly five independent slip systems. The limitations of considering twinning as a slip mode (again, discussed below) help one to understand why Mg does not exhibit such general ductility as more typical engineering materials based upon Fe or Al. The topic of twinning is covered in greater depth in Chapter 3, so the coverage here will be brief. The goal here is to provide a holistic consideration of the low temperature deformation mechanisms of Mg, with the same overall outline used for the slip mechanisms above, and to consider interactions between twinning and slip.

2.5.1 Role in polycrystal deformation

The predominant deformation twinning mechanism in all hcp metals is the {10.2} twin, which produces extension along the c-axis for metals with c/a ratio less than $\sqrt{3}$ and contraction along the c-axis for metals with larger c/a ratio.[141] For Mg and all of its alloys, save Mg-Cd solid solutions,[142] the {10.2} twin produces c-axis extension. For that one exception, high Cd content alloys actually undergo {10.2} contraction twinning. For the precise condition where c/a = $\sqrt{3}$, the {10.2} is inactive and the ductility of the alloy was observed to experience a minimum.

One thing is clear, the main deformation twinning mechanism, the {10.2} extension twin, is responsible for the tension–compression asymmetry that is normally observed in wrought magnesium alloys. The textures induced by most wrought processing (i.e. rolling or extrusion) render magnesium and its alloys softer in compression than in tension, along the prior working direction, i.e. the rolling or extrusion direction. This is because the majority of grains are poorly oriented to accommodate strain in either direction using their softest mechanism, basal slip of <a> dislocations. Therefore, the harder slip and twinning mechanisms must be activated. Because the {10.2} twinning mechanism can only accommodate extension along the c-axis, it is activated when crystals are either pulled in tension along their c-axes or compressed perpendicular to that direction. Again, due to the texture of most wrought magnesium alloys, this condition is satisfied when they are compressed along the prior working direction, and a sigmoidal hardening behavior characteristic of twinning is observed. If pulled in tension along the prior working direction, the material is forced to deform by hard slip and twinning mechanisms, which results in an elevated yield stress.

In situ deformation studies have revealed important insights about twinning, in particular. The initial study by Brown et al.[144] revealed the power of the neutron diffraction technique for revealing the internal stresses at the onset of twinning and the evolution of these internal stresses with twin growth. A more recent study by Muránsky et al.[145] combined the approach with acoustic emission, which allowed them to parse the effects of twin nucleation and growth. One benefit of the neutron diffraction approach is the statistical character of the data. In other words, it provides information about stresses within all the grains (or twins) that contribute to a given diffraction peak. As such, it can provide information from many grains of similar orientation immersed within various grain neighborhoods. This has made the data particularly useful for developing polycrystal plasticity models based upon the mean field self-consistent theory (discussed below). From another perspective, what has just been considered as a strength of the neutron diffraction technique may be considered a weakness. If we need to know the stress level within an individual grain, and/or we want to determine the variation of internal stress of a grain within various grain neighborhoods, the technique has insufficient spatial resolution. A recent study by Aydiner et al.[146] has demonstrated that the tensorial stresses within individual twins and the grains in which they form can be assessed using three-dimensional high energy X-ray diffraction at a synchrotron, such as the Advanced Photon Source at Argonne National Laboratory, Argonne, IL, USA. They examined the internal stress development within a few grains, one of which enabled them to make progress toward their goal of examining the stress at the outset of twinning and its evolution with twin growth. Their results confirm a concept that had been proposed previously, based upon *in situ* neutron diffraction,[143,144] that twins can have a very different stress state than the surrounding grain in which it appears. This difference in stress state between the twin and the parent grain in which it appears represents a sort of 'backstress' that helps to explain why the phenomenon that has come to be known as detwinning is so common in Mg.

Detwinning is the contraction of a prior twinned region (i.e. reversion to the original parent orientation). In one sense it is 'twinning of the twin', back into the parent orientation, but it does not require nucleation of a new twin since it is essentially nothing more than the original twin boundaries sweeping back over the previously twinned region. The detwinning phenomenon was actively researched in shape memory alloys[147] before it was recently discovered in Mg alloys.[148] Cáceres et al.[148] observed a strong pseudoelastic effect during loading, unloading cycles at low plastic strains and found twinning-detwinning to be responsible. Kleiner and Uggowitzer[149] observed the effect at larger plastic strains. Lou et al.[150] performed a rather exhaustive study of the phenomenon at such strains and provided a descriptive model describing the oddly shaped hysteresis loop that develops during repeated twinning–detwinning cycles. It is imagined that detwinning may play an important role during strain path changes,[135] which are typical of metal forming operations. For example, the bend-unbending that sheet metal undergoes during deep drawing will likely induce this phenomenon. Additionally, twinning–detwinning is very active during low cycle fatigue of wrought Mg.[151]

2.5.2 Latent hardening

One of the matters that has perplexed the community is precise prediction of the strain hardening due to twinning. Cases where Mg is twinning extensively lead to exceptional strain hardening rates. It is clear that {10.2} extension twinning frequently reorients the crystal from a soft orientation (at least vis-à-vis twinning) to an orientation that is quite hard, with respect to any soft slip or twinning mode. This reorientation (or texture) effect is indeed heavily responsible for the induced hardening.[57,97] One might call this a 'composite' effect, since the material contains an increasing volume fraction of a very hard phase, as it twins. On the other hand, there appear to be distinctions in both the degree of hardening and the timing of the strong increase in hardening, if one models the hardening as exclusively due to the texture effect.

It has been widely demonstrated that cases where twinning dominates the strain accommodation or cases where there is almost no twinning can be successfully modelled using crystal plasticity approaches[139,152] However, cases where slip is a partial contributor are much less successfully modelled. Even modeling of the former cases has required that *ad hoc* latent hardening corrections be made in order to achieve good agreement with the very high level of strain hardening observed experimentally.[139] These *ad hoc* corrections are designed to account for two additional possible sources of hardening, beyond the textured-based hardening discussed above: Interface hardening (i.e. a Hall–Petch effect on the slip mechanisms due to the presence of twin boundaries) and transmutation effects. An initial attempt to account for the former effect was performed by Proust et al.[153] using a so-called 'composite grain model' that met with some success. However, this approach is probably more appropriate for systems in which prolific twinning of very thin twins occurs (e.g. many low stacking fault energy fcc metals). It is probably not the correct vision for Mg, in which the {10.2} tension twins are observed to readily thicken and grow to the point where they can consume an entire grain. Under such circumstances, the grains are ultimately not significantly refined by the twinning process. In fact, El Kadiri and Oppedal[154] are quick to point out that the strongest grain size refinement is expected to occur at low strains, where the strain hardening rate is low, and that higher strain levels, where the strain hardening rate is rapid, are associated with an expected decrease in interface hardening, since twins are experimentally observed to increasingly consume entire grains. In short, it is pointed out that the twins in Mg are frequently not thin lamellae that occur in ever-increasing numbers, which increasingly refine the microstructure. Rather, {10.2} extension twins tend to occur in rather small numbers, which increasingly grow into one another (hence removing the prior interfaces between them). Oppedal and El Kadiri[155] emphasize the role of transmutation in the high hardening rates associated with twinning. They review the process whereby glissile dislocations within the matrix are transmuted into sessile dislocations within the twin. Their crystal plasticity-based model is shown to be effective for predicting the hardening rates regardless of the level of twin activity.

Recent activities have emphasized the distinction between twin nucleation and twin growth (thickening). It is imagined that nucleation may require high local stresses and may be followed by very rapid growth to a certain size (very frequently a lath encompassing the entire cord subtended by the twinning plane through a grain). Subsequent twin growth is imagined to require less stress and to proceed with much slower kinetics. Evidence for this interpretation is provided by the aforementioned study involving *in situ* neutron diffraction combined with acoustic emission.[145] One area that has undergone rapid development in the past few years is the understanding of twin nucleation. The interested reader is referred to Chapter 3. The techniques of EBSD and atomistic modeling have figured prominently in these advancements.

2.5.3 Solid solution alloy effects

Hauser et al.[74] showed that Li solid solution alloying appeared to reduce the incidence of twinning, if one considered the incidence of twinning as a function of strain. This makes sense, since they also showed that Li alloying promoted prismatic slip, which is competitive with {10.2} extension twinning. In short, since prismatic slip is relatively easier in Mg-Li alloys, less extension twinning will be observed, since the two mechanisms can provide similar types of strain (e.g. compression within the basal plane). However, they go on to show that the incidence of twinning is not reduced if one considers the influence of stress, i.e. they saw the same level of twinning in the alloy as they did in the pure metal at a given stress level.[74] As mentioned in the companion chapter in this volume, 'The single crystal experiments performed by Kelly and Hosford[156] revealed that the addition of Th or Li hardened the normal stress in crystals oriented for twinning (compression perpendicular to c) by an increment approximately one-third of that seen in crystals oriented for compression along the c axis.' One wonders if this is perhaps due to the possible latent hardening effect mentioned earlier in the present chapter. (Recall that a larger than expected hardening of basal slip has been observed for some solute additions, and that this has been attributed to increased forest dislocation density in alloy single crystals, relative to pure Mg.[15]) Since the 1960s, there has been very little detailed study of the effect of solid solution alloying on twinning. Despite numerous qualitative statements in the literature, the present author is not aware of any systematic studies of this effect in single crystals or polycrystals. This would seem to be an area of fundamental scientific importance, since twinning has such a profound influence on the deformation behavior of Mg alloys.

One area where considerable speculation has been made concerns the effect of rare earth element alloying on contraction twinning. Couling *et al.*[157] speculated that dilute additions of rare earth elements may promote {10.1} contraction- and double-twinning, as an explanation for the extensive shear banding observed in the 'cold-rollable' Mg alloys they had discovered. More recently, it was demonstrated, using polycrystal plasticity modeling, that such a mechanism could explain the flow curve and texture evolution of Mg-Y alloys tested in uniaxial compression.[97] On the other hand, a recent EBSD-based study of a Mg-Ce alloy suggested that there is no clear link between the shear banding and double twinning;[158] however, an EBSD and TEM-based examination claimed that all shear bands observed in an Mg-Y alloy are associated with double twinning.[159] Both of these studies revealed that the severity of shear banding was reduced by rare earth solute additions.[159,160] It would appear that this is a specific area worthy of further research. A final example where 'solid solution alloying' has been shown to have a strong effect upon twinning is the LPSO alloys based upon Mg-RE-Zn (e.g. $Mg_{97}Y_2Zn_1$). There, the ordered phase has been shown effective to block the advance of {10.2} twinning, in particular.[160] This is not surprising, given the fact that the ordering precludes twin formation and suggests that the pseudotwin that would be formed is not energetically favorable.

2.5.4 Precipitation effects

The notion of controlling twin formation and growth by precipitation has received enough attention for some general conclusions to be drawn. The aforementioned study Clark performed to determine the effects of the c-axis rod-shaped precipitates that form in Mg-Zn alloys also examined the effect the precipitates had on twinning.[31] He determined that {10.2} twinning is not as strongly affected by the presence of fine, uniformly dispersed precipitates as is dislocation motion. Stanford et al.[161] recently confirmed this conclusion, and added the point that the nucleation of twins actually appears to be enhanced in the presence of such fine precipitates[162] while twin growth is mildly impeded (though less than slip). TEM observations by Singh et al.[163] confirmed this observation in a Mg-Zn-Y alloy, showing that age-hardening increases the tension compression asymmetry of textured wrought alloys because the strength in tension is controlled by the ease of slip, while the strength in compression is controlled by the ease of twinning.

Gharghouri et al.[164,165] performed studies of a Mg-Al alloy using *in situ* neutron diffraction. They concluded, first of all, that twinning was not affected by the hydrostatic pressure but rather the resolved shear stress on the twinning plane and in the twinning direction. Secondly, they show that the precipitates characteristic of the Mg-Al alloy they examined do indeed impede twinning. Jain et al. confirmed this conclusion during their investigations of alloy AZ80.[166] In contrast with the findings for the Mg-Zn system, aging of this Mg-Al-Zn alloy leads to a decrease in the tension compression asymmetry, i.e. $Mg_{17}Al_{12}$ precipitates impede twinning more than slip.

2.5.5 Grain size effects

This subject is extensively detailed in its companion chapter 3 in this volume. Barnett et al.[39] have observed that the tension–compression asymmetry is less pronounced in material with a smaller grain size. In fact, the same study provided strong evidence that the common {10.2} tension twinning mode has a stronger grain size effect than does slip. Thus, one expects the deformation to be increasingly slip-based at finer grain sizes, where twinning appears to be difficult. This has led some researchers to suggest that there is a limiting grain size for twinning. However, others have shown that even when the material exhibits no discernable tension–compression yield stress asymmetry, deformation twinning persists to very small grain sizes.[167] There is now evidence that samples with truly nanocrystalline grain sizes can undergo significant twinning as well[168] so this concept of eliminating twinning via grain size reduction should be revisited. In other materials systems, particularly the pure fcc metals, there has been a great deal of recent research into the increased incidence of twinning at nanocrystalline grain sizes, where nucleation of perfect dislocations is more energetically costly than partial dislocations involved in twin formation, e.g.[169]

Contraction {10.1} twinning also appears to exhibit a strong grain size effect though it is more difficult to diagnose, since contraction twinning does not induce the characteristic changes in the flow curve shape that are observed for {10.2} tension twins. A more direct approach of deforming and characterizing the twin content must be pursued. *In situ* neutron diffraction studies of fine-grained alloys have not provided evidence that compression twinning is an important contributor to strain accommodation.[82,83] As detailed below, this final type of observation is probably a fundamental (and credible) reason why so many researchers have suggested grain refinement as a strategy for reducing twin activity and improving the ductility of Mg and its alloys.[59]

2.5.6 Role in fracture

There is a popular view that if twinning were minimized, the ductility/formability would be improved. The aforementioned results obtained by Stoloff and Gensamer for Mg-Cd alloys[142] challenges that notion, since they observed a decrease in ductility when twinning was eliminated, rather than an increase. A more recent report of Barnett[57] also challenges this notion that twinning is universally deleterious for ductility. Barnett shows that {10.2} extension twinning can actually enhance ductility, since such twinning leads to rapid strain hardening (discussed above), and strain hardening is known to suppress plastic instability. On the other hand, {10.1} compression and {10.1}–{10.2} double twinning have been repeatedly associated with shear localization due to what is known as 'geometric' or 'texture' softening,[58,59] surface relief,[170] and/or microvoid nucleation[171] leading to crack formation and failure.[59]

The fracture of magnesium is very frequently mischaracterized as brittle. Magnesium alloys may be characterized as brittle, from the perspective of area reduction, i.e. they frequently fracture prior to the onset of macroscopic necking. However, the fracture mode is not one of brittle cleavage, nor do Mg alloys often suffer from brittle intergranular fracture. Rather, the fracture mode is one of ductile microvoid nucleation, growth and coalescence. There may be flat, facet-like features on the fracture surface, but these are most frequently highly inclined to the main tensile direction. They are the result of shear-off fractures, which can be associated with deformation twins. Careful inspection of fracture surfaces typically reveals regions of equiaxed dimples as well as elongated dimples on inclined facets. The new technique of high resolution X-ray tomography has already been used to advantage to make connections between microstructure (porosity) and fracture properties of cast Mg alloys.[172–174] It is suggested that could also provide significant insight into the questions of ductile fracture of wrought Mg alloys.

2.5.7 Kink banding

The final aspect discussed in this section on twinning is kink banding. It is a mechanism of deformation involving the cooperative motion of dislocations that leads to arbitrary reorientations of the crystal lattice across kink boundaries.[175] Readers interested in broader studies of kinking, and incipient kinking in particular, may like to see recent papers by Barsoum et al.[176,177] The mechanism was initially observed to occur in hexagonal single crystals of metals such as Cd and Zn, when pulled along the c-axis.[178] Although the basal planes are unfavorably oriented to accommodate slip in such a crystal, the cooperative motion of basal dislocations is observed to reorient a portion of the crystal, thus enabling it to deform further by slip. (Recall that the common {10.2} twin only operates during compression along the c-axis, in such high axial ratio crystals.) Unlike twins, which have their own characteristic twinning dislocations and atomic shuffles, general kink bands are composed by organized motion of conventional lattice dislocations. Kink bands have also been observed in some intermetallic compounds that have limited slip systems, such as NiAl. Some researchers thus contemplated whether kink banding might contribute additional independent 'slip' modes, thus enabling more homogeneous deformation and extending ductility of such brittle phases. However, the fact that they are formed by the motion of conventional lattice dislocations has clearly established that they do not provide additional independent deformation modes.[179]

Layered structure materials such as mica and certain laminated composites are particularly prone to kink banding. In fact, Hess and Barrett[178] proposed a model for which kink banding of axial crystals that demands a minimum c/a ratio of 1.732. Thus, kink banding is not generally expected to occur in Mg or its alloys. The recently discovered LPSO alloys are obviously an exception to this rule, and they have indeed been observed to undergo kink banding.[25,180] Kink bands have been most frequently observed in single crystals and sometimes as accommodation mechanisms in single crystals that are undergoing deformation twinning. However, some researchers have suggested that some twin-like features observed by EBSD within an SEM must be kink bands, since they did not match any twin type known to the authors. However, there are a large number of twin types, including {10.1}, {10.2}, and {10.3}, as well as double-twins composed of the above, especially {10.1}–{10.2} and {10.3}–{10.2}. If one considers all the possible combinations and permutations, a large number of possible twin boundary angles are possible[181, 182]. Thus, it is suggested that many of the features that researchers might previously have identified as due to kink bands[183] may actually be due to twinning.

2.6 Dynamic deformation

The final aspect that concerns the low temperature deformation mechanisms of Mg is their behavior at high rates of deformation. Recent studies of low-cost magnesium alloys, at strain rates relevant to automotive crashworthiness, have been disappointing and forced reconsideration of the use of magnesium in safety critical components such as the crumple zones in the front of a vehicle.[184] On the other hand, it was determined in the same preliminary examinations that the energy absorption during failure (fracture) can be considerable. There have been relatively few studies of the dynamic constitutive behavior of Mg alloys. Most of the published work is due to the groups of Mukai,[185] Gray[186] and Horstemeyer.[187] All of these studies focused on pure Mg or the leading wrought magnesium alloy (AZ31). There was another study on alloy AZ80,[188] the authors of which also compared Al, Mg, and Ti.[189] There has also been some work performed on alloy ZK60.[190] These papers have unanimously emphasized the importance of crystallographic texture and plastic anisotropy at high rates, as was done previously for quasi-static testing conditions (e.g. 10^{-3} s^{-1}). They have also demonstrated the athermal character of mechanical twinning, since the twinning dominated flow curves (compression along the extrusion or rolling direction) yield at essentially the same stress level irrespective of rate, from quasi-static conditions all the way up to rates typical of split Hopkinson pressure bar testing (~10^3 s^{-1}). To the author's knowledge, there is no published analysis of the high strain rate deformation behavior of the newer high RE-content Mg alloys, aside from preliminary investigations of the ballistic behavior of alloys WE43[191] and Elektron 675[192] by the US Army Research Laboratory.

The most widely used class of constitutive models to simulate dynamic metal behavior is generally referred to as the Johnson–Cook Models[193] These models account for strain-rate and temperature dependence and are capable of capturing some of the aspects of shear localization. Zerilli and Armstrong[194] were among the first to note that the functional form of the empirical Johnson–Cook relationship most frequently used to model the constitutive response of metals under dynamic loading conditions fails to account for the differences between metals of different crystal structures, particularly at high levels of plastic strain. They also noted the potential impact that mechanical twinning would have, relative to dislocation motion. Other research groups have demonstrated the utility of a physics-based description of thermally activated slip pioneered by Kocks, Argon and Ashby[195] in their models of the flow strength of metals in the quasi-static through dynamic loading regimes. Follansbee et al.[196,197] and Nemat-Nasser et al.[198–200] have made notable contributions, demonstrating improved success (relative to Johnson-Cook) for modeling the temperature and rate sensitivity of fcc (e.g. Cu), bcc (e.g. Ta), hcp (e.g. Zr) and two-phase mixtures (e.g. α plus β Ti-64).

It is presumed that dislocation-based plasticity in Mg and its alloys at high strain rates will obey the same constitutive rules that have been shown to work for fcc and bcc metals. In some cases, involving primarily basal slip, the rate and temperature dependence may be very similar to fcc metals.[197,199] In other cases, involving more thermally activated non-basal slip, the behavior is expected to be more like that of bcc metals.[200] These expectations have yet to be tested. Attempts to model the dynamic behavior of hcp Ti and its alloys have met with some success, though shortcomings are often attributed to the effects of deformation twinning, and Mg alloys typically undergo more prolific twinning than Ti alloys. Numerous studies have emphasized the athermal character of the most common

{10.2} type twinning, including at rate insensitivity up to dynamic rates, e.g.[187] This suggests that twinning mechanisms will accommodate a more significant fraction of the strain at dynamic rates than during quasi-static loading. As discussed in the section on twinning, it will be very important in distinguishing the roles of different twin types (tension vs compression/contraction). How these effects operate at high strain rates, where twinning may be even more profuse, is entirely unknown.

Magnesium alloys (along with other hexagonal metals, e.g. Ti) are particularly prone to shear failure during primary deformation processing (e.g. rolling) or high strain rate deformation. In comparison with Ti and its alloys, Mg alloys have higher thermal conductivity, which should render it less intrinsically susceptible to adiabatic softening. Other key factors (plastic anisotropy, etc.) that do render Mg prone to shear failure have already been discussed. A paper was recently published that emphasized the presence of very fine grains refined by dynamic recrystallization within shear bands.[201] However, it is mentioned that dynamic recrystallization during low temperature deformation at high rates is generally understood to be a result of shear localization[202] rather than a cause of it, as can be the case during hot deformation.

In summary, if high-performance Mg alloys or processing strategies are to be developed for applications involving dynamic loading, the fundamental structure property relationships addressed by the proposed research must be established. For example, it must be determined whether it is preferable to have fine grains, coarse grains, or if there is some optimal intermediate grain size. It must be determined if a heat-treatable precipitation-strengthened alloy or a solution-strengthened alloy (such as currently employed 5000 series aluminum alloys) approach is preferable from the combined perspectives of strength, ductility, and energy absorption at high rates. Finally, there are numerous first-order mechanistic questions that need to be answered, such as whether or not deformation twinning should be suppressed or enhanced to promote the optimal dynamic response.

2.7   Conclusions

Although the present review is far from comprehensive, it is hoped that the reader is left with the impression that there have already been a large number of studies of the low temperature deformation mechanisms of Mg and its alloys. A great deal is already known, at least qualitatively speaking. However, firm quantitative connections between alloy composition/microstructure and properties are not generally available. One area that seems ripe for development is latent hardening, i.e. the interplay between individual mechanisms, especially between slip and twinning. An area that is rapidly developing is that of atomistic modeling (both molecular dynamics simulation and first principles modeling) of the defects responsible for the observed deformation behavior. One aspect that needs attention is how to translate these fundamental insights into constitutive rules that can be used to predict macroscopic behavior. Although it was not considered in any detail here, some hints regarding the fracture behavior of wrought Mg alloys are already present in the literature. The potential to provide quantitative connections between microstructure and toughness appears to be an attainable goal. Finally, improvement of the dynamic deformation and fracture behavior is a specific area of concern that must be addressed if there are to be broader applications of Mg alloys in safety critical, structural automotive components (as well as applications, which demand resistance to blast and ballistic threats, of interest to the defense industry).

2.8 Acknowledgements

The author would also like to express his gratitude to the National Science Foundation (Grant Number 1121133) and the United States Automotive Materials Partnership for sponsoring this effort to synthesize the state of the art in this area of Mg science and technology. Finally, the author would also like to dedicate this paper to the memory of Professor Fereshteh Ebrahimi, Department of Materials Science and Engineering, University of Florida. She had the reputation of a solid researcher and was a dedicated teacher and mentor. Her students remember her fondly despite the difficult exams she offered. She never hesitated to offer encouragement to the members of the next generation of mechanical metallurgists that seek to follow in her footsteps (the present author included). The recent paper she co-authored on <c+a> dislocations[116] is a landmark work that provides a good overview of prior work on this critical defect and provides new insights that will spark future advances.

2.9 References

1. Raynor GV, *The Physical Metallurgy of Magnesium and Its Alloys*. Pergamon, Oxford, 1959.
2. Roberts CS, *The Deformation of Magnesium, in Magnesium and Its Alloys*. John Wiley & Sons, New York, 1964.
3. Schmid E and Boas W, *Kristallplastizität*. Julius Springer, Berlin, 1935, p. 64.
4. Von Mises R, Mechanics of the ductile form changes of crystals. *Zeitschrift Fur Angewandte Mathematik Und Mechanik* 1928; **8**: 161–185.
5. Taylor GI, Plastic strain in metals. *Journal of the Institute of Metals* 1938; **62**: 307–338.
6. Feltner CE and Laird C, Cyclic stress-strain response of FCC metals and alloys–I Phenomenological experiments. *Acta Metallurgica* 1967; **15**: 1621.
7. Feltner CE and Laird C, Cyclic stress-strain response of FCC metals and alloys–II Dislocation structures and mechanisms. *Acta Metallurgica* 1967; **15**: 1633.
8. Starke EA and Williams JC, in *Fracture Mechanics: Perspective and Directions (Twenty Symp.)*, ASTM STP 1020, Wei RP and Gangloff RP, Eds., ASTM, West Conshohocken, Pennsylvania, 1989, pp. 184–205.

9. Yasi JA, Nogaret T, Trinkle DR, Qi Y, Hector LG et al., Basal and prism dislocation cores in magnesium: comparison of first-principles and embedded-atom-potential methods predictions. *Modeling and Simulation in Materials Science and Engineering* 2009: **17**: 055012 (13pp).
10. Wang Y, Chen LQ, Liu ZK and Mathaudhu SN, First-principles calculations of twin-boundary and stacking-fault energies in magnesium. *Scripta Materialia* 2010; **62**: 646–649.
11. Conrad H and Robertson WD, Effect of temperature on the flow stress and strain-hardening coefficient of magnesium single crystals. *Transactions AIME* 1957; **209**: 503–512.
12. Akhtar A and Teghtsoonian E, Solid solution strengthening of magnesium single crystals-I Alloying behaviour of basal slip. *Acta Metallurgica* 1969; **17**: 1339–1349.
13. Akhtar A and Teghtsoonian E, Substitutional solution hardening of magnesium single-crystals. *Philosophical Magazine* 1972; **25**: 897–916.
14. Suzuki M, Sato H, Maruyama K and Oikawa H, Creep behavior and deformation microstructures of Mg-Y alloys at 550 K. *Materials Science and Engineering A* 1998; **252**: 248.
15. Miura S, Imagawa S, Toyoda T, Ohkubo K and Mohri T, Effect of rare-earth elements Y and Dy on the deformation behavior of Mg alloy single crystals. *Materials Transactions* 2008; **49**: 952–956.
16. Yasi JA, Hector LG and Trinkle DR, First-principles data for solid-solution strengthening of magnesium: From geometry and chemistry to properties. *Acta Materialia* 2010; **58**: 5704–5713.
17. Gao L, Zhou J, Sun ZM, Chen RS and Han EH, Electronic origin of the anomalous solid solution hardening of Y and Gd in Mg: A first-principles study. *Chinese Science Bulletin* 2011; **56**: 1038–1042.
18. Nagamatsu J, Nakagawa N, Muranaka T, Zenitani Y and Akimitsu J, Superconductivity at 39 K in magnesium diboride. *Nature* 2001; **410**: 63–64.
19. Abe E, Kawamura Y, Hayashi K and Inoue A, Long-period ordered structure in a high-strength nanocrystalline Mg-1 at% Zn-2 at% Y alloy studied by atomic-resolution Z-contrast STEM. *Acta Materialia* 2002; **50**: 3845–3857.
20. Yamasaki M, Anan T, Yoshimoto S and Kawamura Y, Mechanical properties of warm-extruded Mg-Zn-Gd alloy with coherent 14H long periodic stacking bordered structure precipitate. *Scripta Materialia* 2005; **53**: 799–803.
21. Kawamura Y, Kasahara T, Izumi S and Yamasaki M, Elevated temperature $Mg_{97}Y_2Cu_1$ alloy with long period ordered structure. *Scripta Materialia* 2006; **55**: 453–456.
22. Zhu YM, Weyland M, Morton AJ, Oh-ishid K, Honod K et al., The building block of long-period structures in Mg-RE-Zn alloys. *Scripta Materialia* 2009; **60**: 980–983.
23. Zhu YM, Morton AJ and Nie JF, The 18R and 14H long-period stacking ordered structures in Mg-Y-Zn alloys. *Acta Materialia* 2010; **58**: 2936–2947.
24. Nie JF, Oh-ishi K, Gao X and Hono K, Solute segregation and precipitation in a creep-resistant Mg-Gd-Zn alloy. *Acta Materialia* 2008; **56**: 6061–6076.
25. Hagihara K, Umakoshi Y and Yokotani N, Plastic deformation behavior of $Mg_{12}YZn$ with 18R long-period stacking ordered structure. *Intermetallics* 2010; **18**: 267–276.
26. Hagihara K, Kinoshita A, Sugino Y, Yamasaki M, Kawamura Y et al., Effect of long-period stacking ordered phase on mechanical properties of $Mg_{97}Zn_1Y_2$ extruded alloy. *Acta Materialia* 2010; **58**: 6282–6293.
27. Nie JF, Effects of precipitate shape and orientation on dispersion strengthening in magnesium alloys. *Scripta Materialia* 2003; **48**: 1009–1015.

28. Geng J, Chun YB, Stanford N, Davies CHJ, Nie JF et al., Processing and properties of Mg–6Gd–1Zn–0.6Zr: Part 2. Mechanical properties and particle twin interactions. *Materials Science and Engineering: A* 2011; **528**: 3659–3665.
29. Singh A, Rosalie JM, Somekawa H and Mukai T, The structure of $\beta'1$ precipitates in Mg-Zn-Y alloys. *Philosophical Magazine* 2010; **90**: 641–651.
30. Rosalie JM, Somekawa H, Singh A and Mukai T. Orientation relationships between icosahedral clusters in hexagonal $MgZn_2$ and monoclinic Mg_4Zn_7 phases in Mg-Zn(-Y) alloys. *Philosophical Magazine* 2011; **91**: 2634–2644.
31. Clark JB, Transmission electron microscopy study of age hardening in a Mg-5 wt% Zn alloy. *Acta Metallurgica* 1965; **13**: 1281–1289.
32. Blankenship CP, Hornbogen E and Starke EA, Predicting slip behaviour in alloys containing shearable and strong particles. *Materials Science and Engineering A* 1993; **169**: 33–41.
33. Muddle BC and Polmear IJ, The precipitate omega-phase in Al-Cu-Mg-Ag alloys. *Acta Metallurgica* 1989; **37**: 777–789.
34. Pekguleryuz MO and Kaya AA, Creep resistant magnesium alloys for powertrain applications, magnesium alloys and their applications. *Deutsche Gesellschaft für Materialkunde (DGM)*. Garmisch, Germany, 2003, pp.74–79.
35. Mendis CL, Bettles CJ, Gibson MA, Gorsse S and Hutchinson CR, Refinement of precipitate distributions in an age-hardenable Mg–Sn alloy through microalloying. *Philosophical Magazine* 2006; **86**: 443–456.
36. Mendis CL, Bettles CJ, Gibson MA, Gorsse S and Hutchinson CR, An enhanced age hardening response in Mg–Sn based alloys containing Zn. *Materials Science and Engineering A* 2006; **435–436**: 163–171.
37. Hauser FE, Landon PR and Dorn JE, Fracture of magnesium alloys at low temperature. *Transactions of the Metallurgical Society of AIME* 1956; **206**: 589–593.
38. Armstrong RW, Theory of the tensile ductile-brittle behavior of poly-crystalline h.c.p. materials, with application to beryllium. *Acta Metallurgica* 1968; **16**: 347–355.
39. Barnett MR, Keshavarz Z, Beer AG and Atwell D, Influence of grain size on the compressive deformation of wrought Mg–3Al–1Zn. *Acta Materialia* 2004; **52**: 5093–5103.
40. Cáceres CH, Mann GE and Griffiths JR, Grain Size hardening in Mg and Mg-Zn solid solutions. *Metallurgical and Materials Transactions A* 2011; **42A**: 1950–1959.
41. Raeisinia B and Agnew SR, Using polycrystal plasticity modeling to determine the effects of grain size and solid solution additions on individual deformation mechanisms in cast Mg alloys. *Scripta Materialia* 2010; **63**: 731–736.
42. Blake AH and Cáceres CH, Solid-solution hardening and softening in Mg–Zn alloys. *Materials Science and Engineering A* 2008; **483–484**: 161–163.
43. Jain A, Duygulu O, Brown DW, Tomé CN and Agnew SR, Grain size effects on the tensile properties and deformation mechanisms of a magnesium alloy, AZ31B, sheet. *Materials Science and Engineering A* 2008; **486**: 545–555.
44. Raeisinia B, Agnew SR and Akhtar A, Incorporation of solid solution alloying effects into polycrystal modeling of Mg alloys. *Metallurgical and Materials Transactions A* 2011; **42A**: 1418–1430.
45. Barnett MR and Hutchinson WB, Effective values of critical resolved shear stress for slip in polycrystalline magnesium and other hcp metals. *Scripta Materialia* 2010; **63**: 737–740.
46. Chapman JA and Wilson DV. The room-temperature ductility of fine-grain magnesium. *Journal of the Institute of Metals* 1962; **91**: 39–40.

47. Wilson DV and Chapman JA, Effects of preferred orientation on the grain size dependence of yield strength in metals. *Philosophical Magazine* 1963; **8**: 1543–1551.
48. Barnett MR, Jacob S, Gerard BF and Mullins JG, Necking and failure at low strains in a coarse-grained wrought Mg alloy. *Scripta Materialia* 2008; **59**: 1035–1038.
49. Barnett MR, Stanford N, Cizek P, Beer A, Xuebin Z *et al.* Deformation mechanisms in Mg alloys and the challenge of extending room-temperature plasticity. *JOM Journal of the Minerals, Metals and Materials Society* 2009; **61**(8): 19–24.
50. Kang DH, Kim DW, Kim S, Bae GT, Kim KH *et al.*, Relationship between stretch formability and work-hardening capacity of twin-roll cast Mg alloys at room temperature. *Scripta Materialia* 2009; **61**: 768–771.
51. Mukai T, Yamanoi M, Watanabe H and Higashi K, Ductility enhancement in AZ31 magnesium alloy by controlling its grain structure. *Scripta Materialia* 2001; **45**: 89–94.
52. Kim WJ, An CW, Kim YS and Hong SI, Mechanical properties and microstructures of an AZ61 Mg Alloy produced by equal channel angular pressing. *Scripta Materialia* 2002; **47**: 39–44.
53. Agnew SR, Horton JA, Lillo TM and Brown DW, Enhanced ductility in strongly textured magnesium produced by equal channel angular (ECA) processing. *Scripta Materialia* 2004; **50**: 377–381.
54. Kim WJ, Lee JB, Kim WY, Jeong HT and Jeong HG, Microstructure and mechanical properties of Mg–Al–Zn alloy sheets severely deformed by asymmetrical rolling. *Scripta Materialia* 2007; **56**: 309–312.
55. Huang XS, Suzuki K, Watazu A, Shigematsu I and Saito N, Mechanical properties of Mg–Al–Zn alloy with a tilted basal texture obtained by differential speed rolling. *Materials Science and Engineering A* 2008; **488**: 214–220.
56. Wang H, Wu PD, Boyle KP and Neale KW, On crystal plasticity formability analysis for magnesium alloy sheets. *International Journal of Solids and Structures* 2011; **48**: 1000–1010.
57. Barnett MR, Twinning and the ductility of magnesium alloys: Part I. 'Tension' twins. *Materials Science and Engineering A* 2007; **464**: 1–7.
58. Jiang L, Jonas JJ, Luo AA, Sachdev AK and Godet S, Twinning-induced softening in polycrystalline AM30 Mg alloy at moderate temperatures. *Scripta Materialia* 2006; **54**: 771–775.
59. Barnett MR, Twinning and the ductility of magnesium alloys: Part II. 'Contraction' twins. *Materials Science and Engineering A* 2007; **464**: 8–16.
60. Ball EA and Prangnell PB, Tensile-compressive yield asymmetries in high strength wrought magnesium alloys. *Scripta Metallurgica et Materialia* 1994; **31**: 111–116.
61. Barnett MR, Nave MD and Bettles CJ, Deformation microstructures and textures of some cold rolled Mg alloys. *Materials Science and Engineering A* 2004; **386**: 205–211.
62. Sandlöbes S, Zaefferer S, Schestakow I, Yi S and Gonzalez-Martinez R, On the role of non-basal deformation mechanisms for the ductility of Mg and Mg–Y alloys. *Acta Materialia* 2011; **59**: 429–439.
63. Senn JW and Agnew SR, Texture randomization of Mg alloys containing rare earth elements. In *Magnesium Technology* 2008, Kaplan HI, Ed., TMS, Warrendale, PA, 2008, pp. 153–159.
64. Mackenzie LFW and Pekguleryuz MO, The recrystallization and texture of magnesium–zinc–cerium alloys. *Scripta Materialia* 2008; **59**: 665–668.
65. Stanford N and Barnett MR, The origin of 'rare earth' texture development in extruded Mg-based alloys and its effect on tensile ductility. *Materials Science and Engineering A* 2008; **496**: 399–408.

66. Stanford N, Sha G, La Fontaine A, Barnett MR and Ringer SP, Atom probe tomography of solute distributions in Mg-based alloys. *Metallurgical and Materials Transactions A* 2009; **40**: 2480–2487.
67. Hadorn JP, Hantzsche K, Yi S, Bohlen J, Letzig D et al., Role of solute in the texture modification during hot deformation of Mg-rare earth alloys, *Metallurgical and Materials Transactions A* 2011, **43**: 1347-1362.
68. Hadorn JP, Hantzsche K, Yi S, Bohlen J, Letzig D and Agnew SR, Effects of solute and second-phase particles on the texture of Nd-containing Mg alloys. *Metallurgical and Materials Transactions A* 2012, 43: 1363-1375.
69. Mishra RK, Gupta AK, Rama RP, Sachdev AK, Kumar AM et al., Influence of cerium on the texture and ductility of magnesium extrusions. *Scripta Materialia* 2008; **59**: 562–565.
70. Chino Y, Kado M and Mabuchi M, Enhancement of tensile ductility and stretch formability of magnesium by addition of 0.2 wt% (0.035 at%) Ce. *Materials Science and Engineering A* 2008; **494**: 343–349.
71. Dreyer CE, Chiu WV, Wagoner RH and Agnew SR, Formability of a more randomly textured magnesium alloy sheet: Application of an improved warm sheet formability test. *Journal of Materials Processing Technology* 2010; **210**: 37–47.
72. Ward Flynn P, Mote J and Dorn JE, On the thermally activated mechanism of prismatic slip in magnesium single crystals. *Transactions of the TMS-AIME* 1961; **221**: 1148–1154.
73. Wonsiewicz BC and Backofen WA, Plasticity of magnesium crystals. *Transactions TMS-AIME* 1967; **239**: 1422–1431.
74. Hauser FE, Landon PR and Dorn JE, Deformation and fracture of alpha solid solutions of lithium in magnesium. *Transactions of the ASM* 1958; **50**: 856–883.
75. Agnew, SR, Plastic anisotropy of magnesium alloy AZ31B sheet. In *Magnesium Technology 2002*, Kaplan H, Ed. TMS, Warrendale, PA, 2002, pp. 169–174.
76. Agnew SR and Duygulu O, Plastic anisotropy and the role of non-basal slip in magnesium alloy AZ31. *International Journal of Plasticity* 2005; **21**: 1161–1193.
77. Styczynski A, Hartig Ch, Bohlen J and Letzig D, Cold rolling textures in AZ31 wrought magnesium alloy. *Scripta Materialia* 2004; **50**: 943–947.
78. Ebeling T, Hartig Ch, Laser T and Bormann R, Material law parameter determination of magnesium alloys. *Materials Science and Engineering A* 2009; **527**: 272–280.
79. Keshavarz Z and Barnett MR, EBSD analysis of deformation modes in Mg–3Al–1Zn. *Scripta Materialia* 2006; **55**: 915–918.
80. Koike J, Enhanced deformation mechanisms by anisotropic plasticity in polycrystalline Mg alloys at room temperature. *Metallurgical and Materials Transactions A* 2005; **36A**: 1689–1696.
81. Agnew SR, Tomé CN, Brown DW, Holden TM and Vogel SC, Study of slip mechanisms in a magnesium alloy by neutron diffraction and modeling. *Scripta Materialia* 2003; **48**: 1003–1008.
82. Agnew SR, Brown DW and Tomé CN, Validating a polycrystal model for the elasto-plastic response of magnesium alloy AZ31 using *in-situ* neutron diffraction. *Acta Materialia* 2006; **54**: 4841–4852.
83. Muránsky O, Carr DG, Barnett MR, Oliver EC and Sittner P, Investigation of deformation mechanisms involved in the plasticity of AZ31 Mg alloy: *In situ* neutron diffraction and EPSC modeling. *Materials Science and Engineering A* 2008; **496**: 14–24.
84. Akhtar A and Teghtsoonian E, Solid solution strengthening of magnesium single crystals-II The effect of solute on the ease of prismatic slip. *Acta Metallurgica* 1969; **17**: 1351–1356.

85. Mathis K, Trojanova Z, Lukac P, Cáceres CH and Lendvai J, Modeling of hardening and softening processes in Mg alloys. *Journal of Alloys and Compounds* 2004; **378**: 176–179.
86. Staroselsky A and Anand L. A constitutive model for hcp materials deforming by slip and twinning: application to magnesium alloy AZ31B. *International Journal of Plasticity* 2003; **19**: 1843–1864.
87. Püschl W, Models for dislocation cross-slip in close-packed crystal structures: a critical review. *Progress in Materials Science* 2002; **47**: 415–461.
88. Couret A and Caillard D, An *in-situ* study of prismatic glide in magnesium—I. The rate controlling mechanism. *Acta Metallurgica* 1985; **33**: 1447–1454.
89. Couret A and Caillard D, An *in-situ* study of prismatic glide in magnesium—II. Microscopic activation parameters. *Acta Metallurgica* 1985; **33**: 1455–1462.
90. Yoshinaga H and Horiuchi R, On the nonbasal slip in magnesium crystals. *Transactions of the Japan Institute of Metals* 1963; **5**: 14–21.
91. Friedel J, in *Internal Stresses and Fatigue in Metals*, Rassweiler GM, Grube WL, Eds. Elsevier, Amsterdam, 1959, p. 220.
92. Groh S, Marin EB, Horstemeyer MF and Bammann DJ, Dislocation motion in magnesium: a study by molecular statics and molecular dynamics. *Modeling and Simulation in Materials Science and Engineering* 2009; **17**: 075009.
93. Liu XY, Adams JB, Ercolessi F and Moriarty JA, EAM potential for magnesium from quantum mechanical forces. *Modeling and Simulation in Materials Science and Engineering* 1996; **4**: 293–303.
94. Sun DY, Mendelev MI, Becker CA, Kudin K, Haxhimali T *et al.*, Crystal-melt interfacial free energies in hcp metals: A molecular dynamics study of Mg. *Physical Review B* 2006; **73**: 024116 (12 pp).
95. Yasi JA, Nogaret T, Trinkle DR, Qi Y, Hector LG *et al.*, Basal and prism dislocation cores in magnesium: comparison of first-principles and embedded-atom-potential methods predictions. *Modeling and Simulation in Materials Science and Engineering* 2009; **17**: 055012.
96. Hauser FE, Landon PR and Dorn JE, Deformation and fracture of alpha solid solutions of lithium in magnesium. *Transactions of the ASM* 1958; **50**: 856–883.
97. Agnew SR, Yoo MH and Tomé CN, Application of texture simulation to understanding mechanical behavior of Mg and solid solution alloys containing Li or Y. *Acta Materialia* 2001; **49**: 4277–4289.
98. Urakami A, Meshii M and Fine ME, Electron microscopic investigation of prismatic slip in Mg-10.5 at.% Li single crystals. *Acta Metallurgica* 1970; **18**: 87–99.
99. Urakami A and Fine ME, Influence of misfit centers on formation of helical dislocations. *Scripta Metallurgica* 1970; **4**: 667–671.
100. Sato A and Meshii M, Solid solution softening and solid solution hardening. *Acta Metallurgica* 1973; **21**:753–768.
101. Gibala R and Mitchell TE, Solid solution softening and hardening. *Scripta Metallurgica* 1973; **7**: 1143–1148.
102. Trinkle DR and Woodward C, The chemistry of deformation: How solutes soften pure metals. *Science* 2005; **310**: 1665–1667.
103. Medvedeva NI, Gornostyrev YN and Freeman AJ, Solid solution softening and hardening in the group-V and group-VI bcc transition metals alloys: First principles calculations and atomistic modeling. *Physical Review B* 2007; **76**: 212104.
104. Yasi JA, Hector LG and Trinkle DR, Prediction of thermal cross-slip stress in magnesium alloys from direct first-principles data. *Acta Materialia* 2011; **59**: 5652–5660.

105. Kobayashi T, Koike J, Mukai T, Suzuki M, Watanabe H et al., Anomalous activity of nonbasal dislocations in AZ31 Mg alloys at room temperature. *Materials Science Forum* 2003; **419–422**: 231–236.
106. Koike J, New deformation mechanisms in fine-grain Mg alloys. *Materials Science Forum* 2003; **419–422**: 189–194.
107. Kobayashi T, Koike J, Yoshida Y, Kamado S, Suzuki M et al., Grain size dependence of active slip systems in an AZ31 magnesium alloy. *Journal of the Japan Institute of Metals* 2003; **67**: 149–152.
108. Jonas JJ, Mu S, Al-Samman T, Gottstein G, Jiang L et al., The role of strain accommodation during the variant selection of primary twins in magnesium. *Acta Materialia* 2011; **59**: 2046–2056.
109. Yoshinaga H and Horiuchi R, Deformation mechanisms in magnesium single crystals compressed in the direction parallel to the hexagonal axis. *Transactions of the Japan Institute of Metals* 1963; **4**: 1–8.
110. Obara T, Yoshinaga H and Morozumi S, $\{11\bar{2}2\}<\bar{1}\bar{1}23>$ slip system in magnesium. *Acta Metallurgica* 1973; **21**: 845–853.
111. Morozumi S, Kikuchi M and Yoshinaga H, Electron microscope observation in and around twins in magnesium. *Transactions of the Japan Institute of Metals* 1976; **17**: 158–164.
112. Agnew SR, Horton JA and Yoo MH, TEM Investigation of <c+a> dislocation structures in Mg and Mg-Li α-solid solution alloys. *Metallurgical and Materials Transactions A* 2002; **33A**: 851–858.
113. Máthis K, Nyilas K, Axt A, Dragomir-Cernatescu I, Ungár T et al., The evolution of non-basal dislocations as a function of deformation temperature in pure magnesium determined by X-ray diffraction. *Acta Materialia* 2004; **52**: 2889–2894.
114. Yoo MH, Agnew SR, Morris JR and Ho KM, Non-basal slip systems in H.C.P. metals and alloys: Source mechanisms. *Materials Science and Engineering A* 2001; **319–321**: 87–92.
115. Kim DH, Ebrahimi F, Manuel MV, Tulenko JS and Phillpot SR, Grain-boundary activated pyramidal dislocations in nano-textured Mg by molecular dynamics simulation. *Materials Science and Engineering A* 2011; **528**: 5411–5420.
116. Niewczas M, personal communication, 2010.
117. Morris JR, Scharff J, Ho KM, Turner DE, Ye YY et al., Prediction of a $\{11\bar{2}2\}$ hcp stacking fault using a modified generalized stacking-fault calculation. *Philosophical Magazine A* 1997; **76**: 1065–1077.
118. Morris JR, Ye YY, Ho KM, Chan CT and Yoo MH, A first-principles study of compression twins in h.c.p. zirconium. *Philosophical Magazine Letters* 1994; **69**: 169.
119. Ando S, Gotoh T and Tonda H, Molecular dynamics simulation of < c+a > dislocation core structure in hexagonal-close-packed metals. *Metallurgical and Materials Transactions A* 2002; **33**: 823–829.
120. Li B and Ma E, Pyramidal slip in magnesium: Dislocations and stacking fault on the $\{1011\}$ plane. *Philosophical Magazine* 2009; **89**: 1223–1235.
121. Nogaret T, Curtin WA, Yasi JA, Hector LG and Trinkle DR, Atomistic study of edge and screw <c + a> dislocations in magnesium. *Acta Materialia* 2010; **58**: 4332–4343.
122. Balk TJ and Hemker KJ, High resolution transmission electron microscopy of dislocation core dissociations in gold and iridium. *Philosophical Magazine A* 2001; **81**: 1507–1531.

123. Mills MJ, Baluc NL and Sarosi PM, HRTEM of Dislocation Cores and Thin-Foil Effects in Metals and Intermetallic Compounds. *Microscopy Research and Technique* 2006; **69**: 317–329.
124. Ando S and Tonda H, Effect of temperature and shear direction on yield stress by $\{11\bar{2}2\}<\bar{1}\bar{1}23>$ slip in HCP metals. *Metallurgical and Materials Transactions A* 2002; **33**: 831–836.
125. Kear BH and Wilsdorf HGF, Dislocation configurations in plastically deformed polycrystalline Cu_3Au alloys. *Transactions of the AIME* 1962; **224**: 382–386.
126. Ando S and Tonda H, Non-basal slip in magnesium-lithium alloy single crystals. *Materials Transactions, Japan Institute of Metals* 2000; **41**: 1188–1191.
127. Ando S, Tanaka M and Tonda H, Pyramidal slip in magnesium alloy single crystals. *Materials Science Forum* 2003; **419–422**: 87–92.
128. Goh CS, Wei J, Lee LC and Gupta M, Ductility improvement and fatigue studies in Mg-CNT nanocomposites. *Composite Science and Technology* 2008; **68**: 1432–1439.
129. Goh CS, Wei J, Lee LC and Gupta M, Properties and deformation behaviour of Mg–Y2O3 nanocomposites, *Acta Materialia* 2007; **55**: 5115–5121.
130. Goh CS, Gupta M, Wei J and Lee LC, The cyclic deformation behavior of Mg-Y_2O_3 nanocomposites. *Composite Science and Technology* 2008; **42**: 2039–2050.
131. Kocks UF, Relation between polycrystal deformation and single crystal deformation. *Metallurgical Transactions* 1970; **1**: 1121–1143.
132. Wu TY, Bassani JL and Laird C, Latent hardening in single crystals. 1. Theory and experiments. *Proceedings of the Royal Society of London* 1991; **435**: 1–19.
133. Bassani JL and Wu TY, Latent hardening in single crystals. 2. Analytical characterization and predictions. *Proceedings of the Royal Society of London* 1991; **435**: 21–41.
134. Rauch EF, Gracio JJ, Barlat F and Vincze G, Modeling the plastic behaviour of metals under complex loading conditions. *Modeling and Simulation in Materials Science and Engineering* 2011; **19**: 035009 (18pp).
135. Laverentev FF and Pokhil YuA, Effect of 'forest' dislocations in the {1122} <1123> system on hardening in Mg single crystals under basal slip. *Physica Status Solidi (a)* 1975; **32**: 227.
136. Monnet G, Devincre B and Kubin LP, Dislocation study of prismatic slip systems and their interactions in hexagonal close packed metals: application to zirconium. *Acta Materialia* 2004; **52**: 4317–4328.
137. Capolungo L, Dislocation junction formation and strength in magnesium. *Acta Materialia* 2011; **59**: 2909–2917.
138. Capolungo L, Beyerlein IJ and Wang ZQ, The role of elastic anisotropy on plasticity in hcp metals: a three-dimensional dislocation dynamics study. *Modeling and Simulation in Materials Science and Engineering* 2011; **18**: 085002.
139. Jain A and Agnew SR, Modeling the temperature dependent effect of twinning on the behavior of magnesium alloy AZ31B sheet. *Materials Science and Engineering A* 2007; **462**: 29–36.
140. Kocks UF and Westlake DG, The importance of twinning for the ductility of CPH polycrystals. *Transactions of the TMS-AIME* 1967; **239**: 1107–1109.
141. Yoo MH, Slip, twinning, and fracture in hexagonal close-packed metals. *Metallurgical Transactions A* 1981; **12A**: 409–418.
142. Stoloff NS and Gensamer M, Deformation and fracture of polycrystalline cadmium. *Transactions of the Metallurgical Society of AIME* 1963; **227**: 70–80.

143. Clausen B, Tomé CN, Brown DW and Agnew SR, Reorientation and stress relaxation due to twinning: modeling and experimental characterization for Mg. *Acta Materialia* 2008: **56**: 2456–2468.
144. Brown DW, Agnew SR, Bourke MAM, Holden TM, Vogel SC *et al.*, Internal strain and texture evolution during deformation twinning in magnesium. *Materials Science and Engineering A* 2005; **399**: 1–12.
145. Muránsky O, Barnett MR, Carr DG, Vogel SC and Oliver EC, Investigation of deformation twinning in a fine-grained and coarse-grained ZM20 Mg alloy: Combined *in situ* neutron diffraction and acoustic emission. *Acta Materialia* 2010: **58**; 1503–1517.
146. Aydiner CC, Bernier JV, Clausen B, Lienert U, Tomé CN *et al.*, Evolution of stress in individual grains and twins in a magnesium alloy aggregate. *Physical Review B* 2009; **80**: 024113.
147. Liu Y, Xie Z, Humbeeck JV and Delaey L, Some results on the detwinning process in NiTi shape memory alloys. *Scripta Materialia* 1999; **41**: 1273–1281.
148. Cáceres CH, Sumitomo T and Veidt M, Pseudoelastic behavior of cast magnesium AZ91 alloy under cyclic loading-unloading. *Acta Materialia* 2003; **51**: 6211–6218.
149. Kleiner S and Uggowitzer PJ, Mechanical anisotropy of extruded Mg–6% Al–1% Zn alloy. *Materials Science and Engineering A* 2004; **379**: 258–263.
150. Lou XY, Li M, Boger RK, Agnew SR and Wagoner RH, Hardening evolution of AZ31 Mg Sheet. *International Journal of Plasticity* 2007; **23**: 44–86.
151. Wu L, Agnew SR, Ren YZ, Brown DW, Clausen B *et al.*, The effects of texture and extension twinning on the low-cycle fatigue behavior of a rolled magnesium alloy, AZ31B. *Materials Science and Engineering A* 2010; **527**: 7057–7067.
152. Graff S, Brocks W and Steglich D, Yielding of magnesium: From single crystal to polycrystalline aggregates. *International Journal of Plasticity* 2007; **23**: 1957–1978.
153. Proust G, Tome CN, Jain A and Agnew SR, Modeling the effect of twinning and detwinning during strain-path changes of magnesium alloy AZ31. *International Journal of Plasticity* 2009; **25**: 861–880.
154. El Kadiri H and Oppedal AL, A crystal plasticity theory for latent hardening by glide twinning through dislocation transmutation and twin accommodation effects. *Journal of the Mechanics and Physics of Solids* 2010; **58**: 613–624.
155. Oppedal AL and El Kadiri H, Transmutation and accommodation effects by glide twinning. in *Magnesium Technology 2010*, Agnew SR *et al.*, Eds. TMS, Warrendale, PA, 2010, pp. 14–18.
156. Kelley EW and Hosford Jr, WF, Plane-strain compression of magnesium and magnesium alloy crystals. *Transactions of the Metallurgical Society of AIME* 1968; **242**: 5.
157. Couling SL, Pashak JF and Sturkey L, Unique deformation and aging characteristics of certain magnesium-base alloys. *Transactions of the American Society of Metals* 1959; **51**: 94–107.
158. Barnett MR, Nave MD and Bettles CJ, Deformation microstructures and textures of some cold rolled Mg alloys. *Materials Science and Engineering A* 2004; **386**: 205–211.
159. Sandlöbes S, Zaefferer S, Schestakow I, Yi S and Gonzalez-Martinez R, On the role of non-basal deformation mechanisms for the ductility of Mg and Mg–Y alloys. *Acta Materialia* 2011; **59**: 429–439.

160. Matsuda M, Li S, Kawamura Y, Ikuhara Y and Nishida M, Interaction between long period stacking order phase and deformation twin in rapidly solidified Mg97Zn1Y2 alloy. *Materials Science and Engineering A* 2004; **386**: 447–452.
161. Stanford N and Barnett MR, Effect of particles on the formation of deformation twins in a magnesium-based alloy. *Materials Science and Engineering: A* 2009; **516**: 226.
162. Robson JD, Stanford N and Barnett MR, Effect of particles in promoting twin nucleation in a Mg-5 wt.% Zn alloy. *Scripta Materialia* 2010; **63**: 823.
163. Singh A, Rosalie JM, Somekawa H and Mukai T, Structure of 1 Precipitates in g-Zn Based Alloys: Co-existence of MgZn2 and Mg4Zn7 phases presented at Magnesium Technology 2010, 14–18 February 2010, Seattle, WA
164. Gharghouri MA, Weatherly GC and Embury JD, The interaction of twins and precipitates in a Mg-7.7 at.% Al alloy. *Philosophical Magazine* 1998; **78**: 1137–1149.
165. Gharghouri MA, Weatherly GC, Embury JD and Root J, Study of the mechanical properties of Mg-7.7at.% Al by *in situ* neutron diffraction. *Philosophical Magazine* 1999; **79**: 1671–1695.
166. Jain J, Poole WJ, Sinclair CW and Gharghouri MA, Reducing the tension–compression yield asymmetry in a Mg–8Al–0.5Zn alloy via precipitation. *Scripta Materialia* 2010; **62**: 301–304.
167. Singh A, Somekawa H and Mukai T, Compressive strength and yield asymmetry in extruded Mg–Zn–Ho alloys containing quasicrystal phase. *Scripta Materialia* 2007; **56**: 935–938.
168. Wu XL, Youssef KM, Koch CC, Mathaudhu SN, Kecske's LJ *et al.*, Deformation twinning in a nanocrystalline hcp Mg alloy. *Scripta Materialia* 2011; **64**: 213–216.
169. Van Swygenhoven H, Derlet PM and Froseth AG, Stacking fault energies and slip in nanocrystalline metals. *Nature Materials* 2004; **3**: 399–403.
170. Ando D and Koike J, Relationship between deformation-induced surface relief and double twinning AZ31 magnesium alloy. *Journal of the Japan Institute of Metals* 2007; **71**: 684–687.
171. Cizek P and Barnett MR, Characteristics of the contraction twins formed close to the fracture surface in Mg-3Al-1Zn alloy deformed in tension. *Scripta Materialia* 2008; **59**: 959–962.
172. Waters AM, Martz HE, Dolan KW, Horstemeyer MF and Green RE, Three dimensional void analysis of AM60B magnesium alloy tensile bars using computed tomography imagery. *Materials Evaluation* 2000; **58**: 1221–1227.
173. Weiler JP, Wood JT, Klassen R, Maire E, Berkmortel R *et al.*, Relationship between internal porosity and fracture strength of die-cast magnesium AM60B alloy. *Materials Science and Engineering A* 2005; **395**: 315–322.
174. Weiler JP and Wood JT, Modeling fracture properties in a die-cast AM60B magnesium alloy II-The effects of the size and location of porosity determined using finite element simulations. *Materials Science and Engineering A* 2009; **527**: 32–37.
175. Orowan E, A type of plastic deformation new in metals. *Nature* 1942; **149**, 643–644.
176. Barsoum MW, Farber L and El-Raghy T, Dislocations, kink bands, and room-temperature plasticity of Ti3SiC2. *Metallurgical and Materials Transactions A* 1999; **30**: 1727–1738.
177. Barsoum MW, Zhen T, Kalidindi SR, Radovic M and Murugaiah A, Fully reversible, dislocation-based compressive deformation of Ti3SiC2 to 1 GPa. *Nature Materials* 2003; **2**: 107–111.
178. Hess JB and Barrett CS, Structure and nature of kink bands in zinc. *Transactions of the AIME* 1949; **185**: 599–606.

179. Pascoe RT and Newey CWA, Deformation modes of the intermediate phase NiAl. *Physica Status Solidi* 1968; **29**: 357–366.
180. Mayama T, Ohashi T, Higashida K and Kawamura Y, Crystal plasticity analysis on compressive loading of Mg with suppression of twinning. In *Magnesium Technology 2011*, Sillekens WH, Agnew SR, Neelameggham NR, Mathaudhu SN, Eds., TMS, Warrendale, PA, 2011, pp. 273–277.
181. Barnett MR, Keshavarz Z, Beer AG and Ma X, Non-Schmid behaviour during secondary twinning in a polycrystalline magnesium alloy. *Acta Materialia* 2008; **56**: 5–15.
182. Martin E, Capolungo L, Jiang L and Jonas JJ, Variant selection during secondary twinning in Mg-3%Al. *Acta Materialia* 2010; **58**: 3970–3983.
183. Zhou AG and Barsoum MW, Kinking nonlinear elasticity and the deformation of magnesium. *Metallurgical and Materials Transactions A* 2009; **40A**: 1741–1756.
184. Luo A, Magnesium Front End Research and Development (MFERD) Project ID 'LM008' AMD 603, 604 and 904, 2011 DOE Merit Review Presentation. http://www1.eere.energy.gov/vehiclesandfuels/pdfs/merit_review_2011/lightweight_materials/lm008_luo_2011_o.pdf, accessed 9/8/2011.
185. Ishikawa K, Watanabe H and Mukai T, High temperature compressive properties over a wide range of strain rates in an AZ31 magnesium alloy. *Journal of Materials Science* 2005; **40**; 1577–1582.
186. Livescu V, Cady CM, Cerreta EK, Henrie BL and Gray GT, The high strain rate deformation behavior of high purity Mg and AZ31B magnesium alloy. In *Magnesium Technology 2006*, Luo AA, Neelameggham NR, Beals RS, Eds., TMS, Warrendale, PA, 2006, pp. 153–158.
187. Tucker MT, Horstemeyer MF, Gullett PM, El Kadiri H and Whittington WR, Anisotropic effects on the strain rate dependence of a wrought magnesium alloy. *Scripta Materialia* 2009; **60**: 182–185.
188. El-Magd E and Abouridouane M, Influence of strain rate and temperature on the flow behaviour of magnesium alloy AZ80. *Zeitschrift Fur Metallkunde* 2001; **92**: 1231–1235.
189. El-Magd E and Abouridouane M, Influence of strain rate and temperature on the compressive ductility of Al, Mg and Ti alloys. *Journal de Physique* IV 2003; 110: 15–20.
190. Xing Z, Li BC, Zhang ZM and Zhi WW, Investigation on deformation in ZK60 at high strain rate. *Materials Science Forum* 2005: **488–489**: 527–529.
191. Cho K, Sano T, Doherty K, Yen C, Gazonas G *et al.*, Magnesium technology and manufacturing for ultra lightweight armored ground vehicles, Army Research Lab-Report-236, 2009.
192. Jones T and Placzankis B, The ballistic and corrosion evaluation of Magnesium Elektron 675 vs baseline magnesium alloy AZ31B and aluminum alloy 5083 for armor applications. Army Research Lab-Technical Report-5565, 2011.
193. Johnson GR and Cook WH, A constitutive model and data for metals subjected to large strains, high strain rates, and high temperatures. *Proceedings of the 7th International Symposium on Ballistics*, Hague, Netherlands, 1983, pp. 1–7.
194. Zerilli FJ and Armstrong RW, Dislocation-mechanics-based constitutive relations for material dynamics calculations. *Journal of Applied Physics* 1987; **61**: 1816–1826.
195. Kocks UF, Argon AS and Ashby MF, Thermodynamics and Kinetics of Slip. *Progress in Materials Science* 1975; **19**: 1–303.
196. Follansbee PS and Kocks UF, A constitutive description of the deformation of copper based on the use of mechanical threshold stress as an internal state variable. *Acta Metallurgica* 1988; **36**: 81–93.

197. Follansbee PS and Gray GT, An analysis of the low temperature low and high strain rate deformation of Ti-6Al-4V. *Metallurgical Transactions* 1989; **20A**: 863.
198. Nemat-Nasser S and Li YF, Flow stress of f.c.c. polycrystals with application to OFHC Cu. *Acta Materialia* 1998; **46**: 565–577.
199. Nemat-Nasser S, Li YF and Isaacs JB, Experimental/computational evaluation of flow stress at high strain rates with application to adiabatic shear banding. *Mechanics of Materials* 1994; **17**: 111–134.
200. Nemat-Nasser S, Guo WG, Nesterenko VF, Indrakanti SS and Gu YB, Dynamic response of conventional and hot isostatically pressed Ti–6Al–4V alloys: Experiments and modeling. *Mechanics of Materials* 2001; **33**: 425–439.
201. Wu XL and Tan CW, Deformation localization of AZ31 magnesium alloy under high strain rate loading. *Rare Metal and Materials Engineering* 2008; **37**: 1111–1113.
202. Xu Y, Zhang J, Bai Y and Meyers MA, Shear localization in dynamic deformation: microstructural evolution. *Metallurgical and Materials Transactions A* 2008; **39A**: 811–843.

3
Twinning and its role in wrought magnesium alloys

M. R. BARNETT, Deakin University, Australia

Abstract: In magnesium and its alloys, twinning plays a critical role in determining many characteristics of the mechanical response. The present chapter provides an overview of twinning in magnesium base alloys with a focus on recent advances made using new techniques such as electron backscattering diffraction, advanced phenomenological models and atomistic modelling. The present chapter is organized to first present the reader with a picture of the fundamental features of twinning. Next, the manner in which twinning affects the mechanical response is examined. Finally, the scope to manipulate twinning with the metallurgical 'tools' of alloying and processing is considered.

Key words: magnesium, twinning, nucleation, mechanical properties, crystallography.

3.1 Introduction

Magnesium twins readily.[1] In so doing it follows the other hexagonal close-packed metals such as zinc, titanium and zirconium. These materials twin simply because twinning can be accomplished at stresses lower or equivalent to those required for the dislocation glide with which twinning competes. In consequence, disk or needle-like regions of twinned crystal are frequently seen in the course of metallographic analysis of magnesium samples. But it is not just the metallographic structure that changes during twinning. In magnesium and its alloys, twinning plays a critical role in determining many characteristics of the mechanical response.

The trace of stress as a function of strain obtained during uniaxial loading is frequently used to indicate key features of a material's mechanical behaviour. Here the effect of twinning in magnesium (and its alloys) is stark. Figure 3.1 shows an example of true stress–strain curves obtained from the most common commercially extruded alloy, AZ31 (3 wt% Al, 1 wt% Zn and ~0.4 wt% Mn). In the case shown, twinning dominates in compression but is virtually absent in tension. It is clear that the compressive yield stress is less than half that in tension and that the apparent work hardening behaviour differs considerably. Not obvious in the figure is the fact that the failure strains are also noticeably dissimilar. Failure strains (established from reduction in area) in tension are often seen to significantly exceed those in compression, contrary to what is observed in most metals.[2] To understand and manage this asymmetry in material response requires that we understand the role played by twinning.

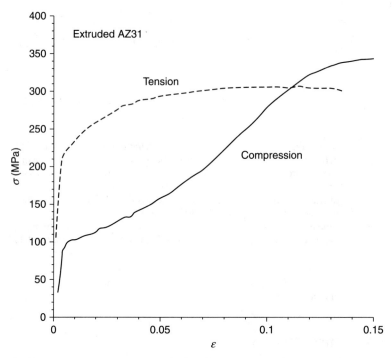

3.1 True stress–strain curves obtained in tension and compression tests performed along the extrusion direction of a bar of magnesium alloy AZ31.

But then we must also understand how to manipulate twinning itself. In the present case of wrought alloys, the metallurgical variables available to be mustered for this end include texture, dislocation density, solute addition, second phase particle dispersion, strain, strain rate, temperature and grain size. By way of example, the last two of these are examined in Fig. 3.2.[3] The plot shows what is commonly understood: twinning is suppressed when the temperature is raised but stimulated when the grain size is increased. Superimposed on Fig. 3.2 are the temperature-grain size 'windows' commonly encountered in practical forming processes. It is clear that for alloy AZ31 twinning will be active in many of the wrought manufacturing processes, but not always. There is scope to decide.

There have been a number of recent advances in understanding the twinning phenomenon in magnesium and its alloys. These have come on the back of new techniques such as electron backscattering diffraction, advanced phenomenological models and atomistic modelling. The present chapter is organized to first present the reader with some highlights of these advances and how they paint a picture of the fundamental features of twinning. Next, the manner in which twinning affects the mechanical response is examined. Finally, the scope to manipulate twinning with the metallurgical 'tools' of alloying and processing is considered.

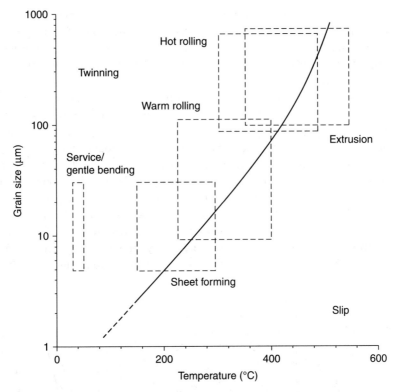

3.2 Plot of the transition between slip and twinning dominated deformation as a function of temperature and grain size for magnesium alloy AZ31. The relation derives from compression tests performed on extrusions.[3] Different deformations and textures may cause slip to dominate in the twinning region but not vice versa. Approximate 'windows' are shown for the main wrought processes.

3.2 Fundamentals of twinning

3.2.1 Phenomenology

Some generalizations can be made with respect to the phenomenon of twinning in magnesium and its alloys (from here on collectively referred to as 'magnesium'). With the application of load, appreciable levels of twinning are seen to arise once the applied stress attains a critical value. In a similar sense to that seen for dislocation glide, the full nature of the stress state can, for most intents and purposes, be ignored. What matters are the components that resolve onto the twin shear plane and in the twin shear direction.[4] That is, twinning in magnesium follows, to a first approximation, a critical resolved shear stress law.[4] However, when examined more closely, it is seen that twin growth honours this statement more than twin nucleation, which is more stochastic in nature.[5] Unlike many cases of glide, the sense of the shear is critical for twinning. Twinning is polar. In magnesium, for example, twinning on the $\{10\bar{1}2\}$ plane only activates in the sense that extends the c-axis. This fact is central to the asymmetry of yielding noted above.

Twinning commences with the formation of a nucleus, nearly always it seems, at a grain boundary;[6,7] though sometimes twin interfaces provide sites for twin formation (e.g. secondary twinning[8]). There is a greater probability of nucleation at a higher stress[9] and consequently more twins are found at higher stress levels. When normalized by the area of grain boundaries an almost monotonic increase in twin density with stress is found in alloy AZ31 tested in compression (Fig. 3.3a).[7] Twin nucleation is typically followed by the very rapid growth of the twinned crystal over the grain in which it formed. The twin thus formed is often wedge, disk or ellipsoidal in shape. Sometimes this growth is accompanied by audible clicks, which testify to the rapidity of the energy release. Acoustic emission has been employed by a number of workers to monitor the progress of

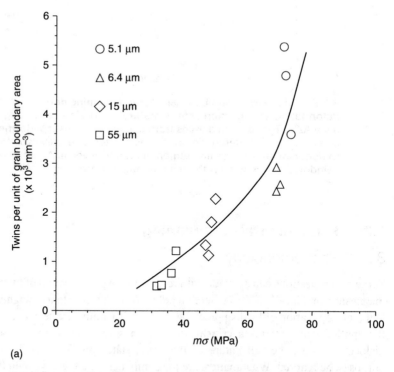

(a)

3.3 Influence of applied stress normalized by the mean maximum Schmid factor for samples of extruded alloy AZ31 of differing grain sizes, tested to strains of up to 1.5% in compression. (a) The number of twins per unit boundary area; (b) the twin aspect ratio.[7]

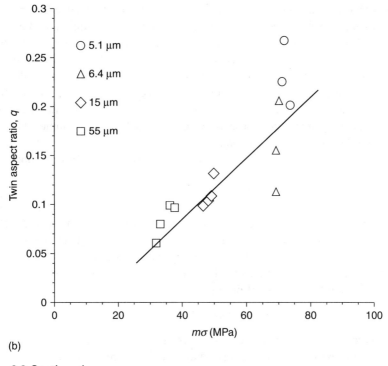

(b)

3.3 Continued.

twin formation in magnesium.[10–12] For extruded material, considerably more intense acoustic activity is seen in compression than in tension, due to profuse twinning in the former. The rapidity of this initial burst of growth is also evidenced by the fact that twins are frequently seen in macrostructural analysis to have impinged upon obstacles such as grain boundaries and other twins.[7] If the initial growth of twins were more 'slow', twins spanning only a portion of their parent grain might be expected to be much more commonly observed; caught, as it were, in the processes of crossing their parent grains.

The twin interiors, which are readily indexed using electron backscatter diffraction (EBSD)[13] in addition to other diffraction techniques, reveal the new orientation of the twinned crystal. The mean trace of twins is usually seen to follow a trace consistent with the relevant habit plane (at least initially). The twin boundaries can be understood to comprise facets of perfect twin interface and, to accommodate local misfit and 'curvature', interfacial line defects ('disconnection' steps in the interface). One class of such defects are zonal twinning dislocations, the shear 'zone' of which extends over more than one atomic plane (the atomic movements within the dislocation zone differ on each plane and include atomic shuffles).[14–18] If these defects are both stable and glissile, their glide over the twin interface advances (or shrinks) the twin. In general the effective Burgers vector of a twinning dislocation is less than that for a lattice dislocation. The Burgers vector divided by the height of the step in the interface that comprises the twinning dislocation gives the characteristic twin shear, s. As alluded to above, the shear is polar in that it operates in one direction only, unlike the slip of lattice dislocations. Once the crystal (grain) is entirely consumed (i.e. reoriented) by the twin, the crystal will have been sheared by s. Partially twinned grains (with twin fraction f), have undergone lower total shears when averaged over the volume. One can write the following expression, which accounts for rotation and shearing, for the macroscopic shear accounted for by twinning in uniaxial deformation:[19]

$$\varepsilon = f\left(\sqrt{1 + 2s \cos \alpha \cos \beta + s^2 \cos^2 \alpha} - 1\right) \quad [3.1]$$

where α is the angle between the twin shear direction and the tension direction and β is the angle between the twin plane normal and the tensile direction. (Note that reference 19 uses the angle between the shear plane and the loading direction.) For small values of s (i.e. $s < \sim 1/3$, which is true for the most prevalent twins in magnesium) Eq. 3.1 simplifies to:

$$\varepsilon \approx fs \cos \alpha \cos \beta = fsm \quad [3.2]$$

where m is the (uniaxial) Schmid factor for the twin system.

In magnesium, the twinning volume fraction increases with stress (and strain) by the nucleation and growth of new twins as well as by the thickening of pre-existing twins (see increasing twin aspect ratio in Fig. 3.3b). This implies that both must proceed under similar levels of applied stress. This is not a trivial observation because it is generally held that twin nucleation occurs under stresses considerably greater than those required for twin growth.[15,20] This apparent paradox can be understood in terms of two points.

1. There may be a dearth of mobile twinning dislocations in the interface, and the generation of new defects may become 'rate' controlling.[20] That is, the exhaustion of sources for twinning dislocations can be expected to lead to a rise in the stress for twin growth.
2. With the introduction of a twin there is a considerable redistribution of local stresses. These can be considered to comprise 'forward' stresses at the twin tip, and 'back' stresses in the twin interior and adjacent surrounding regions.[21] As the twin thickens, these back stresses build up to the point where thickening halts.

To help picture the stress redistribution that accompanies twinning, Fig. 3.4 presents an extract of the results of a finite element simulation performed in a similar manner to that employed by Zhang et al.[21] An ellipsoidal inclusion is defined and allowed to shear a specific amount s (achieved using an artificial thermal strain), while being constrained by its elastic surroundings. This is effectively what is done in the well known Eshelby inclusion analysis.[22] Here, we simply illustrate the concentration of the stress at the twin tips and the relaxation of stresses in the twin and in the regions next to its broad faces. In Fig. 3.4, positive values indicate forward stresses and negative values represent back stresses. These are shear stresses, on the twin plane and in the twin shear direction. The grain in Fig. 3.4 was initially unstressed. For twinning under an applied stress, one simply adds the applied stress to the values shown in the figure.

For an ideal ellipsoidal inclusion, the stress driving twinning dislocations forward at the point of maximum twin thickness is equivalent to the shear stress in the twin interior. As the twin thickens under a constant applied stress, the shear stress in the twin interior drops (because the back stress rises in magnitude). Once the interior stress falls below the critical stress required to move a twinning

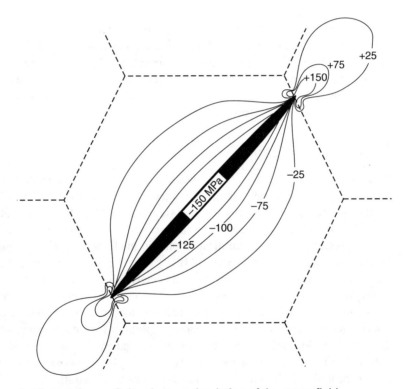

3.4 Extract from a finite element simulation of the stress field surrounding an 'elastic' ellipsoidal twin. Shear stresses on the twin plane in the twin shear direction are shown in MPa in relation to the applied stress level. The dashed lines illustrate possible grain boundary locations (not used in the simulation). Note the stress concentrations at the twin tips that extend into neighbouring grains and the stress relaxation that occurs in the twin and in the parent grain adjacent to the twin. For the simulations, the twin strain was 0.13, Young's modulus was 45 000 MPa and Poisson's ratio 0.3. (Source: Filip Siska, personal communication.)

dislocation, the twin will cease to thicken. A corollary of the effect is that the twin aspect ratio should vary (linearly) with the applied stress.[23] The data in Fig. 3.3b, obtained from extruded AZ31 alloy tested in compression, are reasonably consistent with this.

Turning now to what happens to twins with the release of load, here a detectable drop in the volume fraction of twins is often seen.[24,25] This is presumably accomplished by twins shrinking under the back stresses just described. Complete twin removal upon the release of load is prevented by the presence of dislocation arrays, tangles and debris. These relax forward stresses and shield back stresses. They also create obstacles to twin boundary movement. The finite value of the critical stress required to move twinning dislocations also prevents twinning from being entirely elastic in nature. The strain recovered upon unloading has been termed the anelastic strain.[26]

As one might expect, with load reversal the twin volume fraction drops further. Upon further strain in the reverse direction, complete detwinning of the structure has been reported.[27,28] The phenomenon is not restricted to magnesium and arises due to the relative ease of twin boundary migration[29] in conjunction with the back stress. Thus it is clear that twinning will play an important role in the cyclic deformation and fatigue response of magnesium.

3.2.2 Crystallography

As pointed out in a review of magnesium deformation by Roberts,[30] and of twinning by Christian and Mahajan,[15] magnesium crystals can display twinning on the following planes: $\{10\bar{1}2\}$, $\{30\bar{3}4\}$, $\{10\bar{1}3\}$, $\{11\bar{2}4\}$, $\{10\bar{1}1\}$, $\{10\bar{1}4\}$, $\{10\bar{1}5\}$ and $\{11\bar{2}1\}$. The situation thus appears to be rather complex. However, for the present purposes, simplification can be made on a number of grounds. Firstly, as in many other hcp systems, $\{10\bar{1}2\}$ twinning is the most readily formed and frequently observed (e.g. Roberts[30]). Secondly, when it comes to compression of single crystals along the c-axis, the $\{10\bar{1}1\}$ and $\{10\bar{1}3\}$ modes are the most commonly seen,[31–33] and when sharply textured polycrystals are considered, the $\{10\bar{1}1\}$ mode appears to be the most important for compression near to the c-axis.[13,34] Thirdly, the identification of an apparent $\{30\bar{3}4\}$ twinning mode can be ascribed to a combination of $\{10\bar{1}1\}$ and $\{10\bar{1}2\}$ twinning[35] or by the arrangement of $\{1013\}$ twins.[33] Crocker has shown that it is difficult to rationalize this habit change due to twinning alone.[36] Hartt and Reed-Hill[35,37,38] subsequently concluded that slip on the matrix basal planes cause/allow the double twin band to rotate into a habit nearer to that of $\{30\bar{3}4\}$.

Couling et al.[34] interpreted banded structures seen in their rolled material in terms of $\{10\bar{1}1\}$–$\{10\bar{1}2\}$ double twinning. More recently this twinning mode has been identified using EBSD analysis in pure magnesium[13] and in a number of alloys.[39–46] In these cases a near $\{10\bar{1}1\}$ habit is usually preserved. One important consequence of $\{10\bar{1}1\}$–$\{10\bar{1}2\}$ double twinning is that in many instances it places the basal plane such that it is more favourably aligned for slip; the net rotation of the basal planes through the two twinning modes is 37.5° <$1\bar{2}10$>.[31] This can result in profuse basal slip in the doubly twinned volumes, and the interiors of these twins can take a disproportionately large portion of subsequent plastic strain.[31,47]

The twins just underlined fall into two separate categories based on the sense of the shear provided by the twinning transformation.[48] Twinning on $\{10\bar{1}2\}$ gives a shear that transforms to an extension along the c-axis whereas $\{10\bar{1}1\}$ and $\{10\bar{1}3\}$ twinning give compression along the c direction. Twinning elements and a number of other characteristic parameters for $\{10\bar{1}2\}$ and $\{10\bar{1}1\}$ twinning are given in Table 3.1. In this table, and with reference to the standard twinning elements shown in Fig. 3.5, the values K_1 and K_2 are the first invariant (shear plane) and second invariant (or conjugate) planes respectively. The terms η_1 and η_2 refer to the shear direction and its conjugate, i.e. the vector formed by the intersection of K_2 and the plane of shear. The symbol ω refers to the angle through which the basal planes are rotated around the axis <$hkil$> (which is normal to the plane of shear, see Fig. 3.6). The twinning shear is given by s, the sign of which indicates the sense (negative value gives c-axis extension). Also shown are calculations, made using atomic scale simulations and assuming a dislocation model of the twin, of the twinning dislocation core width (in terms of lattice parameter, a) and the Burgers vector.[49]

A number of important points can be made with respect to some of the relative values in Table 3.1.[49] One is that the magnitudes of the twinning shears are similar for the two common modes though opposite in sign. Another point is that the estimated core widths of the twinning dislocations on the interface and the respective Burgers vector magnitudes are quite different. A $\{10\bar{1}2\}$ twinning dislocation is expected to be more mobile than a $\{10\bar{1}1\}$ twinning dislocation due to the wider core width of the former. The atomic shuffles involved in the propagation of a 'wide' twinning dislocation are lower in magnitude.[49]

These observations have a number of ramifications that are upheld in experimental analysis.[49] One is that the formation of $\{10\bar{1}2\}$ twins might be expected, in general, to occur more readily than that of $\{10\bar{1}1\}$ twins. Secondly, the thickness of $\{10\bar{1}2\}$ twins might be expected to attain, in general, to values greater than that of $\{10\bar{1}1\}$ twins (for reasons of mobility) (see e.g.[46]). Thirdly, thermal activation is more likely to aid mobility in the case of $\{10\bar{1}1\}$ twins than for $\{10\bar{1}2\}$ twins, because the core of the twinning dislocations in the former is narrow (see[31,50]).

Also shown in Table 3.1 are examples of the six double twin variants that can arise when $\{10\bar{1}1\}$ twinning is followed by $\{10\bar{1}2\}$ twinning. Of these, variants IIIa and IIIb create a final double-twinned orientation that can be related to the parent orientation by an equivalent rotation. The same is true for variants IVa and IVb. Thus, when considering the net orientation change accomplished by double twinning, there are four different variants. These can be readily detected using EBSD analysis and it turns out that variant I is by far the most commonly seen.[43]

Table 3.1 Description of the {10 $\bar{1}$ 2} and {10 $\bar{1}$ 1} twinning modes and the relevant interface dislocation.[15,48,49]

	K_1	K_2	η_1	η_2	s	ω	$\langle hkil \rangle$	Width	b (nm)
Primary twinning									
Extension/tensile twin	{10 $\bar{1}$ 2}	{$\bar{1}$012}	<10 $\bar{1}$1>	<$\bar{1}$01 $\bar{1}$>	−0.129	86.3°	<$\bar{1}$2 $\bar{1}$0>	~6a	0.049
Contraction/compression twin	{10 $\bar{1}$ 1}	{10 $\bar{1}$ 3}	<10 $\bar{1}$2>	<$\bar{3}$0 32>	0.138	56°	<$\bar{1}$2 $\bar{1}$0>	~a	0.135
Double twinning									
Primary	(10 $\bar{1}$ 1)		[$\bar{1}$012]						
Secondary I A	(10 $\bar{1}$ 2)		[$\bar{1}$011]			37.5°	<$\bar{1}$2 $\bar{1}$0>		
Secondary II B	($\bar{1}$012)		[10 $\bar{1}$ 1]			30.1°	<$\bar{1}$2 $\bar{1}$0>		
Secondary IIIa D1	(1 $\bar{1}$ 02)		[$\bar{1}$101]			69.9°	<2 $\bar{4}$21> <$\bar{7}$14 $\bar{7}$3>		
Secondary IIIb D2	(01 $\bar{1}$ 2)		[0 $\bar{1}$11]			69.9°	<2 $\bar{4}$21> <$\bar{7}$14 $\bar{7}$3>		
Secondary IVa C1	($\bar{1}$102)		[1 $\bar{1}$01]			66.5°	<5 $\bar{9}$43> <4 $\bar{7}$32>		
Secondary IVb C2	(0 $\bar{1}$12)		[01 $\bar{1}$1]			66.5°	<5 $\bar{9}$43> <4 $\bar{7}$32>		

The value of s is calculated assuming c/a = 1.624. Also shown are the rotations arising from secondary twinning,[42,43] the rotation axes for which differ slightly between these two references. Both are given in the table.

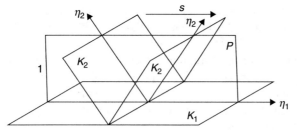

3.5 Twinning elements shown in relation to the plane of shear *P*, which in this image is drawn one unit high. Thus distance *s* gives the twinning shear. The values K_1 and K_2 are the first invariant (shear plane) and second invariant (or conjugate) planes respectively. The terms η_1 and η_2 refer to the shear direction and its conjugate, i.e. the vector formed by the intersection of K_2 and the plane of shear.

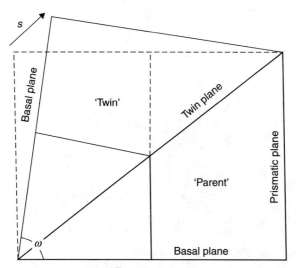

3.6 Schematic of $\{10\bar{1}2\}$ twinning. Shown is the magnesium unit cell seen looking down an <a> or <$1\bar{2}10$> axis, so that the plane of shear is parallel to the page. Illustrated here are the traces of the prismatic and basal planes, which are 'edge on', the angle ω through which the basal plane is rotated during twinning, and the displacement of the η_2 direction (see Fig. 3.4), which defines the twinning shear *s*.

A micrograph illustrating the two main twin types, and the double twin they produce, is presented in Fig. 3.7, which shows an image taken from a rolled sample of pure magnesium compressed along the rolling normal direction.[31] The lenticular twins in the left most grain are $\{10\bar{1}2\}$ twins. The twin at the top right of the grain has undergone considerable thickening and reveals the readiness with which these twins appear to give up coherency (i.e. the ease by which new twinning dislocations are generated at the twin interface). The straight twins are consistent with twins on the $\{10\bar{1}1\}$ plane and this identification is upheld by the detection of fragments of $\{10\bar{1}1\}$ twin boundary in their interiors. The difficulty in indexing these twins and the disruption to the lattice seen in their interior is due to the occurrence of secondary $\{10\bar{1}2\}$ twins and to the profuse basal slip that occurs in the doubly twinned region. The double twin (Type I) boundaries observed between the twinned region and the parent matrix in Fig. 3.7c are also consistent with this interpretation.

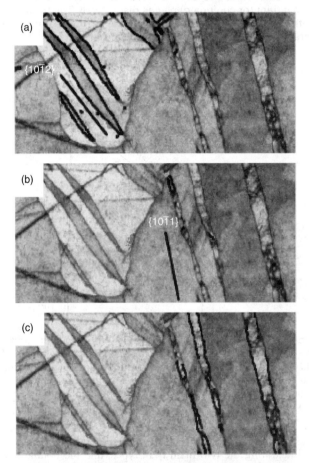

3.7 Twinning in a sample of rolled magnesium compressed in a channel die along the normal direction. (a) $\{10\bar{1}2\}$ twin boundaries, (b) $\{10\bar{1}1\}$ twin boundaries and (c) $\{10\bar{1}1\}$–$\{10\bar{1}2\}$ twin boundaries. Traces of the $\{10\bar{1}2\}$ and $\{10\bar{1}1\}$ twin planes are marked with thick straight black lines in (a) and (b).[13] The grey scale refers to different levels of Kikuchi band contrast. (Reproduced with permission.)

3.2.3 Mechanisms of nucleation and growth

As noted above, twin boundaries can be considered to translate over a crystal by the glide of twinning dislocations in the twin interface. For the $\{10\bar{1}2\}$ twin in hcp metals, this has been studied both experimentally and theoretically by quite a number of workers starting with Thompson and Millard in the 1950s.[49,51–53] As also shown above, the Burgers vector of the step that forms the twinning dislocation is small (~0.05 nm) and its height corresponds to the distance between two $\{10\bar{1}2\}$ lattice planes. It is only those atoms on the second $\{10\bar{1}2\}$ plane that are sheared to their correct positions. The intervening atoms 'shuffle'.[53] Recently, the applicability of this mechanism to magnesium was questioned but this seems to have been in error.[54] What concerns us below is how these dislocations are generated. Growth is considered first, because it is simpler, then nucleation.

For twins to thicken appreciably there must be a constant supply of mobile twinning dislocations. This can be provided by spontaneous thermally activated generation of loops of new twinning dislocations in the interface[55] but this has not received much attention in relation to magnesium. Instead, more consideration has been given to the role of the interaction of lattice dislocations with the twin interface.[52,53] Such a possibility arises in the pole mechanism for twin formation (e.g.[52]). However, recent calculations show that such a mechanism is not necessary[53] nor is it energetically favourable.[56] In 1969, Yoo[52] noted that a lattice <a> dislocation can react with the twin interface to produce a glissile twinning dislocation. This is evidently one mechanism that can provide for twin growth. Recently, Serra and Bacon[53] developed the mechanism further using atomic simulations. In that work, focus was placed on the defect left in the twin interface by the decomposition of a lattice <a> dislocation into a twinning dislocation. Under an applied shear stress, this defect provides a source of new glissile twinning dislocations. The glide of two twinning dislocations away from the defect moves the twin front but leaves the defect in the interface. There, it can continue to provide further twinning dislocations, without the need for additional lattice dislocations.

The nucleation of twins has been a long standing problem in hcp metals but recently, Wang, Hirth and Tomé[56] have advanced a number of mechanisms for $\{10\bar{1}2\}$ twinning based on atomistic simulations and topological analysis. Homogeneous nucleation in the lattice is discounted due to the high stresses required. So, twins nucleate in regions of stress concentration; grain boundaries and other dislocations. But even then the situation is complicated by the details of the atomic nature of the initial defects that give rise to a stable twin nucleus. Wang *et al.* propose two mechanisms, both of which require the simultaneous generation of multiple dislocations. One involves the simultaneous generation of at least three zonal twinning dislocations. The other is the creation of a partial dislocation along with multiple twinning dislocations. The minimum stable nuclei are therefore multiple atomic layers thick; six $\{10\bar{1}2\}$ planes for the first mechanism and 17 for the second mechanism (for magnesium). However, the authors note that thinner twins may be stable under applied shear stresses. Nevertheless, one still requires a mechanism for the simultaneous generation of these defects.

In this regard, it becomes more clear why twin nucleation occurs at grain boundaries. Not only are stresses higher near grain boundaries, but the boundary itself provides a defect structure from which the defects required for a twin nucleus may be obtained.[6,57] It was mentioned above that magnesium frequently displays double twinning – i.e. $\{10\bar{1}1\}$ twinning followed by $\{10\bar{1}2\}$ twinning. The preference for the Type I double twinning variant, even when it is less favoured by the applied stress,[42,43] can be explained in part by restrictions placed by the primary twin boundary on the nucleating 'defects' that can form.[58] That is, it seems that the primary twin boundary dictates what secondary variants can nucleate. It is also found that twin nucleation is favoured at low angle boundaries,[57] and others have noted that particular boundaries are favourable for the transfer of twinning events over them.[59]

The generation of twinning dislocations by the dissociation of dislocations at the head of pile-ups was considered by Capolungo and Beyerlein.[60] Here a pile-up of lattice <a> dislocations was postulated to stimulate the dissociation of the leading dislocation and, under appropriate conditions, the repulsion of a twin loop from the pile-up. Such repulsion was able, in their simulations, to provide for a stable twin nucleus. Naturally, such a mechanism is also likely to occur near grain boundaries, which provide barriers to dislocation motion, causing pile-ups to form. In this instance, it should be mentioned that pile-ups considerably shorter than the grain size are sufficient.

In any case, twins nucleate at grain boundaries and at prior twins. They do this in a stochastic manner; because of fluctuations in internal stress concentrations, variations in incoming fluxes of matrix dislocations and differences in grain boundary character. The average nucleation density, per unit of boundary area, can be rationalized in terms of the stress.[6,7,57,61] The higher the stress, the greater the density of nuclei. The effect can be considered to follow, to a first approximation, a Weibull type distribution in stress,[62,63] which reduces to a power law at low stresses (Fig. 3.3a).[7] See reference 9 for a more detailed statistical analysis. Further discussion of this will be taken up in the sections below on the roles of stress and grain size in producing twins. First, the influence twins have on the mechanical properties is considered.

3.3 The impact of twinning on mechanical response

3.3.1 Yield phenomena

It was noted above in relation to Fig. 3.1 that twinning is responsible for the dramatic yield asymmetry seen in extruded material. This holds true also for rolled magnesium.[64] In these materials the textures are usually very strong. This, in conjunction with ease and the polarity of $\{10\bar{1}2\}$ twinning, gives rise to the yield asymmetry. However, there is also a detectable, though weak, yield asymmetry between compression and tension in randomly textured cast material.[65,66] In tension of a randomly textured sample, the grains that will twin on $\{10\bar{1}2\}$ are those with a c-axis near to parallel with the loading direction. The fraction of grains with a c-axis within α of the loading direction is $1-\cos(\alpha)$.[67] In compression, only the grains with a c-axis close to perpendicular to the loading direction will twin. The fraction of these grains is given by $\sin(\beta)$ where here β is 90° minus the angle between the c-axis and the loading direction. For $\alpha = \beta = 30°$, there are nearly four times as many grains oriented for twinning in compression compared to tension.

The textures of typical extrusions are so sharp and $\{10\bar{1}2\}$ twinning so profuse that the first few percent of compressive strain (when imposed along the extrusion direction) can be entirely attributed to twinning (e.g.[64,68,69]). That is, Eq. 3.1 links the observed twin volume fraction with the imposed strain. In contrast to this, $\{10\bar{1}1\}$ twinning is typically only seen after some strain[46,70] and is therefore not involved in yield or near yield phenomena.

The compressive and tensile tests discussed so far only provide two points on the yield surface. Experimental determination of the yield surface employing the method developed by Backofen[71-74] has been carried out by a number of workers. An example is provided in Fig. 3.8.[71] Twinning dominates in the lower left corner and it can be seen that there is a marked asymmetry. Capturing these effects in a

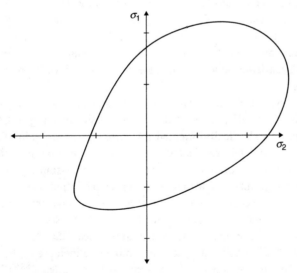

3.8 Example of a typically asymmetric experimental yield surface obtained for alloy ZM61 extruded into sheet. The axes refer to normal stresses in the plane of the sheet (based on data from reference 71). The bottom left corner is dominated by twinning.

mathematical form that is readily utilized in finite element models of plastic response is a major challenge.[75–81] It should be noted though that yield surfaces can be predicted using crystal plasticity modelling and this been examined for magnesium by Jain et al.[82] A simplified version of this approach, once reduced to a few key representative orientation classes, can provide time-efficient finite element simulations.[77]

At finer grain sizes, a yield elongation arises in stress–strain curves of magnesium alloy samples tested such that $\{10\bar{1}2\}$ twinning dominates the deformation (e.g.[3,64]). It appears that under these conditions twinning is autocatalytic, in the sense that twin formation stimulates further twin nucleation and a cascade of twinning spreads over the sample.[7,59] A similar phenomenon is seen in zinc.[83,84] Analysis of stress states using neutron diffraction reveals zero[12,68] and even reversed[85] stresses within twins under these conditions. It is also apparent that the deformation spreads over the sample in a Lüders-like front. The twins formed behind the front largely maintain their number density and size as the front traverses the sample.[59]

3.3.2 Work hardening

In Fig. 3.1, the extent of work hardening seen in the sample undergoing twinning appears to exceed that evident in a sample undergoing deformation by slip. That is, over an equivalent strain of 0.1, the difference between the flow stress and that at which plasticity began is considerably greater in the compressive data. In other metals, twins have been reported to affect hardening by texture change, transformation of dislocations and shortening the mean slip length (e.g.[3,86–89]). This phenomenon in magnesium has been the subject of some discussion (e.g.[90]). One point of contention is whether or not twins in magnesium exert a significant influence on work hardening once appropriate corrections have been made for orientation effects.

A pertinent observation is the stress–strain curves seen in single crystals. A subset of the results obtained by Kelley and Hosford[32] is shown in Fig. 3.9. Curves A and B represent crystals ideally oriented for twinning. By a strain of ~0.06, these crystals have completely reoriented to the twinned orientation which, for crystal A is very close to the orientation of the c-axis compression crystal. It can be seen that the stress–strain curve then takes on a period of rapid work hardening not unlike that seen in the c-axis compression crystal. For equivalent stresses, the hardening in the case of the sample undergoing twinning (A) is seen to marginally exceed that for the c-axis compression sample. It thus appears that the presence of $\{10\bar{1}2\}$ twins and their interfaces may lead to a moderate increase in the rate of work hardening. This is not considered by some to be significant.[88,90] Indeed, crystal plasticity models can reproduce monotonic stress–strain behaviour during twinning reasonably well without introducing any enhanced latent hardening effects associated with twinning,[91] although more sophisticated models that

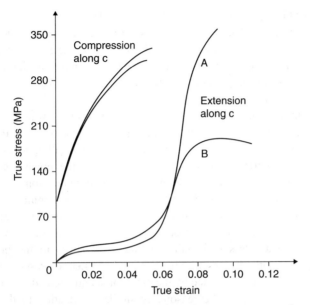

3.9 Stress–strain curves obtained from pure magnesium crystals subjected to compression in a channel die (based on data in reference 32). Curves A and B are compressed along different directions in the basal plane (A, <10$\bar{1}$0> and B, <1$\bar{2}$10>).

capture strain path change effects do require high levels of latent hardening to reproduce experimental observations.[28] This has led others to emphasize the role of transformed dislocations.[92]

Despite the debate, it is important to note that the stark difference in the apparent work hardening between compression and tension in extruded material (Fig. 3.1) does relate to the role of twinning. The key is that twinning changes the texture.[69,93,94] The parent orientations undergo twinning at relatively low stresses while the daughter twinning orientations eventually undergo slip, once the back stress is overcome. The stress required for this slip is high, so as the fraction of daughter orientations rises, so too does the stress for continued plastic deformation. In extruded bar tested in tension (Fig. 3.1), twinning induced texture change does not occur and the plastic deformation begins at considerably higher stresses and only moderate work hardening takes place. In this case, the deformation is dominated by basal and prismatic <a> slip. These observations have practical relevance. If the texture can be altered to encourage twinning to occur in tension, higher work hardening rates may be achieved. A simple illustration of this is the increased work hardening that is seen in extruded material tested in tension perpendicular to the extrusion direction.[95] These samples undergo some twinning and display high uniform elongations. Despite differences in hardening mechanisms, this effect is similar to that seen in some fcc metals that undergo twinning.[19]

Another illustration of the significance of the role {10$\bar{1}$2} exerts on the apparent work hardening is in the change seen in stress–strain curves either side of the twinning transition. This is illustrated in Fig. 3.10 for changing strain rate. There is a marked drop in apparent work hardening once the twinning transition occurs with decreasing rates of testing, at rates between 0.03 s^{-1} and 0.3 s^{-1}.

In cyclic loading carried out to stresses that exceed the yield stress, the occurrence of detwinning plays a role in the stress–strain response.[25,96,97] The most pertinent observation for the present point of discussion is the case where tension is performed following compression along the rolling or extrusion direction. In these situations, the tensile curve displays the upward concave appearance typical of twinning but there are no twin nucleation signatures recorded in acoustic emission (Fig. 3.11).[96] This is because detwinning relies on twin boundary migration, not twin nucleation. A strong Bauschinger effect is also evident in Fig. 3.11a; the macroscopic yielding stresses due to detwinning in the tension cycle are less than half the final flow stress in the previous compression cycle. Another important observation is the near correspondence in Fig. 3.11a of the final stresses attained by the monotonic and cyclic deformations. Here the textures are likely to be reasonably similar in relation to the angle between the c-axis and the loading direction (due to detwinning). Evidently, during twinning and detwinning, the material accumulates a considerable population of dislocations, which cause hardening.

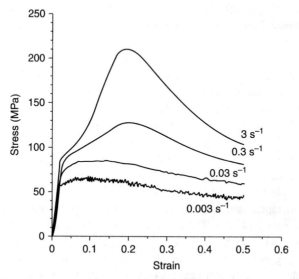

3.10 Stress–strain curves for extruded alloy AZ31 tested in compression at 300 °C. The change in curve shape seen at strain rates between 0.3 s^{-1} and 0.03 s^{-1} corresponds to the twinning transition. Twinning is dominant at higher strain rates.

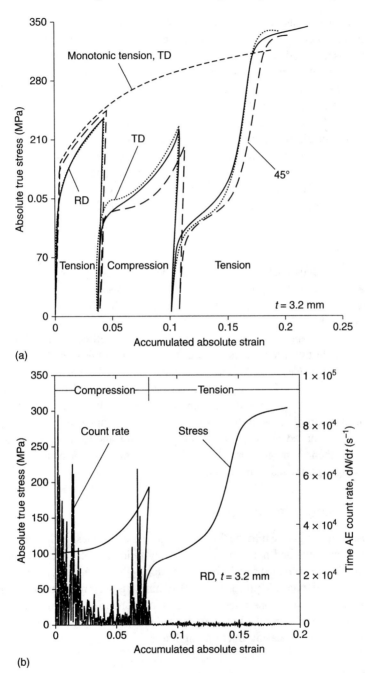

3.11 Stress–strain curves for alloy AZ31 seen during the application of cycles of deformation. (a) Tension-compression-tension; (b) compression followed by tension, showing acoustic emission signals, which indicate twin nucleation. (Reproduced with permission.[96])

Twinning can also create crystallographically 'soft' orientations. The double twinning mechanism described above in Section 3.2 is one example. When $\{10\bar{1}1\}$ twinning is followed by $\{10\bar{1}2\}$ twinning in the twinned volume, the basal plane can end up favourably oriented for slip.[31,38] Slip on the basal plane occurs under quite low stresses in magnesium and in consequence high levels of deformation can concentrate in these doubly-twinned volumes. The net result is that the occurrence of such double twinning can lead to apparent work softening[40,70] although the quantitative details of the mechanism are poorly understood. In contradiction to this, it has been suggested that very fine $\{10\bar{1}1\}$ twins might be the source of high work hardening rates seen in material textured such that the c-axis is subjected to compression.[90] However, such twins have not been observed. It is also apparent that <c+a> slip itself can provide high rates of work hardening.[98]

3.3.3 Ductility

Ductility in metals relates closely to work hardening. So, under conditions where twinning contributes to apparent work hardening – irrespective of the mechanism – extended uniform elongations can be expected.[99] In circumstances where the effective work hardening is lowered by twinning, the deformation will be uniform over a reduced range of strain.[70,95] Where local 'work softening' arises, in freshly formed double twins for instance,[31] the strains can become so high that voids are formed at and inside twins. This has been long known to be a mechanism of failure in single crystals[31,38,100–102] but there is increasing evidence that it is important in polycrystals too.[45–47,70] Thus, if future investigations hold this link to be a strong one, control of twinning is likely to become an important engineering tool in creating alloys with enhanced ductility. Indeed, it is possible that the rise in ductility seen with grain refinement in magnesium[103] is due to the suppression of twinning, although convincing experimental support is lacking.

3.3.4 Fracture toughness

Extensive investigation into the fracture toughness of magnesium has been carried out by Somekawa and colleagues.[104–110] They rationalize the influence of grain size on fracture in part through the influence of grain size on twinning.[110,111] In coarser grained material, profuse $\{10\bar{1}2\}$ twinning is seen near the crack tip. It is supposed that dislocations pile-up at the twin boundaries leading eventually to crack propagation along the boundary.[110] Suppression of these twins is seen as a key principle for increasing fracture toughness.

3.3.5 Fatigue

The polar nature of twinning means that it has an asymmetric effect during cyclic loading.[112] With cyclic loading, twins de-twin (under lower stresses[97]). This is aided by the back stresses that accompany twin formation – due to their arrest at grain boundaries – and is thus largely absent in single crystals.[113] Wu et al.[114] found that when the texture favoured twinning, samples attained much lower stresses and displayed an extended low cycle fatigue life compared with samples in which slip was dominant. The localization of flow in the crystallographically 'soft' regions of double twins can also play a role in fatigue.[115]

Park et al.[116] have been able to rationalize their low cycle fatigue data using an empirical energy density term. When the textures are such that twinning is favoured on the tension rather than the compression cycle, the tensile stresses are lower and consequently the energy density term employed by these workers is also lower, which increases low cycle constant strain amplitude fatigue.[117] However, twins introduced during pre-straining appear to have a detrimental effect on fatigue life.[118]

Koike et al.[115] recently set out to clarify the roles of twinning in fatigue of a coarse-grained sample of AZ31. They conclude that although $\{10\bar{1}2\}$ twinning plays a role, it is seen to occur above and below the fatigue limit, so it is not a controlling role. Instead, prismatic slip appears to be key. The hardening induced by this mode gives rise to stresses sufficient to form double twins of the type discussed above. These in turn create surface ledges, which lead to harmful stress concentrations.

3.3.6 Recrystallization

Recrystallization anneals are often employed in the manufacture of wrought products. During annealing the structure can recrystallize by the nucleation and growth of new strain free grains. These form on the pre-existing defect structure. So, when twinning is prevalent, its affect on the nucleation of recrystallization cannot be ignored.[119-122] In the case of double twins in which enhanced levels of plasticity have occurred, one should also expect there to be greater rates of growth of the recrystallizing grains contained by the twin.

3.4 The impact of structure and processing on twinning

3.4.1 Texture and stress

During the production of wrought alloys, the basal plane rotates to align itself more or less in the plane of material flow. Thus, in rolled plate the c-axes are typically near to the normal direction. In extruded bar the c-axes are commonly found perpendicular to the extrusion direction. Subjecting plate or bar to compression such that extension occurs in the normal direction of a plate, or in the radial direction of a bar, calls for extension along the c-axis. In these situations, $\{10\bar{1}2\}$ twinning is expected.

When single crystals are subjected to compression along the c-axis direction various twinning modes have been observed. As mentioned above, in these studies the occurrence of twins with $\{30\bar{3}4\}$ and/or $\{10\bar{1}1\}$ habits are commonly seen (these planes are 7.3° apart).[31–33,35,37,38,101,123] As also mentioned above, these twins frequently display secondary twinning on $\{10\bar{1}2\}$ in their interiors. A reduced number of workers report the presence of $\{10\bar{1}3\}$ twins and it seems that these twins are only particularly important in single crystals tested at intermediate temperatures, 100–300 °C.[8,33] Thus 'compression' or 'contraction' twinning on $\{10\bar{1}1\}$ arises most in the tension of extrusion and plate along the extrusion or rolling directions. Naturally, it is also seen in the compression of plate along the normal direction.

To make more precise predictions of twin occurrence, one must consider the resolved shear stress and its critical value (critical resolved shear stress – CRSS). In Fig. 3.9, the compressive stresses required for flow for the crystal ideally orientated for $\{10\bar{1}2\}$ twinning (i.e. extension along c) are in the range of 10–15 MPa. These values are approximately 4 times higher than those seen in their crystals that were oriented to favour basal slip.[32] Assuming a Schmid factor of 0.5 gives a CRSS value in the range of 5–10 MPa for $\{10\bar{1}2\}$ twinning. This is similar to the value of ~8 MPa recently reported by Chapuis and Driver.[50] In that case, the twinning CRSS was approximately double the value obtained for basal slip. In their compression tests, Wonsiewicz and Backofen[31] quote a normal stress for $\{10\bar{1}2\}$ twinning in their pure magnesium of ~4 MPa. It appears that the ratio of CRSS values of basal slip to deformation twinning in single crystals falls in the range of 1: 2–4.

Turning now to polycrystalline material, Gharghouri et al.[4] have used neutron diffraction to determine the CRSS for $\{10\bar{1}2\}$ twinning in Mg-7.7at%Al. This was done by noting the resolved shear stresses corresponding to the point at which the texture change due to twinning was first detected. The CRSS was determined to be 70 MPa irrespective of the normal stresses on the $\{10\bar{1}2\}$ plane. In crystal plasticity studies, reported values for alloy AZ31 include: 15 MPa,[66] 18 MPa,[124] 30 MPa,[125] 32 MPa,[126] and 35 MPa.[127] Brown et al.[69] also agree with these values, reporting a CRSS range of 25–35 MPa for AZ31B examined using neutron diffraction.

For typical extrusions, the texture is sufficiently sharp that the pronounced knee in the stress-strain curve that accompanies widespread plastic deformation can be attributed to the onset of twinning. Thus for practical purposes a reasonable estimate of the CRSS for $\{10\bar{1}2\}$ twinning can be made by multiplying the yield point seen in compression of an extrudate by a mean orientation (Schmid) factor. The mean maximum Schmid factor for $\{10\bar{1}2\}$ twinning in typically textured extrusions subject to compression often falls at a value around 0.4–0.45. Thus the CRSS values measured by Muránsky et al.[12] using neutron diffraction for two samples of alloy ZM30, 13 MPa and 28 MPa, will be estimated from their stress–strain curves to be ~15 MPa and ~34 MPa. Finally, it is worth noting that the process of $\{10\bar{1}2\}$ detwinning occurs under stresses approximately one third of those required for twinning during deformation in the opposite sense.[97]

Turning now to the stresses required to initiate $\{10\bar{1}1\}$ twinning; these are considerably higher than those for $\{10\bar{1}2\}$ twinning. This is consistent with the theoretical predictions of low mobility of the $\{10\bar{1}1\}$ twinning dislocation noted above (see Table 3.1). In work on single crystals, Yoshinaga and Horiuchi[123] quote a CRSS of 110 MPa at room temperature for pure magnesium, and Wonsiewicz and Backofen[31] give values over the range 70–150 MPa at room temperature but suggest a Schmid rule might not hold. Recent work by Chapuis and Driver[50] report values of ~65 MPa at 100 °C (with the implication that the value at room temperature will be higher). No values appear to have been measured for polycrystals. (Although Agnew et al. assume a value of 150 MPa for Mg-1Y, six times the value they employ for $\{10\bar{1}2\}$ twinning.[128]) However, we can make a crude estimate here for AZ31. Above, it was seen that the CRSS for $\{10\bar{1}2\}$ twinning in alloy AZ31 fell at values between 10–25 MPa higher than those seen in pure single crystals. Assuming an equivalent hardening increment[129] and a nominal single crystal value of 100 MPa, we estimate the CRSS for $\{10\bar{1}1\}$ twinning in AZ31 to be ~110–125 MPa. A typical extruded bar (m ~0.4) is thus expected to display this twinning mode when the stress exceeds ~290–320 MPa in tension along the extrusion direction. The yield stress for these conditions (which can be understood to be determined largely by prismatic slip[128, 130]) typically lies in the range 170–200 MPa. Thus only once considerable work hardening has occurred, are we likely to see these twins. This is indeed the case in practice.[70]

Armed with an idea of the magnitudes of the CRSS values and knowledge of the twin plane normal and shear directions, a number of workers have set out to rationalize local twin formation seen in EBSD.[5,13,43,44] In these studies, a number of interesting findings have come to light. Some twins are seen when the applied stress resolves onto the twin planes in the 'wrong' sense.[44] However, this is not common and the majority of twins are seen only in grains within which the Schmid factor based on the applied stress state is favourable. Nevertheless, twins[5] are seen when the resolved stress is low and not seen in grains where it is high. The twins found in a grain are not always those most heavily stressed – in terms of the applied stress. This may seem to contradict the general idea that a CRSS law holds for twinning. Indeed, the stochastic nature of twin nucleation was pointed out above.

In this regard, it is worth noting that as the twin frequency rises, the perturbation of the local stress states is likely to increase. This arises from the forward stresses illustrated in Fig. 3.4 and can give rise to Lüders-like twin cascades. It may also generate 'anomalous' twins on planes and in grains where they might not be otherwise expected.[24,44,131] With higher twin frequencies, subsequent twinning events will relate more to the local perturbed stress and less to the applied stress. Conversely, one might therefore expect twins that form initially to be more sensitive to the applied stress and the Schmid factor. In one study, a subset of grains in a lightly twinned sample was examined that contained only one twin per grain.[43] In these cases, the active twin was seen to differ only marginally in Schmid factor from the most heavily stressed twin within the grain. Similarly, the work by Gharghouri et al.[4] pertains to the earliest detection of twinning in a particular subset of grain orientations. Of course, such detection is still far from a measure of nucleation stresses – it averages the entire volume of many grains, and requires a significant twinned fraction to have formed.

Recently, Jonas et al.[132] have suggested that $\{10\bar{1}1\}$ twin variants can only be rationalized once compatibility issues with neighbouring grains are resolved. For this they believe that the orientation of prismatic planes in adjacent grains is important. This, along with the point made above that it is not possible to use applied stresses to understand double twinning,[42,43,133] shows that prediction of the nucleation stresses and conditions for twinning remains an important area for further study.

However, although twins may form in consequence of the local stress state, they will only adopt an appreciable thickness if favoured by the applied stress. That is, twin thickening seems to obey a CRSS law. This has been born out in a statistical study carried out by Beyerlein et al.[5] which showed a strong correlation between twin thickness and the Schmid factor. Thus, there will be an increased likelihood that 'CRSS-favoured' twins will register in bulk analyses such as those that employ neutron diffraction.[4,69,85]

An interesting point can now be made with respect to twin thickening. The dependence of twin thickening on applied stress[5] corresponds, for twins halted at grain boundaries, to an increase in twin aspect ratio with applied stress (Fig. 3.3b). It was pointed out above in relation to Fig. 3.4 that a back stress is imposed on the twin by the surrounding material. If these back stresses are not relaxed by slip, increasing the twin aspect ratio by twin thickening will increase their magnitude. One can imagine that a point will eventually be reached where the back stress rises to such an extent that the stress 'felt' by the twinning dislocations will become too low to move them. Twin thickening should then cease. A further rise in applied stress will allow further thickening. Hence a CRSS law arises. If the internal stress in the twin is uniform it should be equivalent in value to the stress required to move the twinning dislocations over its surface. However, in practice lower stresses are seen.[12,85] One example of this has been obtained using a reconstruction of synchrotron data by Aydiner et al.[134] and their results are reproduced in Fig. 3.12. The freshly formed twins display an internal shear stress so low that it is opposite in sense to that present before the twin formed.

The appearance of shear and normal stresses in twins opposite in sign to those that caused the twin to form have been ascribed to an 'overshoot' that occurs as part of the rapid drop in internal energy that accompanies twinning.[85,134] It should be pointed out that this does not mean the material has adopted an unfavourably high energy state. Indeed, the simulations by Zhang et al.[21] show that the optimal energy state – determined by the trade-off between elastic relaxation and plastic dissipation (ignoring twin interfacial energies) – corresponds to a twin configuration where the twin possesses 'reversed' stresses. The difficult point to

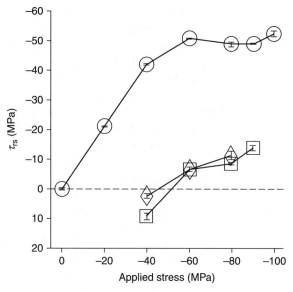

3.12 Shear stresses on the {10$\bar{1}$2} twin plane in the twin direction measured for a single grain in a sample of alloy AZ31. Circles refer to the parent, and diamonds and squares refer to daughter twins. (Reproduced with permission.)[134]

rationalize is how the twinning dislocations can remain 'positively' stressed as they take the twin into a 'negative' stress state. One explanation is that the twins do not remain as ideal ellipsoids and so the stress in their interior does not remain uniform. So, in the steep stress gradient that can be expected near the twin periphery, stresses favourable for twin dislocation movement may well occur while those deeper in the twin interior attain more negative values.

Finally, *in situ* tensile testing of alloy AZ31 has revealed the formation of twins on a free surface during unloading.[24] It is most likely that this is due to asymmetries in the growth and contraction of pre-existing twins rather than the nucleation of new twins upon unloading. However, Wonsiewicz and Backofen[31] compressed a crystal in a vice while observing the surface, and twins were observed to form during unloading. Reed-Hill and Robertson[8] proposed that, in their samples, tensile stresses were set up by the accommodation of {30$\bar{3}$4} twins at the surface of single crystals and that these lead to transient {10$\bar{1}$2} twin formation upon unloading. This phenomenon reflects the important role of local and residual stresses in the formation twins.

3.4.2 Effect of strain and dislocation density

The high mobility of {10$\bar{1}$2} twin boundaries means that these twins readily fill the crystals (grains) in which they form. If no slip accompanies the twinning reaction, twinning would be expected to be exhausted once the macroscopic shear strain reaches the twinning shear strain, 0.129 in the present case. In an ideally oriented crystal, with a Schmid factor of 0.5, the normal strain at the completion of twinning should be of the order of 0.065. This is consistent with the flow curve for the crystal oriented for twinning in Fig. 3.1.[32] In polycrystals of wrought AZ31 (extruded and rolled) $\{10\bar{1}2\}$ twinning has been reported, based on bulk texture measurements, to exhaust at strains of around 0.08–0.1,[69,135] prolonged presumably due to the concurrent occurrence of dislocation glide. EBSD analysis suggests that the reaction might actually persist to higher strains. It does, at least in the tube samples examined by Jiang et al.[41,93] Higher exhaustion strains have been employed in analytical modelling work and this is probably a reflection, at least in part, of the assumption made in these models of an iso-strain law of mixtures.[94,127]

Insight into the role of residual lattice dislocations on twinning can be gained by considering the stresses needed for twinning in pre-strained samples. Lou et al.[96] and Proust et al.[28] examined cases where a pre-strain in tension was employed to initiate ~90 MPa and ~130 MPa of hardening respectively. Upon re-loading in compression, the compressive stresses were found to have increased by ~40 MPa in both cases. To a first approximation, this suggests that the dislocation network generated in the pre-strain is one third to one half as effective when it comes to strengthening against twinning. Of course the situation is more complex. Back stresses may play a role and pre-existing dislocations are likely to impact on twin nucleation. The degree of elastic interaction between twinning and lattice dislocations is also lessened by the small Burgers vector of the twinning dislocations.

3.4.3 Strain rate

Generally, twinning becomes more prevalent at higher rates of deformation[15] and this holds for $\{10\bar{1}2\}$ twinning in magnesium.[136] A simple rationale is that the twinning stress is only marginally sensitive to strain rate compared to slip. This can be understood, in part at least, to be a consequence of the insensitivity of moving twinning dislocations to the obstacles that impede the glide of lattice dislocations.[15] Above it was shown that the core of $\{10\bar{1}2\}$ zonal twinning dislocations is quite wide. Thermal activation and the concomitant role of time and hence strain rate is therefore not expected.

Klimanek[137] carried out shock loading and observed little change in twinning frequency compared to static testing. Indeed, the relative insensitivity of the twinning stress to strain rate can be seen in the yield stresses of the stress-strain curves in Fig. 3.10. In these curves rate sensitivity is only seen at higher strains and lower strain rates, where slip is involved more in the deformation.

3.4.4 Temperature

Meyers et al.[63] show that temperature has a minor effect on the twinning stress for a wide cross-section of materials. However, in magnesium the influence of temperature on the CRSS for twinning in single crystals has only been measured recently.[50] It is seen that the effect of temperature on the stress for $\{10\bar{1}2\}$ twinning is indeed small. Over the temperature range of room temperature to 300 °C no systematic change in CRSS was detected. As mentioned above, this is quite consistent with the wide core width predicted for $\{10\bar{1}2\}$ twinning (Table 3.1). Crystal plasticity simulations[126] and Hall–Petch parameters[136] also show a minor effect of temperature on $\{10\bar{1}2\}$ twinning. However, there has been a suggestion made in recent crystal plasticity analysis that the CRSS in AZ31 may decrease with temperature.[138] And Bakarian and Mathewson have reported increased $\{10\bar{1}2\}$ twin volume fraction and thickness in single crystals compressed at 300 °C.[139]

Unlike $\{10\bar{1}2\}$ twins, the stress required for $\{10\bar{1}1\}$ twinning seems to drop dramatically with temperature.[31,50,123] The sensitivity to temperature has been ascribed to the requirement of some non-basal slip to generate a local rise in the stress.[31] This idea predicts a similarity in the temperature dependencies of $\{10\bar{1}1\}$ twinning and non-basal slip. However, the calculations of Serra et al.[49] (low core width in Table 3.1) and their analysis predict that the glide of the localized steps involved in the propagation of $\{10\bar{1}1\}$ twin boundaries should be thermally activated. The drop in twinning stress with temperature in AZ31 polycrystals must be less than the drop in slip stress otherwise there would be no suppression of twinning with increasing temperature, which is a characteristic feature in this material.[70]

3.4.5 Grain size

In general, the values of the Hall–Petch slope when twinning dominates the deformation are greater than when slip controls the flow.[63] This has been shown to be true for $\{10\bar{1}2\}$ twinning in magnesium alloy AZ31[136] and the results of that study are reproduced in Fig. 3.13. At intermediate temperatures, it can be seen that twinning is suppressed by grain refinement. The transition from slip- to twinning-dominated flow with changing grain size and temperature was shown in Fig. 3.2.[136] At room temperature it appears, from extrapolation of the plot in Fig. 3.2, that grain sizes less than 1 μm would be required to suppress $\{10\bar{1}2\}$ twinning in this alloy. Twinning can also persist to quite high temperatures, if the grain sizes are coarse enough.[140]

An interesting phenomenon that accompanies the suppression of twinning with grain refinement is a 'reverse' Hall–Petch trend in the flow stress. This is shown in Fig. 3.13b. The phenomenological rationalization is that with the suppression of twinning by grain refinement comes also a suppression of the texture change that provides additional hardening. Without the texture change due to twinning, the samples continue to deform by slip, which, at intermediate temperatures, occurs with only small amounts of hardening.

Now it remains to consider why it is that the grain size appears to exert a greater impact on twinning than it does on slip. One important part of the answer lies in

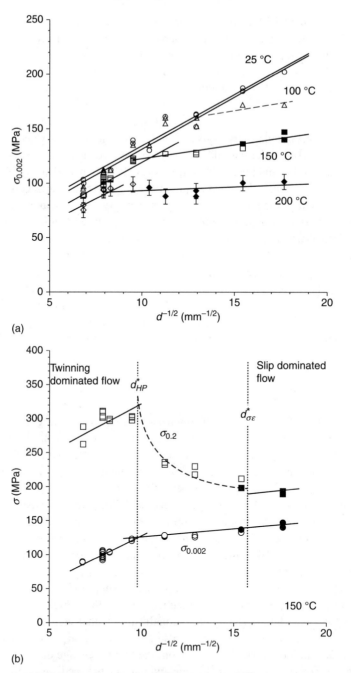

3.13 Influence of grain size on (a) yield and (b) flow stress in magnesium AZ31 extrusions tested in compression at various temperatures. These samples deform at room temperature by mechanical twinning, which is characterized by a concave up-curve (i.e. $d^2\sigma/d\varepsilon^2 < 0$). Filled symbols: $d^2\sigma/d\varepsilon^2 < 0$. (Reproduced with permission.)[3]

dominant role that grain boundaries play in twin nucleation.[9,141] Thus, with grain refinement, we might expect greater densities, N_v, of twin nucleation, per volume. This is quite obvious, but it has some less obvious implications. Take, for example, two samples with differing grain sizes, D and d, but with the same number densities, N_{Sv}, of twins formed per unit of grain boundary area, S_v. The finer grained sample will contain more twins per volume, by the factor D/d. However, the finer grained sample will contain fewer twins per each grain; coarser grains contain more surface area according to the square of their size. In a cursory metallographic analysis, this will appear as a considerable effect. For the same density of twins per unit area of grain boundary, a sample with a grain size of 100 µm with one twin per grain will contrast strikingly with a 10 µm grain-sized sample, with only one twin for each 100 grains.

Now consider the role that twins play in the onset of widespread plastic deformation. To a first approximation, we can assume that yielding (at some offset strain) occurs by $\{10\bar{1}2\}$ twinning once a critical volume fraction f^* of twins is attained. Assume that twins form on grain boundaries (surface area per volume $4/d$, where d is the grain size) as a function N of the applied stress τ. Imagine also that they form as oblate spheroids (volume $= \pi d^2 t/6$ where t is the twin thickness, which is also a function of stress). The volume fraction f of twins can thus be given by:

$$f \propto N(\tau) t(\tau) d \qquad [3.4]$$

If the grain size is reduced, higher values of $t(\tau)$ and/or $N(\tau)$ are required for f to attain its critical value f^* for widespread yielding. To do this requires an increase in stress. We have thus identified a simple grain size dependency term for the yield stress that is not seen for slip. If it is assumed that this term operates in addition to any grain size sensitivity of slip, then it can be seen that twinning will tend to display greater sensitivities to grain size than slip.

3.4.6 Alloying

Here we concern ourselves with the direct roles of solutes and precipitates upon twinning. It should be noted though that alloying addition can also play an important indirect role in twin formation through its influence on grain size and texture.

We have already noted that the CRSS for twinning rises with Al level (CRSS = 70 MPa in AZ80 and ~25 MPa in AZ31) but the most striking effect of alloying addition reported so far is that seen when yttrium is added. In an alloy containing 9% by weight of yttrium $\{10\bar{1}2\}$ twinning was nearly completely suppressed.[64,142] The reason for the effect is unclear. The solubility of yttrium in magnesium is high at ~2 wt% at 200 °C and it is not yet clear if twin suppression is attributable to solute or particles effects, or both. Indeed, there appears to have been little study done into the role of solute addition on twinning in magnesium. The single crystal experiments performed by Kelle and Hosford[32] revealed that the addition of thorium and lithium hardened the normal stress in crystals oriented for twinning (compression perpendicular to c) by an increment approximately one-third of that seen in crystals oriented for compression along the c-axis.

Long period stacking order (LPSO) phases in yttrium-containing alloys have been shown to interfere with twin propagation.[142] This was only seen to occur in regions of high density of plates of an 18 atomic-layer-thick phase. Otherwise the plates were sheared by the twin, changing their habit plane by 3.8°.[142] The shearing of γ′ precipitates by twinning in a Mg-6Gd-1Zn-0.5Zr alloy has also recently been reported.[143] In this instance little hardening was seen and the particles adopted a new plane of habit in the twin. This coincided approximately with the prismatic plane and made an angle of ~3.2° to the original basal habit.

In his transmission electron microscope study on a magnesium-5% zinc alloy, Clark[144] observed the interaction between dislocations, twins and MgZn′ particles. The particle phase is a transition structure with a high coherency with the magnesium lattice. Basal dislocations were not able to shear the precipitates, but twins did. Sheared precipitates were seen to be 'engulfed' by the twin. It is interesting that the same obstacle was not sheared by slip dislocations. Clark ascribed this to two effects:

- The relatively small shear associated with the twinning dislocations.
- The supposed difficulty for a twin dislocation to bow around a particle due to its confinement in the twin interface.

However, there appears to be little evidence to support the latter.

The impact of MgZn′ on twinning in alloy Mg-5Zn has also more recently been studied by Stanford et al.[61,145] These rod-shaped precipitates raise the twinning stress while simultaneously increasing the twin number density. It appears that in this case the precipitates exert a retarding force on twin growth but do not so much hinder twin nucleation. (The aged alloys displayed a precipitate-free zone near the grain boundaries.) Thus, twins propagated under higher stresses. These higher stresses stimulated further twin nucleation. With smaller particles this trend breaks down and samples in the peak and over-aged conditions do not show the same additional twin nucleation from the same stress increase.[146] Evidently the precipitates can also interfere with twin nucleation under these circumstances.

In the same study, the stress increment upon aging was consistently greater in tension than in compression, except for the underaged condition, where only small amounts of hardening were seen in both samples. Put simply, this indicates that twinning is hardened less than slip in the case of rod-shaped precipitates in Mg-5Zn. A fuller analysis is required to elucidate this further. Nevertheless it is interesting to note that the peak aged condition differed for compression and tension (of extruded bar). The peak aged condition for compression, where twinning dominates, was obtained for particles 120 nm long, 20 nm in diameter. For the tensile tests, optimal strength was obtained for particles 10 nm in diameter and 150 nm long. This raises the possibility that the finer precipitates may have been sheared, as reported by Clark.

Gharghouri et al.[147] studied the interaction between twins and precipitates in Mg-7.7Al, which forms plate-shaped particles upon aging. In their study twins were observed to be held up by particles, to engulf particles and in some instances to bypass particles. Important for the present consideration is the fact that Gharghouri et al. maintain that although the precipitates were engulfed by their twins they were *not* sheared. This conclusion appears to be based largely on their determination of the yield stress of the precipitate, which was found to be in the order of 1 GPa, approximately an order of magnitude higher than the yield stress of the matrix. The engulfed precipitates were seen to be inclined ~4° to their initial orientation in the direction of the twinning shear.

As would be expected, the particles engulfed by twins should carry considerable elastic stresses. Gharghouri et al. measured lattice strains using neutron diffraction. Precipitates in their sample deformed by twinning carried 180 MPa when the macroscopic stress was 125 MPa. The plastic strain corresponding to this condition was ~0.004 so the twin fraction in their strongly textured extruded material is likely to have been roughly 5–10%. To a first approximation, we can assume that it is only the precipitates that are engulfed by the twins that carry additional stresses. This provides an upper estimate. We can then calculate that in the material examined by Gharghouri et al. the 'twinned' particles carried normal axial stresses approximating to 1 GPa. This seems an extremely high value. Precipitates outside of twins will undoubtedly also carry an enhanced load. Nevertheless, this points to the importance of back stresses in non-deforming precipitates that become engulfed by twinned (i.e. sheared) material.

These back stresses have been calculated by Robson et al. for a number of cases[148] along with Orowan hardening stresses. It is seen that the low Burgers vector for twinning and the need to accommodate twins by slip are important in understanding the impact of precipitates on twinning. The stress required to bow a twinning dislocation between two obstacles is lower than that required to bow a slip dislocation by a factor equivalent to the ratio of the Burgers vectors. However, it is clear that the high elastic stresses born by non-shearing particles are relaxed by slip in the twinned matrix. It is likely that the impact of precipitates on this slip, as well as the slip occurring in a plastic zone around the twin, must be taken into account. Indeed, it is seen that the stresses required for twinning in alloy AZ91, containing plate-shaped precipitates, are considerably higher than those expected from Orowan looping.[149] Non-deforming $Mg_{17}Al_{12}$ precipitates that form on the basal plane of Mg-Al alloys are most effective in hardening against twinning.

The transition from twinning to slip seen with increasing temperature has been studied as a function of aging by Beladi and Barnett[150] in alloy WE54. In the age-hardened condition, an increase in the transition temperature was seen. This can be ascribed to precipitates hardening against slip more than they do against twinning in this system. This meant that twinning was able to persist as a dominating deformation mode up to temperatures where it would have otherwise been replaced by slip.

3.5 Conclusion

It has been shown that magnesium undergoes profuse twinning and that the influences of this on the mechanical response are considerable. It has also been shown that our understanding of how twinning may be controlled using alloying is rudimentary at best. Thus a greater endeavour is warranted towards understanding how mechanical objectives such as strengthening can be achieved when twinning is dominant. Although recent stochastic modelling has proven fruitful there will be a continuing need to capture new understandings in a useful mathematical framework.

3.6 Acknowledgements

The author would like to acknowledge Bevis Hutchinson, Sean Agnew and Carlos Tomé for their useful comments on the text.

3.7 References

1. Schmid E and Boas W. *Plasticity of Crystals: With Special Reference to Metals*. London: Chapman and Hall, 1935.
2. Beck, A. *The Technology of Magnesium and its Alloys*. London: Hughes, 1943.
3. Barnett MR, Keshavarz Z, Beer AG and Atwell D. Influence of grain size on the compressive deformation of wrought Mg–3Al–1Zn. *Acta Materialia* 2004; **52**: 5093.
4. Gharghouri MA, Weatherly GC, Embury JD and Root J. Study of mechanical properties of Mg-7.7at.%Al by *in situ* neutron diffraction alloy. *Philosophical Magazine A* 1999; **79**: 1671.
5. Beyerlein IJ, Capolungo L, Marshall PE, McCabe RJ and Tomé CN. Statistical analyses of deformation twinning in magnesium. *Philosophical Magazine* 2010; **90**: 2161.
6. Wang J, Beyerlein IJ and Tomé CN. An atomic and probabilistic perspective on twin nucleation in Mg. *Scripta Materialia* 2010; **63**: 741.
7. Ghaderi A and Barnett MR. Sensitivity of deformation twinning to grain size in titanium and magnesium. *Acta Materialia*, 2011; **59**: 7824–7839.
8. Reed-Hill RE and Robertson WD. Additional modes of deformation twinning in magnesium. *Acta Metallurgical* 1957; **5**: 717.
9. Beyerlein IJ and Tomé CN. A probabilistic twin nucleation model for HCP polycrystalline metals. *Proceedings of the Royal Society A: Mathematical, Physical and Engineering Sciences* 2010; **466**: 2517.
10. Bohlen J, Chmelik F, Dobron P, Kaiser F, Letzig D *et al*. Orientation effects on acoustic emission during tensile deformation of hot rolled magnesium alloy AZ31. *Journal of Alloys and Compounds* 2004; **378**: 207.
11. Mathis KC, Trojanova Z, Lukac P and Lendvai J. Investigation of some magnesium alloys by use of the acoustic emission technique. *Materials Science and Engineering A* 2004; **387–389**: 331.

12. Muránsky O, Barnett MR, Carr DG, Vogel SC and Oliver EC. Investigation of deformation twinning in a fine-grained and coarse-grained ZM20 Mg alloy: Combined *in situ* neutron diffraction and acoustic emission. *Acta Materialia* 2010; **58**: 1503.
13. Nave MD and Barnett MR. Microstructures and textures of pure magnesium deformed in plane-strain compression. *Scripta Materialia* 2004; **51**: 881.
14. Yoo MH. Reply to 'Zonal dislocations and dislocation reactions with twins in hexagonal-close-packed metals'. *Scripta Metallurgica* 1970; **4**: 9.
15. Christian JW and Mahajan S. Deformation twinning. *Progress in Materials Science* 1995; **39**: 1.
16. Hull D and Bacon DJ. *Introduction to Dislocations*. Oxford: Butterworth-Heinemann, 2001.
17. Pond RC, Serra A and Bacon DJ. Dislocations in interfaces in the h.c.p. metals—II. Mechanisms of defect mobility under stress. *Acta Materialia* 1999; **47**: 1441.
18. Kronberg ML. Atom movements and dislocation structures in some common crystals. *Acta Metallurgica* 1961; **9**: 970.
19. Reed-Hill RE. Role of deformation twinning in determining the mechanical properties of metals. *Inhomogeneity of Plastic Deformation*. Metals Park, OH: AMS, 1973.
20. Cahn RW. Survey of recent progress in the field of deformation twinning. In: Reed-Hill RE, Hirth JP, Rogers HC, editors. *TMS-AIME Conf., Deformation Twinning*, vol. 25. Gainesville, Florida: American Institute of Mining, Metallurgical and Petroleum Engineers, INC., Printed in Great Britain, 1964, p.1.
21. Zhang RY, Daymond MR and Holt RA. A finite element model of deformation twinning in zirconium. *Materials Science and Engineering: A* 2008; **473**: 139.
22. Lebensohn RA and Tomé CN. A study of the stress state associated with twin nucleation and propagation in anisotropic materials. *Philosophical Magazine A* 1993; **67**: 187.
23. Friedel J. *Dislocations*. Oxford: Pergamon, 1964.
24. Barnett MR, Nave M and Keshavarz Z. Microstructural features of rolled Mg-3Al-1Zn. *Metallurgical and Materials Transactions A* 2004; **36A**: 1697.
25. Muránsky O, Carr DG, Sittner P and Oliver EC. *In situ* neutron diffraction investigation of deformation twinning and pseudoelastic-like behaviour of extruded AZ31 magnesium alloy. *International Journal of Plasticity* 2009; **25**: 1107.
26. Cáceres CH, Sumitomo T, and Veidt M. Pseudoelastic behaviour of cast magnesium AZ91 alloy under cyclic loading-unloading. *Acta Materialia* 2003; **51**: 6211.
27. Uota T, Suzu T, Fukumoto S and Yamamoto A. EBSD Observation for reversible behavior of deformation twins in AZ31B magnesium alloy. *Materials Transactions* 2009; **50**: 2118.
28. Proust G, Tomé CN, Jain A and Agnew SR. Modeling the effect of twinning and detwinning during strain-path changes of magnesium alloy AZ31. *International Journal of Plasticity* 2009; **25**: 861.
29. Proust G, Kaschner GC, Beyerlein IJ, Clausen B, Brown DW *et al*. Detwinning of high-purity zirconium: *In Situ* neutron diffraction experiments. *Experimental Mechanics* 2010; **50**: 125.
30. Roberts CS. *The Deformation of Magnesium. Magnesium and Its Alloys*. New York: John Wiley & Sons, Inc., 1960, p. 81.
31. Wonsiewicz BC and Backofen WA. Plasticity of magnesium crystals. *Transactions of the Metallurgical Society of AIME* 1967; **239**: 1422.
32. Kelley EW and Hosford Jr, WF. Plane-strain compression of magnesium and magnesium alloy crystals. *Transactions of the Metallurgical Society of AIME* 1968; **242**: 5.

33. Yoshinaga H, Obara T and Morozumi S. Twinning deformation in magnesium compressed along the c-axis. *Materials Science and Engineering* 1973; **12**: 255.
34. Couling SL, Pashak JF and Sturkey L. Unique deformation and aging characteristic of certain magnesium-base alloys. *Transactions of the ASM* 1959; **51**: 94.
35. Hartt WH and Reed-Hill RE. The Irrational Habit of Second-order $\{10\bar{1}1\}$–$\{10\bar{1}2\}$ Twins in Magnesium. *Transactions of the Metallurgical Society of AIME* 1967; **239**: 1511.
36. Crocker AG. Double twinning. *Philosophical Magazine* 1962; **4**: 1901.
37. Hartt WH and Reed-Hill RE. Shear accommodation kinking at second order $\{10\bar{1}1\}$–$\{10\bar{1}2\}$ twins in magnesium. *Transactions of the Metallurgical Society of AIME* 1968; **242**: 2207.
38. Hartt WH and Reed-Hill RE. Internal deformation and fracture of second-order $\{10\bar{1}1\}$–$\{10\bar{1}2\}$ twins in magnesium. *Transactions of the Metallurgical Society of AIME* 1968; **242**: 1127.
39. Barnett MR, Nave MD and Keshavarz Z. Microstructural features of rolled Mg-3Al-1Zn. *Metallurgical Transactions A* 2005; **36A**: 1697.
40. Jiang L, Jonas J, Luo A, Sachdev A and Godet S. Twinning-induced softening in polycrystalline AM30 Mg alloy at moderate temperatures. *Scripta Materialia* 2006; **54**: 771.
41. Jiang L, Jonas JJ, Mishra RK, Luo AA, Sachdev AK *et al.* Twinning and texture development in two Mg alloys subjected to loading along three different strain paths. *Acta Materialia* 2007; **55**: 3899.
42. Martin É, Capolungo L, Jiang L and Jonas JJ. Variant selection during secondary twinning in Mg-3%Al. *Acta Materialia* 2010; **58**: 3970.
43. Barnett MR, Keshavarz Z, Beer AG and Ma X. Non-Schmid behaviour during secondary twinning in a polycrystalline magnesium alloy. *Acta Materialia* 2008; **56**: 5.
44. Barnett MR, Nave MD and Bettles CJ. Deformation microstructures and textures of some cold rolled Mg alloys. *Materials Science and Engineering* A 2004; **386**: 205.
45. Ando D, Koike J and Sutou Y. Relationship between deformation twinning and surface step formation in AZ31 magnesium alloys. *Acta Materialia* 2011; **58**: 4316.
46. Koike J. Enhanced deformation mechanisms by anisotropic plasticity in polycrystalline Mg alloys at room temperature. *Metallurgical and Materials Transactions A* 2005; **36A**: 1689.
47. Barnett MR, Jacob S, Gerard BF and Mullins JG. Necking and failure at low strains in a coarse-grained wrought Mg alloy. *Scripta Materialia* 2008; **59**: 1035.
48. Yoo MH. Slip, Twinning and fracture in hexagonal close-packed metals. *Metallurgical Transactions A* 1981; **A12**: 409.
49. Serra A, Bacon DJ and Pond RC. Computer simulation of the structure and mobility of twinning dislocations in H.C.P. metals. *Acta Metallurgica et Materialia* 1991; **39**: 1469.
50. Chapuis A and Driver JH. Temperature dependency of slip and twinning in plane strain compressed magnesium single crystals. *Acta Materialia* 2010; **59**: 1986.
51. Thompson N and Millard DJ. *Philosophical Magazine* 1952; **43**: 422.
52. Yoo MH. Interaction of slip dislocations with twins in HCP metals. *Transactions of the Metallurgical Society of AIME* 1969; **245**: 2051.
53. Serra A and Bacon DJ. A new model for $\{1012\}$ twin growth in HCP metal. *Philosophical Magazine A* 1996; **73**: 333.
54. Serra A, Bacon DJ and Pond RC. Comment on 'atomic shuffling dominated mechanism for deformation twinning in magnesium'. *Physical Review Letters* 2010; 104, 2, 029603.

55. Christian JW. *Mechanical twinning. The Theory of Transformations in Metals and Alloys – An Advanced Textbook in Physical Metallurgy*. Oxford: Pergamon Press, 1965, p. 48.
56. Wang J, Hirth JP and Tomé CN. Twinning nucleation mechanisms in hexagonal-close-packed crystals. *Acta Materialia* 2009; **57**: 5521.
57. Beyerlein IJ, McCabe RJ and Tomé CN. Effect of microstructure on the nucleation of deformation twins in polycrystalline high-purity magnesium: A multi-scale modeling study. *Journal of the Mechanics and Physics of Solids* 2011; **59**: 988.
58. Beyerlein IJ, Wang J, Tomé CN and Barnett MR. Nucleation of secondary twins in magnesium. Submitted to *Proceedings of the Royal Society*.
59. Barnett MR and Ghaderi A. submitted to *Acta Materialia* 2011.
60. Capolungo L and Beyerlein IJ. Nucleation and stability of twins in hcp metals. *Physical Review B – Condensed Matter and Materials Physics* 2008; 78, 024117.
61. Robson JD, Stanford N and Barnett MR. Effect of particles in promoting twin nucleation in a Mg-5 wt.% Zn alloy. *Scripta Materialia* 2010; **63**: 823.
62. Mann, G.E Sumitomo, T, Griffiths JR and Cáceres CH. Reversible plastic strain during cyclic loading-unloading of Mg and Mg-Zn alloys. *Materials Science and Engineering: A* 2007; **456**(1–2): 138.
63. Meyers MA, Vohringer O and Lubarda VA. The onset of twinning in metals: a constitutive description. *Acta Materialia* 2001; **49**: 4025.
64. Eckelmeyer KH and Hertzberg RW. Deformation of wrought Mg-9Wt Pct Y. *Metallurgical Transactions A* 1970; **1**: 3411.
65. Mann G, Griffiths JR and Cáceres CH. Hall–Petch parameters in tension and compression in cast Mg-2Zn alloys. *Journal of Alloys and Compounds* 2004; **378**: 188.
66. Agnew SR, Yoo MH and Tomé CN. Application of texture simulation to understanding mechanical behaviour of Mg and solid solution alloys containing Li or Y. *Acta Materialia* 2001; **49**: 4277.
67. Hosford WF. *The Mechanics of Crystals and Textured Polycrystals*. The Oxford Engineering Science Series. New York: Oxford University Press, 1993.
68. Muránsky O, Barnett MR, Luzin V and Vogel S. On the correlation between deformation twinning and Lüders-like deformation in an extruded Mg alloy: *In situ* neutron diffraction and EPSC.4 modelling. *Materials Science and Engineering: A* 2010; **527**: 1383.
69. Brown DW, Agnew SR, Bourke MAM, Holden TM, Vogel M and Tomé C. Internal strain and texture evolution during deformation twinning in magnesium. *Materials Science and Engineering A* 2005; **399**: 1.
70. Barnett MR. Twinning and the ductility of magnesium alloys Part II: 'Contraction' twins. *Materials Science and Engineering* 2007; **464**: 8.
71. Safi-Naqvi SH, Hutchinson WB and Barnett MR. Texture and mechanical anisotropy in three extruded magnesium alloys. *Materials Science and Technology* 2008; **24**: 1283.
72. Dillamore IL, Hadden P and Stratford DJ. Texture control and the yield anisotropy of plane strain magnesium extrusions. *Texture* 1972; **1**: 17.
73. Kelley EW and Hosford Jr, WF. The deformation characteristic of textured magnesium. *Transactions of the Metallurgical Society of AIME* 1968; **242**: 654.
74. Barnett MR, Keshavarz Z and Ma X. A semianalytical Sachs model for the flow stress of a magnesium alloy. *Metallurgical and Materials Transactions A* 2006; **37A** (2006): 2283.
75. Banabic D and Hussnatter W. Modeling the material behavior of magnesium alloy AZ31 using different yield criteria. *International Journal of Advanced Manufacturing Technology* 2009; **44**: 969.

76. Cazacu O, Ionescu IR and Yoon JW. Orthotropic strain rate potential for the description of anisotropy in tension and compression of metals. *International Journal of Plasticity* 2010; **26**: 887.
77. Li M, Lou XY, Kim JH and Wagoner RH. An efficient constitutive model for room-temperature, low-rate plasticity of annealed Mg AZ31B sheet. *International Journal of Plasticity* 2010; **26**: 820.
78. Mayama T, Aizawa K, Tadano Y and Kuroda M. Influence of twinning deformation and lattice rotation on strength differential effect in polycrystalline pure magnesium with rolling texture. *Computational Materials Science* 2009; **47**: 448.
79. Naka T, Uemori T, Hino R, Kohzu M, Higashi K *et al*. Effects of strain rate, temperature and sheet thickness on yield locus of AZ31 magnesium alloy sheet. *Journal of Materials Processing Technology* 2008; **201**: 395.
80. Plunkett B, Cazacu O and Barlat F. Orthotropic yield criteria for description of the anisotropy in tension and compression of sheet metals. *International Journal of Plasticity* 2008; **24**: 847.
81. Plunkett B, Lebensohn RA, Cazacu O and Barlat F. Anisotropic yield function of hexagonal materials taking into account texture development and anisotropic hardening. *Acta Materialia* 2006; **54**: 4159.
82. Jain A and Agnew SR. Measuring the temperature dependence of the flow surface of magnesium alloy sheet. *Magnesium Technology* 2005; Wrought Magnesium Alloys I: 71.
83. Ecob N and Ralph B. Effect of grain size on flow stress of textured Zn alloy. *Metal Science* 1983; **17**: 317.
84. Ecob N and Ralph B. The effect of grain size on deformation twinning in a textured zinc alloy. *Journal of Materials Science* 1983; **18**: 2419.
85. Clausen B, Tomé CN, Brown DW and Agnew SR. Reorientation and stress relaxation due to twinning: Modeling and experimental characterization for Mg. *Acta Materialia* 2008; **56**: 2456.
86. Karaman I, Sehitoglu H, Beaudoin AJ, Chumlyakov YI, Maier HJ *et al*. Modeling the deformation behavior of Hadfield steel single and polycrystals due to twinning and slip. *Acta Materialia* 2000; **48**: 2031.
87. Cáceres CH and Blake AH. On the strain hardening behaviour of magnesium at room temperature. *Materials Science and Engineering A* 2007; **462**: 193.
88. Cáceres CH, Lukac P and Blake A. Strain hardening due to {1012} twinning in pure magnesium. *Philosophical Magazine* 2008; **88**: 991.
89. Salem AA, Kalidindi SR, Doherty RD and Semiatin SL. Strain hardening due to deformation twinning in alpha-titanium: Mechanisms. *Metallurgical and Materials Transactions A* 2006; **37A**: 259.
90. Knezevic M, Levinson A, Harris R, Mishra RK, Doherty RD *et al*. Deformation twinning in AZ31: Influence on strain hardening and texture evolution. *Acta Materialia* 2010; **58**: 6230.
91. Wang H, Raeisinia B, Wu PD, Agnew SR, and Tomé CN. Evaluation of self-consistent polycrystal plasticity models for magnesium alloy AZ31B sheet. *International Journal of Solids and Structures* 2010; **47**: 2905.
92. Oppedal AL, El Kadiri H, Tomé CN, Baird JC, Vogel SC *et al*. Limitation of current hardening models in predicting anisotropy by twinning in hcp metals: Application to a rod textured AM30 magnesium alloy. In: Sillekens WH, Agnew SR, Neelameggham NR, Mathaudu SN, editors. *Magnesium Technology* 2011; **2011**: 313.

93. Jiang L, Jonas JJ, Luo AA, Sachdev AK and Godet S. Influence of {10–12} extension twinning on the flow behavior of AZ31 Mg alloy. *Materials Science and Engineering: A* 2007; **445–446**: 302.
94. Barnett MR, Davies CHJ and Ma X. An analytical constitutive law for twinning dominated flow in magnesium. *Scripta Materialia* 2005; **52**: 627.
95. Barnett MR. Twinning and the ductility of magnesium Alloys Part I: Experiments and 'tension' twins. *Materials Science and Engineering* 2007; **464**: 1.
96. Lou XY, Li M, Boger RK, Agnew SR and Wagoner RH. Hardening evolution of AZ31 Mg Sheet. *International Journal of Plasticity* 2007; **23**: 44.
97. Wu L, Jain A, Brown DW, Stoica GM, Agnew SR et al. Twinning-detwinning behavior during the strain-controlled low-cycle fatigue testing of a wrought magnesium alloy, ZK60A. *Acta Materialia* 2008; **56**: 688.
98. Jones IP and Hutchinson WB. Stress-state dependence of slip in Titanium-6Al-4V and other H.C.P. metals. *Acta Metallurgica* 1981; **29**: 951.
99. Reed-Hill RE. Role of deformation twinning in the plastic deformation of a polycrystalline anisotropic metal. In: Reed-Hill RE, Hirth JP, Rogers HC, editors. TMS-AIME Conf., *Deformation Twinning*, vol. 25. Gainesville, Florida: American Institute of Mining, Metallurgical and Petroleum Engineers, INC., Printed in Great Britain, 1964, p. 295.
100. Reed-Hill RE. Written discussion. *Transactions of the ASM* 1959; **51**: 105.
101. Reed-Hill RE. A study of the {10 1} and {10 3} Twinning modes in magnesium. *Transactions of the Metallurgical Society of AIME* 1960; **218**: 554.
102. Reed-Hill RE and Robertson WD. The crystallographic characteristics of fracture in magnesium single crystals. *Acta Metallurgica* 1957; **5**: 728.
103. Wilson DV. Plastic anisotropy in sheet metals. *Journal of the Institute of Metals* 1966; **94**: 84.
104. Somekawa H and Higashi K. Diffusion bonding in commercial superplastic AZ31 magnesium alloy sheets. *Key Engineering Materials* 2003; **233–236**: 857.
105. Somekawa H, Hosokawa H, Watanabe H and Higashi K. Diffusion bonding in superplastic magnesium alloys. *Materials Science and Engineering A* 2003; **339**: 328.
106. Somekawa H and Mukai T. Fracture toughness in a rolled AZ31 magnesium alloy. *Journal of Alloys and Compounds* 2006; **417**: 209.
107. Somekawa H and Mukai T. Fracture toughness in an extruded ZK60 magnesium alloy. *Materials Transactions* 2006; **47**: 995.
108. Somekawa H and Mukai T. Fracture toughness in Mg-Al-Zn alloy processed by equal-channel-angular extrusion. *Scripta Materialia* 2006; **54**: 633.
109. Somekawa H, Osawa Y and Mukai T. Effect of solid-solution strengthening on fracture toughness in extruded Mg-Zn alloys. *Scripta Materialia* 2006; **55**: 593.
110. Somekawa H, Singh A and Mukai T. Fracture mechanism of a coarse-grained magnesium alloy during fracture toughness testing. *Philosophical Magazine Letters* 2009; **89**: 2.
111. Somekawa H, Nakajima K, Singh A and Mukai T. Ductile fracture mechanism in fine-grained magnesium alloy. *Philosophical Magazine Letters* 2010; **90**: 831.
112. Zhang J, Yu Q, Jiang Y and Li Q. An experimental study of cyclic deformation of extruded AZ61A magnesium alloy. *International Journal of Plasticity* 2010, in press.
113. Stevenson R and Sande JBV. The cyclic deformation of magnesium single crystals. *Acta Metallurgica* 1974; **22**: 1079.
114. Wu YJ, Zhu R, Wang JT and Ji WQ. Role of twinning and slip in cyclic deformation of extruded Mg-3%Al-1%Zn alloys. *Scripta Materialia* 2009; **63**: 1077.

115. Koike J, Fujiyama N, Ando D and Sutou Y. Roles of deformation twinning and dislocation slip in the fatigue failure mechanism of AZ31 Mg alloys. *Scripta Materialia* 2010; **63**: 747.
116. Hyuk Park S, Hong S-G, Ho Lee B, Bang W and Soo Lee C. Low-cycle fatigue characteristics of rolled Mg-3Al-1Zn alloy. *International Journal of Fatigue* 2010; **32**: 1835.
117. Park SH, Hong S-G, Bang W and Lee CS. Effect of anisotropy on the low-cycle fatigue behavior of rolled AZ31 magnesium alloy. *Materials Science and Engineering: A* 2010; **527**: 417.
118. Park SH, Hong S-G and Lee CS. Role of initial $\{10\bar{1}2\}$ twin in the fatigue behavior of rolled Mg-3Al-1Zn alloy. *Scripta Materialia* 2010; **62**: 666.
119. Martin E, Jiang L, Godet S and Jonas JJ. The combined effect of static recrystallization and twinning on texture in magnesium alloys AM30 and AZ31. *International Journal of Materials Research* 2009; **100**: 576.
120. Yi S, Schestakow I and Zaefferer S. Twinning-related microstructural evolution during hot rolling and subsequent annealing of pure magnesium. *Materials Science and Engineering A* 2009; **516**: 58.
121. Beer AG and Barnett MR. The influence of twinning on the hot working flow stress and microstructural evolution of magnesium alloy AZ31. *Materials Science Forum* 2005; **488–489** (2005): 611.
122. Beer A and Barnett MR. Microstructural development during hot working of Mg-3Al-1Zn. *Metallurgical and Materials Transactions A* 2007; **38**: 1856.
123. Yoshinaga H and Horiuchi R. Deformation mechanisms in magnesium single crystals in the direction parallel to hexagonal axis. *Trans. JIM* 1963; **4**: 1.
124. Staroselsky A and Anand L. A constitutive model for hcp materials deforming by slip and twinning: application to magnesium alloy AZ31B. *International Journal of Plasticity* 2003; **19**: 1843.
125. Agnew SR, Tomé CN, Brown DW, Holden TM and Vogel SC. Study of slip mechanisms in a magnesium alloy by neutron diffraction and modeling. *Scripta Materialia* 2003; **48**: 1003.
126. Barnett MR. A Taylor model based description of the proof stress of magnesium AZ31 during hot working. *Metallurgical and Materials Transactions A* 2003; **34A**: 1799.
127. Barnett MR, Keshavarz Z, and Ma Z. A Semi-analytical Sachs model for the flow stress of a magnesium alloy. *Metallurgical and Materials Transactions A* 2006; **37A**: 2283.
128. Agnew SR, Brown DW, and Tomé CN. Validating a polycrystal model for the elastoplastic response of magnesium alloy AZ31 using *in situ* neutron diffraction. *Acta Materialia* 2006; **54**: 4841.
129. Hutchinson WB and Barnett MR. Effective values of critical resolved shear stress for slip in polycrystalline magnesium and other hcp metals. *Scripta Materialia* 2010; **63**: 737.
130. Muránsky O, Carr DG, Barnett MR, Oliver EC and Sittner P. Investigation of deformation mechanisms involved in the plasticity of AZ31 Mg alloy: *In situ* neutron diffraction and EPSC modelling. *Materials Science and Engineering: A* 2008; **496**: 14.
131. Jiang J, Godfrey A, Liu W and Liu Q. Identification and analysis of twinning variants during compression of a Mg-Al-Zn alloy. *Scripta Materialia* 2008; **58**: 122.
132. Jonas JJ, Mu S, Al-Samman T, Gottstein G, Jiang L *et al.* The role of strain accommodation during the variant selection of primary twins in magnesium. *Acta Materialia* 2011; **59**: 2046.

133. Cizek P and Barnett MR. Characteristics of the contraction twins formed close to the fracture surface in Mg-3Al-1Zn alloy deformed in tension. *Scripta Materialia* 2008; **59**: 959.
134. Aydiner CC, Bernier JV, Clausen B, Lienert U, Tomé CN et al. Evolution of stress in individual grains and twins in a magnesium alloy aggregate. *Physical Review B – Condensed Matter and Materials Physics* 2009; **80**: 024113.
135. Brown DW, Agnew SR, Abeln SP, Blumenthal WR, Bourke MAM et al. The role of texture, temperature and strain rate in the activity of deformation twinning. *Materials Science Forum* 2005; **495–497**: 1037.
136. Barnett MR, Keshavarz Z, Beer AG and Atwell D. Influence of grain size on the compressive deformation of wrought Mg-3Al-1Zn. *Acta Materialia* 2004; **52**: 5093.
137. Klimanek P and Pötzsch A. Microstructure evolution under compressive plastic deformation of magnesium at different temperatures and strain rates. *Materials Science and Engineering* 2002; **A324**: 145.
138. Jain A and Agnew SR. Modeling the temperature dependent effect of twinning on the behavior of magnesium alloy AZ31B sheet. *Materials Science and Engineering A* 2007; **462**(1–2): 29.
139. Bakarian PW and Mathewson CH. Slip and twinning in magnesium single crystals at elevated temperatures. *AIME Transactions* 1943; **152**: 226.
140. Myshlyaev MM, McQueen HJ, Mwembela A and Konopleva E. Twinning, dynamic recovery and recrystallization in hot worked Mg-Al-Zn alloy. *Materials Science and Engineering A* 2002; **A337**: 121.
141. Barnett MR. A rationale for the strong dependence of mechanical twinning on grain size. *Scripta Materialia* 2008; **59**: 696.
142. Matsuda M, Ii S, Kawamura Y, Ikuhara Y and Nishida M. Interaction between long period stacking order phase and deformation twin in rapidly solidified Mg97Zn1Y2 alloy. *Materials Science and Engineering A* 2004; **386**: 447.
143. Geng J, Chun YB, Stanford N, Davies CHJ, Nie JF et al. Processing and properties of Mg-6Gd-1Zn-0.6Zr Part 2 – Mechanical properties and particle twin interactions. *Materials Science and Engineering A* 2011; **528**: 3659.
144. Clark JB. Transmission electron microscopy study of age hardening in a Mg-5 wt.% Zn alloy. *Acta Metallurgica* 1965; **13**: 1281.
145. Stanford N and Barnett MR. Effect of particles on the formation of deformation twins in a magnesium-based alloy. *Materials Science and Engineering A* 2009; **516**: 226.
146. Barnett MR, Stanford N, Geng J and Robson JD. On the impact of second phase particles on twinning in magnesium alloys. In press, *Magnesium Technology 2011*. San Diego, US: TMS Warrendale, 2011.
147. Gharghouri MA, Weatherly GC and Embury JD. The interaction of twins and precipitates in a Mg-7.7at.%Al alloy. *Philosophical Magazine A* 1998; **78**: 1137.
148. Robson JD, Stanford N and Barnett MR. Effect of precipitate shape on slip and twinning in magnesium alloys. *Acta Materialia* 2011; **59**: 1945.
149. Stanford N, Geng J, Chun YB, Davies CJH, Nie JF et al. Effect of plate shaped particle distributions on the deformation behaviour of magnesium alloy AZ91 in tension and compression. submitted to *Acta Materialia*.
150. Beladi H and Barnett MR. Influence of aging pre-treatment on the compressive deformation of WE54 alloy. *Materials Science and Engineering: A* 2007; **452–453**: 306.

4
Superplasticity in magnesium alloys by severe plastic deformation

R. LAPOVOK and Y. ESTRIN, Monash University, Australia

Abstract: This chapter discusses superplasticity of wrought magnesium alloys, with an emphasis on enhancement of superplastic properties by severe plastic deformation (SPD). A review of literature addressing superplastic properties of severely deformed magnesium alloys is given. Based on the research of the authors on two important structural alloys, ZK60 and AZ31, the mechanisms underlying superplasticity of Mg alloys and the role of grain refinement by SPD processing are elucidated. The bi-modality of the grain structure produced by SPD is presented as an important microstructural aspect of enhanced superplasticity.

Key words: superplasticity, severe plastic deformation (SPD), equal channel angular pressing, recrystallisation, twinning, ZK60, AZ31, bi-modal grain structure.

4.1 Introduction

Superplasticity was observed as early as in 1934 by Pearson[1] and was extensively investigated after World War II in the former Soviet Union where the term 'superplasticity' was coined. Superplasticity was described broadly as the ability of a material to sustain a large uniform tensile strain (higher than at least 200%) without local necking prior to failure in certain intervals of temperature and strain rate.[2,3] This definition is admittedly somewhat diffuse. More precise definitions were introduced over the years, and the most recent one was suggested by Langdon.[4] It reads verbatim as follows:

> Superplasticity is the ability of a polycrystalline material to exhibit, in a generally isotropic manner, very high elongations prior to failure. The measured elongations in superplasticity are generally at least 400% and the measured strain rate sensitivities are close to 0.5.

It is commonly accepted that metallic materials exhibit superplastic behaviour at high temperatures above half the melting temperature, $0.5\, T_m$, where the diffusion mechanisms of plasticity prevail, and at low strain rates below or around $10^{-2}\, s^{-1}$. Superplasticity also implies a high strain rate sensitivity of the flow stress ($\partial \ln\sigma/\partial \ln\dot{\varepsilon} \approx 0.5$), which is the case when the grain size of the material is sufficiently small (typically less than 10 μm).[5] As superplastic forming is well accepted in industry as a process for manufacturing parts with complex shapes, it is, of course, desirable to move the range of superplasticity to lower temperatures and higher strain rates. An excellent overview of the principles of superplasticity, the history of the area and some new ways of achieving it through grain refinement by means of severe plastic deformation (SPD) was recently given.[4] In the present chapter we focus specifically on superplasticity in magnesium wrought alloys, the emphasis being on the effect of severe plastic deformation on the mechanisms and characteristics of superplasticity. The most popular SPD techniques used to impart superplastic properties on Mg alloys are equal channel angular pressing (ECAP) and high-pressure torsion (HPT).[6]

Manufacturing of wrought magnesium products has been shrinking for many years[7] despite their superior properties, such as doubled strength and ductility,[8] as compared to cast products. The reason why wrought products have been pushed out of the market was the high cost of production associated with the limited ductility of magnesium alloys.[7–11] Hot working operations used for production of magnesium parts, including forging, extrusion and rolling, could be performed only at rather high temperatures (300–500 °C) and relatively low strain rates. The recently discovered capability of wrought magnesium alloys to deform superplastically at lower temperatures and reasonably high strain rates, which will be discussed in this chapter, promises increased applications in automotive,[9] aerospace[10] and construction[11] industries.

Although superplasticity of metallic materials was discovered about 75 years ago (see chronology[4]), it was not until the late 1960s that a systematic investigation of superplasticity of magnesium alloys began, largely initiated by the work of Backofen et al.[12–14] Research activities in this area have been growing ever since. A potential impact of superplastic forming of magnesium wrought alloys on industry and the need to develop a knowledge base in this area have led to a substantial number of publications over the last two decades (Table 4.1). It can be seen that among the wrought Mg alloys, ZK60 and AZ31 are the most widely

Table 4.1 Published data on superplasticity of magnesium wrought alloys

Alloy	Number of publications	Some representative references
WE54	2	15, 16
WE43	2	17, 18
ZC71	1	19
ZW3	1	20
AZM	1	21
AZ80	1	22
AZ61	30	23–28
AZ31	113	29–33
ZK40	2	34, 35
ZK60	98	36–41

investigated ones. However, even for these two alloys there is no consensus about the optimum thermo-mechanical processing route needed to enhanced tensile ductility. There is still no agreement on what type of microstructure is most favourable for increased elongation-to-failure. While this chapter touches on some general issues relevant to many wrought magnesium alloys, the focus is on the two most common alloys, ZK60 and AZ31. The effect of SPD processing on superplasticity of these alloys will be highlighted and possible mechanisms underlying the improved superplastic properties observed will be discussed.

Among the well-known 'brands' of superplasticity, such as fine-structure superplasticity (FSS), internal-stress superplasticity (ISS), high strain rate superplasticity (HSRS) and others,[1] it is FSS that is most commonly associated with superplastic behaviour of wrought alloys, including magnesium wrought alloys. A uniform fine-grained microstructure is usually considered as a prerequisite for superplastic behaviour. However, early studies of constitutive behaviour of magnesium alloys[42-44] have shown that the process of grain refinement in magnesium alloys differs from that in fcc metals (e.g. aluminium alloys),[45] and that a variety of different microstructures can be formed as a result of grain refinement. The types of microstructures produced may range from uniform fine-grained to bi-modal or, for that matter, multi-modal structures of different morphology.

Such diversity is caused by the hexagonal crystal lattice of magnesium alloys, with different mechanisms playing the dominant role at different temperatures. Below about 150 °C, slip is mainly limited to the hexagonal basal planes, as the critical yield stress on prismatic or pyramidal planes for single crystals is about ten times higher, while at temperatures around 400 °C this ratio drops to about two.[46] The deficiency in slip systems at low temperatures leads, at least for a sufficiently large grain size, to activation of twinning, as will be discussed below.

At elevated temperatures, thermally activated pyramidal or prismatic slip enhances dynamic recovery. Dislocations arrange themselves into low-energy subgrain boundaries,[44,47] while the high dislocation density regions trigger dynamic recrystallisation resulting in new randomly oriented grains.[47] At lower temperatures recrystallisation may also be initiated at deformation twins, and this interplay of deformation by dislocation glide and twinning under thermo-mechanical treatment (TMT) may lead to rich microstructures, some of which may be conducive for superplasticity, as will be shown in the following section.

The concurrent processes of deformation and dynamic recrystallisation result in characteristic stress-strain curves of wrought magnesium alloys with a pronounced strain softening part of the curve after a peak stress is reached. The plastic flow characteristics of ZK60 and AZ31 alloys were widely investigated and reported in the literature.[42] There is also extensive literature on aspects of deformation behaviour relevant specifically to superplasticity of Mg alloys.[15-41]

Methods employed for production of commercial superplastic alloys include thermomechanical processing, rapid solidification routes, powder metallurgy and mechanical alloying.[1] Very good results have been obtained by combining heat treatment and plastic deformation in TMT of aluminium alloys, and suitable regimes of TMT have been developed to produce fine grain sizes.[48] However, only a few papers on TMT of magnesium alloys, such as AZ91D and Mg-4Y-3Re[49,50] have been published.

The success in achieving superplastic behaviour has been based on grain refinement using conventional deformation processes at elevated temperatures, such as extrusion, forging and rolling.[28,37,39,42] The new opportunities in grain refinement and manipulating the microstructures in magnesium wrought alloys have come with the advent of severe plastic deformation methods, such as HPT, accumulated roll bonding (ARB) and ECAP.[51] The application of these processes alone[18,32–35,52,53] or in combination with conventional processes[54–56] has yielded record values of elongation-to-failure in ZK60 and AZ31.[41,33,57] This promising route to achieving superplastic behaviour of Mg alloys will be discussed in this chapter.

4.2 Microstructure evolution during thermomechanical processing

Various methods, both conventional (such as extrusion or rolling) and SPD-based (including ECAP and HPT), as well as their combinations (e.g. extrusion + ECAP or rolling + ECAP) have been used for grain refinement in magnesium wrought alloys. The resulting range of microstructures is responsible for the observed diversity in superplastic properties. A summary of the results pertaining to superplastic behaviour published to date in relation to the type of microstructure, mechanical testing conditions and strain at fracture is given in Table 4.2 for alloys ZK60 and AZ31.

It is seen that the microstructure produced by these methods can be either uniform, with the grain size ranging from 0.7–6 μm, or bi-modal with two populations of grains of different size and morphology, and that the bi-modal one is more common in severely deformed specimens. The multiplicity of deformation mechanisms in hcp metals results in a far greater complexity of microstructure formation by SPD processing than in fcc metals. In magnesium alloys the mechanisms of grain refinement and the final grain size depend on the temperature of processing, the initial grain size, the strain rate, the texture, the occurrence of twinning, softening due to dynamic recrystallisation and other factors.

The analysis of strategies and principles of grain refinement in magnesium alloys with particular reference to ECAP are described.[45] As discussed therein, if the initial grain size is smaller than a critical grain size, d_c, a uniform ultrafine-grained structure can be obtained as a result of severe plastic deformation. By contrast, if the initial grain size is larger than d_c, a bi-modal grain structure, consisting of large deformed and small recrystallised grains, is formed. The final grain size and the volume fraction of both grain populations depend on the initial

Table 4.2 Published data on superplasticity of magnesium alloys ZK60 and AZ31

Grain size (µm)	Engineering strain at fracture (%)	Tensile test conditions		Grain refinement technique used	References
		Temperature (°C)	Strain rate (s^{-1})		
ZK60					
1	810	260	3×10^{-3}	ECAP (R_B)	58
1	960	260	7×10^{-4}	ECAP (R_B)	59
1.4	1083	200	1×10^{-5}	ECAP (R_B) + anneal.	60
1.6	570	250	6×10^{-4}	ECAP (R_B)	61
2.1	680	250	6×10^{-4}	ECAP (R_C)	61
3.1	480	250	6×10^{-4}	ECAP (R_A)	61
2.4	730	300	4×10^{-4}	Extrusion	62
3.3	544	323	1×10^{-2}	Extrusion	36
6.5	440	200	3×10^{-6}	Extrusion	63
1, 15	1200	270	3×10^{-3}	Extrusion	12, 13
2, 25	220	180	1×10^{-5}	Extrusion	63
2.5, 10	140	350	5×10^{-3}	ECAP (R_B)	18
3.7	1330	250	1.4×10^{-4}	Extr. + compr. + rolling	37
1.5, 12	1400	220	3×10^{-3}	ECAP (R_B)	57
1.5, 12	2040	220	3×10^{-4}	ECAP (R_B)	57
0.8	3050	200	1×10^{-4}	Extr. + ECAP (R_B)	58
AZ31					
3	600	200	1×10^{-4}	Extrusion	65
3	900	300	1×10^{-4}	Extrusion	65
5	608	325	1×10^{-4}	Extrusion	29
1	400	250	3×10^{-3}	ECAP (R_B)	60
0.7	460	150	1×10^{-4}	Extr. + ECAP (R_B)	30
6	265	450	2×10^{-4}	Rolling	66
5	360	400	1.4×10^{-3}	Roll. + extr. + anneal.	67
2.2	1030	350–450	$10^{-5} - 10^{-4}$	Extr. + ECAP (R_B)	32
1, 17	1215	350	1×10^{-4}	ECAP (R_B)	33, 68

grain size, the amount of strain imposed on the alloy, the strain rate and the temperature of processing. It is also influenced by the propensity for twinning, as will be discussed later in this section.

4.2.1 Twinning and critical grain size

Grain size, d, is one of the variables that govern the amount of deformation twinning in polycrystalline metals and alloys. Often, but not always, the lower yield stress, determined by either the dislocation slip or the twinning mechanism, obeys a Hall–Petch type relation.[68] The effect of grain size is somewhat obscured by the possibility that grains of mixed sizes, as commonly observed after thermo-mechanical processing of Mg alloys, may not behave according to a law of mixtures, but may rather exhibit superior properties, as reported by Ma's group[69] with reference to Cu. Many factors, including temperature, strain rate, amount of pre-strain, initial grain size, composition and the presence of dispersed phases, may influence twinning,[68,70] so that only general trends can be established.

In the case of polycrystalline Mg and its alloys, Taylor's requirement that five independent slip modes be activated for plastic compatibility between the grains to be maintained is satisfied by the activation of non-basal slip and deformation twinning.[71] Along with non-basal dislocation slip associated with a Burgers vector having a component along the c-axis, twinning can ensure deformation along the c-axis, as well. It has also been suggested that compatible deformation can be achieved by four independent slip modes and local stress relief by twinning.[72]

In view of the competition between dislocation slip and twinning, it is of interest to assess the range of grain sizes where twinning prevails. The critical grain size limiting twinning can be estimated by comparing the grain-size dependent stresses for twinning and dislocation glide. We use a relation between the applied stress s and the dimensions of a twin in equilibrium considered by Friedel:[73]

$$\sigma \cong \frac{\mu}{2} s \cdot \frac{h}{L} \qquad [4.1]$$

Here μ is the shear modulus, s is the shear produced by the twin (of the order of 0.1–1),[73] L is the length of the twin and h its thickness. If the stress required for twinning, $\sigma_{TWINNING}$, is assessed on the basis of this equation, the most favourable condition for twinning corresponds to L being identified with the grain size d:

$$\sigma_{TWINNING} \cong \frac{\mu}{2} s \cdot \frac{h}{d}. \qquad [4.2]$$

The $1/d$ dependence of the twinning stress is known from the experimental literature,[65] although a Hall–Petch-like $1/\sqrt{d}$ dependence was also reported.[65] As the stress required for dislocation glide controlled plastic deformation follows a $1/\sqrt{d}$ dependence, it is obvious that for sufficiently small d, the twinning stress given by Eq. 4.2 will become larger than the stress required for dislocation glide, σ_{GLIDE}. Obviously, twinning will not be the preferred deformation mode then. As an exemplary case, an estimation will be made for common wrought Mg alloy ZK60. The grain size dependence of strength for coarse-grained ZK60 compiled from the literature,[74] assumed to be associated with dislocation glide, obeys a Hall–Petch type relation, which can be written for ZK60:[65]

$$\sigma_{GLIDE} = 197\,\text{MPa} + 99\,\text{MPa} \cdot \mu\text{m}^{1/2}/\sqrt{d} \qquad [4.3]$$

(d in μm). The dependence of the twinning stress, Eq. 4.2, and the dislocation glide stress, Eq. 4.3, on the average grain size is shown on Fig. 4.1. The twinning stress was estimated for $s = 1$ (which is somewhat higher than the value of 0.13 associated

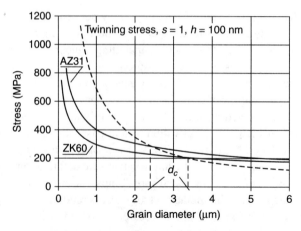

4.1 Grain size dependence of the twinning stress (dashed line) and the dislocation glide stress for ZK60 and AZ31 (solid lines).

with the most common form of twinning in Mg) and $h = 0.1$ μm. (The latter value is suggested by the order of magnitude of the twin width observed in the micrographs of ZK60 structures deformed by ECAP; twins can be wider in more conventionally deformed microstructures). The diagram shows that twinning is prevalent for large d and that it gives way to dislocation glide in the range of small d. The critical grain size d_c at which this cross-over occurs is seen to be about 3–4 μm. Of course, this particular value depends on the magnitude of the parameters chosen and is indicative only. A similar estimate for alloy AZ31 is also presented. The grain size dependence of the stress associated with dislocation glide was adopted from:[64]

$$\sigma_{GLIDE} = 56 \text{MPa} + 348 \text{MPa} \cdot \text{μm}^{1/2}/\sqrt{d},$$ [4.4]

where d is to be given in μm. It should be noted that, in addition to the uncertainties with regard to the values of the parameters in Eq. 4.2, the critical conditions for the occurrence of twinning depend on temperature, strain rate, and texture. Hence, the above estimates are only indicative, rather than numerically accurate.

It will be seen below that twinning is important for the formation of a bi-modal grain structure; when the size of recrystallised grains falls below d_c, no twinning occurs and a uniform ultra-fine grained (UFG) structure is formed. Despite the crudeness of the above estimation of the magnitude of the critical grain size for both alloys, the prediction of a lower value of d_c for AZ31 is valid, as was also confirmed by experiment.[33]

A further argument for the existence of a critical grain size below which twinning should not occur was put forward by Koike.[75,76] It is based on the idea that despite the preference for basal slip in the interior of the grains in Mg alloys, multiple slip is promoted and twinning is suppressed in the regions near grain boundaries. Following Koike it can thus be argued that for sufficiently small grain size the zone where deformation occurs only by multiple slip may cover the entire grain. Following Koike's arguments it can be expected that ductility of the material is enhanced in this 'sub-critical' grain size range (P. Uggowitzer, private communication).

4.2.2 Twinning and bi-modal grain structure

The role of twinning in the formation of a bi-modal grain structure was discussed.[77,78] It was shown that twins act as sites where dynamic recrystallisation starts (Fig. 4.2). Indeed, transmission electron microscopy (TEM) provides evidence of the occurrence of small recrystallised grains that appear to have nucleated at the twin boundary seen in the micrograph (Fig. 4.2b). Therefore, not only old grain boundaries but also twins appear to have acted as sites where new ultrafine grains nucleate. In a bi-modal population, there are distinct families of coarse and fine grains. The greater the twin fraction during low temperature deformation, the smaller is the average grain size of the coarser population, as the initial grains must have been fragmented and are surrounded by a necklace of dynamically recrystallised grains.

Lapovok et al.[67,77] studied twinning and bi-modal structure formation in both ZK60 and AZ31 that underwent ECAP. The materials were available in the form of continuously cast billets. Samples of 20 × 10 mm² cross-section and 70 mm length were machined from ZK60 and subjected to different routes of ECAP

(a)   (b)

4.2 Microstructure of ZK60 after warm ECAP (300 °C). (a) Profuse twinning within large grains (optical image). (Source: Japanese Institute of Metals.[78]) (b) Small grains (indicated in the photo by arrows) resulting from recrystallisation that was initiated at a twin crossing a big grain. (Source: courtesy Springer-Verlag GmbH.[77])

(route A and route C[77]) at three temperatures (200 °C, 250 °C and 300 °C). A 120° equal channel die[77] was used and the ram speed was 0.5 mm/s. Prior to processing, samples were homogenised for four hours at 460 °C in alumina powder to reduce surface oxidation. No grain growth occurred at this homogenisation temperature. The initial grain structure was equiaxed with average grain diameter of 45.5 µm (Fig. 4.3). Isolated zinc-zirconium particles dispersed within the grains or occurring in clusters were observed. The initial grain size was well above d_c and, therefore, twinning occurred immediately in the first ECAP pass.

Deformation twins in large grains were clearly recognisable at all three temperatures. As could be expected,[68] the occurrence of twins was markedly reduced with increasing ECAP temperature (Fig. 4.4). Their number density in large grains (defined as the number of twins per unit area of large grains) dropped almost to zero as the temperature of warm ECAP was raised from 200 °C to 300 °C (Fig. 4.5a). Concurrently, the area fraction of large grains rapidly decreased with temperature as they were replaced with small dynamically recrystallised grains (Fig. 4.5b).

The degree of grain refinement is seen to increase with the number of ECAP passes. After severe deformation introduced by eight ECAP passes, the mean size of large grains (in the transverse cross-section) dropped from 45.5 ± 15.2 µm (Fig. 4.3) to 26.7 ± 2.5 µm for route A and to 29.7 ± 4.1 µm for route C. The temperature of warm ECAP was almost immaterial. At the same time, a new population of small recrystallised grains with average diameter ranging from 2 µm (ECAP at 200 °C) to 4 µm (ECAP at 300 °C) was formed. As already mentioned, the area fraction of large grains was found to be smaller for higher ECAP temperatures, and only a very small number density of large grains was

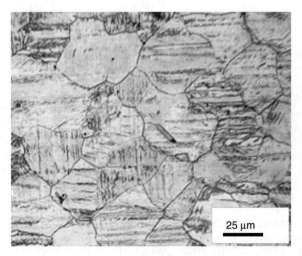

4.3 Microstructure of ZK60 after homogenisation. (Source: courtesy Springer-Verlag GmbH.[77])

Superplasticity in magnesium alloys by severe plastic deformation 153

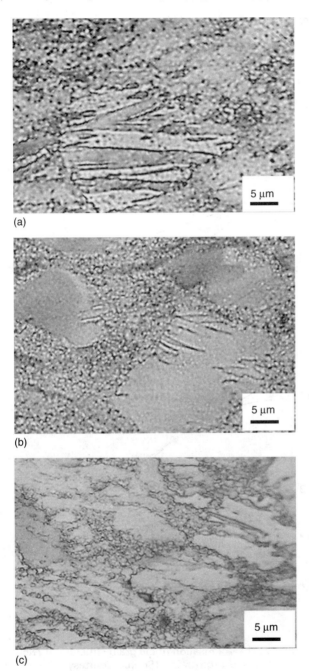

4.4 Optical images of large grains in the microstructure of ZK60 showing twins after eight passes of warm ECAP (Route A) at:
(a) 200 °C (source: courtesy Springer-Verlag GmbH[77]); (b) 250 °C (source: courtesy Springer-Verlag GmbH[77]); (c) 300 °C.

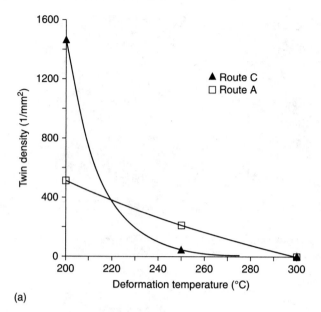

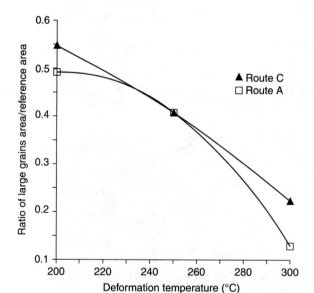

4.5 Quantitative characterisation of the grain structure of ZK60. (a) Density of twins in large grains. (Source: courtesy Springer-Verlag GmbH.[77]) (b) Area fraction of large grains after eight passes as a function of the ECAP temperature. (Source: courtesy Japanese Institute of Metals.[78])

retained after pressing at 300 °C. Upon the increase in the deformation temperature a uniform fine grain structure was formed, with the average grain size ranging from 1–5 μm. For route B_C the population of larger grains (with an average size of about 12 μm) was much smaller. With increasing temperature and/or number of passes the area fraction of large grains and their average size decreased for all routes tested. It was observed that higher temperatures promoted dynamic recrystallisation in that a lower strain or a smaller number of ECAP passes was required to trigger the process. The volume fraction of dynamically recrystallised grains also increased with temperature.

The number density of twins acquired within the first two warm ECAP passes either remained unchanged with further ECAP processing, as in the case of ECAP at 200 °C, or exhibited an apparent decrease, as in the other two cases. This decrease can be attributed to a decline in the fraction of large grains or a drop in size of the small grains below the critical size.

The number density of twins detected after two ECAP passes (about 3000 and 4500 mm^{-2} for routes A and C, respectively) was nearly the same for 200 °C and 250 °C, while for 300 °C, drastically lower values, about two orders of magnitude smaller, were recorded. There was a direct correlation between the occurrence of twinning and the bi-modality of the grain structure formed. It appears that twins promote dynamic recrystallisation providing nucleation sites for it.[79] However, this twin-induced dynamic recrystallisation tended to be incomplete due to the relatively low temperatures used, and some of the initial grains were retained as coarse islands within the fine recrystallised structure.

Quite universally, for all routes and all temperatures used, warm ECAP resulted in a bi-modal microstructure with two populations of grains distinctly different in size, where large elongated grains are embedded in regions of small recrystallised grains. Representative microstructures obtained after eight ECAP passes at different temperatures are shown in Fig. 4.6. Although the processing conditions for AZ31 were quite different from those used for ZK60,[77] the trends reported above are also relevant for AZ31: the area fraction of surviving large grains and their average diameter, as well as the diameter of small recrystallised grains, were the smallest for the lowest possible temperature of ECAP. However, the microstructure of AZ31 samples processed by ECAP at a temperature of 200 °C and above tended to exhibit a more uniform grain size with the growing number of ECAP passes. A significant grain refinement was observed, with the grain size ranging between 1.3–7.5 μm (Fig. 4.7a). No bi-modal structure was observed for these temperatures, and the large variation in the grain size was attributed to the effect of non-uniformity of initial grain structure and dynamic recrystallisation.

To obtain a bi-modal structure in AZ31 (Fig. 4.7b), the temperature of ECAP had to be lowered below 200 °C. Due to poor ductility of AZ31 in this temperature range, this was only possible by virtue of back-pressure.[80] Route B_C ECAP was performed at the ram velocity of 15 mm/s in a 90° die with a sharp corner. Multi-pass ECAP (1, 2, 4, 6 or 8 passes) was carried out in the temperature range of

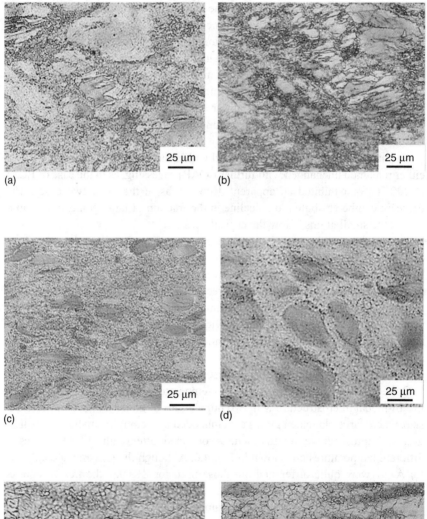

4.6 Optical images of microstructure of ZK60 after eight passes of warm ECAP at: (a, b) 200 °C; (c, d) 250 °C; (e, f) 300 °C; (a, c, e) Route A; (b, d, f) Route C. (Source: courtesy Springer-Verlag GmbH.[77])

Superplasticity in magnesium alloys by severe plastic deformation

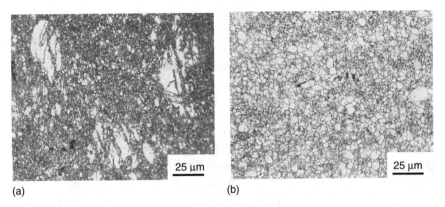

4.7 Optical images of microstructure of AZ31 after six passes of warm ECAP at: (a) 150 °C; (b) 200 °C. (Source: courtesy Wiley-VCH Verlag GmbH & Co.[33])

150–250 °C. The level of back-pressure was varied from 44 MPa at 200 °C to a maximum value of 260 MPa at the lowest ECAP temperature of 150 °C.

Our studies of microstructure formation by ECAP processing of both wrought Mg alloys, ZK60 and AZ31, showed[67,77] that grain refinement is likely to result from two concurrent effects: Formation of a large population of small recrystallised grains at old grain boundaries and fragmentation of large grains by deformation twins serving as sites at which new recrystallised grains formed. ECAP conducted at the low end of the temperature interval studied resulted in profuse twinning in the first two passes triggering dynamic recrystallisation at twin boundaries. The twinning activity then decreased over the subsequent passes. The fraction and the grain size of the population of large grains and the average grain size of the small-grain population can be manipulated by appropriately choosing the temperature and strain rate of processing and the right extent of strain. For example, by increasing the process temperature and the level of strain (i.e. the number of ECAP passes) the large-grain population can be completely eliminated resulting in a uniform fine grain structure.

The observed microstructures will now be shown to give rise to enhanced superplastic properties.

4.3 Superplastic behaviour

A study of elongation-to-failure of ZK60 and AZ31 samples processed by ECAP with back-pressure has shown unexpectedly high values of this quantity in several cases. This extraordinary superplastic ductility was recorded for samples with bi-modal grain structures, as distinct from much more modest values of elongation-to-failure usually observed for uniform fine-grained structures, see summary in Table 4.2. An exception is the extremely high superplastic ductility observed for ECAP-processed Mg alloys with uniform fine-grained microstructure.[57] While not questioning the exceptionally good superplastic ductility reported,[57] we note that the magnitude of strain-to-failure generally depends on the ratio of the gauge length to the diameter or width of the sample and it cannot be directly compared with the strain-to-failure obtained under standard conditions.[81–83] Thorough experiments[82] have shown that dropping this ratio from six to less than three may lead to a significant error in the values of the strain determined from the tensile curves. That is why, in industry practice, specially modified standards for sample dimensions corresponding to the length to diameter ratio equal to four are used.[83] The results on superplasticity of magnesium alloys ZK60 and AZ31 with bi-modal grain structure reported in our publications,[33,41,56,67] which are presented below, were obtained for the samples with the gauge length to diameter ratio equal to four.

In Fig. 4.8a the tensile samples of ZK60 are shown in the initial geometry and after superplastic tensile deformation with different strain rates at 220 °C. Before a tensile test, the homogenised ZK60 billet with initial grain size of 45.5 μm was subjected to six passes of ECAP (route B_C) using a 90° die heated to 200 °C. As a result, a bi-modal grain structure was formed. It comprised two distinctly different populations of grains: Small grains with the grain sizes in the range below 4 μm, and relatively large grains with the size of about 12.5 ± 2.2 μm (Fig. 4.9a).

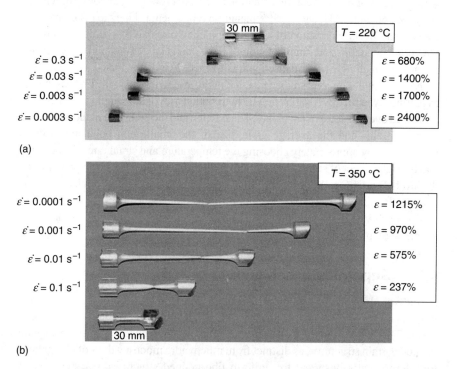

4.8 Tensile samples shown in the initial geometry and after superplastic tensile deformation. (a) ZK60. (Source: courtesy Cambridge University Press.[41]) (b) AZ31. (Source: courtesy Carl Hanser Verlag GmbH & Co.[94])

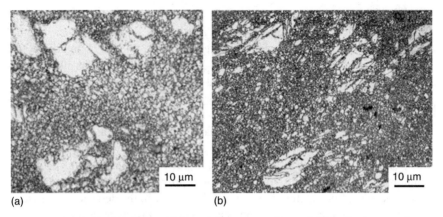

4.9 Optical images of bi-modal structure formed after six passes of ECAP. (a) ZK60 at 200 °C. (Source: courtesy Cambridge University Press.[41]) (b) AZ31 at 150 °C. (Source: courtesy Carl Hanser Verlag GmbH & Co.[94])

A similar processing schedule was also applied to alloy AZ31, only that the ECAP temperature was 150 °C.[33] Figure 4.8b shows tensile samples of ECAP-processed AZ31 in the initial geometry and after superplastic tensile deformation at 350 °C with different strain rates. As in the case of ZK60, a bi-modal grain structure with two distinct grain fractions was formed. The grain sizes of the fine-grain population were in the range below 3 μm, while those of the coarse-grain population were about 16.5 ± 3.5 μm (Fig. 4.9b).

The general trend observed for both alloys was that the area fraction of surviving large grains and their average diameter, as well as the diameter of small recrystallised grains, were the smallest for the lowest temperature at which ECAP processing was still possible. It should be noted that the temperature at which the maximum strain-to-failure was obtained was different for the two alloys studied. Furthermore, this optimum temperature was not related to the temperature of ECAP processing leading to bi-modality, which was lower for AZ31 than for ZK60.

It is remarkable that the tensile curves for the samples shown in Fig. 4.8 do not exhibit plateau-like portions typical of classical superplastic behaviour. Such a phenomenon is typically associated with steady-state plastic flow. This is illustrated by Fig. 4.10a, b. By contrast, a pronounced strain-hardening portion, followed by a continual drop-off in stress, was observed at all strain rates and temperatures investigated. A plausible explanation for this behaviour is seen in the bi-modality of the microstructure of the ECAP-processed material. The ascending branch of a deformation curve may be associated with the presence of a population of large non-recrystallised grains deforming by dislocation glide, thus giving rise to strain hardening. The reason for the occurrence of a descending branch may be seen in a continual increase of the population of small grains due to ongoing dynamic recrystallisation under uniaxial tensile deformation. Provided that the small grains do not give rise to strain hardening, as they deform by grain boundary sliding or another mechanism involving diffusion, an increase in their volume fraction naturally leads to overall softening. The role of bi-modality of microstructure in the large tensile elongation achieved by ECAP was discussed.[69]

The TEM study of bi-modality revealed even more complex microstructures emerging due to continuous dynamic recrystallisation involving different sites and times of grain nucleation, such as the nucleation at a twin boundary (Fig. 4.2b), nucleation at the boundary of a large grain (Fig. 4.11a, b), or nucleation at a triple junction of grain boundaries (Fig. 4.12a). It is a well-known fact that grain boundaries can act as nucleation sites for recrystallised grains; profuse twinning in the beginning of deformation provides increased numbers of nucleation sites. Recrystallised grains that form early in the deformation process are part of the small grain population, but are comparatively large, about 4 μm in size (Fig. 4.11c, d, Fig. 4.12b). They continue to store dislocations and can be

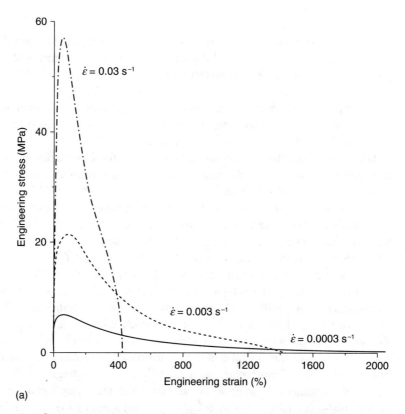

(a)

4.10 Stress vs strain curves (at different strain rates) shown for the temperature corresponding to maximum superplastic ductility.
(a) ZK60, T = 220°C. (Source: courtesy Cambridge University Press.[41])
(b) AZ31, T = 350 °C. (Source: courtesy Springer-Verlag GmbH.[67])

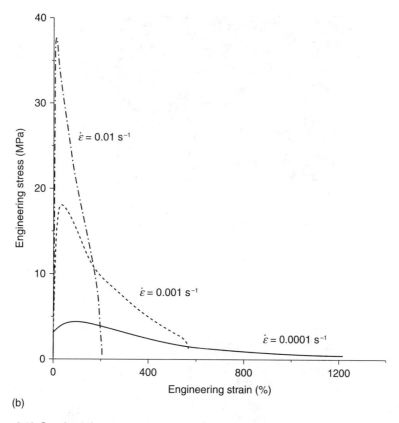

(b)

4.10 Continued.

further fragmented into dislocation cells (Fig. 4.11d, Fig. 4.12b). During continued deformation these cells can transform to new small grains due to rotation and increasing misorientation.[84] Concurrently, new recrystallised grains may emerge. The boundaries of the grains of this second grain population and dislocation walls/sub-boundaries may act as new sites for further grain nucleation. The occurrence of ultrafine grains about 0.4 μm in diameter, which are devoid of dislocations, may be a result of the latter process. In this sense, the small grain population itself can be viewed as a bi-modal one, so that the grain structure produced by ECAP processing can be considered as effectively tri-modal.

At a phenomenological level, the Hart criterion[85] for strain localisation under tensile loading,

$$\theta/\sigma + m \leq 1 \tag{4.5}$$

can be used to rationalise the peaked stress-strain curves seen in Fig. 4.10 and the concomitant extraordinary superplasticity of the bi-modal grain structure. In

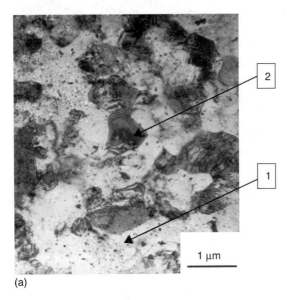

(a)

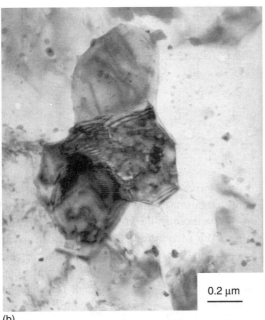

(b)

4.11 TEM images of the microstructure of ZK60 processed by six ECAP passes (route B_c) at 200 °C. (a) Part of a large grain ~10 μm in diameter (1) surrounded by small recrystallised grains (2). (b) Small grains ~0.5–0.6 μm in diameter with visibly different dislocation structure. (c) Two fractions within small grain population: relatively large ~4 μm (1) and small ~0.4 μm (2) grains. (d) Fragmentation of a 4 μm grain into dislocation cells. (Source: courtesy Cambridge University Press.[41])

Superplasticity in magnesium alloys by severe plastic deformation

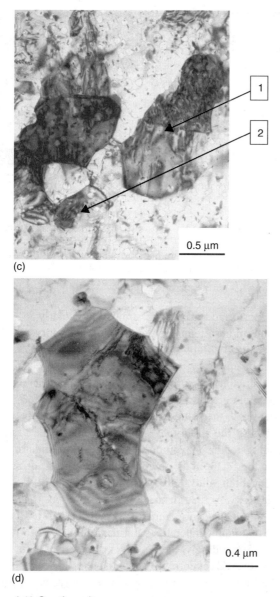

(c)

(d)

4.11 Continued.

addition to the strain hardening coefficient θ normalised by stress σ, Hart's criterion involves the strain-rate sensitivity of stress represented by the parameter

$$m = \left(\partial \ln \sigma / \partial \ln \dot{\varepsilon}\right)_\varepsilon \qquad [4.6]$$

where ε and $\dot{\varepsilon}$ denote the strain and the strain rate, respectively. The strain rate sensitivity parameter was measured for both alloys using strain rate jump tests.

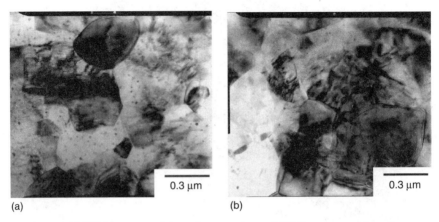

4.12 TEM images of the microstructure of AZ31 processed by six ECAP passes (Route B_c) at 150 °C. (a) Recrystallised small grain at a triple junction. (b) Freshly recrystallised small grains and formation of dislocation cells in bigger grains. (Source: courtesy Wiley-VCH Verlag GmbH & Co.[33])

Strain rate jumps were performed at various strains, particularly in the strain hardening, stress peak and strain softening regions of the tensile curves. The strain rate in these tests was changed abruptly between 3×10^{-4} s^{-1} and 3×10^{-1} s^{-1} several times along a tensile curve.

At 220 °C the measured values of m for ZK60, which were close to 0.5, were fairly insensitive to the strain at which a strain rate jump was performed. They were also quite consistent with the values of strain-rate sensitivity obtained near the peak stress. The plots of the strain-rate sensitivity versus strain rate at the temperature corresponding to the maximum elongation-to-failure for the two alloys are presented in Fig. 4.13. It is obvious that the values of strain rate sensitivity parameter for both alloys are well within the range commonly associated with superplasticity. At first glance, the strain-rate sensitivity of stress for ZK60 is less strain-rate-dependent compared to that for AZ31. However, it should be noted that the optimum temperatures for maximum strain-to-failure are different for these alloys. Accordingly, the dependence m versus $\dot{\varepsilon}$ for ZK60 is shown for 220 °C, while for AZ31 it corresponds to the temperature of 350 °C.

The magnitude of the strain-rate sensitivity parameter, m, typical for classical superplastic behaviour that is believed to involve grain boundary sliding[86] should be at least 0.3. In the strain rate range studied, this condition is satisfied for both alloys, with a distinct trend for larger values of m for ZK60.

To study the behaviour of the same alloy with a uniform ultrafine grain structure, i.e. in a condition when no bi-modality effects are involved, ECAP of alloy ZK60 was conducted at a higher temperature or by two-step processing (rolling followed by ECAP). It should be mentioned that the microstructure after the two-step processing was similar to the one obtained by ECAP with an increased number of passes.

Superplasticity in magnesium alloys by severe plastic deformation 165

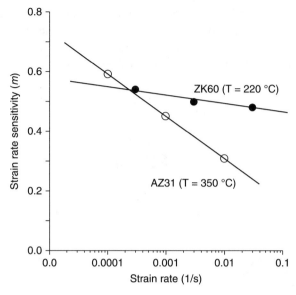

4.13 Strain rate sensitivity parameter, *m*, vs strain rate for AZ31 and ZK60 at the respective temperatures of maximum elongation-to-failure.

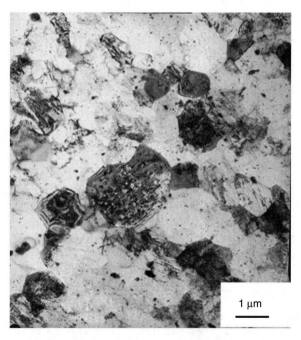

4.14 Microstructure of ZK60 sample after rolling followed by six passes of ECAP at 200 °C. (Source: courtesy Elsevier.[56])

A comparison of the tensile curves obtained in the temperature range of 200–300 °C at the strain rates of 3×10^{-4} s^{-1} to 3×10^{-2} s^{-1} for samples processed by six passes of ECAP proper and by the two-step procedure (Fig. 4.15), shows that samples with a uniform ultra-fine-grained structure exhibit a lower elongation-to-

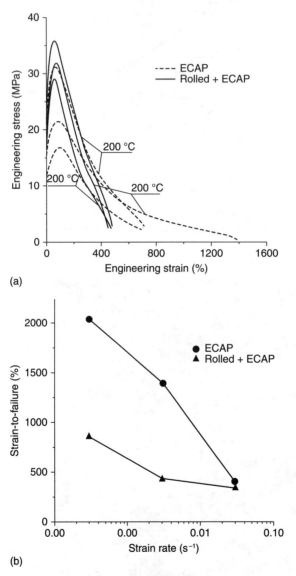

4.15 Tensile tests for samples of ZK60 processed by one-step deformation (ECAP) and two-step deformation (rolling + ECAP). (a) Stress-strain curves at strain rate of 3×10^{-3} s^{-1}. (b) Strain-to-failure vs strain rate for tensile tests performed at 220 °C. (Source: courtesy Elsevier.[56])

failure and a less pronounced softening stage associated with dynamic recrystallisation.

Similarly, AZ31 samples rolled at the same conditions prior to six passes of ECAP demonstrated superplastic behaviour, albeit with lower values of strain-to-failure (Fig. 4.16).

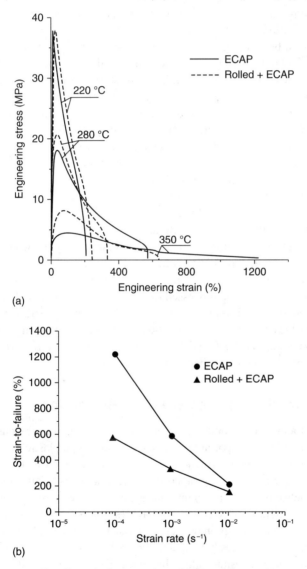

4.16 Tensile tests for samples of AZ31 processed by one-step deformation (ECAP) and two-step deformation (rolling + ECAP). (a) Stress-strain curves at strain rate of 1×10^{-4} s^{-1}. (b) Strain-to-failure vs strain rate for tensile tests performed at 350 °C.

In summary, after processing leading to uniform grain structure both alloys exhibited superplastic behaviour at similar temperatures and strain rates, characterised by high-tensile ductility and large strain rate sensitivity values ($m \approx 0.4$–0.5) in the same range as their bi-modal equivalents. However, the strain-to-failure did not reach the extraordinarily high levels as in the case of bi-modal structure. It can be concluded that bi-modality of grain structure is not a prerequisite of superplastic behaviour but is rather a factor promoting higher levels of superplastic ductility.

The benefits of a bi-modal grain structure can be rationalised in terms of the mentioned Hart instability criterion.[85] According to this criterion, two factors stabilise plastic flow against tensile instability: The strain hardening coefficient θ and the strain rate sensitivity parameter m (Eq. 4.5). Retention of relatively large, non-recrystallised grains after ECAP provides for the strain hardening seen in the ascending branch of the stress-strain curves in Fig. 4.10. Loss of the strain hardening capability is associated with the gradual decrease in the fraction of the coarse-grained population (due to dynamic recrystallisation under tensile loading), but it is compensated for by the growth of the volume fraction of fine grains and the increasing role of the strain rate sensitivity. That is why a bi-modal grain structure is believed to be particularly beneficial for good superplastic properties. Still, failure will eventually occur when saturation of m below unity for a vanishing value of θ is reached. A similar scenario (strain hardening associated with dislocation glide-controlled plasticity followed by softening during dynamic recrystallisation), albeit without the grain boundary sliding component, may apply to the case of the two-step processed material, see Fig. 4.15 and 4.16.

The above phenomenological picture provides only a crude general scenario of failure. Further clues to the role of bi-modal structure in obtaining superior superplastic elongation can be found by investigating the fracture surfaces of tensile samples that failed under superplastic conditions.

4.4 Fracture during superplastic deformation

The capability of a material to undergo superplastic deformation is obviously not unlimited. One of the most important processes that limit superplastic ductility is cavitation occurring during deformation. A commonly accepted picture of the processes underlying fracture during superplastic deformation is that in most cases it is caused by nucleation, growth and interlinkage or coalescence of cavities – the aforementioned cavitation. Nucleation of cavities normally occurs at grain boundaries at two-grain, three-grain or four-grain junctions (spheroidic voids) and at the inclusion-matrix interfaces (non-spheroidic voids). The mechanism of cavity or void nucleation in magnesium alloys has been described in detail.[87] It was shown that stress concentration at grain junctions caused by grain boundary sliding during superplastic deformation, should it not be relaxed by grain boundary diffusion and/or intragranular dislocation movement, results in nucleation of cavities.[88] Under the assumption that the stress concentration relief is controlled by grain boundary diffusion and using the expression given in[89] for the incubation time for cavity nucleation by vacancy condensation, the critical condition for cavitation to occur takes the form[87]

$$d \geq d_{crit} \qquad [4.7]$$

where

$$d_{crit} = \left(\frac{26 F_v E \Omega}{kT}\right)^{1/3} \frac{\gamma}{2\sigma} \qquad [4.8]$$

Here F_v is the cavity shape factor (of the order of unity), E is Young's modulus, Ω is the atomic volume, k is the Boltzmann constant, T is the absolute temperature, σ is the applied stress and γ is the surface energy.

Using Eq. 4.8 with the parameter values $F_v = 1$, $E = 45$ GPa, $\Omega = 2.33 \cdot 10^{-28}$ m^3, and $\gamma = 0.74$ J m^{-2} the critical grain size above which cavitation occurs at 350 °C is estimated at 1.38 µm. This value is below the average grain size for the large grain population but well above the average over the small grain population. Therefore, it can be expected that during the tensile test cavitation will be delayed until the small grains coarsen to the extent that the average grain size exceeds the critical value d_{crit}.

The above condition makes it possible to draw a 'cavitation map', defining areas on the grain size-stress plane where cavitation is to be expected. The above condition also shows that simultaneous increase of grain size and stress results in greater propensity for cavity nucleation. Therefore, if the ultrafine grain structure is not thermally stable at the temperature of the tensile test and grain growth occurs *in situ*, cavity nucleation will be promoted. In thermally stable ultrafine-grained materials cavitation will be delayed and will start at a higher strain than in their coarse-grained counterparts. An example of this has been shown for AZ31.[90] High resistance to cavitation was also found in other alloys that underwent grain refinement by ECAP.[91]

Models and mechanisms of cavity growth during superplastic tensile deformation of magnesium alloys were considered in detail[88, 92] and in many other publications. Three distinct mechanisms of cavity growth have been identified:

- Diffusion with transport of vacancies to cavities.
- Superplastic diffusion with multiple diffusion paths via grain boundaries.
- Plasticity-controlled growth of cavities by deformation of ligaments between cavities.

Each of the three mechanisms is described by a kinetic equation for the variation of the cavity radius, r, with the plastic strain as summarized.[88] For plasticity-controlled cavity growth the following equation holds:

$$\frac{dr}{d\varepsilon} = r - \frac{3\gamma}{2\sigma} \qquad [4.9]$$

Cavity growth due to plastic deformation results in a shape change of a void from spheroidic to ellipsoidal, the major axis being aligned with the direction of tensile stress. The aspect ratio of such voids for magnesium alloy AZ31 deformed in superplastic tensile tests before the collapse of inter-cavity ligaments increases. The observed values of the aspect ratio can be as high as 2.5.[93] Obviously, in such a case the cavity radius r in Eq. 4.9 can only be considered as an average radius of curvature. Depending on the prevalent mechanism of cavity growth, a prediction of failure can be made using Eq. 4.9. However, in the case under consideration, cavities with extremely large aspect ratios of up to 6 were observed. Equation 4.9 cannot be applied in this case.

Fracture of magnesium wrought alloys with bi-modal microstructure in superplastic conditions is exemplified by the behaviour of alloy AZ31.[94] Failure of the specimens under superplastic tensile deformation was studied with regard to the appearance of the fracture surface and the outer surface of tensile specimens and the microstructural features of fracture. The failure behaviour described below refers to the samples tested at a temperature of 350 °C and a strain rate of 10^{-4} s^{-1}, under which conditions the highest superplastic ductility was obtained.

The failed specimens exhibited a number of characteristic features (Fig. 4.17). It can be seen that the specimen tip at the fracture site has an asymmetrical shape (Fig. 4.17b). This is not unexpected given the type of texture produced by ECAP, known to preferentially orient the basal plane at 45° to the pressing direction.[95–97] An unexpected feature is the occurrence of unusually deep voids observed at the dimpled fracture surface (Fig. 4.17a). The void size was comparable with the size of large $Mg_{17}Al_{12}$ particles, which were partly fractured during ECAP processing, as seen in an optical micrograph (Fig. 4.17d). Fracture of some particles and their rotation during ECAP obviously led to de-bonding, and large voids were formed at particle/matrix interfaces. During the subsequent tensile tests these voids grew and coalesced to voids of sizable dimensions. It was found that void growth occurred in two directions: parallel to the longitudinal axis of the sample and at an angle to the axis. Void growth at an angle to the tensile loading direction, which is known to be about 50°,[98] resulted in cavities occurring at the outer surface of tensile sample (Fig. 4.17c) (indicated by black arrows in the photograph).

To determine the depth of longitudinally elongated voids, the tip of the failed tensile sample was cut lengthwise by spark-erosion. Voids uncovered in this way (Fig. 4.17d), show a depth of about 250 μm. This confirms the assumption that the voids were formed at an early stage of the tensile test, most probably due to pull-out of relatively large precipitates at the sites where de-bonding occurred. The fact that the ligaments between the dimples have survived large strains during the tensile test suggests that their load-bearing capacity was high enough to carry the load as the voids were elongated together with the sample. It should be noted that the aspect ratio of the observed voids reached values in excess of six.

The ruptured surface exhibited two different length scales of ligaments. In addition to the 'primary' ligaments separated by voids, which were described

Superplasticity in magnesium alloys by severe plastic deformation 171

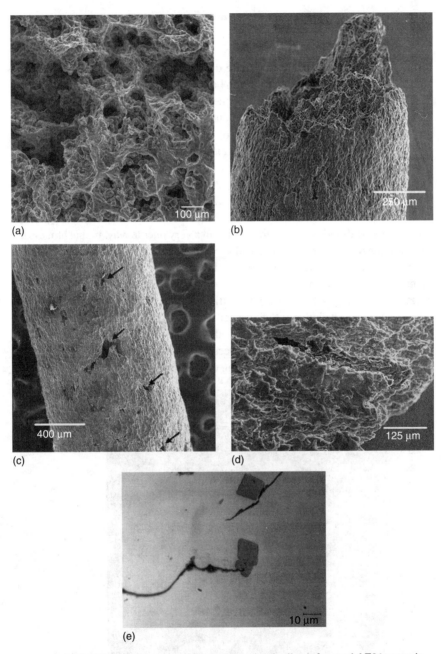

4.17 SEM micrographs of the superplastically deformed AZ31 sample. (a) Fracture surface (top view); (b) fracture surface (side view); (c) sample surface with visible pores; (d) pore opened by spark-erosion cutter. (Source: courtesy Carl Hanser Verlag GmbH & Co.[94]) (e) $Mg_{17}Al_{12}$ particle at which a crack was initiated. (Source: courtesy Springer-Verlag GmbH.[67])

above, smaller 'secondary' fibres were discovered on their surfaces, as well. The spacing between these secondary fibres was comparable with the size of small Mn_5Al_8 precipitates (Fig. 4.18). Apparently, the bonds between the matrix and these smaller-size precipitates are more difficult to break than those between the matrix and the $Mg_{17}Al_{12}$ particles. It can be conjectured that these small secondary fibres were formed on the primary ligaments at late stages of the tensile test and did not play a decisive role in tensile ductility. It can further be surmised from these findings that it is the presence of large $Mg_{17}Al_{12}$ particles that controls tensile ductility. Hence, a strategy to raising superplastic ductility even further would be to eliminate these large particles by suitable heat treatment.

Scanning electron microscopy (SEM) evidence suggests that a fibrous structure forms during superplastic deformation, which is promoted by extension of voids nucleated at failed $Mg_{17}Al_{12}$ particles. The character of fracture of ligaments between the voids indicates that they failed in a very ductile way. In this process the cross-section of ligaments was reduced to 50–100 nm at the fracture point, which is typical for ductile rupture. After superplastic deformation the inspected part of the sample had a lamellar structure composed of fibres or ligaments oriented parallel to the tensile axis (Fig. 4.19a). There was little variation in the width of the ligaments and their majority had a very high aspect ratio (Fig. 4.19b, c). Large areas with a pronounced fibre structure were observed. It was suggested that the fibres were formed during the tensile elongation process and were extended until some of them broke and de-lamination occurred leading eventually to overall failure.

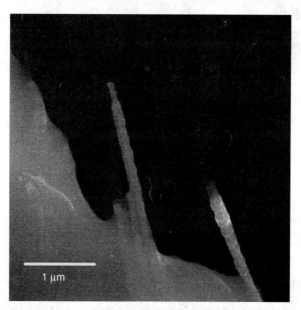

4.18 Secondary fibres with a spacing (~1 μm) commensurate with the Mn_5Al_8 precipitate size. (Source: courtesy Carl Hanser Verlag GmbH & Co.[94])

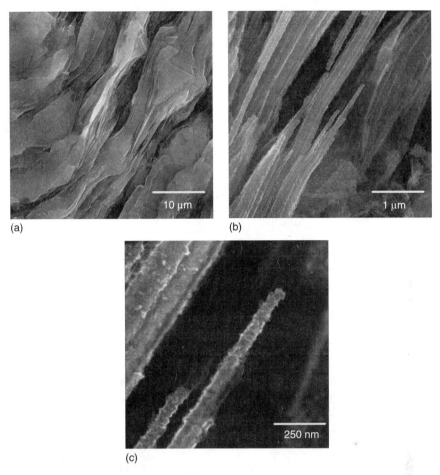

4.19 Micrographs showing a fibrous structure and ductile rupture of fibres. (Source: courtesy Carl Hanser Verlag GmbH & Co.[94])

While so far the emphasis in the discussion of enhanced superplastic ductility has been on the microstructural effects, the role of ECAP-induced texture should be mentioned as well. As with other Mg alloys, the ECAP-processed AZ31 develops a pronounced texture, with basal planes oriented at 45° to the ECAP direction,[95–97] (Fig. 4.20). Activation of basal slip is favoured in this case and becomes a predominant mode of deformation. This texture promotes tensile ductility in general[99] and is conducive for enhanced superplastic ductility in particular.

The above considerations regarding the texture are consistent with the morphology of slip in the ECAP processed AZ31 under uniaxial tensile loading.[67] The slip morphology was observed by SEM using argon plasma etching of the specimens. Tensile test samples were placed in a GATAN PECS (precision etching/coating system) and slope-cut at 45° to the longitudinal axis for several hours.

174   Advances in wrought magnesium alloys

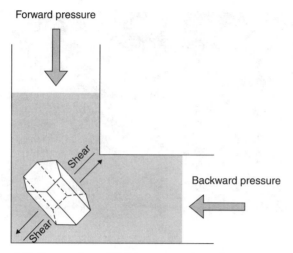

4.20 Texture in an ECAP processed specimen: orientation of basal plane at 45° to the pressing direction, after reference 95 (Source: courtesy Carl Hanser Verlag GmbH & Co.[94])

A normal view of a cut at 45° to the tensile axis of the tensile specimen produced by plasma etching is seen in Fig. 4.21. The geometry of the cut and the direction of view are indicated by a schematic sketch included in the figure. A terrace-like pattern observed in many regions of the sample is a signature of the crystallographic slip on the basal plane of the hexagonal Mg structure retained until failure. The

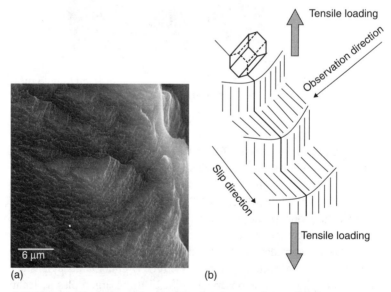

4.21 Terrace-like morphology of the fracture surface reflecting crystallographic slip. (a) SEM micrograph. (b) Schematic drawing. (Source: courtesy Carl Hanser Verlag GmbH & Co.[94])

examination of the fracture surface and the outer surface of the failed tensile specimens revealed unusually deep voids, presumably originating, as mentioned above, from pull-out of $Mg_{17}Al_{12}$ particles at an early stage of the tensile test. These voids had an unexpectedly high aspect ratio and their direction of growth was inclined to the tensile direction. The ductility and load-bearing capacity of the ligaments between the voids were extremely high, and the ligaments eventually failed in a characteristically ductile manner. This confirms the soundness of a strategy of enhancing tensile ductility based on elimination of coarse $Mg_{17}Al_{12}$ particles,[67] thus capitalising on the high ductility of the matrix material.

4.5 Mechanisms and models

A comprehensive review of phenomenological correlations and physically based constitutive models pertaining to superplasticity has been given.[3] A classical relation between the plastic strain rate $\dot{\varepsilon}$, temperature T, and stress σ, which holds under superplastic conditions,[1–3] is given by

$$\dot{\varepsilon} = A \frac{D_0 G b}{kT} \left(\frac{b}{d}\right)^p \left(\frac{\sigma - \sigma_0}{G}\right)^{1/m} \exp\left(-\frac{Q}{kT}\right) \quad [4.10]$$

Other quantities entering Eq. 4.10 are the activation energy Q, for the rate controlling mechanism, the shear modulus G, the magnitude of the Burgers vector b, and the pre-exponential factor in the diffusivity D_0. The product $D = D_0 \exp(-Q/kT)$ gives the relevant diffusivity, which has the meaning of D_{gb} for grain boundary diffusion, D_L for lattice (bulk) diffusion or D_{eff} for more complex combinations of diffusion mechanisms.[2] A is a constant proportionality coefficient.

The dependence of the strain rate on the grain size d, represented by a power law with an exponent p is essential, as is the power-law stress dependence with an exponent $1/m$. The quantity σ_0 is interpreted as a 'threshold stress' below which the rate of plastic flow is considered to be negligible. In most cases, however, this 'threshold stress' is set to zero.[100,101] The parameters m, p and Q are determined by the underlying deformation mechanism governing superplastic behaviour. Typical values of these parameters, along with the associated thermally activated mechanisms of plastic flow are presented in Table 4.3.

Table 4.3 Parameters characterising deformation mechanisms in superplasticity

m	p	Q (eV)	D (m² s⁻¹)	Deformation mechanism
0.33	0	1.40	D_{eff}	Dislocation climb-controlled creep
0.50	2	1.40	D_L	Grain boundary sliding accommodated by dislocation climb and lattice diffusion
0.50	3	0.95	D_{gb}	Grain boundary sliding accommodated by grain boundary diffusion

The most interesting ranges of $1/m$, which are of relevance to superplasticity of severely deformed Mg alloys, are around 0.33 and 0.5. These values correspond to the dislocation climb control and the grain boundary sliding mechanisms of superplastic deformation, respectively (Table 4.3). As seen from the table, the mechanism underlying superplasticity is characterised not just by the magnitude of m: Both the power p in the power-law dependence of the plastic strain rate on the grain size and the activation energy for the rate controlling process provide a clear signature of the underlying deformation mechanism.

Equation 4.10 has an obvious semblance with the relation describing diffusion-controlled creep, such as the Nabarro–Herring or the Coble creep, and a close connection between superplasticity and diffusion creep has been established.[102]

It should be noted that in the case of dislocation-climb controlled plastic flow the grain size dependence vanishes and $p = 0$ is found (Table 4.3).

Grain boundary sliding (GBS) is considered to be the principal mechanism of superplastic deformation. It is promoted when the temperature is raised and/or the strain rate is lowered. GBS occurs by virtue of grain boundary diffusion in conjunction with intragranular dislocation motion, at least at local scale, in the vicinity of grain boundaries. As GBS is facilitated by grain size reduction, grain refinement by SPD can promote superplastic properties, as was first suggested.[103] A first confirmation of this idea[104] was followed by a stream of publications substantiating it.

Grain-boundary sliding can be detected by scribing a line on a polished and etched surface and observing a shear offset where the line crosses a grain boundary. Evidence of grain-boundary sliding for ZK60 by this technique is presented in Fig. 4.22.

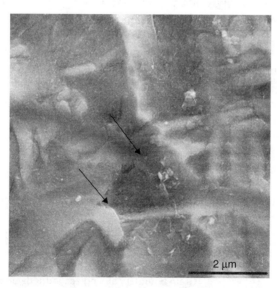

4.22 Scribed line on the surface of a ZK60 sample after superplastic deformation. A scribe line, which was initially straight, has been transformed to a Z-shaped one after shear deformation. The arrows indicate the corners of the sheared scribe line.

Superplastic behaviour of coarse-grained[12,13,61,63] and ultrafine-grained[49–59] alloys ZK60 and AZ31 was analysed using the above mathematical relation. The values of the activation energy Q were obtained from a plot of $\ln \dot{\varepsilon}$ versus $1/T$ and, based on these values together with the measured strain rate sensitivity parameter m and the exponent p, conclusions regarding a most plausible deformation mechanism were made.

In Fig. 4.23 a modified Arrhenius plot for determination of the activation energy for AZ31 at $\dot{\varepsilon} = 0.0001 \ s^{-1}$ and for ZK60 at $\dot{\varepsilon} = 0.003 \ s^{-1}$ is shown. Commonly an Arrhenius plot is represented in terms of a $\ln \dot{\varepsilon}$ versus $1/T$ diagram. Based on the above equation for superplastic deformation and setting σ_0 to zero, an 'Arrhenius plot' in the coordinates $\ln \sigma$ versus $1/T$ can be constructed for a fixed strain rate. The activation energy can be obtained from the slope of a $\ln \sigma$ versus $1/T$ curve as follows:

$$Q = \frac{k}{m} \frac{\ln(\sigma_1/\sigma_2)}{\frac{1}{T_1} - \frac{1}{T_2}} \qquad [4.11]$$

Setting the value of the strain rate sensitivity parameter m at 0.4, the values of the activation energy for AZ31 and ZK60 are obtained. They turn out to be close to each other and to the magnitude of the activation energy for grain boundary diffusion in Mg (about 0.95 eV).

Despite the fact that thermal stability of bi-modal microstructure was different for the two alloys, the deformation mechanism controlling superplastic

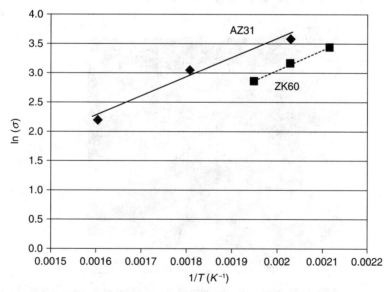

4.23 Arrhenius plot for determination of activation energy for AZ31 at $\dot{\varepsilon} = 0.0001 \ s^{-1}$ and for ZK60 at $\dot{\varepsilon} = 0.003 \ s^{-1}$.

deformation appears to be the same for both alloys. Alloy AZ31, which exhibited the highest strain-to-failure at the temperature of 350 °C and the strain rate of 10^{-4} s^{-1}, had a completely recrystallised microstructure with grain sizes ranging from 3–17 µm (Fig. 4.24). The microstructure of ZK60, which exhibited the highest strain-to-failure at the temperature of 220 °C and the strain rate of 3×10^{-3} s^{-1}, also had a fully recrystallised microstructure with uniform grains being slightly coarser than the initial ones, but lying below 4 µm (Fig. 4.25). Static annealing experiments of one hour duration showed that no grain growth appeared at the temperatures below 250 °C.

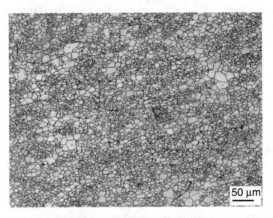

4.24 Optical microscopy image of microstructure of AZ31 after tensile test, $T = 350$ °C, strain rate 10^{-4} s^{-1}.

4.25 TEM image of microstructure of ZK60 after tensile test, $T = 220$ °C, strain rate 3×10^{-3} s^{-1}.

4.6 Conclusions

The focus in the present chapter has been on the effect of grain refinement by severe plastic deformation on superplastic properties of Mg alloys, as exemplified by alloys ZK60 and AZ31. It has been shown that the grain structure produced by ECAP is conducive for establishing superior superplastic characteristics. The most favourable ECAP processing conditions leading to retention of a population of large non-recrystallised grains embedded in a distribution of recrystallised fine grains were found. Achieving such structures was possible due to the use of back pressure during ECAP processing, which permitted reduction of the processing temperature. The ensuing incomplete dynamic recrystallisation giving rise to a bi-modal grain structure turned out to result in excellent superplastic properties. An alternative, two-step rolling + ECAP process was seen to lead to superplasticity of ZK60 and AZ31 for a uniform fine-grained structure, without bi-modality. However, the materials with a bi-modal (or tri-modal) grain structure produced by low-temperature ECAP with back-pressure showed significantly better superplastic properties.

A phenomenological interpretation of delayed tensile failure appears to be possible on the basis of Hart's instability criterion, which relates the resistance of a material against strain localisation to the strain hardening capability and the strain rate sensitivity of the flow stress. Fractography studies by SEM provided deeper insights into the nature of post-localisation deformation leading to failure. Cavitation at interfaces between the matrix and coarse incoherent $Mg_{17}Al_{12}$ particles in alloy AZ31 was identified as a critical process that limits strain-to-failure. Indeed, the ligaments between the voids were found to still possess a substantial resource of ductility. This means that by suppressing decohesion between the particles and the matrix even higher superplastic ductility levels could be achieved in principle.

As a general conclusion, it can be stated that processing by the severe plastic deformation technique of equal-channel angular pressing has a great potential in the area of superplastic forming of magnesium alloys.

4.7 References

1. C.E. Pearson, The viscous properties of extruded eutectic alloys of Lead-Tin and Bismuth-Tin, *Journal Institute of Metals*, 1934, **54**: 111-123.
2. T.G. Nieh, J. Wadsworth, and O.D. Sherby, *Superplasticity in Metals and Ceramics*, Cambridge University Press, 1997, p. 273.
3. J. Pilling and N. Ridley, *Superplasticity in Crystalline Solids*, USA Institute of Metals, 1989, p. 214.
4. T.G. Langdon, Seventy-five years of superplasticity: Historic developments and new opportunities, *Journal of Material Science*, 2009, **44**: 22, 5998–6010.
5. K.A. Padmanabhan, R.A. Vasin and F.U. Enikeev, *Superplastic Flow: Phenomenology and Mechanics*, Springer-Verlag, Berlin, Heidelberg, 2001, p. 363.
6. R.Z. Valiev, Y Estrin, Z. Horita, T.G. Langdon, M.J. Zehetbauer *et al.*, Producing bulk ultrafine-grained materials by severe plastic deformation, *JOM*, 2006, **58**:4, 33–39.

7. David J Lewis and Associates Pty Ltd, *Magnesium: Opportunities in Australia*, Industry Science Resources, Canberra, Dept. of Industry, Science and Resources, 1999, p. 121.
8. F.H. Frost, D. Eliezer and E. Aghion, The science, technology and applications of magnesium, *JOM*, 1998, **50**:9, 30–34.
9. L. Gaines, R. Cuenca, F. Stodolsky and S. Wu, *Potential Automotive Uses of Wrought Magnesium Alloys*, Technical Paper Argon National Laboratory, 1996, p. 130.
10. E. Colasanti, Design and development of a lightweight seat frame using magnesium extrusions and stampings, *SAE, Journal of Materials and Manufacturing*, 1995, **103**:5, 206–212.
11. E. Aghion and B. Bronfin, Magnesium alloys development towards the 21st century, *Materials Science Forum*, 2000, **350**: 19–28.
12. W.A. Backofen, G.S. Murty and S.W. Zehr, Evidence for diffusional creep with low strain rate sensitivity, *Metallurgical Society of American Institute of Mining, Metallurgical and Petroleum Engineers – Transactions*, 1968, **242**:2, 329–331.
13. A.U. Karim and W.A. Backofen, Grain-size dependence of strain-rate hardening behaviour in a Mg-Zn-Zr alloy, *Material Science and Engineering*, 1969, **3**:5, 306–307.
14. A.U. Karim and W.A. Backofen, Some observations of diffusional flow in a superplastic alloy, *Materials Transactions*, 1972, **3**:3, 709–712.
15. Y. Xiao, X.-M. Zhang, J.-M. Chen and H. Jiang, Microstructures and mechanical properties of extruded Mg-9Gd-4Y-0.6Zr-T5 at elevated temperatures, *Chinese Journal of Nonferrous Metals*, 2006, **16**:4, 709–714.
16. X. Liu, R. Chen and E. Han, High temperature deformations of Mg-Y-Nd alloys fabricated by different routes, *Materials Science and Engineering A*, 2008, **497**:1–2, 326–32.
17. H. Watanabe, T. Mukai, K. Ishikawa, T. Mohri, M. Mabuchi et al., Superplasticity of a particle-strengthened WE43 magnesium alloy, *Materials Transactions*, 2001, **42**:1, 157–162.
18. S.R. Agnew, G.M. Stoica, L.J. Chen, T.M. Lillo, J. Macheret et al., Equal channel angular processing of magnesium alloys, *Ultrafine Grained Materials II. Proceedings. TMS Annual Meeting*, 2002, 643–652.
19. K.P. Park, M.J. Birt, and K.J.A. Mawella, Near net shape magnesium alloy components by superplastic forming and thixoforming, *Advanced Performance Materials*, 1996, **3**:3–4, 365–375.
20. M.M. Tilman and L.A. Neumeier, *Superplasticity in commercial and experimental compositions of magnesium alloy sheet*, Bur. Mines, Washington, DC, USA, 1982, p. 29.
21. T. Chen and M.L. Mecartney, Comparison of the high-temperature deformation of alumina-zirconia and alumina-zirconia-mullite composites, *Journal of Materials Research*, 2005, **20**:1, 13–17.
22. W.J. Kim and Y.G. Lee, Enhanced superplasticity of 1 wt.%Ca-AZ80 Mg alloy with ultrafine grains, *Materials Letters*, 2010, **64**:16, 1759–1762.
23. Y. Wang and J.C. Huang, Superplasticity enhanced by two-stage deformation in a hot-extruded AZ61 magnesium alloy, *Journal of Materials Science and Technology*, 2005, **21**:1, 71–74.
24. T.G. Langdon, Y. Miyahara and Z. Horita, Exceptional superplasticity in an AZ61 magnesium alloy processed by extrusion and ECAP, *Materials Science and Engineering A*, 2006, **420**:1–2, 240–244.

25. W.J. Kim, S.W. Chung, C.W. An and K. Higashi, Superplasticity in a relatively coarse-grained AZ61 magnesium alloy, *Journal of Materials Science Letters*, 2001, **20**:17, 1635–1637.
26. Y. Yoshida, K. Arai, S. Itoh, S. Kamado and Y. Kojima, Superplastic deformation of AZ61 magnesium alloy having fine grains, *Materials Transactions*, 2004, **45**:8, 2537–2541.
27. Tsutsui, H. Watanabe, T. Mukai, M. Kohzu, S. Tanabe *et al.*, Superplastic deformation behaviour in commercial magnesium alloy AZ61, *Materials Transactions, JIM*, 1999, **40**:9, 931–934.
28. M.T. Pérez-Prado, J.A. Del Valle and O.A. Ruano, Superplastic behaviour of a fine grained AZ61 alloy processed by large strain hot rolling, *Materials Science Forum*, 2004, **447–448**: 221–226.
29. H. Watanabe, J.-K. Morinomiya, T. Mukai, K. Ishikawa, Y. Okanda *et al.*, Superplastic characteristics in an extruded AZ31 magnesium alloy, *Journal of Japan Institute of Light Metals*, 1999, **49**:8, 401–404 (in Japanese).
30. H.K. Lin, J.C. Huang and T.G. Langdon, Relationship between texture and low temperature superplasticity in an extruded AZ31 Mg alloy processed by ECAP, *Materials Science and Engineering A*, 2005, **402**:1–2, 250–257.
31. J. Xing, X. Yang, H. Miura and T. Sakai, Superplasticity of fine-grained magnesium alloy AZ31 processed by multi-directional forging, *Materials Transactions*, 2007, **48**:6, 1406–1411.
32. R.B. Figueiredo and T.G. Langdon, Developing superplasticity in a magnesium AZ31 alloy by ECAP, *Journal of Materials Science*, 2008, **43**:23–24, 7366–7371.
33. R. Lapovok, Y. Estrin, M.V. Popov and T.G. Langdon, Enhanced superplasticity in a magnesium alloy processed by equal-channel angular pressing with a back-pressure, *Advanced Engineering Materials*, 2008, **10**:5, 429–433.
34. L. Lin, L. Chen and Z. Liu, An investigation of low temperature superplasticity of ZK40 magnesium alloy subjected to equal channel angular pressing, *Materials Science Forum, Proceedings of the International Conference on Magnesium – Science, Technology and Applications*, 2005, **488–489**: 581–584.
35. L. Yang, X.M. Yang, T. Liu, S.D. Wu and L.J. Chen, Superplasticity of magnesium alloy ZK40 processed by equal channel angular pressing, *Materials Science Forum, Proceedings of the International Conference on Magnesium – Science, Technology and Applications*, 2005, **488–489**: 575–580.
36. H. Watanabe, T. Mukai and K. Higashi, Superplasticity in a ZK60 magnesium alloy at low temperatures, *Scripta Materialia*, 1999, **40**:4, 477–484.
37. A. Galiyev and R. Kaibyshev, Superplasticity in a magnesium alloy subjected to isothermal rolling, *Scripta Materialia*, 2004, **51**:2, 89–93.
38. A. Ben-Artzy, A. Shtechman, A. Bussiba, Y. Salah, S. Ifergan *et al.*, Low temperature super-plasticity response of AZ31B magnesium alloy with severe plastic deformation, *Magnesium Technology 2003 Symposium*, 2003, 259–263.
39. W.J. Kim, M.J. Kim and J.Y. Wang, Superplastic behaviour of a fine-grained ZK60 magnesium alloy processed by high-ratio differential speed rolling, *Materials Science and Engineering A*, 2009, **527**:1–2, 322–327.
40. R.B. Figueiredo and T.G. Langdon, Factors influencing superplastic behaviour in a magnesium ZK60 alloy processed by equal-channel angular pressing, *Materials Science and Engineering A*, 2009, **503**:1–2, 141–144.
41. R. Lapovok, R. Cottam, P.F. Thomson and Y. Estrin, Extraordinary superplastic ductility of magnesium alloy ZK60, *Journal of Materials Research*, 2005, **20**:6, 1375–1378.

42. H.J. McQueen, M. Myshlaev, M. Sauerborn and M. Mwembela, Flow stress, microstructures and modeling in hot extrusion of magnesium Alloys, In *Proceeding of the Conference: Magnesium Technology, 2000*, Eds. H.I. Kaplan, J. Hryn and B. Clow, The Mineral, Metals and Materials Society, 2000, pp. 355–362.
43. H.J. McQueen, A. Mwembela, E.V. Konopleva and M. Myshlaev, Hot working characteristics of Mg-2.8Al-0.9Zn, In *Proceeding of the Conference: Magnesium Alloys and their Applications*, Eds. B.L. Mordike and K.U. Kainer, 1998, Germany, 201–208.
44. A. Mwembela and H.J. McQueen, Hot workability of magnesium alloy ZK60, In *Proceeding of the Conference: Hot Workability of Steels and Light Alloys Composites'* Eds. H.J. McQueen, E.V. Konopleva and N.D. Ryan, Met. Soc., CIM, 1996, pp. 181–188.
45. R.B. Figueiredo and T.G. Langdon, Principles of grain refinement in magnesium alloys processed by equal-channel angular pressing, *Journal of Materials Science*, 2009, **44**: 4758–4762.
46. H.J. McQueen and D.L. Bourell, Hot workability of metals and alloys, *Journal of Metals*, 1987, **39**:9, 28–35.
47. A. Mwembela, E.V. Konopleva and H.J. McQueen, Microstructural Development in Mg Alloy AZ31 during Hot Working, *Scripta Materialia*, 1997, **37**:11, 1789–1795.
48. D.J. Lloyd and D.M. Moore, Aluminum alloys design for superplasticity, in *Proc. Symp. Superplastic Forming of Structural Alloys*, TMS, 1982, pp. 147–172.
49. K. Sekihara, S. Ohnishi, S. Kamado and Y. Kojima, Semi-solid forming of strain-induced AZ91D magnesium alloy, *J. Japan Inst. Light Metals*, 1995, **45**: 560–565.
50. T. Mohri, M. Mabuchi, N. Saito and M. Nakamura, Microstructure and mechanical properties of a Mg-4Y-3RE alloy processed by thermo-mechanical treatment, *J. Materials Science and Engineering A*, 1998, **257**:2, 287–294.
51. R.Z. Valiev, Y. Estrin, Z. Horita, T.G. Langdon, M.J. Zehetbauer *et al.*, Producing bulk ultrafine-grained materials by severe plastic deformation, *JOM*, **58**:4, 33–39.
52. L. Ceschini, L. Balloni, I. Boromei, M. El Mehtedi and A. Morri, Superplastic behaviour of the AZ31 magnesium alloy produced by twin roll casting, *Metallurgia Italiana*, 2007, **9**: 5–11.
53. Q.F. Wang, X.P. Xiao, X.J. Chen and W. Chen, Superplasticity in ultrafine grained magnesium alloy AZ31 prepared by accumulative roll bonding, *Materials Science Forum*, 2007, **551–552**: 249–254.
54. H. Akamatsu, T. Fujinami, Z. Horita and T.G. Langdon, Influence of rolling on the superplastic behavior of an Al-Mg-Se alloy after ECAP, *Scripta Materialia*, 2001, **44**:5, 759–764.
55. Z. Horita, K. Matsubara, K. Makii and T.G. Langdon, A two-step processing route for achieving a superplastic forming capability in dilute magnesium alloys, *Scripta Materialia*, 2002, **47**: 255–260.
56. R. Lapovok, P.F. Thomson, R. Cottam and Y. Estrin, Processing routes leading to superplastic behaviour of magnesium alloy ZK60, *Materials Science and Engineering A*, 2005, **410–411**: 390–393.
57. R.B. Figueiredo and T.G. Langdon, Record superplastic ductility in a magnesium alloy processed by equal-channel angular pressing, *Advanced Engineering Materials*, 2008, **10**:1–2, 37–40.
58. V.N. Chuvil'deev, T.G. Nieh, M.Y. Gryaznov, V.I. Kopylov and A.N. Sysoev, Superplasticity and internal friction in microcrystalline AZ91 and ZK60 magnesium alloys processed by equal-channel angular pressing, *Journal of Alloys and Compounds*, **378**:1–2 (2004), 253–257.

59. V.N. Chuvil'deev, V.I. Kopylov, M.Y. Gryaznov and A.N. Sysoev, Low temperature superplasticity of microcrystalline high-strength magnesium alloys processes by equate-channel angular pressing, *Doklady Akademii Nauk* (in Russian), 2003, **391**:1, 47–51.
60. H. Watanabe, T. Mukai, K. Ishikawa and K. Higashi, Superplastic behavior of an ECAE processed ZK60 magnesium alloy, *Materials Science Forum*, 2003, **419–422**:1, 557–562.
61. A. Galiyev, R. Kaibyshev and T. Sakai, Continuous dynamic recrystallization in magnesium alloy, *Materials Science Forum*, 2003, **419–422**:1, 509–514.
62. M. Nakanishi, M. Mabuchi, N. Saito, M Nakamura and K. Higashi, Tensile properties of the ZK60 magnesium alloy produced by hot extrusion of machined chip, *Journal of Materials Science Letters*, 1998, **17**:23, 2003–2005.
63. A. Bussiba, A. Ben Artzy, A. Shtechman, S. Ifergan and M. Kupiec, Grain refinement of AZ31 and ZK60 Mg alloys – Towards superplasticity studies, *Materials Science and Engineering A*, 2001, **302**:1, 56–62.
64. H.K. Lin and J.C. Huang, High strain rate and/or low temperature superplasticity in AZ31 Mg alloys processed by simple high-ratio extrusion methods, *Materials Transactions*, 2002, **43**:10, 2424–2432.
65. J.C. Tan and M.J. Tan, Superplasticity and grain boundary sliding characteristics in two stage deformation of Mg-3Al-1Zn alloy sheet, *Materials Science and Engineering A*, 2003, **339**:1–2, 81–89.
66. D.L. Yin, K.F. Zhang, G.F. Wang and W.B. Han, Superplasticity and cavitation in AZ31 Mg alloy at elevated temperatures, *Materials Letters*, 2005, **59**:14–15, 1714–1718.
67. R. Lapovok, Y. Estrin, M.V. Popov, S. Rundell and T. Williams, Enhanced superplasticity of magnesium alloy AZ31 obtained through ECAP with back-pressure, *J. Material Science*, 2008, **43**: 7372–7378.
68. J.W. Christian and S. Mahajan, Deformation twinning, *Progress in Materials Science*, 195, **39**: 1–157.
69. Y. Wang, M. Chen, F. Zhou and E. Ma, High tensile ductility in a nanostructured material, *Nature*, 2002, **419**: 912–915.
70. R.E. Reed-Hill, Role of deformation twinning in the plastic deformation of a polycrystalline anisotropic metal, *Deformation Twinning, Proc. Conf.*, R.E. Reed-Hill, J.P. Hirth and H.C. Rogers (eds), Gainesville, Florida, 21–22 March, 1963, Gordon and Breach, New York, pp. 295–320.
71. M.H. Yoo, Slip, twinning and fracture in hexagonal close-packed metals, *Materials Transactions A*, 1981, **12A**: 409–417.
72. F. Kocks and D.G. Westlake, *Trans. TMS – AIME*, 1967, **239**: 1107–1109.
73. J. Friedel, *Dislocations*, Pergamon, Oxford, UK, 1964, p.178.
74. Yugang, M., Mabuchi, M., Shimojima, K., Yamada, Y., Wen, C.E. et al., High strength and high strain rate superplasticity in magnesium alloys, *Materials Science Forum*, Trans. Tech Publications, Switzerland, 2001, **357–359**: 327–332.
75. J. Koike, Dislocation plasticity and complementary deformation mechanisms in polycrystalline Mg alloys, *Materials Science Forum*, 2004, **449–452**: 665–668.
76. J. Koike, Enhanced deformation mechanisms by anisotropic plasticity in polycrystalline Mg Alloys at room temperature, *Metallurgical and Materials Transactions*, 2005, **36A**: 1689–1696.
77. R. Lapovok, P.F. Thomson, R. Cottam and Y. Estrin, The effect of grain refinement by warm equal channel angular extrusion on room temperature twinning in magnesium alloy ZK60, *Journal of Materials Science*, 2005, **40**:7, 1699–1708.

78. R. Lapovok, P.F. Thomson, R. Cottam and Y. Estrin, The effect of warm equal channel angular extrusion on ductility and twinning in magnesium alloy ZK60, *Materials Transactions*, 2004, **45**:7 2192–2199.
79. O. Sitdikov, R. Kaibyshev and T. Sakai, Dynamic recrystallisation based on twinning in coarse-grained Mg, *Materials Science Forum*, Trans. Tech Publications, Switzerland, 2003, **419–422**: 521–526.
80. R. Lapovok, The role of Back-Pressure in Equal Channel Angular Extrusion, *Journal of Materials Science*, 2005, **40**:2: 341–346.
81. F. Abu-Farha, M. Nazzal and R. Curtis, Optimum specimen geometry for accurate tensile testing of superplastic metallic materials, *Experimental Mechanics*, 2010, 1–15 (published on line 6 August 2010).
82. M.A. Khaleel, K.I. Johnson, C.A. Lavender, M.T. Smith and C.H. Hamilton, Specimen geometry effect on the accuracy of constitutive relations in a superplastic 5083 Aluminum alloy, *Scripta Materialia*, 1996, **34**:9, 1417–1423.
83. P.N. Comley, ASTM E2448 – a unified test for determining SPF properties, *Journal of Materials Engineering and Performance*, 2008, **17**:2, 183–186.
84. L. S. Tóth, Y. Estrin, R. Lapovok and C. Gu, A model of grain fragmentation based on lattice curvature, *Acta Materialia*, 2010, **58**:5, 1782–1794.
85. E.W. Hart, Theory of the tensile test, *Acta Metallurgica*, 1967, **15**:2, 351–355.
86. R.Z. Valiev, I.V. Alexandrov, Y.T. Zhu and T.C. Lowe, Paradox of strength and ductility in metals processed by severe plastic deformation, *J. Material Research*, 2002, **17**: 5–8.
87. R. Raj and M.F. Ashby, Intergranular fracture at elevated temperature, *Acta Metallurgica*, 1975, **23**:6, 653–666.
88. A.H. Chokshi, Cavity nucleation and growth in superplasticity, *Materials Science and Engineering A*, 2005, **410–411**, 95–99.
89. R. Raj, Nucleation of cavities at second phase particles in grain boundaries, *Acta Metallurgica*, 1978, **26**:6, 995–1006.
90. H. Somekawa and T. Mukai, Effect of dominant diffusion process on cavitation behaviour in superplastic Mg-Al-Zn alloy, *Scripta Materialia*, 2007, **57**: 1008–1011.
91. A. Mussi, J.J. Blandin, L. Salvo and E.F. Rauch, Resistance to strain-induced damage of an ultrafine-grained magnesium alloy deformed in superplastic conditions, *Acta Materialia*, 2006, **54**: 3801–3809.
92. A.H. Chokshi and T.G. Langdon, A model for diffusional cavity groth in superplasticity, *Acta Metallurgica*, 1987, **35**:5, 1089–1101.
93. C.J. Lee and J.C. Huang, Cavitation characteristics in AZ31 Mg alloy during LTSP or HSRSP, *Acta Materialia*, 2004, **52**: 3111–3122.
94. R. Lapovok, T. Williams and Y. Estrin, Superplastic failure mode in ultrafine grained magnesium alloy International, *Journal of Materials Research*, 2009, **100**:4, 609–613.
95. T. Mukai, M. Yamanoi, H. Watanabe and K. Higashi, Ductility enhancement in AZ31 magnesium alloy by controlling its grain structure, *Scripta Materialia*, 2001, **45**:1, 89–94.
96. S. Suwas, G. Gottstein and R. Kumar, Evolution of crystallographic texture during equal channel angular extrusion (ECAE) and its effects on secondary processing of magnesium, *Materials Science and Engineering A*, 2007, **471**:1–2, 1–14.
97. Y. Estrin, S.B. Yi, H.-G. Brokmeier, Z. Zúberová, S.C. Yoon, H.S. Kim *et al.*, Microstructure, texture and mechanical properties of the magnesium alloy AZ31 processed by ECAP, *International Journal of Materials Research*, 2008, **99**:1, 50–55.
98. L.E. Miller and G.C. Smith, Tensile fractures in carbon steels, *Journal of Iron Steel Institute (London)*, 1970, **208**:11, 998–1005.

99. Z. Zúberová, I. Sabirov and Y. Estrin, The effect of deformation processing on tensile ductility of magnesium alloy AZ31, *Kovove Materialy*, 2011, **49**: 29–36.
100. A. Mukherjee, Plastic deformation and fracture of materials, in: R.W. Cahn, P. Haasen and E.J. Kramer, Eds., *Materials Science and Technology*, Vol. 6, Weinheim, NY, 1993, pp. 407–460.
101. R. Valiev, R. Islamgaliev, I. Semenova and N. Yunusova, New trends in superplasticity in SPD-processed nanostructured materials, *International Journal of Materials Research*, 2007, **98**:4, 314–319.
102. H.F. Frost and M.F. Ashby, *Deformation-Mechanism Maps, The Plasticity and Creep of Metals and Ceramics*, 1982, Pergamon Press, pp. 166.
103. Y. Ma, M. Furukawa, Z. Horita, M. Nemoto, R.Z. Valiev and T.G. Langdon, Significance of microstructural control for superplastic deformation and forming, *Materials Transactions, JIM*, 1996, **37**:3, 336–339.
104. R.Z. Valiev, D.A. Salimonenko, N.K. Tsenev, P.B. Berbon and T.G. Langdon, Observations of high strain rate superplasticity in commercial aluminum alloys with ultrafine grain sizes, *Scripta Materialia*, 1997, **37**:12, 1945–1950.

5
Dynamic recrystallization in magnesium alloys

R. KAIBYSHEV, Belgorod State University, Russia

Abstract: This chapter discusses grain refinement in magnesium and its alloys through dynamic recrystallization (DRX). It is shown that DRX in magnesium alloys occurs in a highly specific manner due to the strong relationship between DRX mechanisms and deformation mechanisms. The effect of chemical composition on DRX in magnesium alloys is also attributed to this relationship. Three types of DRX mechanisms operating in magnesium and its alloys are considered in detail. The necessity to use intense plastic straining (IPS) techniques to produce submicrometer scale grains in magnesium alloys is not obvious due to the high rate of DRX even at low temperatures.

Key words: magnesium, dynamic recrystallization, deformation mechanisms, intense plastic straining.

5.1 Introduction

Magnesium alloys exhibit poor workability at ambient temperature due to the fact that basal slip and twinning are dominant deformation mechanisms. As a result, elongation-to-failure of low alloy magnesium at room temperature usually does not exceed 15% (Pérez-Prado et al., 2005, Sakai and Miura, 2011). Ductility of magnesium alloys containing high amounts of alloying elements and exhibiting relatively high stress is significantly less. In practice, useful improvements in ductility and toughness of magnesium alloys can be attained by extensive grain refinement (Somekawa et al., 2009, Singh et al., 2011, Xu et al., 2009a, Kang et al., 2010, Wang et al., 2008). In addition, strength at ambient temperature can be improved significantly through grain refinement in accordance with the Hall–Petch relationship. No phase transformation of the magnesium matrix takes place. As a result, fine grain sizes can only be produced through recrystallization processes.

Static recrystallization (SRX) is widely used in the rolling of magnesium alloys to provide softening between passes. Numerous low reduction (<15%) rolling passes, each followed by a recrystallization anneal, are used to produce thin sheets from low alloy magnesium billets. The use of repetitive rolling passes compensates for the low workability of magnesium alloys at room temperature. Unfortunately, in order to produce a uniform microstructure the total imposed strain should be higher than 40% (Somekawa et al., 2009). In addition, SRX can only provide grain sizes ranging from 8–25 μm. For significant enhancement of mechanical properties in magnesium alloys, grain refinement to the micron or submicron level is required (Somekawa et al., 2009, Singh et al., 2011).

Therefore, extensive grain refinement in magnesium alloys in commercial practice is better attained through the occurrence of dynamic recrystallization (DRX). Thermomechanical processing of magnesium alloys is frequently carried out at intermediate and/or high temperatures. Under these conditions, prismatic and pyramidal slip systems are activated and the ductility is raised. As a result, at higher temperatures, magnesium alloys have sufficient plasticity to be processed into an ultra-fine grained (UFG) state. This can be achieved using conventional deformation techniques.

It should be noted that DRX in magnesium alloys occurs in a highly specific manner due to the strong relationship between DRX mechanisms and deformation mechanisms (Xu et al., 2009a). In this regard magnesium is a unique material in which it is possible to observe a range of 'exotic' forms of DRX.

5.2 Dynamic recrystallization (DRX) mechanisms operating in magnesium alloys

There exist numerous works that deal with studies of the DRX mechanisms in magnesium and its alloys (Kang et al., 2010, Humphreys and Hatherly, 2004, Ion et al., 1982, Galiyev et al., 2001, Sitdikov and Kaibyshev, 2001, Galiyev et al., 2003a, Sitdikov et al., 2003, Sun et al., 2010, Watanabe et al., 2009, Xu et al., 2009b, Al-Samman and Gottstein, 2008, Ravi Kumar et al., 2003, Beer and Barnett, 2006, 2007, 2008, Martin and Jonas, 2010, Xu et al., 2010, Valle and Ruano, 2008). Like any recrystallization process, DRX proceeds by nucleation and growth (Humphreys and Hatherly, 2004). Inspection of experimental data shows that usually in magnesium and magnesium alloys, nucleation is the slowest process and, therefore, the rate-controlling one (Sivakesavam et al., 1993). For example, if nucleation occurs due to dislocation rearrangement, the rate-controlling process for dislocation rearrangement is the rate-controlling process for nucleation. Numerous deformation mechanisms are operative in magnesium and its alloys and these include basal and pyramidal slip (1st and 2nd order), cross-slip, dislocation climb and twinning. Each mechanism makes its own contribution to the nucleation process and, as a result, different DRX mechanisms operate under different conditions (Galiyev et al., 2001, Kaibyshev and Sitdikov, 2000, Galiyev and Kaibyshev, 2001, Martin and Jonas, 2010).

One important factor involved in the DRX of magnesium and its alloys is that there exists a strong temperature dependence of the active deformation mechanisms (Galiyev et al., 2001, Kaibyshev and Sitdikov, 2000, Martin and Jonas, 2010). Thus, in contrast with cubic metals (Kang et al., 2010), different DRX nucleation mechanisms are found to be active during hot, warm and even cold deformation (Xu et al., 2009a, Galiyev et al., 2001, 2003a, Galiyev and Kaibyshev, 2001, Sitdikov and Kaibyshev, 2001, Sitdikov et al., 2003, Sun et al., 2010, Watanabe et al., 2009, Xu et al., 2009b, Al-Samman and Gottstein, 2008, Ravi Kumar et al., 2003, Beer and Barnett, 2006, 2007, 2008, Martin and Jonas,

2010, Xu *et al.*, 2010, Valle and Ruano, 2008, Sivakesavam *et al.*, 1993). Three main DRX mechanisms are now understood to be operative in magnesium and its alloys:

- Continuous DRX (CDRX), which includes the formation of stable three-dimensional arrays of deformation low-angle boundaries (LABs) followed by their gradual transformation into high-angle grain boundaries (HABs) upon straining (Sivakesavam *et al.*, 1993). New grains are formed progressively within the deformed original grains from the continuous increase in misorientation across deformation-induced boundaries (Humphreys and Hatherly, 2004, Beer and Barnett, 2007, Xia *et al.*, 2005).
- Discontinuous DRX (DDRX), which involves the development of HABs via the nucleation and growth of new grains. Nuclei evolve on original HABs due to the operation of a bulging mechanism (Sitdikov and Kaibyshev, 2001, Kaibyshev and Sitdikov, 1994a, Sun *et al.*, 2010, Beer and Barnett, 2007). DDRX usually occurs in materials with relatively low stacking fault energies (Humphreys and Hatherly, 2004, Kaibyshev and Sitdikov, 1994a, Beer and Barnett, 2007). The local migration, i.e. bulging, of grain boundaries leads to the formation of nuclei, which then grow out and consume a deformed matrix, resulting in decreased dislocation density, and providing strain softening. Thus this mechanism involves the development of high-angle grain boundaries via the nucleation and growth of new grains. It is closely related to strain-induced migration of initial boundaries. Local migration of boundaries of large DRX grains resulting in repetitive DDRX is rarely observed.
- DRX mechanism associated with twinning (TDRX), in which twinning leads to the formation of coarse lamellae surrounded by special grain boundaries (Muránskya *et al.*, 2008). There are at least three processes by which this can occur: Mutual intersection of primary twins, the occurrence of secondary twinning within the coarse lamella, and coarse twin lamellae can be subdivided by deformation-induced LABs that transform into conventional HABs upon further straining and provide chains of DRX grains (Kaibyshev and Sitdikov, 2000, 1994b, Sitdikov and Kaibyshev, 2001, Sitdikov *et al.*, 2003).

It is worth noting that CDRX is a recovery process and proceeds by continuous accumulation of lattice dislocations in LABs without change in the sub-grain size (Humphreys and Hatherly, 2004). As a result, these accumulated dislocations take no part in strain hardening. Numerous authors consider this mechanism as being a strong recovery process (Al-Samman and Gottstein, 2008) rather than a classical recrystallization phenomenon. In contrast, DDRX is characterized by nucleation and growth by migration of HABs. This process provides absorption of lattice dislocations by migrating HABs. As a result, these two mechanisms affect differently the mechanical behavior (Beer and Barnett, 2007). The TDRX mechanism is considered here to fall between these two extremes. In what follows, we consider these mechanisms in more detail.

5.2.1 CDRX in magnesium and magnesium alloys

In magnesium alloys, CDRX is strongly dependent on the operative deformation mechanisms (Galiyev et al., 2001). Non-basal and (**a+c**) slip play an important role due to thermal activation. The formation of three-dimensional (3D) arrays of LABs in the vicinity of initial boundaries requires mutual reactions of dislocations of two different Burgers vectors (Humphreys and Hatherly, 2004) followed by a rearrangement of the dislocations. Therefore, it is obvious that (**a+c**) dislocations are required in addition to (**a**) dislocations for the formation of 3D recrystallization nuclei and CDRX occurrence. In magnesium alloys the (**a**) dislocations can glide in basal, prismatic and first order pyramidal planes. However, the generation of (**a+c**) dislocations in second order pyramidal planes is a necessary condition for the formation of the 3D nuclei (Galiyev et al., 2001). The Burgers vector of (**a+c**) dislocations is significantly larger than that for (**a**) dislocations. As a result, the critical resolved shear stresses (CRSS) for non-basal slip are significantly higher that that for basal slip (Biswas et al., 2010). The CRSS values for (**a**) first order pyramidal glide and (**a+c**) second-order pyramidal glide are higher by factors of 8 and 6, respectively, than the value for (**a**) basal glide for pure magnesium at 250 °C. However, the difference in CRSS values tends to decrease with increasing temperature (Beer and Barnett, 2006). That is, the generation of (**a+c**) dislocations requires significant thermal activation and can be highly facilitated by an increase in deformation temperature. Higher temperatures also facilitate (**a**) dislocation glide on non-basal planes.

It was recently shown (Muránsky et al., 2008, Koike et al., 2003) that three types of dislocation glide are operative in magnesium at ambient temperatures: Slip of (**a**) dislocations on the $\{0001\}$ basal planes; slip of (**a**) dislocations on the first-order prismatic planes; slip of (**a+c**) dislocations on the second-order pyramidal planes. Therefore, the necessary prerequisite slip conditions for CDRX occurrence are fulfilled. However, initial structure, chemical and phase composition, and deformation rate can affect the relative contributions of these three types of dislocation glide to total strain and, therefore, the rate of CDRX and its contribution to overall recrystallization process.

The formation of LABs occurs by rearrangement of the (**a**) lattice dislocations forming pile-ups in the vicinity of initial boundaries (Galiyev et al., 2001). It is known (Couret et al., 1991) that the value of the stacking fault energy in a non-basal plane (SFE~125–145 mJ m^{-2}) is higher by a factor of about four in comparison with the value for basal plane stacking faults (SFE = 36 mJ m^{-2}). Some authors (Xia et al., 2005) have reported even higher values of SFE for non-basal planes. As a result, edge and screw (**a**) dislocations lying in non-basal planes can readily climb and rearrange themselves by cross-slip. In contrast, (**a**) dislocations lying in a basal plane can arrange only by cross-slip mechanisms that require significant thermal activation. Thus, the formation of arrays of LABs is strongly dependent on temperature; initiation of non-basal slip is a prerequisite condition for CDRX occurrence. Therefore, the initial orientation plays an important role in initiation of CDRX, crystallographic planes should be tailored such that the Schmid factor is maximized to facilitate non-basal slip. This is why there exists a strong effect of initial texture on DRX in magnesium alloys (Kaibyshev and Sokolov, 1992, Kaibyshev et al., 1994b). This phenomenon will be considered below.

All features of CDRX and the recrystallized structure originating from this mechanism of DRX can be linked to the non-basal slip operating in magnesium and its alloys. The shear stress for non-basal slip is significantly higher than that for basal slip. As a result, initiation of non-basal slip takes place near initial boundaries where extensive accumulation of lattice dislocation occurs (Galiyev et al., 2001, Zaripov et al., 1987). Due to strain incompatibility, the local internal stress in the vicinity of initial grain boundaries exceeds the critical resolved shear stress for non-basal slip. Thus non-basal slip is initiated near the grain boundaries. This is why CDRX occurs in the vicinity of initial boundaries and results in the formation of recrystallized 'mantle'.

The ratio between the contributions of basal and non-basal slip to total strain controls the shape of the subgrains evolved in the first stage of CDRX. If basal slip is dominant, the formation of lamellar arrays of LABs takes place (see Fig. 5.1a) (Sitdikov and Kaibyshev, 2002). This is typical for materials with restricted numbers of slip systems. The formation of this layered structure can be interpreted in terms of microband (MB) formation (Martin and Jonas, 2010). The formation of the extended longitudinal boundaries with low-to-moderate misorientation is attributed to the collection of basal dislocations in these boundaries. The boundaries are relatively long, straight and perpendicular to the loading axis (Martin and Jonas, 2010). They delineate MBs and transform into HABs with a high rate during deformation due to high density of accumulated (**a**) dislocations lying in a basal plane. Some authors (Martin and Jonas, 2010) consider these boundaries to be geometrically necessity boundaries (GNB). The formation of transverse LABs within MBs subdivides the lamellar structure into separate subgrains and leads to the evolution of 3D arrays of LABs, which facilitate CDRX. (Sub)grains bounded partly by LABs (transverse boundaries) and partly by HABs (longitudinal boundaries) are evolved (see Fig. 5.1b). This (sub)grain represents a small volume of relatively perfect material, which is at least partly bounded by a HAB (see Fig. 5.1b) and can be considered to be a recrystallization nucleus (Kaibyshev et al., 2005). These crystallites acquire equiaxed shapes due to the simple transition of triple junctions of LABs with contact angles of ~90° to HABs with triple junction angles of ~120°. Dislocation glide occurs within these (sub) grain interiors (see Fig. 5.1c). Mobile dislocations move across the (sub)grains and are trapped by LABs resulting in an increase in their misorientation (see Fig. 5.1c). As a result, LABs eventually convert to true HABs. Therefore, the formation of microbands followed by their subdivision by transverse LABs is a specific feature of magnesium and its alloys caused by the dominance of basal slip.

5.1 Schematic representation of CDRX mechanism with basal slip being predominant. (a) The formation of lamellar arrays of subgrains. (b) A nucleus. (c) Interaction of low-angle boundaries with lattice dislocations resulting in progressive increase in their misorientation.

(*Continued*)

(c)

5.1 Continued.

If the contribution of non-basal slip and basal slip to the total strain is essentially the same, the formation of subgrains having an equiaxed shape takes place initially (see Fig. 5.2a). Extensive trapping of lattice dislocation by stable LABs takes place during deformation (see Fig. 5.2b). These boundaries consequently increase their misorientation under strain. The critical condition for the high rate of this process is a low mobility of deformation-induced LABs. The latter is required to hinder the collision of LABs consisting of dislocations of opposite Burgers vector, which would cause mutual annihilation (Biberger and Blum, 1992). If arrays of LABs are stable, the major process of increasing misorientation of deformation-induced boundaries is trapping of lattice dislocations (Kaibyshev *et al.*, 2005). 3D arrays of sub-boundaries can be stabilized by the presence of secondary phase particles, which effectively pin LABs in magnesium alloys containing nano-scale dispersoids, or low temperatures suppressing migration of LABs. In the latter case a strain-induced continuous reaction, which will be considered below in detail, providing the formation of grains with size less than 1 μm, is observed in pure Mg at ambient temperature (Kaibyshev and Sitdikov, 1994b, Edalati *et al.*, 2011, Biswas *et al.*, 2010). Stability of arrays of LABs is an important requirement for CDRX (Kaibyshev *et al.*, 2005, Dudova *et al.*, 2010).

Initially, the boundary between a subgrain and one of its neighbors may acquire a high-angle misorientation. However, the transformation of a nucleus into a grain occurs by the conversion of all its LAB facets into HABs. In addition to the process depicted in Fig. 5.1c, the rate of transformation of LABs to HABs is

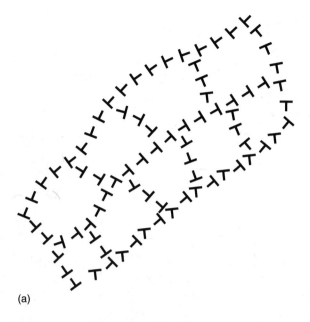

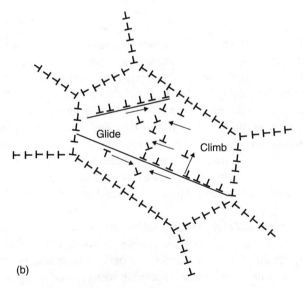

5.2 Schematic representation of CDRX mechanism operating with balanced contribution of basal and non-basal slips to total strain. (a) The formation of arrays of subgrains having equiaxed shape. (b) Interaction of low-angle boundaries with lattice dislocations resulting in progressive increase in their misorientation. (c) Rotation of (sub)grains facilitates the transformation of low-angle boundaries into high-angle boundaries.

(*Continued*)

5.2 Continued.

accelerated by the operation of grain boundary sliding (GBS) (Kaibyshev et al., 2005, 2002). GBS starts to operate along the isolated segments of HABs. However, GBS is not restricted to HAB segments, but can also occur on adjacent LABs. Continuity between the sliding of LABs is achieved by the generation and absorption of dislocations (Dougherty et al., 2003); GBS occurring along separate segments of HABs in partially recrystallized (see Fig. 5.2c) and layered (see Fig. 5.1) structures is accommodated by the rotation of the LABs attached to them. Extensive (sub)grain rotation highly accelerates the increase in the misorientation of LABs with strain (Kaibyshev et al., 2005, 2002). In addition, grains retain their equiaxed shape despite extensive dislocation glide within their interiors.

A necklace of dynamically recrystallized grains can also evolve on former twin boundaries. In contrast with the TDRX mechanism, twinning in this case simply provides additional HABs that act like grain boundaries (Xu et al., 2009a). Twin boundaries can originate from either annealing twins or deformation twins (Kaibyshev and Sokolov, 1992, Kaibyshev et al., 1994b, 1999, Kaibyshev and Sitdikov, 2000, Sitdikov and Kaibyshev, 2001, Sun et al., 2010, Watanabe et al., 2009, Xu et al., 2009b, Al-Samman and Gottstein, 2008). Twin boundaries transform into boundaries of common type due the formation of walls of mismatch dislocations during dislocation movement through these boundaries (Kaibyshev and Sitdikov, 2000, Sitdikov et al., 2003). There is not a great difference between the role of conventional HABs and former twin boundaries in CDRX.

The necklaces of CDRX grains decorate the pre-existing grains, and the CDRX grains evolved in vicinity of former twin boundaries decorate twins.

CDRX serves as a restoration mechanism. Initial strain hardening leads to the accumulation of lattice dislocations in the vicinity of initial boundaries (Zaripov et al., 1987). Strain softening after the peak stress may be due to the first stage of CDRX. However, it is very difficult to reveal its exact role in mechanical behavior due to the fact that usually several mechanisms of DRX are operative during plastic deformation of Mg and its alloys (Beer and Barnett, 2006, 2007, Martin and Jonas, 2010). In addition, the formation of MBs and extensive twinning can provide significant texture softening by re-orientation of crystal lattice from hard slip to softer slip directions. Thus, numerous factors can be responsible for the flow softening. It is also known (Humphreys and Hatherly, 2004, Beer and Barnett, 2007) that CDRX involves very little migration of both LABs and HABs; though deformation-induced LABs and HABs do serve as dislocation sinks. As mentioned above, the lattice dislocations trapped by LABs increase their misorientation with strain and give no contribution to strain hardening; lattice dislocations trapped by HABs are eliminated by absorption (Valiev et al., 1984).

We can consider the effect of temperature on σ–ε curves using Fig. 5.3 for a ZK60A alloy. It is seen that at $T \leq 350$ °C a well-defined peak stress is observed. The flow stress rises to a maximum at strains ranging from 15–22%; strain softening then occurs and finally a steady state is attained. At high temperatures, the steady state is attained after small strain; no peak stress is observed. At these temperatures, the major contribution to strain softening is provided by texture softening. For instance, orientation of MBs (see Fig. 5.1a) may change due to rotation of basal planes. As a result of such rotations, the basal planes within the MBs become more favorably oriented for glide, leading to geometric softening (Martin and Jonas, 2010). These rotations are attributed to glide of the basal dislocations within the MB interiors (see Fig. 5.1b). At high temperatures, CDRX provides equilibrium between the number of dislocations emitted by dislocation sources and the number of dislocations adsorbed by deformation-induced boundaries or mutually annihilated.

As LABs transform into HABs, a dynamic equilibrium is established between the number of emitted dislocations and dislocations trapped by deformation-induced boundaries. The LABs transforming into HABs provide an effective sink for mobile dislocations. As a result, no significant strain hardening takes place in regions adjacent to 3D arrays of deformation-induced boundaries. The grain mantle region becomes the soft region in which the localization of plastic deformation occurs (Beer and Barnett, 2006, 2007, Zaripov et al., 1987). The grain core becomes the hard region in which accumulation of lattice dislocations takes place.

As it was mentioned above, the formation of necklaces of recrystallized grains along initial boundaries leads to initiation of extensive GBS that highly accelerates

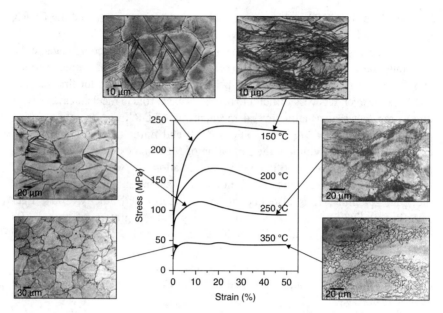

5.3 Flow curves of ZK60 alloy at various temperatures and an initial strain rate of 2.8×10^{-3} s^{-1} and deformation microstructures evolved.

the adsorption of lattice dislocations by grain boundaries (Kaibyshev *et al.*, 2002). As a result, the accumulation of lattice dislocations in the vicinity of the DRX grain necklaces slows and the rate of DRX becomes slower. The contribution of GBS to the total strain is strongly dependent on grain size. In magnesium alloys containing nanoscale dispersoids the grain size is sufficiently fine for the material to become superplastic, even after the recrystallized volume fraction attains 40% (Zaripov *et al.*, 1987, Zaripov and Kaibyshev, 1988). Extensive localization of plastic deformation into the grain mantle region takes place due to the fact that the flow stress for superplastic deformation (i.e. GBS) is significantly less than that for dislocation glide. In pure magnesium and magnesium alloys without nanoscale dispersoids the recrystallized grains are relatively coarse and there is no well-defined transition from hot deformation to superplastic deformation. In this instance, a strong localization of plastic deformation in the grain mantle region is observed.

There is a specific mechanism of CDRX that operates in magnesium and its alloys at temperatures between ambient and 200 °C (Galiyev and Kaibyshev, 2001, Sitdikov and Kaibyshev, 2001, Kaibyshev and Sitdikov, 1994b, Edalati *et al.*, 2011, Biswas *et al.*, 2010, Al-Samman and Gottstein, 2008, Martin and Jonas, 2010). This mechanism of extensive grain refinement termed low-temperature DRX (LTDRX) is not strictly a recrystallization process due to the

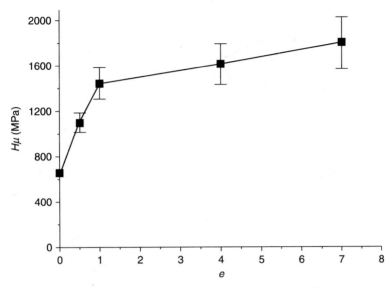

5.4 Strain dependence of microhardness for ZK60 alloy deformed at room temperature.

fact that it provides no restoration. The formation of new grains entirely delimited by HABs yields extensive strain hardening (see Fig. 5.4) (Galiyev and Kaibyshev, 2001, Sitdikov and Kaibyshev, 2001, Kaibyshev and Sitdikov, 1994b, Edalati *et al.*, 2011, Biswas *et al.*, 2010, Xia *et al.*, 2005, Jin *et al.*, 2005, Belyakov *et al.*, 2000a, Liu *et al.*, 2009). The extensive grain refinement during cold deformation is a strain-induced continuous reaction (Belyakov *et al.*, 2000a, 1998). Details of this continuous reaction are still unknown. It was established that it involves the evolution of a very high density of uniformly distributed dislocations that provide very high strain hardening (see Fig. 5.5a). The establishment of large long-range internal stress fields causes the initiation of non-basal slip (Muránskya *et al.*, 2008, Koike *et al.*, 2003) (see Fig. 5.5b). High-density dislocation pile-ups subdivide the interior of original grains into micro-regions in which different secondary dislocation systems operate (see Fig. 5.5c). Interaction of dislocations belonging to non-basal systems with basal pile-up dislocations leads to the formation of deformation-induced low-to-moderate angle boundaries (see Fig. 5.5b). In addition, low-to-moderate angle boundaries increase their misorientation by 'consuming' lattice dislocations. Both of these processes provide restoration. However, HABs are not susceptible for absorption of lattice dislocations at ambient temperature. As a result, lattice dislocations trapped by HABs are accumulated within HABs as grain boundary dislocations (Mabuchi *et al.*, 1999). HABs containing a very

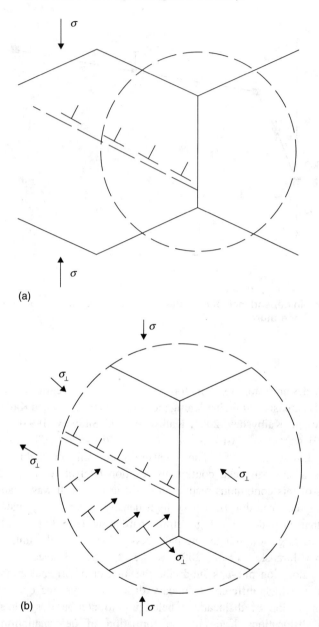

5.5 Schematic representation of a continuous reaction operating at low temperatures and resulting in the formation of submicrometer scale grains. (a) The formation of high-dense (**a**) dislocation pile-ups yielding high long-range stress fields in basal plane. (b) Initiation of (**a+c**) non-basal slip under high applied and internal stress fields. (c) The formation of dislocation-induced boundary due to operation of different slip systems within different areas.

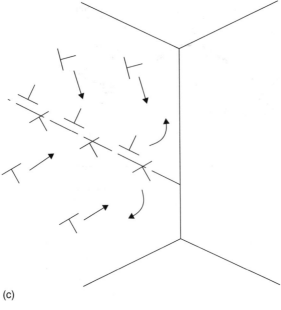

(c)

5.5 Continued.

high density of grain boundary dislocations produce large long-range elastic stress fields (Belyakov *et al.*, 2000b). This process is responsible for kinematic strain hardening.

5.2.2 DDRX in magnesium and magnesium alloys

The mechanism of DDRX is considered in the literature (Sitdikov and Kaibyshev, 2001, Kaibyshev and Sitdikov, 1994a, Sun *et al.*, 2010, Beer and Barnett, 2007, Barnett *et al.*, 2007a, b) and can be presented as follows (see Fig. 5.6). A recrystallized grain is first developed from a grain boundary (see Fig. 5.6a). Thus, bulging of grain boundaries is a prelude to DDRX (Humphreys and Hatherly, 2004), and is related to strain-induced grain boundary migration. Under DDRX conditions the initial boundary will migrate in a direction dictated by the forces acting on it. One force is provided by the stored energy difference over the boundary (see Fig. 5.6b) (Humphreys and Hatherly, 2004), and another arises by the interaction of lattice dislocations with the grain boundary. When lattice dislocations interact with a grain boundary, they will dissociate and be absorbed by the grain boundary. Before the dissociation process completes, a stress field builds up, which exerts a force on the grain boundary (see Fig. 5.6c). At high temperatures, rearrangement of lattice dislocations occurs easily within dislocation pile-ups and the force acting on the first dislocation trapped by a HAB (see Fig. 5.6c) becomes small. As a result,

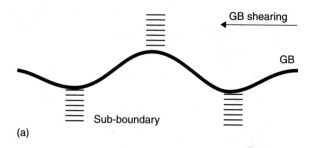
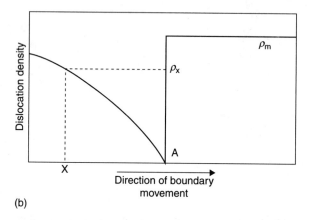
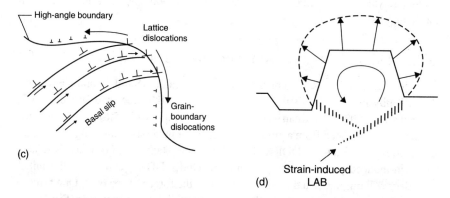

5.6 Schematic representation of DDRX. (a) The formation of LABs anchors local bulging. (b) Appearance of driving force for migration of HABs under deformation and local migration toward increased dislocation density. (c) The formation of dislocation pile-ups within bulging interiors promotes local migration toward decreased dislocation density. Dissociation of leading lattice dislocations trapped by the grain boundary interact with grain boundary dislocations. (d) Grain boundary sliding facilitates the formation of a LAB subdivided bulged area from a parent grain.

the boundary bulges towards the grain with the higher dislocation density, under the first force described above. At low temperatures, the second force is dominant due to the restricted rearrangement of lattice dislocations within pile-ups (see Fig. 5.6c). When a bulge is formed, it is still a part of the original grain (see Fig. 5.6a). The bulged configuration of the initial boundary is unstable and is usually anchored by the subgrains evolved perpendicular to the initial HAB (see Fig. 5.6a). In magnesium and its alloys these LABs usually consist of (**a**) dislocations located in basal planes.

After a grain boundary segment starts to bulge, a bridging dislocation wall forms and anchors the bulged grain boundary (see Fig. 5.6d). This stage is clearly demonstrated in the literature (Beer and Barnett, 2007, Yang *et al.*, 2003) (see Fig. 5.7). Yang *et al.* (2003) suggested that the places N_1, N_2 (see Fig. 5.7) were evolved by formation of kink bands following the grain boundary bulging. The

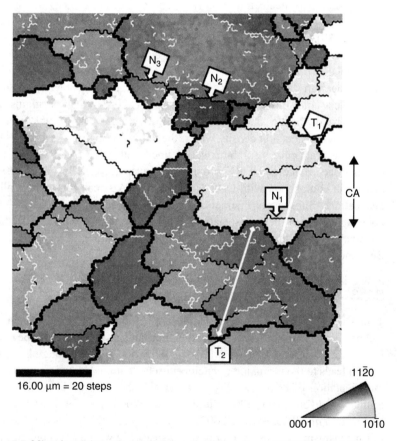

5.7 Misorientation map indicating the formation of DDRX nuclei (N_1, N_2, N_3) depicted in Fig. 5.6d, AZ31 strained at 400 °C up to 10%; CA = compression axis (Yang *et al.*, 2003).

formation of a DDRX nucleus can occur in the following sequence. The formation of LABs subdivides the bulged region from the 'parent' grain. This LAB becomes the border between the bulged region and the parent grain (Sun *et al.*, 2010); a DDRX nucleus evolves. This process is facilitated by GBS along the initial boundary. GBS exerts a local stress at the base of the bulge (see Fig. 5.6e). These extra elastic stresses are relieved by the emission of lattice dislocations from grain boundary sources. Next, the emitted dislocations form a transversal LAB by mechanisms described above for lamellar LABs. It is worth noting that, in general, this LAB can be formed by lattice dislocations belonging to one slip system. However, multiple slip highly facilitates the formation of a bridging dislocation wall across the bulge. During further deformation, the misorientation of this bridging wall gradually increases by trapping lattice dislocations (see Fig. 5.6e). It thereby transforms into a HAB, and a DRXed grain forms. Thus, the process that controls the rearrangement of lattice dislocations controls both deformation and nucleation (Galiyev *et al.*, 2001). Hence, the controlling deformation mechanism becomes the climb of non-basal dislocations (Galiyev *et al.*, 2003b).

As discussed above, DDRX grains grow and consume lattice dislocations. As a result, DDRX provides significant strain softening, especially at intermediate and low temperatures (Galiyev *et al.*, 2001, Sitdikov and Kaibyshev, 2001, Sun *et al.*, 2010, Beer and Barnett, 2007, Yang *et al.*, 2003). DDRX grains form a recrystallized mantle on initial grains providing a significant decrease in accumulated dislocation density within the deformed matrix. As a result, the true stress-true strain curves exhibit a well-defined peak stress (see Fig. 5.8). It seems that the occurrence of DDRX provides a higher decrease in lattice dislocation density in comparison with CDRX; migrating HABs can trap almost all lattice dislocations. Subsequent interaction of trapped lattice dislocations with a grain boundary leads to their dissociation followed by their absorption by the grain boundary. CDRX involves very little boundary migration (Galiyev *et al.*, 2001, Beer and Barnett, 2007).

5.2.3 TDRX in magnesium and magnesium alloys

TDRX is a specific DRX mechanism that was found to be unique to magnesium and its alloys. Three different nucleation mechanisms were found to be operative. First, mutual intersection of primary contraction twins belonging to $\{10\bar{1}1\}$ system leads to the formation of nuclei entirely delimited by twin boundaries with misorientation of 86.3° (see Fig. 5.9a) (Kaibyshev and Sitdikov, 2000, 1994b, Sitdikov and Kaibyshev, 2001, Sitdikov *et al.*, 2003, Yoo, 1981, Ma *et al.*, 2011, Knezevic *et al.*, 2010, Barnett, 2007a, b, Zhang *et al.*, 2010). It is worth noting that this mechanism was observed only with as-cast Mg with an average grain size of about 2 mm. The large grain size is necessary to initiate the operation of five or even more twin system belonging to $\{10\bar{1}1\}$ family. This number of non-coplanar

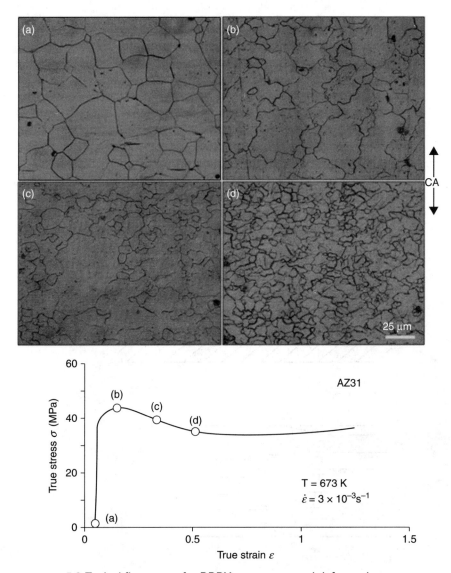

5.8 Typical flow curve for DDRX occurrence and deformed microstructure; AZ31 strained at 400 °C and an initial strain rate of 3×10^{-3} s^{-1} (Yang *et al.*, 2003).

twin systems is required (Yoo, 1981) to provide their mutual intersection. Increasing grain size facilitates twins and, therefore, this mechanism can be operative only in magnesium with very coarse initial grains.

Second, original grains can be segmented by the primary twins providing the formation of extended twin boundaries with misorientation close to 86.3° (Sitdikov *et al.*, 2003, Watanabe *et al.*, 2009, Xu *et al.*, 2009b, Al-Samman and

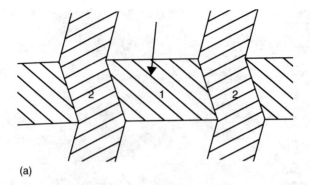

(a)

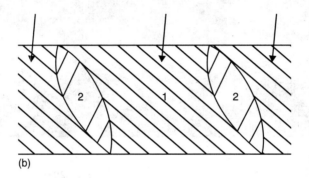

(b)

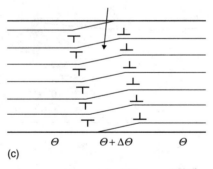

(c)

5.9 Schematic representation of TDRX mechanism. (a) Nucleation by mutual intersection by primary twins. (b) Nucleation by subdivision of coarse lamellae (1) of primary twins by fine secondary twins (2). (c) Nucleation by subdivision of coarse lamellae of primary twins by transverse low-angle boundaries. (d) The formation of orientation misfit dislocations with b_3 and b_4 Burgers vector on twin boundaries due to the passage of lattice dislocations with b_1 Burgers vector through a twin (this dislocation acquires b_2 Burgers vector within twin lamellae).

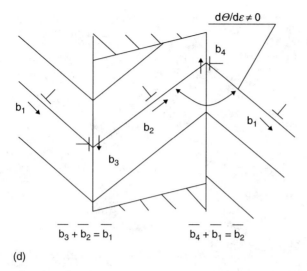

(d)

5.9 Continued.

Gottstein, 2008). With further strain the secondary contraction $\{10\bar{1}2\}$ twinning starts to be operative within primary coarse twins belonging to the $\{10\bar{1}1\}$ family and nuclei entirely bounded by twin boundaries (see Fig. 5.9b) (Watanabe *et al.*, 2009, Xu *et al.*, 2009b, Al-Samman and Gottstein, 2008). Such a nucleus is depicted in Fig. 5.10a (Al-Samman and Gottstein, 2008).

However, the most probable nucleation mechanism in TDRX is the subdivision of the primary twin lamellae by LABs forming within interiors of these lamellae (see Figs. 5.9c and 5.10b). Upon further strain these transversal LABs increase their misorientation and transform into HABs by a CDRX mechanism (see Fig. 5.1c). Thus a nucleus bounded partly by LABs and partly by twin boundaries is formed (see Fig. 5.9c).

At the second stage of TDRX, nuclei delimited by twin boundaries transform into recrystallized grains due to transformation of special twin boundaries into random HABs. This process occurs once the deviation of misorientation from the ideal coincident site lattice (CSL) relationship exceeds the value given by the standard Brandon criterion. The deviation in misorientation originates from the superposition of the misorientation corresponding to a wall of misfit dislocations onto the misorientation of a twin boundary (see Fig. 5.9d). Misfit dislocations result from interaction between mobile dislocations and a twin boundary (see Fig. 5.9d) (Sitdikov and Kaibyshev, 2001). A change in direction of dislocation glide takes place with the passage of a lattice dislocation through the twin boundary. As a result, a sessile misfit dislocation, the Burgers vector b_3, compensates for the change in the Burgers vector b_1 of the gliding lattice dislocation when it passes through the twin boundary ($b_1 \rightarrow b_2$) (see Fig. 5.9d).

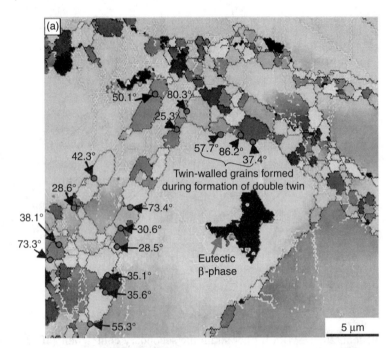
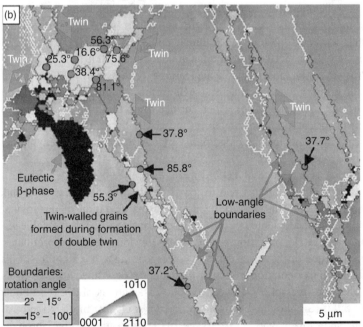

5.10 TDRX nucleation mechanisms operating in an AZ91 alloy strained at 300 °C. (a) Nucleation process depicted in Fig. 5.9b (Xu *et al.*, 2009a). (b) Nucleation process depicted in Fig. 5.9c (Xu *et al.*, 2009a).

Taking into account the very high misorientation associated with primary twins (~86.3°) and, therefore, the high value of the Burgers vector of the misfit dislocation, high misorientations can be quickly established. As a result, the misorientation of the twin boundary deviates from special misorientation and this boundary becomes a random HAB. This results in non-transparency of former twin boundaries for gliding lattice dislocations. Accumulation of lattice dislocation in the vicinity of the former twin boundary thus starts to occur. It is worth noting that the arrays of twins and transverse subgrain boundaries are very stable under hot deformation conditions (Watanabe *et al.*, 2009, Xu *et al.*, 2009b, Al-Samman and Gottstein, 2008); no significant growth of such nuclei has been observed.

A third stage of TDRX occurs once the grains having irregular shape acquire an equiaxed shape due to the limited migration needed to establish 120° triple junctions. This migration eliminates increased dislocation density in vicinity of the former twin boundaries and provides minor levels of strain softening.

5.3 Effect of initial structure on DRX

In contrast with materials with an fcc lattice (Humphreys and Hatherly, 2004) the initial structure of magnesium and its alloys impacts on the characteristics of the final recrystallized structure and the rate of DRX. A number of authors (Kaibyshev *et al.*, 1994a, Beer and Barnett, 2007, 2008) have shown that the size of DRX grains evolved in the wrought state of ZK60 and AZ31 alloys is less than those evolved in as-cast states of these alloys. For AZ31 alloy, this difference is attributed to a difference in flow stress (Beer and Barnett, 2007). It is known that the DRX grain size is related to the flow stress at steady-state. For ZK60 it was assumed (Kaibyshev *et al.*, 1994a) that this difference is caused by the effect of grain size on dislocation glide and, therefore, on the rate-controlling process for nucleation. It was shown (Kaibyshev *et al.*, 1994a, Koike *et al.*, 2003) that grain refinement highly facilitates (**a**) and (**a+c**) non-basal dislocation glide, respectively and, therefore, CDRX occurs easily in a magnesium alloy with a fine initial structure. The aforementioned dependence of DRX grain size on initial grain size is attributed to a strong dependence of non-basal slip on grain size, and therefore, the initial structure affects the operating DRX mechanism. This is why the dependence of DRX grain size on initial grain size, which is not observed in other materials (Sakai and Jonas, 1984), is a typical feature of magnesium alloys.

This assumption was supported by the literature (Sitdikov and Kaibyshev, 2001, Kaibyshev *et al.*, 1994a), which shows a linear dependence of the DRX grain size on the normalized steady-state flow stress, σ/G. This is observed only for restricted temperature ranges in which the rate-controlling process for nucleation remains unchanged. Changes in these processes and, therefore, in DRX mechanisms results in changes in this dependence. Each DRX mechanism considered above provides the formation of grains with a specific size. Operation of several DRX mechanisms leads to the appearance of a strain dependence of

DRX grain size (Sitdikov and Kaibyshev, 2001, Kaibyshev and Sitdikov, 1994b, Yang et al., 2003), which is also a unique feature of DRX in magnesium and its alloys (Humphreys and Hatherly, 2004). Concurrent operation of DDRX and CDRX mechanisms yields a weak strain dependence of DRX grain size (Yang et al., 2003). In contrast, TDRX occurrence leads to a strong strain dependence of DRX grain size (Sitdikov and Kaibyshev, 2001, Kaibyshev and Sitdikov, 1994b). It is worth noting that TDRX always occurs concurrently with other DRX mechanisms (Beer and Barnett, 2008). As a result, a strong strain dependence of DRX grain size is a unique feature of TDRX. Refinement of initial structure suppresses TDRX so this provides a strong effect of initial grain size on both the morphology and size of DRX grains and the rate of DRX as well.

A strong relationship between the mechanisms of plastic deformation and DRX results in a strong effect of initial crystallographic texture on DRX kinetics and the size of DRX grains (Kaibyshev and Sokolov, 1992, Kaibyshev et al., 1994b, Kaibyshev et al., 1999). Samples were cut from a hot extruded bar of ZK60 alloy having a sharp axial <1120> texture under the angles of 0° (state 1), 45° (state 2), and 90° (state 3) to the extrusion axis (see Fig. 5.11). Compression test at $T = 300$ °C and a strain rate of 2.8×10^{-3} s^{-1} showed that state 3 exhibits the highest steady-state flow stress, and that state 2 exhibits the lowest flow stress. A well-defined peak stress could be observed in the stress–strain curves of states 1 and 3 (see Fig. 5.12). Analysis of deformation behavior of the three states for alloy ZK60 showed that state 2 exhibits an apparent activation energy and stress exponent of 192 kJ/mol and 6.2, respectively. The other states of the ZK60 alloy showed the values of apparent activation energy and stress exponents as ~135 kJ/mol and ~5, respectively. Deformation behavior of state 2 was interpreted in terms of thermoactivated cross-slip by Friedel mechanism. The Friedel–Escaig mechanism of cross slip of (**a**) dislocations located in non-basal planes was believed to operate in the other samples. Microstructural characterization showed that state 3 demonstrates the highest rate of DRX; state 2 demonstrates lowest rate. The rate of DRX in state 1 is slightly less than that in state 3 (see Fig. 5.13). Lowest size of DRX grains was attained in state 2 despite the fact that this state of the ZK60 alloy shows the lowest steady-state flow stress. The DRX grain size correlates with the Zener–Hollomon parameter, Z, according to a power-law relationship. For state 2 the dependence on Z differs from that seen for the other states (see Fig. 5.14). It is worth noting that it was found that the CDRX mechanism is the dominant DRX mechanism for the deformation conditions used for the ZK60 alloy in this study.

Inspection of texture evolution, topographic observations and Burgers vector analysis (Kaibyshev and Sokolov, 1992, Kaibyshev et al., 1994b, 1999) shows that a strong effect of initial texture on CDRX behavior is attributed to its effect on deformation mechanisms (see Fig. 5.13). In State 1, the slip of (**a**) dislocations on prismatic and first-order pyramidal planes occurs in the early stages of plastic flow ($\varepsilon < 15\%$). Contribution of basal slip to the total strain is minor. In the strain

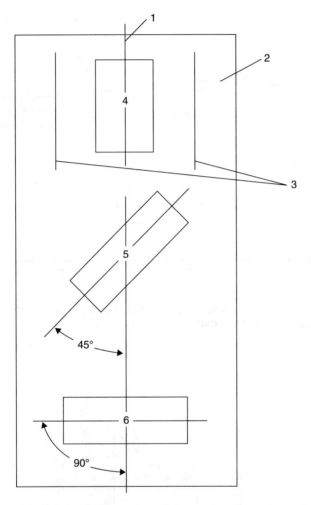

5.11 Schematic illustration of the cutting of specimens from a rod.
1: Extrusion, 2: rod, 3: basal plane, 4: state-1 specimens, 5: state-2 specimens, 6: state-3 specimens.

interval ε = 15–50%, rotation of crystallographic planes provides opportunity for operation of basal slip. At higher strains ($\varepsilon > 50\%$) the rotation of crystallographic planes suppresses the basal slip and initiates gliding of (**a+c**) dislocations.

In state 3, the greatest number of slip systems is operative at the early stage of plastic flow. As a result, the highest rate with which the percentage CDRX increases is observed. Therefore, multiple slip with operation of (**a+c**) dislocations glide provides favorable conditions for CDRX occurrence. In contrast, in state 2 the basal slip is highly facilitated due to highest value of Schmid factor; no CDRX occurs at low strains. Rotation of crystallographic planes makes possible the

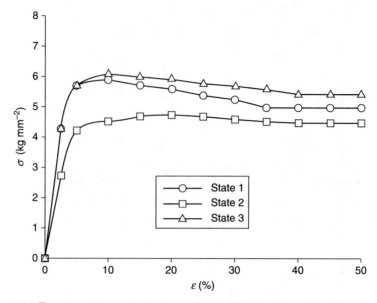

5.12 True stress–true strain curves for the ZK60A alloy with different initial crystallographic texture. States of the ZK60A alloy are shown in Fig. 5.11.

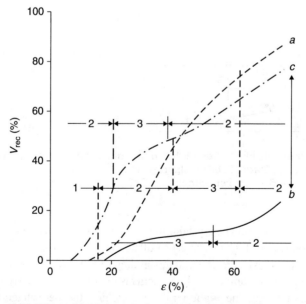

5.13 Effect of initial crystallographic texture on the percentage DRX and dependence of operation system of dislocation glide on strain. 1: Region of minor contribution of basal slip to total strain, non-basal slip systems are predominant. 2: Region of balanced basal and non-basal slip. 3: Region where basal slip system is predominant. *a*: State 1. *b*: State 2. *c*: State 3.

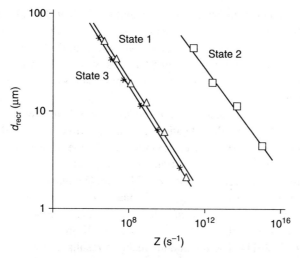

5.14 The effect of the Zener–Hollomon parameter, Z, on the plateau grain size in samples of ZK60 alloy with different initial texture.

operation of non-basal slip, including (**a+c**) dislocation glide. As a result, at $\varepsilon \geq 25\%$, CDRX starts to occur. Thus, the effect of initial texture on CDRX in magnesium alloys (Kaibyshev and Sokolov, 1992, Kaibyshev *et al.*, 1994b, 1999, Valle and Ruano, 2008) is caused by the strong effect of dislocation glide on the nucleation mechanism of CDRX.

5.4 DRX in different magnesium alloys

5.4.1 DRX in pure magnesium

DRX in pure magnesium in a wide temperature range was examined in the literature (Sitdikov and Kaibyshev, 2001, Beer and Barnett, 2006, 2008, Kaibyshev and Sitdikov, 1992, Biswas *et al.*, 2010, Edalati *et al.*, 2011, Galiyev *et al.*, 1995). In as-cast magnesium with an initial coarse-grained structure, all three DRX mechanisms are operative; their contributions to overall DRX change with strain. As a result, a strong effect of strain on DRX grain size was found. It is worth noting that this effect is attributed mainly to the operation of the TDRX mechanism. A strong decrease in grain size due to the use of wrought magnesium as a starting material provides a significant decrease in the contribution of TDRX to the overall DRX process (Biswas *et al.*, 2010). This diminishes the effect of strain on the DRX grain size.

At all temperatures, basal slip is favored in Mg (Galiyev *et al.*, 2003b). As a result, a mantle of recrystallized grains along initial boundaries grows at a low rate with increasing strain. DDRX makes a significant contribution to the overall recrystallization process. Deformation-induced HABs, which are not pinned by nanoscale particles or solutes (Humphreys and Hatherly, 2004), exhibit high mobility even at ambient temperature (Sitdikov and Kaibyshev, 2001, Kaibyshev and Sitdikov, 1994b) consuming lattice dislocations. As a result, at intermediate temperatures, high peak stresses are observed and the dynamically recrystallized structure is unstable during interpass holding (Biswas et al., 2010). Conventional DRX resulting in significant levels of strain softening can occur in Mg even at ambient temperature.

DRX was reported in magnesium subjected to high-pressure torsion (HPT) (Kaibyshev and Sitdikov, 1994b, Edalati et al., 2011, Galiyev et al., 1995). At ambient temperature under HPT, in coarse-grained Mg the TDRX mechanism occurs initially resulting in very high strain softening and the formation of an ultra-fine structure with an average grain size of 4–6 μm (Sitdikov and Kaibyshev, 2001, Kaibyshev and Sitdikov, 1994b). The formation of UFG structures highly facilitates non-basal slip (Koike et al., 2003); and the continuous reaction described above occurs (Sitdikov and Kaibyshev, 2001, Kaibyshev and Sitdikov, 1994b, Edalati et al., 2011). The continuous reaction results in the formation of micron-scale grains and high strain softening.

5.4.2 DRX in alloys belonging to Mg-Al-Zn system

Magnesium alloys belonging to the AZ series are most widely used among the structural magnesium alloys. As a result, numerous studies dealt with examination of DRX phenomena in these alloys (Xu et al., 2009a, Sun et al., 2010, Ravi Kumar et al., 2003, Beer and Barnett, 2006, 2007, 2008, Martin and Jonas, 2010, Valle and Ruano, 2008, Yang et al., 2003, Xing et al., 2005, Ding et al., 2007, Fatemi-Varzaneh et al., 2007, Maksouda et al., 2009, Laser et al., 2003, Somekawa and Mukai, 2009). Features of DRX behavior in alloys belonging to the Mg-Al-Zn system are attributed to two factors. Although basal slip occurs in these alloys (Agnew et al., 2005), they exhibit balanced secondary slip of non-basal (**a**) and (**a+c**) dislocations. However, the contribution of non-basal slip to total strain is minor even at high temperatures. Increasing aluminum content results in insignificant changes in character of dislocation glide (Sun et al., 2010, Agnew et al., 2005), because the additional Al is involved in second-phase particles. Therefore, the DRX behavior of AZ alloys and pure magnesium is nearly the same. As a result, the DDRX mechanism is predominant at intermediate and high temperatures (Sun et al., 2010, Yang et al., 2003, Ding et al., 2007, Fatemi-Varzaneh et al., 2007); a well-defined peak stress is observed in s–e curves under compression (Sun et al., 2010, Xu et al., 2009b, Ravi Kumar et al., 2003, Beer and Barnett, 2007, Martin and Jonas, 2010, Ding et al., 2007). DDRX occurrence leads to significant strain softening due to extensive migration of HABs. Also, TDRX was found to be operative at 300 °C in as-cast AZ91 alloy (Xu et al., 2009a) as in Mg.

Second, coarse particles of β-phase ($Mg_{17}Al_{12}$) do not exert a remarkable drag pressure (Humphreys and Hatherly, 2004) to suppress migration of deformation-induced boundaries. It is worth noting that in AZ31 alloy these precipitates dissolve into the matrix at $T \geq 200$ °C, and therefore, play no role in impeding grain boundary migration during DRX. It seems that β-phase particles play no role in DRX processes. As a result, the recrystallized grain size in Mg-Al-Zn alloys is independent on aluminum content and is less than that in pure Mg by a factor of two or even less. This difference tends to decrease with increasing temperature. We can assume that the decreased size of DRX grains in AZ alloys in comparison with pure magnesium is attributed to the presence of nanoscale particles of Mn and Al_4Mn in AZ alloys. These particles tend to coarsen with increasing temperature and even may dissolve at $T > 400$ °C. In addition, the volume fraction of these particles is low. As a result, no great difference between DRX grain size in AZ alloys and pure magnesium was found. There exists a bilinear relationship between DRX grains and the Zener–Hollomon parameter or normalized stress in AZ31 alloy (Laser et al., 2003). This relationship reflects the change in the structural mechanisms responsible for DRX grain development under different processing conditions. It was shown (Somekawa and Mukai, 2009) that at 200 °C in a coarse-grained (~50 µm) AZ31 alloy the occurrence of TDRX takes place in the same manner as in pure Mg at ambient temperature.

Effect of Ca on DRX in Mg-Al alloys

Ca additions to AZ alloys have no effect on deformation mechanisms (Masoudpanah and Mahmudi, 2009). Textures after extrusion and equal channel angular pressing (ECAP) in AZ31 with and without Ca were almost the same. However, these additives provide the formation of the following particles: $(Mg,Al)_2Ca$ (C36, dihexagonal), Mg_2Ca (C14, hexagonal) or Al_2Ca (C15, cubic) (Kim et al., 2009a). These phases can form nanoscale particles within the magnesium matrix. In addition, Ca may dissolve in the β-phase ($Mg_{17}Al_{12}$), making it resistant to dissolution (Masoudpanah and Mahmudi, 2009). A strong effect of Ca additives on DRX behavior of Mg-Al alloys is attributed to this effect of Ca additions on a dispersion of secondary phases. The β-phase additionally alloyed by Ca acquires a more equiaxed shape (Hakamada et al., 2010). This phase is located on matrix grain boundaries in an AZ91 alloy with Ca additives. Particle stimulated nucleation (PSN) observed in Ca modified AZ alloys is associated with this β-phase. As a result, a strong localization of DRX takes place in the Ca modified AZ alloy as in ZK60. It is worth noting that no evidence for PSN in conventional AZ alloys was reported (Ding et al., 2007). It seems that this localization is also attributed to the aforementioned operation of GBS within the recrystallized mantle (Zaripov et al., 1987, Zaripov and Kaibyshev, 1988) due to the fact that nanoscale phases containing Ca effectively pin deformation-induced HABs providing a grain size lower than that in a ZK60 alloy (Hakamada et al., 2010). As a result, a Mg-Al-Ca alloy exhibits superplasticity (Kim et al., 2009a). In general, Ca additives provide a two-fold decrease in DRX grain size in comparison with AZ alloys (Xu et al., 2009b, Xu et al., 2010, Hakamada et al.,

2010, Kim et al., 2009a, Masoudpanah and Mahmudi, 2009) due to the additional Zener drag force originated from aforementioned nanoscale particles formed by Ca. Decrease in mobility of HABs in Ca-bearing alloys in comparison with base AZ31 alloys leads to decreased strain softening (Kim et al., 2009a). This is caused by the restricted ability of deformation-induced HABs to consume lattice dislocations in these alloys. In addition, Ca additives serve as a strong refiner of the as-cast structure. As a result, (**a+c**) non-basal slip is facilitated (Kim et al., 2009a) and this highly accelerates DRX, specifically at low temperatures. A fully recrystallized structure is evolved at strains at which the recrystallized volume fraction in AZ31 is negligible. In addition, Ca additives also suppress twining.

5.4.3 DRX in alloys belonging to Mg-Zn-Zr system

In contrast to AZ alloys, the ZK alloys appear to favor non-basal (**a+c**) slip (Galiyev et al., 2001, 2003b, Agnew et al., 2005). In addition to extensive (**a+c**) slip, prismatic and first order pyramidal (**a**) slip are operative. Although basal slip is still dominant, the contribution of non-basal slip systems to the total strain is significantly higher than in alloy AZ31. Cross-slip is also active (Galiyev et al., 2003b). These features of the deformation facilitate the occurrence of CDRX.

ZK alloys differ from AZ alloys in that they form nanoscale dispersoids that are uniformly distributed within the magnesium matrix (Kim et al., 2009b, Gao and Nie, 2007, Shahzad and Wagner, 2009). In addition, these dispersoids are highly stable against coarsening under high temperature annealing and hot deformation conditions due the fact that Zr forms a supersaturated solid solution (Shahzad et al., 2009). Zn is an important alloying element in Mg alloys and provides significant hardening by two mechanisms. First, zinc is a solid solution strengthener. Second, Zn provides precipitation hardening. Two metastable phases precipitate in ZK alloys during ageing. β''-phase ($MgZn_2$) particles are formed and these have a hexagonal structure ($a = 0.520$ nm, $c = 0.857$ nm); this phase is a Laves phase having disc-shaped morphology at intermediate and high temperature and plate-like shape at low temperatures (Kim et al., 2009b, Gao and Nie, 2007). It was found (Shahzad et al., 2009) that Zr may enrich this phase providing high thermal stability. It is apparent that at high temperatures, nanoscale particles of β''-phase enriched by Zr are observed within the magnesium matrix rather than a Zn-Zr phase.

β'-phase also forms and this phase has a base-centered monoclinic structure ($a = 2.596$ nm, $b = 1.428$ nm, $c = 0.524$ nm, $\gamma = 102.5°$) and chemical composition of Mg_4Zn_7 (Gao and Nie, 2007). Particles having a rod shape of this phase precipitate along the [0001] matrix direction and effectively suppress glide of basal dislocations. As a result, at low temperatures, contribution of non–basal dislocation glide to the total strain increases significantly. This accelerates the formation of DRX grains due to the operation of the aforementioned continuous reaction (Galiyev and Kaibyshev, 2001). Thus, prior ageing promotes DRX in ZK alloys due to the retardation of basal slip.

Both metastable particles exert a high Zener drag force restricting mobility of deformation induced boundaries. It was shown (Kaibyshev et al., 1994a) that at $T \leq 250$ °C, DRX occurrence leads to the formation of dispersoids having an equiaxed shape situated on boundaries of DRX grains. At $T \geq 300$ °C, migration of deformation-induced boundaries is hindered by equiaxed particles of β'''-phase. High Zener drag force provides a near two-fold decrease in size of DRX grains in comparison with AZ alloys. This difference has a maximum in the temperature interval 150–250 °C in which a dispersion of β'-phase exerts a high Zener drag force. In the temperature interval 300–350 °C this difference diminishes due to dissolution of the β'-phase. At $T \geq 400$ °C, this difference tends to increase with increasing temperature due to the fact that the β'''-phase enriched by Zr is more stable under high temperature deformation conditions than Mn or $MnAl_4$ phase in AZ alloys. Therefore, the secondary phases in ZK alloys provide finer sized DRX grains in comparison with AZ alloys.

Thus, there exist two types of magnesium alloys distinguished by their DRX behavior. This difference is attributed mainly to the difference in operating deformation mechanisms. AZ alloys belong to the first type, which can be termed as 'pure metal' type due to the fact that the DRX behavior of these alloys and pure magnesium is essentially the same. ZK alloys belong to the second type, which can be termed as 'alloy' type. The DRX behavior of these alloys is different due to the significant effect of alloying elements and secondary phases precipitated in these alloys on deformation mechanisms.

5.5 DRX during severe plastic deformation

In magnesium and most of its alloys, basal slip is the predominant deformation mechanism, as shown above. The SFE in the basal plane is low. As a result, (a) basal dislocations have limited ability for rearrangements by cross-slip and climb (Galiyev et al., 2003b); annihilation of lattice dislocations with opposite Burgers vectors plays an unimportant role in restoration under deformation. Extensive accumulation of lattice dislocation occurs in magnesium and its alloys with strain. As a result, magnesium exhibits a high rate of DRX with strain (Sitdikov and Kaibyshev, 2001, Kaibyshev et al., 1994a, Beer and Barnett, 2006, 2007). A fully recrystallized structure is evolved at strains ranging from 1 to 2 under unidirectional compression, which is the most popular technique for plastic deformation of Mg and its alloys to achieve grain refinement. At intermediate and high temperatures, these strains can be easily achieved via conventional forging techniques, and, therefore, there is no limitation on the overall strains that have to be imposed in Mg and its alloys to produce a fine-grained structure. Thus, there is no reason to apply intense plastic straining (IPS) techniques (Valiev and Langdon, 2006, Zhilyaev and Langdon, 2008, Nandan et al., 2008, Mishra and Ma, 2005), where extremely high strains are imposed in a material without incurring any concomitant changes in the cross-sectional dimensions of the billets, to produce UFG structure in magnesium alloys. However, there are two exceptions. First, there are some reasons to use IPS techniques to produce submicrometer scale grains in semi-finished products from magnesium alloys (Valiev and Langdon, 2006, Zhilyaev and Langdon, 2008). At low temperatures the strains imposed in conventional processing are not sufficient to introduce DRX grains with size less than 1 μm because of the low workability of magnesium and its alloys at these temperatures. Second, application of friction stir welding (FSW) is highly suitable for magnesium alloys because it reduces the formation of the toxic fumes that occur in arc welding and provides welds with good mechanical properties.

DRX is the main process of microstructural evolution occurring under IPS. We will consider DRX during ECAP and FSW in detail. It is worth noting that the literature that deals with DRX phenomenon during multiple forging (Xing et al., 2005) and HPT (Galiyev and Kaibyshev, 2001, Sitdikov and Kaibyshev, 2001, Kaibyshev and Sitdikov, 1994b, Edalati et al., 2011, Galiyev et al., 1995, Zhilyaev and Langdon, 2008) are very limited. In addition, we would like only to say that application of HPT at room temperature for processing of ZK60 and AZ61 alloys (Harai et al., 2008) leads to essentially the same resulting grain size; extensive grain refinement is accompanied by extensive strain hardening of both alloys.

5.5.1 DRX under equal channel angular pressing

Microstructural evolution during ECAP of magnesium and its alloys was considered in numerous studies (Kang et al., 2010, Valle and Ruano, 2008, Biswas et al., 2010, Xia et al., 2005, Jin et al., 2005, Agnew et al., 2005, Masoudpanah and Mahmudi, 2009, Valiev and Langdon, 2006, Ding et al., 2009, Janecek et al., 2007, Yan et al., 2011, Zhao et al., 2011, Matsubara et al., 2004) at $T \geq 200$ °C. Inspection of these data shows that simple application of ECAP for processing of different magnesium alloys or Mg has no advantage in comparison to unidirectional compression. The size of DRX grains evolved in one alloy subjected to ECAP and compression at similar temperatures is essentially the same; the percentage of recrystallized fraction is higher in compressed samples in comparison with samples of the same alloy subjected to ECAP at the similar temperature up to nearly the same strain rate. At 200 °C, the formation of new grains in an AZ31 alloy occurs through the occurrence of CDRX (Ding et al., 2009, Janecek et al., 2007) rather than DDRX. Evidence for extensive non-basal slip after first ECAP pass could be detected (Janecek et al., 2007); the formation of recrystallized mantle along initial grains takes place. In general, magnesium alloys initially processed by extrusion are subjected to subsequent ECAP. That is, the alloy with refined microstructure is processed by ECAP to provide sufficient workability (Valiev and Langdon, 2006). As a result, DRX grain size remains virtually unchanged with increasing strain after the first pass of ECAP (Ding et al., 2009); a linear dependence between DRX grain size and the Zener–Hollomon parameter is observed. No significant effects of ECAP (Valiev and Langdon, 2006) and geometry of ECAP die on DRX grain size were found (Ding et al., 2009).

Exceptionally high yield stresses of 372 MPa with sufficient ductility was attained in a AZ31 alloy subjected to ECAP with decreasing temperature from 200 °C to 115 °C (Ding et al., 2009, 2008). It is apparent that the sequence of DRX process in these works was essentially similar to that reported for pure Mg. At 200 °C, the occurrence of CDRX under ECAP resulted in the formation of a uniform structure with an average grain size of about 1.8 µm. This extensive grain refinement highly facilitate non-basal slip at low temperatures (Muránskya et al., 2008, Koike et al., 2003, Ding et al., 2008). This fact provided sufficient workability to deform the AZ31 alloy in the temperature interval 115–125 °C and facilitates the occurrence of the continuous reaction providing the formation of grains with an average size of 0.37 µm and extensive strain hardening. Thus, it is obvious that submicrometer grained structure in magnesium alloy can be produced by plastic deformation if the temperature is decreased from pass to pass. The high efficiency of this approach is attributed to a key role of grain boundaries in initiation of non-basal slip (Muránskya et al., 2008, Koike et al., 2003).

Significant advantages of ECAP processing in comparison with compression in DRX occurrence can be obtained if this processing is optimized for operation of non-basal slip (Kang et al., 2010, Valle and Ruano, 2008, Ding et al., 2008, Kim et al., 2009c, Foley et al., 2011). This can be accomplished by the use of optimal initial crystallographic textures for CDRX, and application of back pressure, which facilitates non-basal slip and, as a result, promotes DRX occurrence. Taking into account a necessity to activate non-basal slip at low temperature, ECAP deformation of AZ31 alloy allows the achievement of very high yield stress and ductility via extensive grain refinement (Kang et al., 2010, Valle and Ruano, 2008, Ding et al., 2008, Kim et al., 2009c, Foley et al., 2011).

5.5.2 DRX under friction stir welding

During FSW process, the material undergoes intense plastic deformation at elevated temperature, resulting in the generation of fine and equiaxed recrystallized grains due to the occurrence of DRX (Nandan et al., 2008, Mishra and Ma, 2005). The temperature and strain rate of plastic deformation is a strong function of FSW parameters. It was shown that FSW produces recrystallized structures in the stir zone (SZ) in magnesium alloys. The sizes of recrystallized grains range from 100 nm (Chang et al., 2007) to 6–8 µm by varying these parameters (Suhuddin et al., 2009). Generally, successful FSW process of magnesium alloys refines the microstructure of these alloys down to 1–5 µm (Suhuddin et al., 2009, Freeney and Mishra, 2010, Ma et al., 2008, Zhang et al., 2005, Dobriyal et al., 2008, Esparza et al., 2002, Mironov et al., 2009, 2007, Xie et al., 2007, Feng and Ma, 2007, Xunhong and Kuaishe, 2006, Chang et al., 2004). Achieving submicrometer scale grains (Chang et al., 2007, Zhang et al., 2005) can be attained by using FSW regimes with a low heating temperature and a high strain rate in the SZ. The characteristics of DRX under FSW (Chang et al., 2004) are almost the same as those under compression (Suhuddin et al., 2009, Ma et al., 2008, Esparza et al., 2002, Mironov et al., 2009, 2007, Xie et al., 2007). All aforementioned DRX mechanisms are operative under FSW. Microstructures distinctly distinguished by DRX grain size in the recrystallized fraction are evolved in different areas of welded samples. There exists the strong dependence of strain, strain rate and temperature on the distance from the central line of the SZ.

The main feature of DRX occurring under FSW is attributed to superposition of huge changes in phase composition and distribution of secondary phases on microstructural evolution. FSW leads to dissolution of the β-phase ($Mg_{17}Al_{12}$) (Ma et al., 2008, Feng and Ma, 2007, Xunhong and Kuaishe, 2006); FSW produced a supersaturated solid solution. It is worth noting that most of the particles of β-phase have a eutectic origin. As a result, significant grain growth can take place under DRX. However, precipitation of very fine dispersoids of β-phase occurs under or after FSW resulting in significant strengthening of AZ alloys (Feng and Ma, 2007, Xunhong and Kuaishe, 2006). The formation of a dispersion of β-phase in AZ alloys within the SZ is a unique feature of the FSW process.

In ZK60 and Mg-Zn-Y-Zr alloys FSW produces dissolution of the coarse primary phase precipitate (Mironov et al., 2007, Xie et al., 2007); coarse particles of stable Mg_3Zn_2 phase located along initial boundaries completely dissolve under FSW. Subsequent precipitation of metastable phases from supersaturated solid solution provides a high dense dispersion of β''-phase ($MgZn_2$). In Mg–6%Al–3%Ca–0.5%(Ce+La)–0.2%Mn alloy (Zhang et al., 2005) the coarse eutectic particles of Al_2Ca phase break up under FSW and this leads to refinement of these particles. Thus FSW highly refines the secondary phases almost eliminating eutectic coarse particles. This is expected to increase the Zener drag force and decrease the DRX grain size. However, it is not possible to distinguish the effect of phase composition changes, size and distribution of secondary phases on DRX grain size from the same effect of temperature and strain rate.

5.6 Conclusions

At present, the diversity of DRX phenomenon in magnesium and its alloys are relatively well-recognized. The main DRX mechanisms are closely related to the operating deformation mechanisms that control the nucleation process and, therefore, strongly affect the rate of DRX, DRX grain size and the distribution of recrystallized grains within the material. This is a unique feature of the DRX behavior of magnesium and its alloys. Future trends in research and development activity should be focused on the development of a commercial technique for high-volume production of semi-finished products including sheets from magnesium alloys with UFG structures. To achieve this goal the new compositions of wrought magnesium alloys have to be developed taking into account the facilitation of producing a UFG structure via thermomechanical processing based on the DRX phenomenon discussed here.

5.7 Acknowledgements

The author would like to thank Prof. Matthew Barnett for many valuable suggestions and Ms E. Lashina for their assistance in preparation of this chapter. The author is indebted to Prof. T. Sakai for his discussion and help in taking micrographs.

5.8 References

Agnew, S. R., Mehrotra, P., Lillo, T. M., Stoica, G. M. and Liaw, P. K. (2005) 'Texture evolution of five wrought magnesium alloys during route A equal channel angular extrusion: Experiments and simulations', *Acta Materialia*, **53**, 3135–3146.

Al-Samman, T. and Gottstein, G. (2008) 'Dynamic recrystallization during high temperature deformation of magnesium', *Materials Science and Engineering*, A**490**, 411–420.

Barnett, M. R. (2007a) 'Twinning and the ductility of magnesium alloys Part I: "Tension" twins', *Material Science Engineering*, A**464**, 1–7.

Barnett, M. R. (2007b) 'Twinning and the ductility of magnesium alloys Part II. "Contraction" twins', *Material Science Engineering*, A**464**, 8–16.

Barnett, M. R., Keshavarz, Z. and Nave, M. D. (2005) 'Microstructural features of rolled Mg-3Al-1Zn', *Metallurgical Material Transaction*, **36A**, 1697–1704.

Beer, A. G. and Barnett, M. R. (2006) 'Influence of initial microstructure on the hot working flow stress of Mg-3Al-1Zn', *Materials Science and Engineering*, A**423**, 292–299.

Beer, A. G. and Barnett, M. R. (2007) 'Microstructural development during hot working of Mg-3Al-1Zn', *Metallurgical Material Transaction*, **38A**, 1856–1867.

Beer, A. G. and Barnett, M. R. (2008) 'Microstructure evolution in hot worked and annealed magnesium alloy AZ31', *Materials Science and Engineering*, A**485**, 318–324.

Belyakov, A., Gao, W., Miura, H., and Sakai, T. (1998) 'Strain induced grain evolution in polycrystalline copper during warm deformation', *Metallurgical Material Transaction*, **29A**, 2957–2965.

Belyakov, A., Sakai, T. and Miura, H. (2000a) 'Fine-grained structure formation in austenitic stainless steel under multiple deformation at 0.5 T_m', *Material Transaction JIM*, **41**, 476–484.

Belyakov, A., Sakai, T., Miura, H. and Kaibyshev, R. (2000b) 'Substructures and internal stresses developed under warm severe deformation of austenitic stainless steel', *Scripta Materialia*, **42**, no 4, 319–325.

Biberger, M. and Blum, W. (1992) 'Subgrain Boundary Migration during Creep of LiF: I. Recombination of Subgrain boundaries', *Philosophical. Magazine A*, **65**, 757–770.

Biswas, S., Dhinwal, S. S. and Suwas, S. (2010) 'Room-temperature equal channel angular extrusion of pure magnesium', *Acta Materialia*, **58**, 3247–3261.

Chang, C. I., Lee, C. J. and Huang, J. C. (2004) 'Relationship between grain size and Zener–Hollomon parameter during friction stir processing in AZ31 Mg alloys', *Scripta Materialia*, **51**, 509–514.

Chang, C. I., Dua, X. H. and Huang, J. C. (2007) 'Achieving ultrafine grain size in Mg–Al–Zn alloy by friction stir processing', *Scripta Materialia*, **57**, 209–212.

Couret, A., Caillard, D., Puschl, W. and Schoeck, G. (1991) 'Prismatic glide in divalent h.c.p. metals', *Philosophical Magazine A*, **63**, 1045–1057.

del Valle, J. A. and Ruano, O. A. (2008) 'Influence of texture on dynamic recrystallization and deformation mechanisms in rolled or ECAPed AZ31 magnesium alloy', *Materials Science and Engineering*, A**487**, 473–480.

Ding Hanlin, Liu Liufa, Kamado Shigeharu, Ding Wenjiang and Kojima Yo (2007) 'Evolution of microstructure and texture of AZ91 alloy during hot compression', *Materials Science and Engineering*, A**452–453**, 503–507.

Ding, S. X., Lee, W. T., Chang, C. P., Chang, L. W. and Kao, P. W. (2008) 'Improvement of strength of magnesium alloy processed by equal channel angular extrusion', *Scripta Materialia*, **59**, 1006–1009.

Ding, S. X., Chang, C. P. and Kao, P. W. (2009) 'Effects of Processing Parameters on the Grain Refinement of Magnesium Alloy by Equal-Channel Angular Extrusion', *Metallurgical Material Transaction*, **40**, 415–424.

Dobriyal, R. P., Dhindawa, B. K., Muthukumaran, S. and Mukherjee, S. K. (2008) 'Microstructure and properties of friction stir butt-welded AE42 magnesium alloy', *Materials Science and Engineering*, A**477**, 243–249.

Dougherty, L. M., Robertson, I. M. and Vetrano, J. S. (2003) 'Direct observation of the behavior of grain boundaries during continuous dynamic recrystallization in an Al–4Mg–0.3Sc alloy', *Acta Materialia*, **51**, 4367–4378.

Dudova, N., Belyakov, A., Sakai, T. and Kaibyshev, R. (2010) 'Dynamic recrystallization mechanisms operating in a Ni–20%Cr alloy under hot-to-warm working', *Acta Materialia*, **58**, 3624–3632.

Edalati, K., Yamamoto, A., Horita, Z. and Ishihara, T. (2011) 'High-pressure torsion of pure magnesium: Evolution of mechanical properties, microstructures and hydrogen storage capacity with equivalent strain', *Scripta Materialia*, **64**, 880–883.

Esparza, J. A., Davis, W. C., Trillo, E. A. and Murr, L. E. (2002) 'Friction-stir welding of magnesium alloy AZ31B', *J Material Science Letters*, **21**, 917–920.

Fatemi-Varzaneh, S. M., Zarei-Hanzaki, A. and Beladi, H. (2007), 'Dynamic recrystallization in AZ31 magnesium alloy', *Materials Science and Engineering*, A**456**, 52–57.

Feng, A. H. and Ma, Z. Y. (2007) 'Enhanced mechanical properties of Mg–Al–Zn cast alloy via friction stir processing', *Scripta Materialia*, **56**, 397–400.

Foley, D. C., Al-Maharbi, M., Hartwig, K. T., Karaman, I., Kecskes, L. J. et al. (2011) 'Grain refinement vs. crystallographic texture: Mechanical anisotropy in a magnesium alloy', *Scripta Materialia*, **64**, 193–196.

Freeney, T. A. and Mishra, R. S. (2010) 'Effect of friction stir processing on microstructure and mechanical properties of a cast-magnesium–rare earth alloy', *Metallurgical Materials Transactions*, **41**, 73–84.

Galiyev, A. M., Kaibyshev, R. O. and Sitdikov, O. (1995) 'On the possibility of producing a nano-crystalline structure in magnesium and magnesium alloys', *Nano Structured Materials*, **6**, no 5–8, 621–624.

Galiyev, A. and Kaibyshev, R. (2001) 'Microstructural evolution in ZK60 magnesium alloy during severe plastic deformation', *Materials Transactions*, **42**, no 7, 1190–1199.

Galiyev, A., Kaibyshev, R. and Gottstein, G. (2001) 'Correlation of plastic deformation and dynamic recrystallization in magnesium alloy ZK60', *Acta Materialia*, **49**, no 7, 1199–1207.

Galiyev, A., Kaibyshev, R. and Sakai, T. (2003a) 'Continuous dynamic recrystallization in magnesium alloy', *Materials Science Forum*, **419–422**, 509–514.

Galiyev, A., Sitdikov, O. and Kaibyshev, R. (2003b) 'Deformation behavior and controlling mechanisms for plastic flow of magnesium and magnesium alloy', *Materials Transactions*, **43**, no 4, 426–435.

Gao, X. and Nie, J. F. (2007) 'Characterization of strengthening precipitate phases in a Mg–Zn alloy', *Scripta Materialia*, **56**, 645–648.

Hakamada, M., Watazu, A., Saito, N. and Iwasaki, H. (2010) 'Dynamic recrystallization during hot compression of as-cast and homogenized noncombustible Mg–9Al–1Zn–1Ca (in mass %) alloys', *Materials Science and Engineering*, A**527**, 7143–7146.

Harai Yosuke, Kai Masaaki, Kaneko Kenji, Horita Zenji, Langdon Terence G. (2008) 'Microstructural and mechanical characteristics of AZ61 Magnesium alloy processed by high-pressure torsion', *Materials Transactions*, **49**, no 1, 76–83.

Humphreys, F. J. and Hatherly, M. (2004) *Recrystallization and Related Annealing Phenomena*, 2nd ed., Elsevier, Oxford, 285–318.

Ion, S. E., Humphreys, F. J. and White, S. H. (1982) 'Dynamic recrystallization and the development of microstucture during the high temperature deformation of magnesium', *Acta Metallurgica*, **30**, no10, 1909–1919.

Janecek, M., Popov, M., Krieger, M. G., Hellmig, R. J. and Estrin, Y. (2007) 'Mechanical properties and microstructure of a Mg alloy AZ31 prepared by equal-channel angular pressing', *Materials Science and Engineering*, A**462**, 116–120.

Jin, L., Lin, D. T., Mao, D., Zeng, X. and Ding, W. (2005) 'Mechanical properties and microstructure of AZ31 Mg alloy processed by two-step equal channel angular extrusion', *Mater Lett*, **59**, 2267.

Kaibyshev, R. O. and Sitdikov, O. Sh. (1992) 'Structural changes during plastic deformation of pure magnesium', *The Physics of Metals and Metallography*, **73**, no 6, 635–642.

Kaibyshev, R. O. and Sokolov, B. K. (1992) 'The influence of a crystallographic texture on slipping and dynamic recrystallization in a magnesium alloy', *The Physics of Metals and Metallography*, **74/1**, 72–78.

Kaibyshev, R. O., Galiyev, A. M. and Sokolov, B. K. (1994a) 'The influence of grain size on plastic deformation and dynamic recrystallization of magnesium alloy', *The Physics of Metals and Metallography*, **78/2**, 126–139.

Kaibyshev, R. O., Galiyev, A. M. and Sokolov, B. K. (1994b) 'The influence of plastic deformation on texture, crystallographic sliding and structural changes in the magnesium alloy', *The Physics of Metals and Metallography*, **78/2**, 145–158.

Kaibyshev, R. O. and Sitdikov, O. Sh. (1994) 'The crystallographic sliding and dynamic recrystallization caused by local grain boundary migration, Part 1, experimental results', *The Physics of Metals and Metallography*, **78**, no 4, 372–383.

Kaibyshev, R. and Sitdikov, O. (1994) 'Dynamic recrystallization of magnesium at ambient temperature', *Z. Metallkunde*, B **85**, no 10, 738–743.

Kaibyshev, R., Sokolov, B. and Galiyev, A. (1999) 'The influence of crystallographic texture on dynamic recrystallization', *Texture and Microstructures*, **32**, 47–60.

Kaibyshev, R. and Sitdikov, O. (2000) 'On the role of twinning in dynamic recrystallization', *The Physics of Metals and Metallography*, **89**, no 4, 384–390.

Kaibyshev, R., Goloborodko, A., Musin, F., Nikulin, I. and Sakai, T. (2002) 'The role of grain boundary sliding in microstructural evolution during superplastic deformation of a 7055 aluminum alloy', *Materials Transaction*, **43**, no 10, 2408–2414.

Kaibyshev, R., Shipilova, K., Musin, F. and Motohashi, Y. (2005) 'Continuous dynamic recrystallization in an Al-Li-Mg-Sc alloy during equal-channel angular extrusion', *Material Science Engineering*, **396**, no 1–2, 341–351.

Kang, F., Liu, J. Q., Wang, J. T. and Zhao, X. (2010) 'Equal Channel Angular Pressing of a g–3Al–1Zn Alloy with Back Pressure', *Advanced Engineering Materials*, **12**, no 8, 730–734.

Kim, J. H., Kang, N. E., Yim, Ch. D. and Kim, B. K. (2009a) 'Effect of calcium content on the microstructural evolution and mechanical properties of wrought Mg–3Al–1Zn alloy', *Materials Science and Engineering*, A**525**, 18–29.

Kim, W. J., Kim, M. J. and Wang, J. Y. (2009b) 'Superplastic behavior of a fine-grained ZK60 magnesium alloy processed by high-ratio differential speed rolling', *Materials Science and Engineering*, A**527**, 322–327.

Kim, W. J., Yoo, S. J., Chen, Z. H. and Jeong, H. T. (2009c) 'Grain size and texture control of Mg–3Al–1Zn alloy sheet using a combination of equal-channel angular rolling and high-speed-ratio differential speed-rolling processes', *Scripta Materialia*, **60**, 897–900.

Knezevic, M., Levinson, A., Harris, R., Mishra, R., Doherty, R. D. and Kalidindi, S. R. (2010) 'Deformation twinning in AZ31: Influence on strain hardening and texture evolution', *Acta Materialia*, **58**, 6230–6242.

Koike, J.-I., Kobayashi, T., Mukai, T., Watanabe, H., Suzuki, M., Maruyama, K. and Higashi, K. (2003) 'The activity of non-basal slip systems and dynamic recovery at room temperature in fine-grained AZ31B magnesium alloys', *Acta Materialia*, **51**, 2055–2065.

Laser, T., Hartig, Ch., Bormann, R., Bohlen, J. and Letzig, D. (2003) 'Dynamic Recrystallization of Mg–3Al–1Zn', Magnesium, *Proceedings of the 6th International Conference 'Magnesium Alloys and Their Applications'*, ed. Kainer K. U., Wiley-VCH Verlag, Weinheim, pp. 266–271.

Liu Zhiyi, Bai Song and Kan Suk Bong (2009) 'Low-temperature dynamic recrystallization occurring a high deformation temperature during hot compression of twin-roll-cast Mg–5.51Zn–0.49Zr alloy', *Scripta Materialia*, **60**, 403–406.

Ma, Q., Kadiri, H. El., Oppedal, A. L., Baird, J. C., Horstemeyer, M. F. and Cherkaoui, M., (2011) 'Twinning and double twinning upon compression of prismatic textures in an AM30 magnesium alloy', *Scripta Materialia*, **64**, 813–816.

Ma, Z. Y., Pilchak, A. L., Juhas, M. C. and Williams, J. C. (2008) 'Microstructural refinement and property enhancement of cast light alloys via friction stir processing', *Scripta Materialia*, **58**, 361–366.

Mabuchi, M., Ameyama, K., Iwasaki, H. and Higashi, K. (1999) 'Low temperature superplasticity of AZ91 magnesium alloy with non-equilibrium grain boundaries', *Acta Materialia*, **47**, 2047–2057.

Maksouda Ismael Abdel, Ahmed H. and Röde Johannes (2009) 'Investigation of the effect of strain rate and temperature on the deformability and microstructure evolution of AZ31 magnesium alloy', *Materials Science and Engineering*, A**504**, 40–48.

Martin, E. and Jonas, J. J. (2010) 'Evolution of microstructure and microtexture during the hot deformation of Mg–3% Al', *Acta Materialia*, **58**, 4253–4266.

Masoudpanah, S. M. and Mahmudi, R. (2009) 'Effects of rare-earth elements and Ca additions on the microstructure and mechanical properties of AZ31 magnesium alloy processed by ECAP', *Materials Science and Engineering*, A**526**, 22–30.

Matsubara, K., Miyahara, Y., Horita, Z. and Langdon, T. G. (2004) 'Achieving enhanced ductility in a dilute magnesium alloy through severe plastic deformation', *Metallurgical Material Transaction*, **35**, 1735–1744.

Mironov Sergey, Motohashi Yoshinobu, Ito Tsutomu, Goloborodko Alexandre, Funami Kunio and Kaibyshev Rustam (2007) 'Feasibility of Friction Stir Welding for Joining and Microstructure Refinement in a ZK60 Magnesium Alloy', *Materials Transactions*, **48**, no 12, 3140–3148.

Mironov Sergey, Motohashi Yoshinobu, Kaibyshev Rustam, Somekawa Hidetoshi, Mukai Toshiji and Tsuzaki Kaneaki (2009) 'Development of fine-grained structure caused by friction stir welding process of a ZK60A magnesium alloy', *Materials Transactions*, **50**, no 3, 610–617.

Mishra, R. S. and Ma, Z. Y. (2005) 'Friction stir welding and processing', *Materials Science and Engineering*, R**50**, 1–78.

Muránskya, O., Carra, D. G., Barnett, M. R., Oliver, E. C. and Sittner, P. (2008) 'Investigation of deformation mechanisms involved in the plasticity of AZ31 Mg alloy: In situ neutron diffraction and EPSC modeling', *Materials Science and Engineering*, A**496**, 14–24.

Nandan, R., DebRoy, T. and Bhadeshia, H. K. D. H. (2008) 'Recent advances in friction-stir welding – Process, weldment structure and properties', *Progress in Materials Science*, **53**, 980–1023.

Pérez-Prado, M. T., del Valle, J. A. and Ruano, O. A. (2005) 'Achieving high strength in commercial Mg cast alloys through large strain rolling', *Materials Letters*, **59**, 3299–3303.

Ravi Kumar, N. V., Blandin, J. J., Desrayaud, C., Montheillet, F. and Suery, M. (2003) 'Grain refinement in AZ91 magnesium alloy during thermomechanical processing', *Materials and Engineering*, A**359**, 150–157.

Sakai, T. and Jonas, J. J. (1984) 'Dynamic recrystallization. mechanical and microstructural considerations', *Acta Meterialia*, **32**, no 2, 189–209.

Sakai, T. and Miura, H., Mechanical properties of fine-grained magnesium alloys processed by severe plastic forging, in *Magnesium Alloys-Design, Processing and Properties*, ed. F. Czerwinski, (2011, InTeck), 219–244.

Shahzad, M. and Wagner, L. (2009) 'Microstructure development during extrusion in a wrought Mg–Zn–Zr alloy', *Scripta Materialia*, **60**, 536–538.

Singh, A., Osawa, Y., Somekawa, H., Mukai, T. (2011) 'Ultra-fine grain size and isotropic very high strength by direct extrusion of chill-cast Mg–Zn–Y alloys containing quasicrystal phase', *Scripta Materialia*, **64**, 661–664.

Sitdikov, O. and Kaibyshev, R. (2001) 'Dynamic Recrystallization in Pure Magnesium', *Materials Transaction*, **42**, no 9, 1928–1937.

Sitdikov, O. and Kaibyshev, R. (2002) 'Dislocation glide and dynamic recrystallization in LiF single crystals', *Materials Science and Engineering*, **328**, no 1–2, 147–155.

Sitdikov, O., Kaibyshev, R. and Sakai, T. (2003) 'Dynamic recrystallization based on twinning in coarse-grained Mg', *Materials Science Forum*, **419–422**, 521–526.

Sivakesavam, O., Rao, I. S. and Prasad, Y. V. R. K. (1993) 'Processing Map for Hot Working of as Cast Magnesium', *Material Science Technology*, **9**, 805.

Somekawa, H. and Mukai, T. (2009) 'Microstructure evolution of Mg–Al–Zn alloys during compression test at low strain and temperature', *Materials Science and Engineering*, A**527**, 370–375.

Somekawa, H., Singh, A. and Mukai, T. (2009) 'Microstructure evolution of Mg–Zn binary alloy during a direct extrusion process', *Scripta Materialia*, **60**, 411–414.

Suhuddin, U. F. H. R., Mironov, S., Sato, Y. S., Kokawa, H. and Lee, C.–W. (2009) 'Grain structure evolution during friction-stir welding of AZ31 magnesium alloy', *Acta Materialia*, **57**, 5406–5418.

Sun, D. K., Chang, C. P. and Kao, P. W. (2010) 'Microstructural aspects of grain boundary bulge in a dynamically recrystallized Mg-Al-Zn alloy', *Metallurgical Material Transaction*, **41**A, 1864–1870.

Valiev, R. Z., Gertsman, V. Yu., Kaibyshev, O. A. and Khananov, Sh. Kh. (1984) 'Non-equilibrium and recovery of grain boundary structure', *Physics Status Solid*, **77**, no 1, 97–105.

Valiev, R. Z. and Langdon, T. G. (2006) 'Principles of equal-channel angular pressing as a processing tool for grain refinement', *Progress in Materials Science*, **51**, 881–981.

Wang Jing Tao, Yin De Liang, Liu Jin Qiang, Tao Jun, Sua Yan Ling and Zhao Xiang (2008) 'Effect of grain size on mechanical property of Mg–3Al–1Zn alloy', *Scripta Materialia*, **59**, 63–66.

Watanabe, H., Fukusumi, M., Somekawa, H. and Mukai, T. (2009) 'Texture and mechanical properties of a superplastically deformed Mg–Al–Zn alloy sheet', *Scripta Materialia*, **61**, 883–886.

Xia, K., Wang, J. T., Xu, X., Chen, G. and Gurvan, M. (2005) 'Equal channel angular pressing of magnesium alloy AZ31', *Material Science Engineering*, **A410–411**, 324–327.

Xie, G. M., Ma, Z. Y., Geng, L. and Chen, R. S. (2007) 'Microstructural evolution and mechanical properties of friction stir welded Mg–Zn–Y–Zr alloy', *Materials Science and Engineering*, A**471**, 63–68.

Xing, J., Soda, H., Yang, X., Miura, H. and Sakai, T. (2005) 'Ultra-fine grain development in an AZ31 magnesium alloy during multi-directional forging under decreasing temperature conditions', *Materials Transactions*, **46**, no 7, 1646–1650.

Xu, S. W., Kamado, S., Matsumoto, N., Honma, T. and Kojima, Y. (2009a) Recrystallization mechanism of as-cast AZ91 magnesium alloy during hot compressive deformation', *Materials Science and Engineering*, A **527**, 52–60.

Xu, S. W., Matsumoto, N., Kamado, S., Honma, T. and Kojima, Y. (2009b) 'Dynamic microstructural changes in Mg–9Al–1Zn alloy during hot compression', *Scripta Materialia*, **61**, 249–252.

Xu, S. W., Kamado, S. and Honma, T. (2010) 'Recrystallization mechanism and the relationship between grain size and ZenerHollomon parameter of MgAlZnCa alloys during hot compression', *Scripta Materialia*, **63**, 293–296.

Xunhong Wang and Kuaishe Wang (2006) 'Microstructure and properties of friction stir butt-welded AZ31 magnesium alloy', *Materials Science and Engineering*, A**431**, 114–117.

Yan Kai, Sun Yang-Shan, Bai Jing and Xue Feng (2011) 'Microstructure and mechanical properties of ZA62 Mg alloy by equal-channel angular pressing', *Materials Science and Engineering*, A**528**, 1149–1153.

Yang Xuyue, Miura Hiromi and Sakai Taku (2003) 'Dynamic Evolution of New Grains in Magnesium Alloy AZ31 during Hot Deformation', *Materials Transactions*, **44**, no 1, 197–203.

Yoo, M. N. (1981) 'Slip, Twinning and fracture in hexagonal close-packed metals', *Metallurgical Transaction*, **12A**, no 3, 409–418.

Zaripov, N. G., Vagapov, A. R. and Kaibyshev, R. O. (1987) 'Dynamic recrystallization of a magnesium alloy', *The Physics of Metals and Metallography*, **63**, no 4, 774–781.

Zaripov, N. G. and Kaibyshev, R. O. (1988) 'Dynamic recrystallization and superplasticity of the magnesium alloys', Superplasticity and Superplastic Forming, TMS, 91–95.

Zhang Datong, Suzuki Mayumi and Maruyama Kouichi (2005) 'Microstructural evolution of a heat-resistant magnesium alloy due to friction stir welding', *Scripta Materialia*, **52**, 899–903.

Zhao Zude, Chen Qiang, Chao Hongying, Hu Chuankai and Huang Shuhai (2011) 'Influence of equal channel angular extrusion processing parameters on the microstructure and mechanical properties of Mg–Al–Y–Zn alloy', *Materials and Design*, **32**, 575–583.

Zhang, Z., Wang, M., Jiang, N. and Li, S., (2010) 'Orientation analyses for twinning behavior in small-strain hot-rolling process of twin-roll cast AZ31B sheet', *Material Science Engineering*, A527, 6467–6473.

Zhilyaev, A. P. and Langdon, T. G. (2008) 'Using high-pressure torsion for metal processing: Fundamentals and applications', *Progress in Materials Science*, **53**, 893–979.

Part II
Processing of magnesium alloys

6
Direct chill casting of magnesium extrusion billet and rolling slab

J. F. GRANDFIELD, Grandfield Technology Pty Ltd, Australia

Abstract: The direct chill continuous casting process used to produce magnesium extrusion billets and rolling slabs is described. The physics of the process including the heat and fluid flow and mechanical behaviour during casting are explained. The history and current practice of the technology and engineering of the process is outlined. Defect formation and control, particularly cracking defects, are discussed.

Key words: direct chill casting, water cooling, cracking, hot tearing, extrusion billet, rolling slab.

6.1 Introduction

Direct chill (DC) casting is a widely used continuous casting process, producing non-ferrous alloy ingots for remelt, extrusion and rolling. Aluminium alloys are by far the greatest tonnage DC cast and the bulk of the DC casting literature refers to aluminium (Emley 1976; Granger 1989; Katgerman 1991; Grandfield and McGlade 1996; Grandfield 1997; Schneider 2002; Eskin 2008). The aluminium DC casting literature forms a useful resource to understand the process when applied to magnesium but there are important differences when DC casting magnesium that will be highlighted here. Magnesium DC casting has been partially reviewed and contrasted to aluminium (Baker and McGlade 2001). A more comprehensive review of the history and current practice of DC casting of magnesium is given in this chapter.

Direct chill casting is a type of continuous casting developed in the 1930s in Germany (Roth 1936; Zunkel 1939) and in the 1940s in the USA (Ennor 1942) as an improvement over casting extrusion and rolling ingots in permanent moulds. DC casting is characterised by a direct water spray impinging on the ingot as it continuously emerges from a water-cooled mould (Fig. 6.1). Earlier continuous casting processes such as that of Junghaus continuously withdrew the ingot from a long water-cooled mould but had no direct water quench on the ingot (Junghaus 1938).

The Junghaus process was used for magnesium from 1937 onward (Lippert 1944) making 200 mm diameter extrusion billets and slab up to 600 mm × 100 mm. DC casting of magnesium soon followed in the 1940s at Alcoa in the USA (Lippert 1944), the Magnesium Elektron Limited, Manchester plant in 1944 (Wilkinson and Hirst 1952) and at the Vereinigte Leichtmetall Werken, Germany

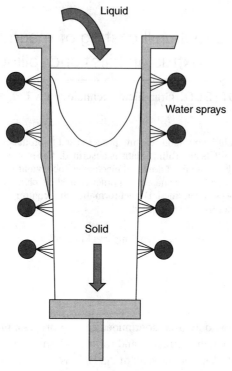

6.1 Early open mould VDC casting for aluminium or magnesium.

(Roth and Weisse 1944) (Fig. 6.2). The main departures from the practice with aluminium were the use of cover gas (SO_2) to stop the magnesium melt from burning and use of steel as the melt delivery system material as opposed to refractory, as in the case of aluminium.

To commence DC casting a starting head is positioned in the mould. Liquid is then delivered into the mould and, after some initial solid forms, the starting head or 'dummy block' is withdrawn, taking the solid part of the ingot with it. The dummy is lowered either on a platen on a hydraulic ram, a suspended platen on a chain, or a screw-driven platen into a pit in the ground (up to 11 m in the case of aluminium) or from a raised platform (generally less than a few metres high). The lowering rate sets the casting speed. Optimum casting speed depends on alloy thermal properties and ingot size, and varies from 50–1000 mm/min. This is the semi-continuous vertical DC process; the cast is stopped once maximum travel of the platen is reached. DC casting can also be conducted horizontally in a fully continuous mode.

In early moulds, the water jets were produced from holes in ring pipes around the product. In modern moulds the emerging ingot is chilled by water jets coming from the mould, spraying directly onto the ingot, which then forms a continuously

Direct chill casting of magnesium extrusion billet and rolling slab

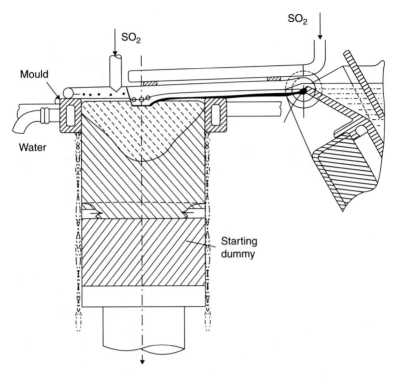

6.2 DC casting of magnesium with steel trough and tilting furnace. (Roth 1943.)

falling film down the ingot. The mould serves two functions: it contains the liquid metal, and acts as the water spray delivery system. The boiling mechanisms of the water spray have a significant effect on the heat transfer and solidification and stresses in the ingot, and are discussed further below.

Liquid metal is continually fed to the liquid pool in the mould to maintain a specified height in the mould. Mould lubrication is essential to stop the ingot shell sticking to the mould. Early embodiments used fat or grease applied to the mould before the start of casting. Modern configurations use continuous lubricant feed of organic oils such as castor and rapeseed, or mineral oils. Early mould designs were simple tubes that suffered distortion (Fig. 6.3).

For aluminium DC casting, aluminium moulds are commonly used, sometimes with a carbon liner. For magnesium, copper moulds are more common but aluminium moulds are also used.

Modern moulds tend to have a box construction with interior baffles for rigidity and control of water flow distribution around the perimeter of the ingot. Mould technology is discussed in more detail below, and several variants have been used over the years for magnesium. A variety of ingot sizes (20–3000 mm) and shapes (rounds, rectangles, T-bar) can be produced by DC casting.

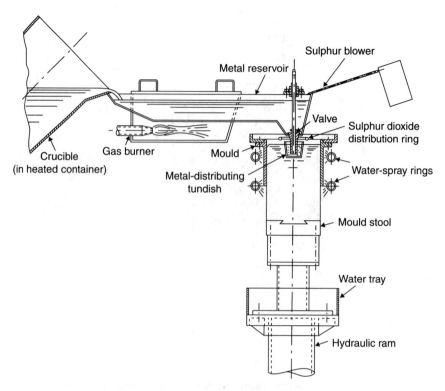

6.3 1950s magnesium DC casting technology. (Wilkinson and Hirst 1952.)

Semi-continuous vertical direct chill (VDC) casting remains the more common version of the process as opposed to horizontal direct chill (HDC) casting. Fully continuous VDC casting is achieved with a flying saw below the mould cutting the ingots to the desired length. DOW pioneered this approach at Madison, Illinois (Brown R. E., personal communication, 2010). This process was used for aluminium and magnesium and continues to be used for magnesium and copper today (Paddock 1954) (Fig. 6.4).

HDC casting with a flying saw was developed in France during the 1960s (Angleys 1960, 1964) and was adopted for small aluminium alloy remelt ingot, small diameter forging stock (Yoh 1989), rolling slab, 10 mm × 1000 mm strip (Moritz and Dietz 1979), T-bar (Gariepy et al. 2005) and billet production (Zeillinger and Beevis 1997; Niedermair 2001). Various mould designs and materials and lubrication delivery systems have been used. A typical configuration is shown in Fig. 6.5. Several attempts have been made to apply HDC casting to magnesium including DOW, Hydro and AMC (Kittilsen and Oistad 1996; McGlade 1999; McGlade and Ricketts 2001); however, due to technical difficulties (examined in detail in Grandfield's PhD thesis (Grandfield 2001)) this process has

Direct chill casting of magnesium extrusion billet and rolling slab 233

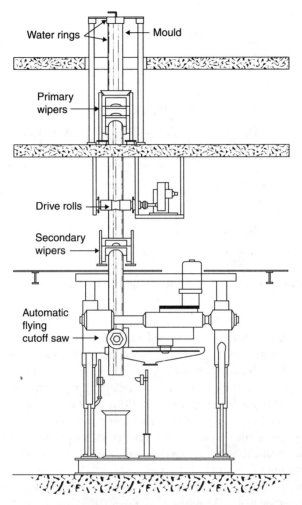

6.4 Process for fully continuous DC casting of magnesium used for many years by DOW Midland plant and still practised at the MEL Madison, Illinois plant.

not been used commercially for magnesium but continues to be investigated (Bong Sun You *et al.* 2005).

Aluminium extrusion billets are not scalped prior to extrusion and therefore control of the size of the surface mould chill zone region has been a strong driving force in the evolution of DC mould technology as applied to aluminium. The small volumes of magnesium extrusion billet have led to a general practice of scalping these ingots before extrusion and therefore there has been less emphasis on adopting those new mould technologies developed for aluminium.

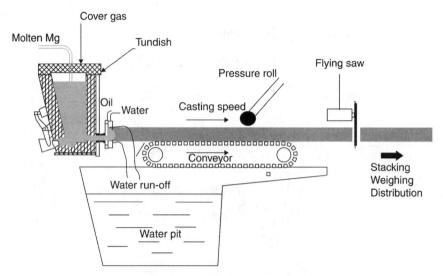

6.5 Schematic of HDC magnesium caster used at CSIRO, Brisbane in 1999.

Aluminium DC casting technology is relatively standardised and open around the world; however magnesium DC casting practice varies considerably from plant to plant and the technology tends to be rather guarded. Tens of millions of tonnes of aluminium are DC cast every year and typical capacities for aluminium VDC units are around 100 000 tonnes per annum. These installations are generally highly automated. The low volume of wrought magnesium that is DC cast (less than 20 000 tonnes per annum) is on a much smaller scale to aluminium and is a 'boutique', almost hand-craft activity, and has precluded extensive investment in large-scale magnesium DC casting technology.

The production of primary magnesium in the Americas and Europe has declined as China's primary production has increased. As a result, the tonnage of magnesium that is DC cast has also significantly reduced. Previously, Western companies used DC casting to make large format T-bar remelt ingot. The newer Chinese smelters are mainly producing small remelt ingots cast into open moulds. In the late 1990s about 45 000 tonnes of magnesium was being DC cast (Baker 1997), much of it remelt product; however today there is likely to be less than 20 000 tonnes per annum DC cast. DC casting facilities in the West that have closed include those of DOW, Hydro and Timminco. A large proportion of the DC cast ingots were large remelt T-bar ingots; a preferred format for the aluminium industry. US Magnesium and AVISMA in Russia continue to produce T-bar. Aluminium DC casting has continuously evolved and improved to meet increasingly stringent downstream customer requirements; however, the technology as applied to magnesium in many cases is exactly as that used in the 1940s, although there are small laboratory scale, highly automated units such as those in Korea, Taiwan China, Israel and Austria (Kettner *et al.* 2006).

6.2 Heat and fluid flow

DC casting heat flow is generally considered in two phases: the start of the process when the temperature field is changing, and the steady-state phase where the temperature distribution is relatively stable with the heat input balanced by the heat extraction rate. The process heat flow determines the temperature distribution and controls the solidification rate, solidification defects and stress development, which determines ingot deformation, final shape and the formation of cracks. It is thus essential to have a good understanding of the heat flow to minimise scrap and to control ingot microstructural quality.

Experimental methods (reviewed by Grandfield (2000)), such as thermocouple implant temperature measurements and liquid doping have revealed that a liquid pool forms during casting (Fig. 6.6). Where the metal first touches the mould, mould cooling causes solidification of a thin shell. The water spray has an upstream cooling effect into the mould and a thicker region of solid forms near the mould exit. In VDC castings the length of solid formed above the water spray is often known as the upstream conduction distance (UCD) (Harrington and Groce 1971; Devadas and Grandfield 1991; Flood et al. 1995; Grandfield et al. 1997a).

The solid in the mould consists of two components: the thin shell formed by the mould cooling, and the solid formed by the upstream effect of the water spray.

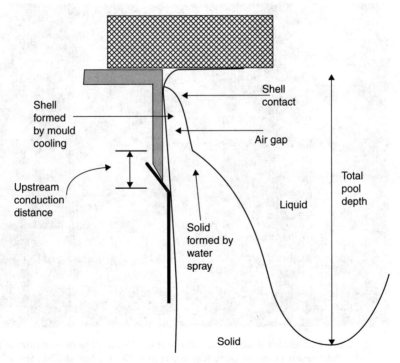

6.6 Typical shape of solidification front in hot top VDC casting.

In the case of a refractory hot top mould, the physical mould length differs from the shell length due to the meniscus determining the first point of metal contact where the shell begins to form and the water spray impact point not being exactly at the end of the mould. The spray is angled and the metal has contracted away from the mould at the mould exit. Shell formation for magnesium was studied in detail for the case of HDC casting (Grandfield 2000) and the shell shape is clear from Zn doping additions (Fig. 6.7).

During the steady-state part of the cast the temperature distribution is determined by a balance between the convective heat input (set by casting speed V, density ρ, ingot size R, specific heat C_p and latent heat L), heat extraction by diffusion (determined by diffusion path length, i.e. the ingot size R and thermal conductivity k) and convection cooling (described by the heat transfer coefficient h). For example, as the casting speed or ingot size is increased or the thermal conductivity of the alloy decreases then the pool depth increases. Two non-dimensional numbers have been used to characterise the heat balance (Flood *et al.* 1995; Hakonsen and Mihr 1995; Grandfield *et al.* 1997a): the Peclet number, $Pe = \rho C_p VR/k$ (the ratio of convective to diffusive heat flow) and the Biot number $Bi = hR/k$ (the ratio of resistance to heat flow from conduction to that from surface convection).

6.7 Sump and shell shape from doping the melt with zinc during HDC magnesium casting pure Mg 100 mm × 80 mm bar at 350 mm/min with a 35 mm long mould. (Source: Grandfield 2000.)

Typical non-dimensional number values for commercial aluminium VDC casting are: $1.8 < Pe < 4.5$ and $2 < Bi < 60$ (Flood et al. 1995). It can be observed that diffusion and convection are both strong in this process. In contrast, continuous casting of steel has much higher casting speeds and lower thermal conductivity values, giving larger Peclet numbers. The diffusive heat flow in the casting direction is very small compared to convection in the case of steel, resulting in large pool depths for given ingot dimensions and little axial heat flow. The low Pe for aluminium also explains why the pool depth is relatively small; typically the order of the diameter or width of the ingot.

Early studies (Roth and Weisse 1942; Dobatkin 1948; Adenis et al. 1962–1963) showed that pool depths in both aluminium and magnesium increase linearly with casting speed. The normalised pool depth Δ_{ss} (absolute pool depth divided by radius) was found to be linear with the Peclet number in a modelling study by Flood et al. (1995), i.e.

$$\Delta_{ss} = (a_1 + b_1 Pe)(c_1 + Bi^{-d_1})$$ [6.1]

where a_1, b_1, c_1, and d_1 are constants. This was later confirmed experimentally (Grandfield et al. 1997a).

The water spray in DC casting results in intense cooling (~6 MW/m^2 (Jensen et al. 1986; Grandfield and Baker 1987) and high Biot numbers. Approximately 90–95% of the heat is typically extracted by the direct water spray (Grandfield and Baker 1987; Hakonsen and Mortensen 1995) with the remainder going into the mould. For aluminium much work has been conducted to characterise the water spray heat transfer coefficients as a function of the boiling mode and spray parameters of the cooling water, such as impact velocity, jet angle and water composition (Weckman and Niessen 1982; Jensen et al. 1986; Grandfield and Baker 1987; Fjaer et al. 1992; Grandfield et al. 1997a, 1997b). This is further discussed below including recent data on Mg DC casting water sprays.

One tends to find that casting speeds used (often set empirically) for a given billet diameter and alloy tend to follow an inverse relationship, i.e. $V_c = \frac{F}{R}$ where V_c is the casting speed, R the radius and F some alloy-dependent factor. If one casts too fast then the ingot develops internal cracks. Although it has been known since the 1940s (Wilkinson and Hirst 1952–1953) that slower speeds must be used for larger diameter billet to prevent cracking it is perhaps not widely appreciated within the industry that the speed at which cracking starts for a given alloy is also found to follow an inverse relationship with diameter.

Cooling rates at the solidification front affect the size of the dendrite arm spacing and grain size. Small ingots naturally have higher cooling rates than large ingots. Typical DC cooling rates are in the order of 1–10 °C/s. The cooling rate is also higher near the surface of the ingot and drops toward the centre as a function of the pool shape and casting speed.

At cast start, the stable temperature distribution has not been established. Initially, when the mould is filled and the starter bar is in place but not moving, more heat leaves the system than comes in and a solid base builds up. Once the platen starts to descend and the ingot emerges into the water cooling the liquid pool starts to develop. Because there is a delay for the water cooling effect to act in the centre of the ingot there tends to be an overshoot of the pool depth before it reverts to the steady-state pool depth if a constant cast start speed is used (Fig. 6.8). Hot tear defects can often occur at this time (Jensen and Schneider 1990; Schneider and Jensen 1990).

In practice, two methods are used to avoid the overshoot of the pool depth and the possible occurrence of hot tears (Grandfield and Wang 2004):

- using an initial slow casting speed, which is then ramped up to the desired run speed after the stage of pool overshoot, or
- having a raised central section in the starting head.

The mould chill zone (MCZ) region has an unacceptable microstructure with surface segregation and a coarse grain region. If the size of this zone is minimised it allows smaller butt discard during extrusion and smaller scalping losses in the case of rolling ingot. Much of the evolution of DC mould technology has been driven by the intention of eliminating shell reheating by reducing mould cooling (Emley 1976; Grandfield and McGlade 1996). The length of the UCD is found to vary inversely with casting speed and alloy thermal conductivity (Harrington and Groce 1971; Flood et al. 1995; Grandfield et al. 1997a) and is generally around 10–20 mm. Surface defect formation is further discussed below.

Fluid flow has an important influence on the heat flow and temperature distribution and on macrosegregation, and has been widely studied (Davidson and

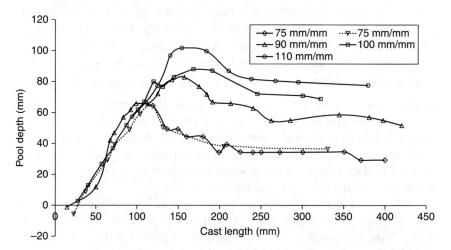

6.8 Development of pool depth measured by dip rod as a function of cast speed for 228 mm diameter 6061 alloy billet. The pool depth goes through a maximum before reaching equilibrium. (Source: Grandfield et al. 1997.)

Flood 1994; Hakonsen and Mortensen 1995; Reese 1997; Jones *et al.* 1998). Flow patterns have been found to affect the temperature distribution within the liquid pool, pool profile and shell temperature (Raffourt *et al.* 1990; Brochu *et al.* 1993). The distribution systems used to deliver the liquid into the mould are therefore very important in controlling surface defects, macrosegregation and internal structure.

Mathematical modelling of the process has developed since Roth made limited thermal calculations (Roth 1943). Heat flow in DC casting was first modelled with a digital computer and numerical methods in the 1960s (Adenis *et al.* 1962–1963); this study was on magnesium DC casting. Weckman reviewed mathematical modelling of DC casting in 1984 (Weckman and Niessen 1984b) but the field has seen rapid development and application to aluminium DC casting. Fully coupled heat and fluid flow stress analysis models are available (Gruen and Schneider 1996; Gruen *et al.* 2000; M'hamdi *et al.* 2000, 2002a, 2002b, 2003, 2004, 2006; M'Hamdi and Mo 2002a, b, 2005, 2008; M'Hamdi and Hakonsen 2003; Mortensen *et al.* 2008). These have been extensively verified against experimental measurements. These models also include macrosegregation, defect and microstructure prediction (Drezet *et al.* 1995; Drezet and Rappaz 1996, 1997).

Prediction of strains and ingot deformation have been well verified but residual stress measurement has been difficult. Residual stresses in DC cast product were measured by removing successive inner layers from round ingots while measuring the displacement in the diameter and length (Roth *et al.* 1942). These results were used for comparison with a stress model (Bohmer and Jordan 1995). The hole drilling strain gauge method often used to measure residual stresses in welds has also been used to measure residual stresses in DC product (Inoue and Ju 1989). Residual stresses in a 126 mm diameter billet were also measured using X-rays (Moriceau 1975). Displacements and final ingot shape are easier to measure than stresses and have been used more widely to verify predictions from stress/strain models. Measurement of the displacement of the butt of the ingot at cast start have been made (Droste and Schneider 1990).

The flatness of rolling ingot can be measured using the mould as a reference after casting (Drezet *et al.* 1995). The displacement of the shell in the mould and the air gap formation was confirmed experimentally in the same study. The amount of contraction between the final ingot size and the mould opening varies according to ingot size. It is greater than the linear thermal contraction.

Little work was done on modelling DC casting of magnesium after the initial work by Adenis *et al.* (1962–63); however in the last ten years there has been a surge of activity. Hibbins (1998) examined water spray heat transfer during Mg DC casting by implanting thermocouples during casting of rolling slab and using a 3D thermal model to deduce position-dependent heat transfer coefficients. Values of the heat transfer coefficient varied from 7000–12 000 W/m^2.

Grandfield conducted thermal modelling and an examination of mould heat transfer coefficients during HDC casting of AZ91 and pure magnesium. Similar mould heat transfer coefficients as those found in aluminium DC casting were found (Grandfield 2000). Grandfield et al. (2009) examined measured hot tearing defects at cast start and compared the incidence to the pressures predicted in the mush using the fully coupled Alsim simulation package developed by IFE in Norway.

Hao et al. modelled stresses (Hao et al. 2005) in AZ31 billet casting and later attempted hot tearing prediction with this model (Hao et al. 2010). Recently, Turski et al. applied a fully coupled thermal, fluid and mechanical model to the problem of hot cracking during casting of high-strength WE43 alloy and developed improved practices to eliminate cracking (Turski et al. 2010). A key feature of this work was the extensive measurement of thermal and mechanical properties and model validation using implant thermocouple measurements. This model was also compared to residual stress measurements (Turski et al. 2010).

6.2.1 Mould cooling

DC mould cooling determines the formation of the initial ingot shell and has a large effect on surface microstructure and surface defects. When the liquid metal first touches the mould the degree of contact is high and the measured heat transfer coefficients are of the order of several 1000 W/m^2 K and a solid shell rapidly starts to form. However, because the ingot below the mould experiences thermal contraction due to the cooling of the water sprays, the shell inside the mould is pulled away from the mould once it is sufficiently strong. Thus, an air gap forms with a correspondingly low heat transfer rate causing reheating of the shell to the extent that it becomes semi-solid.

The range of typical mould heat transfer coefficient values (Ho and Pehlke 1985) for both aluminium and magnesium, are around 1000–2000 W/m^2 K during mould contact and 100–500 W/m^2 K for the air gap (Adenis et al. 1962–1963; Jensen 1984; Weckman and Niessen 1984a; Baker and Grandfield 1987). The exact values for the contact region are thought to depend on the pressure between the metal and mould, the lubrication conditions and the mould roughness. As with air gap formation in permanent mould casting, the size of the gap and the thermal conductivity of the gas in the gap is believed to control the heat transfer (Ho and Pehlke 1985; Nishida et al. 1986; Muojekwu et al. 1995). In permanent moulds with rapeseed oil lubricant, the gas has been shown to be a mixture of lubricant decomposition and combustion products, i.e. water vapour, methane and air (Muojekwu et al. 1995). The size of the air gap during DC casting was measured by Drezet and Rappaz (1997) using displacement sensors and shown to be 1.5 mm at the mould exit for a 510 mm thick 3004 alloy rolling ingot. Recently, shell temperature measurements inside the mould have been made showing that as mould heat flux increases, the shell temperature drops and *vice versa* (Bainbridge and Grandfield 2007).

The reheating of the shell gives rise to exudation out of the surface of the solute-rich liquid in the semi-solid shell region under the driving force of the metallostatic head pressure, giving rise to compositional variation at the surface and surface bumps. The reheating can give rise to periodic behaviour in shell temperatures as the shell becomes weak on reheating and the metallostatic head pushes it back against the mould causing cooling. The details of formation of this mould chill zone (MCZ) region are discussed in a number of studies (Bachowski and Spear 1975; Buxmann 1982; Ohm and Engler 1989a, b; Grandfield *et al.* 2005). The air gap causes a reduced solidification rate, remelting of the shell and formation of a coarse DAS region below the surface. Under the influence of the metallostatic head pressure, the solute rich liquid exudes out of the surface (Mo *et al.* 1997). This exudate can appear as raised areas of the order of 1–10 mm diameter or as sweat bands, depending on the alloy. Oscillatory behaviour can arise due to the shell reheating causing loss of shell strength and collapse back against the mould. This has been observed in mould temperature measurements (Bakken and Bergstrom 1986; Fjaer *et al.* 2000) and periodic composition variation on the ingot surface in the casting direction (Siebel *et al.* 1953).

An interesting early comment by Roth and Weisse (1942) is that surface exudation is reduced by using an inwardly tapered mould (0.5–1% of diameter over the length of the mould) to reduce air gap formation. The mechanism proposed for this improvement was that the reheating of the shell can be limited by limiting the air gap width, thus reducing exudation. Tapered moulds are not used in non-ferrous DC casting, possibly because tearing problems are largely controlled by good lubrication practice. However, it is common practice in ferrous continuous casting.

Mould temperature measurements at various points within open top moulds (Muto *et al.* 1996; Hamadaiid 2004) and hot top moulds have been made (Jensen 1984; Instone *et al.* 2003; Bainbridge and Grandfield 2007) showing the effects of casting speed, mould lubrication and alloy composition on mould heat fluxes. The gas-pressurised hot top moulds show much reduced heat fluxes. Mould heat transfer coefficients and the formation of the air gap have been inferred from mould temperature measurements.

Grandfield (2001) measured mould temperatures during HDC casting of pure magnesium and AZ91 alloy at between 40–110 °C depending on casting conditions. Calculation of the mould heat flux from the pure magnesium pool profiles, for the 100 mm × 100 mm mould casting 99.9% Mg alloy at 350 mm/min, gave values of ~1200–1800 kW/m^2 in the shell contact region and a low value in the air gap of ~50–160 kW/m^2. These calculated shell contact region values fall within the range of values calculated from the measured mould temperatures for the position 5 mm from the refractory. These varied between 500–2000 kW/m^2 within a given cast and generally averaged around 600 kW/m^2. Heat fluxes calculated from mould temperatures for pure magnesium cast with the 80 mm × 100 mm mould were around 200–900 kW/m^2, slightly changing with cast length but for AZ91, mould heat fluxes were much higher around 1000–3000 kW/m^2 (Grandfield and Dahle 2000) Thermocouples inserted in the shell during casting showed that for alloy AZ91 the shell is semi-solid for most of the time in the mould (Fig. 6.9).

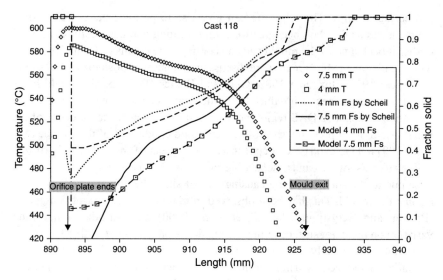

6.9 Measured temperatures in the shell region; AZ91, 80 mm × 100 mm mould, 250 mm/min. cast speed. Note shell is semi-solid for most of the mould length. Fs, fraction solid. (Source: Grandfield 2001.)

Lubrication type and practice and mould material have significant effects on mould heat flow and shell and surface defect formation. There has been a great deal of development of the mould technology to control mould heat flow, and there are many variants. These are discussed below, including the important breakthrough of gas-pressurised hot top casting.

6.2.2 Water cooling

The DC water cooling has a profound effect on the process. The direct chill water spray results in a high heat flux of the order of 2–10 MW/m^2. The heat flux depends on the water boiling mode on the surface of the ingot. As with other water cooling processes, nucleate, unstable film and film boiling can occur in DC casting. Ingot surface temperature, water composition, water temperature and the physical characteristics of the water jet and falling film (velocity flow rate per perimeter, angle of the jet) all affect the boiling mode and the heat transfer coefficients.

The boiling regimes are best described in terms of a boiling curve showing heat flux or heat transfer coefficient as a function of ingot surface temperature. Below ~100 °C convection cooling occurs, above that nucleate boiling takes place where steam bubbles nucleate in non-wetted crevices on the ingot surface (Fig. 6.10). The release of these bubbles into the water generates stirring in the water and reduction of the boundary layer and a high heat flux. These bubbles collapse in the water. If the ingot surface temperature is higher than some critical temperature

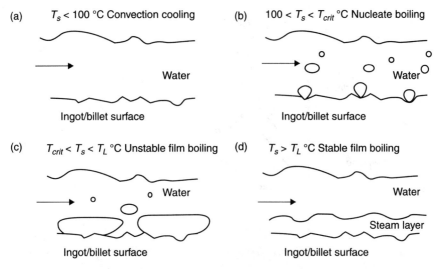

6.10 Schematic of boiling regimes that can occur in the DC casting water cooling. (a) $T_s < 100$ °C, convection cooling. (b) $100 < T_s < T_{crit}$ °C, nucleate boiling. (c) $T_{crit} < T_s < T_L$ °C, unstable film boiling. (d) $T_s < T_L$ °C, convection cooling.

there are so many steam bubbles that they start to agglomerate and form a periodically collapsing steam layer on the ingot. This boiling regime is known as unstable film boiling where the heat flux is reduced by the low thermal conductivity of the steam layer. At temperatures above the Leidenfrost temperature the steam layers become stable and fully stable film boiling occurs.

The water cooling can be divided into the impingement zone where the water jet first impacts the ingot and the falling film below the impingement zone (Fig. 6.11). Typically, there is some unstable film boiling in the impingement zone where ingot surface temperatures are around 350 °C but the ingot surface temperature drops rapidly below the critical temperature and nucleate boiling occurs in the falling film. The surface temperatures drop rapidly and most of the falling film cools by convection cooling.

Many studies of DC casting water cooling have been made (Yu 1985; Bamberger et al. 1979; Bakken and Bergstrom 1986; Fjaer and Mo 1990; Watanabe and Hayashi 1996; Grandfield et al. 1997; Maenner et al. 1997; Caron et al. 2005). By freezing in thermocouples and using inverse calculation, Rappaz et al. (1995) found the peak heat flux in the water spray was 4 MW/m^2, similar to other published values. Grandfield and Baker (1987) found that under normal water flows for a 150 mm ingot, the surface temperature at the impact point was around 250–300 °C and nucleate boiling occurred with a peak h of 40kW/m below a critical temperature of around 150 °C. If the impact velocity was reduced to 1 m/s by increasing the outlet area of the water slot (flow rate per perimeter kept constant) film boiling then occurred and surface temperatures remained high.

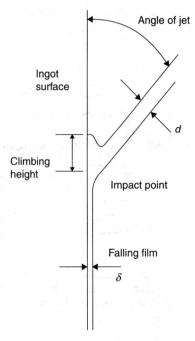

6.11 Configuration and key features of water spray in vertical DC casting.

In most cases nucleate boiling is the desired boiling regime but there are cases where film boiling is deliberately induced to achieve a lower cooling rate. For example, at the start of slab casting the initial butt formed lifts up at the ends due to thermal stresses causing cracking at cast start.

This can also cause problems with liquid break outs and/or the ingot can jam in the mould, resulting in mould damage or catastrophic release of the ingot from the mould. It has been found that by reducing the intensity of water cooling at the start of the process by inducing film boiling, the amount of butt curl can be reduced. Film boiling can be induced by dissolving CO_2 in the water (Yu 1985), pulsing the water on and off so the surface temperature stays above the Leidenfrost temperature, and using a low water flow rate.

Because most plants use recycled evaporative cooling water systems the water composition varies over time. The water lost during evaporation must be made up with fresh water. The dissolved salts build up in the water also due to evaporation, and water is periodically dumped to control the total dissolved solids (TDS) concentration. The higher TDS level, i.e. the harder the water, the higher the Leidenfrost temperature (Grandfield *et al.* 1997) thus variation in water cooling behaviour can arise due to composition variation in the water. Casting lubricant also finds its way into the water system and most be removed both for environmental reasons and because it encourages film boiling. In practice, well run DC casting

Direct chill casting of magnesium extrusion billet and rolling slab

installations pay careful attention to management of the water cooling system and water composition variation.

Hao et al. (2004) made a comparison of measured pool depths from doping and thermocouple insertion during magnesium DC casting with a thermal model. Pool depth was linear with casting speed. Alternative heat transfer coefficient (htc) (T) boiling curves taken from the literature related to aluminium casting were examined and used as a basis. Caron et al. examined water spray heat transfer coefficients for DC casting of magnesium using a quench rig apparatus in tests with AZ31 (Caron et al. 2005). Similar htc values were found as in other studies (Etienne et al. 2005).

6.3 Magnesium direct chill (DC) casting technology and engineering

6.3.1 Melt delivery systems for DC casting

Aluminium DC casting units are fed with liquid along refractory-lined troughs from the furnace to the casting unit via gravity. The liquid then either flows into a flooded refractory table in the case of hot top casting or via downspouts into the mould in the case of rolling ingot casting. The flow through the spouts is usually controlled by automatically positioned pins.

Magnesium is far more reactive to refractories than aluminium but on the other hand is far less reactive to steel. Steel-based systems have therefore historically been used to deliver magnesium liquid to the DC casting units. Delivery can either be gravity fed from a tilting furnace along a trough and down a spout or pumped via steel tubes into the mould.

6.3.2 Cover gas systems for DC casting

Unlike aluminium, magnesium requires some form of protective cover gas to prevent spontaneous combustion. Varying gases and delivery systems have been used in DC casting systems including sulphur, SO_2, air/SF_6 and R134a.

6.3.3 Open top mould casting

Until the late 1970s aluminium extrusion billet DC casting used float and spout open top moulds (Fig. 6.12) and then largely switched to level pour flooded table hot top moulds (Fig. 6.13) (Emley 1976). However, because magnesium tends to react with many refractories there has been far slower uptake of hot top technologies, and magnesium billet casting has used open moulds with the level manually controlled either by the tilt rate on the furnace, if the mould is fed by a steel trough, or by controlling the pumping rate if fed by a pump.

With the open top moulds the metal level was controlled by a float in the case of billet casting. This necessitated relatively high metal levels in the mould, which

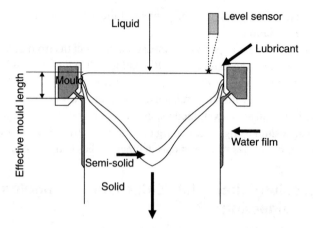

6.12 Typical open top casting set up for magnesium.

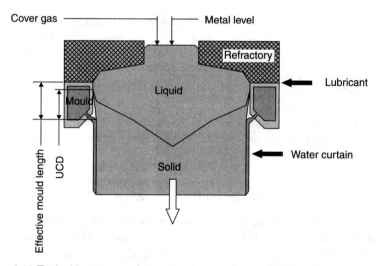

6.13 Typical hot top configuration for magnesium VDC casting.

in turn led to a large degree of shell reheating and poor surface quality. It was well known that if a low metal level was used then a smaller mould chill zone and less surface segregation resulted, as the initial metal contact point approached the extent of the upstream water cooling effect. However, these low levels cause difficulties at cast start and it is not possible to change the level with a float technology (one system known as IsoCAST moved the whole mould table up on screws to lower the metal level). An alternative approach is to use a hot top mould where a refractory top determines the point of initial metal contact against the mould. These moulds produced substantially better surface quality than open top moulds.

Direct chill casting of magnesium extrusion billet and rolling slab

The use of open top moulds to cast AZ31 billets was described in detail by Grandfield *et al.* (2003). Metal was delivered to a single strand VDC unit using a mechanical pump, the level being manually controlled. Lubricant supply was by gravity feed from a reservoir. Cover gas was supplied by a ring around the perimeter of the mould.

Very simple copper open top moulds positioned below the furnace are used in China to produce small VDC cast magnesium alloy slabs.

The use of carbon-lined low-level moulds allows lubrication usage to be minimised (Wagstaff and Bowles 1995). These moulds have been used to produce magnesium slab and T-bar.

6.3.4 Hot top mould development

Initially, hot top moulds were simple open top moulds with a refractory liner (Moritz 1961) still with a float, but later the refractory became an integral part of the mould and was used to deliver the liquid metal via a refractory pan on top of a water box containing the moulds (Furness and Harvey 1975). This allowed level pour delivery from a tilting furnace with reduced dross generation. These systems have relatively high metal head pressures of 100–200 mm head, and this results in the metal being pushed into the corner below the refractory and some cold folding as the meniscus freezes.

Hot top casting has not been as widely used for magnesium with open top moulds remaining more common.

Multi-strand hot top casting of T-bar and billet was conducted at Hydro's Becancour plant (Kittilsen and Pinfold 1996).

Baker *et al.* developed multi-strand hot top mould technology for large diameter billet casting of magnesium alloys at Timminco (McGlade and Baker 2001). Because magnesium has a lower volumetric latent heat than aluminium it is more prone to freezing in the corners below the hot top (see Grandfield *et al.* 2007). This was also a problem for HDC casting (Grandfield 2001).

6.3.5 Gas-pressurised hot top moulds

In the late 1970s Showa Aluminium substantially improved the performance of hot top moulds with the invention of the gas-pressurised hot top mould (Mitamura *et al.* 1978; Mitamura and Itoh 1979; Yanagimoto and Mitamura 1984). In the Showa configuration, air was injected through small grooves on a metal plate just below the refractory at a pressure equal to the metallostatic head pressure resulting in a larger, more stable meniscus, low mould heat fluxes and elimination of shell reheating. The gas-pressurised moulds produce billets with very smooth surfaces and almost no surface segregation.

Many other companies followed the Showa development with variants having air injected at different points and via different delivery systems, e.g. through porous graphite mould liners along with continuous lubricant delivery (Wagstaff et al. 1986; Steen et al. 1997; Apostolou and Armaos 1988; Schneider and Kramer 1994; Bainbridge et al. 2005).

At one time, Hydro used gas-pressurised hot top moulds to produce magnesium billet (McGlade and Baker 2001).

Direct observation of the meniscus has shown it to be the same size as an open-to-mould meniscus with no metal head (Ekenes and Peterson 1990) and can be readily predicted (Baker and Grandfield 2001). The mould heat flow is very low compared to a hot top without gas pressure (Baker and Grandfield 1987; Bainbridge and Grandfield 2007) due to the fact that the gas pressure moves the base of the meniscus to the point where the solid shell formed by the water spray is being pulled away from the mould. The meniscus is stabilised whereby the base is near the point where the shell formed by the water spray is being pulled away from the mould. There is thus essentially no mould contact and therefore very little mould heat flow, and the shell re-heating mechanism, which normally gives rise to macrosegregation, is eliminated. Some inverse segregation can still occur (Benum et al. 1999).

It is important to realise that the mould length of a gas-pressurised hot top mould is a key factor (Bainbridge and Grandfield 2007). If the mould is too long then gas-pressurised mode cannot be obtained at normal casting speed and, if too short, cold folding will occur at normal speeds.

The gas-pressurised hot top technology has continued to evolve with improved gas pressure control systems.

The gas-pressurised hot top process has been applied to magnesium using the Showa mould in 1987 (Yanagimoto 1987) and more recently (Grandfield et al. 2007) with a design specifically tailored to magnesium. However, the process is not in commercial production.

6.3.6 Rolling slab casting

The evolution of aluminium rolling slab casting mould technology has followed a different path to billet casting technology. In order to get better metal level control during casting to achieve smaller mould chill zones and scalping losses, float level control was superseded by the use of a downspout coming from a trough over the moulds with a pin inside the spout. The pin position was controlled by a lever arrangement with a refractory float on the melt. This 'steady eddy' system enables manual adjustment of the level during casting via positioning of a counter-weight on the lever. It is also essential to use a flow distribution system in the metal pool for slab casting to ensure an even melt temperature distribution. The short sides being further from the inlet are otherwise too cold and the long sides too hot. In practice a fibreglass permeable cloth distributor system is used. Designs vary enormously from plant to plant, and optimum distributor design for rolling slab casting is the subject of ongoing research (Tremblay and Lapointe 2002; Hasan and Ragel 2009).

Conversion of slab moulds to hot top casting has been done using the simple refractory liner system but a flooded table hot top system has not been adopted due to practical difficulties, e.g. the large amount of butt curl at the start of casting can drive the slab up into and destroy the refractory hot top.

6.3.7 Electromagnetic DC casting

Electromagnetic DC casting (EMC) was developed in the1960s in Samara, Russia to reduce mould cooling and achieve improved surface quality (Getselev *et al.* 1969). A high-frequency AC current is passed through the mould creating induced currents in the melt and a levitating force balancing the metallostatic head pressure to hold the liquid away from the mould. In this configuration all heat flows to the water spray and there is no shell reheating effect. Excellent cast surface quality is obtained as a result. Although EMC was applied to aluminium DC casting in the West through the 1970s and 1980s (Pritchett 1973; Bergmann 1975; Dunn 1979; Nagae *et al.* 1988; Hudault *et al.* 1989; Goodrich and Tarapore 1991) it has all but disappeared due to its high capital cost and marginal benefits, now that the conventional process has been improved via automation of metal level control. EMC has also been applied to DC casting of magnesium; in this case a low-frequency current has been used to induce stirring in the melt and achieve grain refinement (Zhang *et al.* 2008).

6.4 Solidified structures and defect formation

6.4.1 Grain structure

A major difference between aluminium and magnesium wrought alloy DC casting is the use of grain refiner. For aluminium DC casting, ingots destined for extrusion and rolling are nearly always grain-refined titanium and titanium boride to reduce grain growth and to act as nuclei to achieve small grain sizes (Easton and St John 1999a, 1999b), which reduces the incidence of cracking. It also eliminates any as-cast crystallographic texture (Granger 1989). For MgAl alloys there is no similarly reliable technology.

Various methods around producing carbides by adding wax, etc., are used for magnesium but these methods remain inconsistent. Consequently cracking is a major problem during DC casting of MgAl alloys. Bainbridge and Grandfield found that in order to avoid a condition resulting in freezing against the refractory hot top causing surface tears, higher casting speeds were required, but at these speeds cracking occurred (Grandfield *et al.* 2007). ZrB_2 offers some promise as a nucleant but being such a heavy particle it tends to sink rapidly.

6.4.2 Typical surface defects

A variety of defects can form on the surface of DC castings depending on the mould technology, casting practice and alloy. Cold folding defects form when the meniscus is frozen, then pulled down and new liquid laps over the previously frozen material. They may be eliminated by increasing casting speed and or using gas-pressurised or electromagnetic moulds, which push the liquid meniscus away from the mould. Because magnesium has a lower volumetric latent heat than aluminium it is much more prone to cold folding, thus an extended discussion of cold folding is given below.

In 1954, Porro and Lombardi discussed cold folding of aluminium during DC casting (Porro and Lombardi 1954). Waters conducted tests with aluminium, copper, zinc and lead in 1953 and found that the surface quality was improved as casting speed increased (Waters 1954). The cold fold spacing was found to decrease as casting speed increased. These experiments were performed with 1 inch diameter unlubricated moulds, one of which was a Pyrex mould to enable visual recording and direct observation of the meniscus freezing. The cold fold spacing for the different metals at a constant cast speed was also successfully related to the surface tension-to-density ratio of the metals. Surface tension-to-density ratio determines the height of the meniscus.

In the early 1970s, Bergmann (1970, 1973, 1975) showed clearly that for the case of VDC casting (similar to the HDC case shown in Fig. 6.14), the mechanism of cold folding involves these steps:

1 the meniscus freezes,
2 it is pulled away from the refractory in the casting direction,
3 solidification proceeds into the ingot and then opposite to the casting direction until,
4 the diffusion path is too long and liquid breaks through to fill the gap, and
5 the cycle repeats.

The steps in the formation of a cold fold during HDC casting can be seen in Fig. 6.14:

1 In Fig. 6.14a, liquid has just flooded into the void from the previous cold fold.
2 In Fig. 6.14b, the liquid is solidified and heat loss to the mould is interface-controlled, resulting in a linear growth rate from the mould. Combined with the removal at the casting speed, this results in a straight solid surface with angle θ.
3 In Fig. 6.14c, as the solidification front gets further away from the mould and the diffusion path is longer, heat flow becomes diffusion-controlled and parabolic with distance and thus, the solidification front slows down and curves up.
4 Finally, in Fig. 6.14d, when the fold reaches a depth d, the horizontal solidification rate is less than the casting speed and solid is no longer in contact with the refractory and surface tension holds the liquid in place. When the gap becomes too big, liquid floods into the void, repeating the cycle again.

Bergmann also measured the temperature in the refractory orifice plate (OP) where it overhangs the mould. The cyclic nature of the folding process was

Direct chill casting of magnesium extrusion billet and rolling slab 251

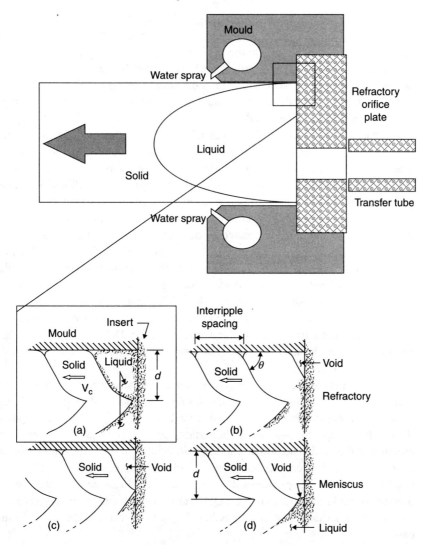

6.14 Steps in formation of a cold fold during HDC casting. (Source: Weckman and Niessen 1984c.)

evident in the cycling of OP temperature due to liquid periodically contacting the refractory. Formation of folds was found to be controlled by the mould heat transfer, the casting speed and the geometry of the refractory overhang.

Weckman and Niessen (1984a) confirmed the previous VDC result that the cold fold spacing becomes smaller as casting speed increased in HDC casting of Al, Zn, and Pb. As the heat input increases with increasing casting speed, the solidification front does not move as far along the meniscus before the thermal diffusion path length becomes too long, allowing the liquid to flood against the mould again. The ratio of the rate of the solidification front movement perpendicular to the mould and the casting speed sets the angle of the cold fold relative to the mould face. Weckman and Niessen (1984) assumed that the growth rate is initially interface-controlled (a good assumption given the Biot number) and were then able to calculate the mould heat transfer coefficient from the angle of the fold and the casting speed.

Weckman also modelled the effect of changes in the mould heat transfer coefficient (Weckman and Niessen 1983). The results showed that an increased mould heat transfer enables the solidification front to extend further into the product, resulting in deeper cold folding. Both Bergmann (1970) and Weckman and Niessen (1984d) studied the effect of metal head. The severity of cold folds increased as the head was increased. Weckman and Niessen (1984a) found that the mould heat transfer increased as the head was increased and proposed that this was caused by a greater degree of contact between the hot semi-solid shell and the mould. It was further proposed that the friction with the mould also increased with head, sometimes causing tearing of the shell. This is one reason why HDC casting can be prone to cold folding and surface tears compared to VDC as the metal head pressures are higher (usually about 400–500 mm for HDC compared to 100–200 mm for vertical hot top DC casting). The reason for the difference in head between HDC and VDC is more a matter of general practice rather than necessity.

Weckman proposed that not only does the metal head affect the mould heat transfer but that the fold also collapses sooner as the head is increased. The differential pressure across the material at the final stage of the fold development (Fig. 6.14d) was proposed to be important, i.e. the head pressure minus the pressure in the void. The pressure in the void is influenced by the amount of gas evolved due to lubricant decomposition and whether gas is injected into the mould, as in the gas-pressurised processes. Therefore, the lubricant supply pressure should be greater than the metallostatic head pressure to reduce cold fold formation. Other defects such as gas bubbles in the surface might occur in this case.

The geometry of the OP has also been found to affect cold folding. Several different arrangements of the OP are possible (Fig. 6.15). They can be quantified by the amount of overhang and the angle of the chamfer. Bergman (1970) found that the depth of the cold folds was reduced when the overhang was reduced. He also found that the cold folds were eliminated when the draft angle exceeded 45° to the horizontal (in the VDC case). Weckman and Niessen (1984) further studied the effect of the overhang, using the key assessment parameter of whether the overhang was less than the cold fold depth (d in Fig. 6.14a). In the case where the overhang is large, the cold fold proceeds as illustrated in Fig. 6.14. If the overhang is about the same as the cold fold depth then solidification proceeds up onto the refractory (Fig. 6.16).

Direct chill casting of magnesium extrusion billet and rolling slab

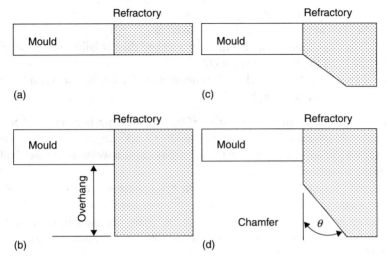

6.15 Possible orifice plate geometry arrangements.

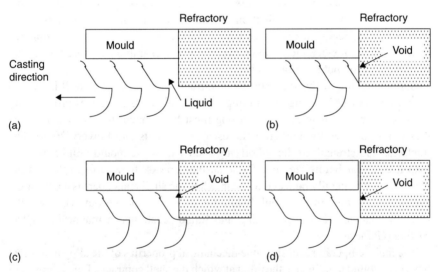

6.16 Steps in formation of a cold fold in HDC, when the OP overhang is about the same as the depth of the cold fold. (a) Liquid has just flooded into the void. (b) Solidification proceeds along the overhang. (c) When the limit of the overhang is reached then solidification proceeds onto the horizontal surface of the OP. (d) The point is reached where the solidification path is too long and the liquid is about to break through.

Weckman and Niessen (1984) found experimentally that cold folds are eliminated if the overhang is essentially zero but that there is the risk of solidifying material into the refractory and causing tears. Also, if the refractory is not flush with the mould and recessed behind the mould face, solid can form on the lip and cause tearing:

1. Liquid has just flooded into the void.
2. Solidification proceeds along the overhang.
3. When the limit of the overhang is reached then solidification proceeds onto the horizontal surface of the OP.
4. The point is reached where the solidification path is too long and the liquid is about to break through.

Cold folds may cause shell tearing if the cold folding is extensive because the solid formed may jam inside the mould (Weckman and Niessen 1984a). In addition, the interface between the folds may cause a weak spot in the shell where surface tearing could initiate.

Cold folding tends to be worse with magnesium due to the lower volumetric latent heat compared to aluminium. Freezing of metal against either the refractory hot top in the case of VDC casting or the transition plate in the case of HDC casting is a common problem (see Grandfield 2001; Bainbridge and Grandfield 2007).

Bergmann (1973) found that graphite as a mould material reduced cold folding compared to a metallic mould. This is probably because of the lower thermal conductivity of graphite. Conduction of heat through the mould is generally not the controlling heat flow resistance with a metallic mould but it may become the controlling factor when the conductivity is low. For example with a carbon mould the cold folds do not penetrate as deep.

Surface tearing of the shell and subsequent liquid break out is well known to practitioners of DC casting and an important problem, as it reduces productivity causing scrap and in some cases casting must be stopped. It is also a potential safety hazard when the molten metal escapes (Roberts and Lowery 2009). The thickness and strength of the solidifying shell within the mould will control the load it can bear before rupture. These surface cracks, transverse to the casting direction, are generally believed to occur when the shell tears open as a part of the shell remains stuck in the mould while the product is pulled out. This was the main mechanism of surface tearing and liquid breakouts in magnesium HDC casting (Grandfield 2001).

The high-temperature mushy zone mechanical properties of the alloy being cast have been found to influence the point at which the shell contracts. Pure aluminium was found to pull away quickly while less rigid wide freezing range alloys such as 5182 and 7075 maintain contact with the mould for longer periods (Bachowski and Spear 1975). It has been concluded that wide freezing range alloys are more prone to tearing because their shells maintain contact with the mould longer than dilute alloys, resulting in greater friction forces acting on the shell (Emley 1976; Ohm and Engler 1989). It was further argued that the more dilute alloys obtain sufficient strength at higher temperatures and are able to pull away from the mould at a higher shell temperature. It was also concluded that higher casting speeds not only increased friction on the shell, but also increased the shell temperature and reduced its strength. Lubrication theory also suggests that higher speeds will generate greater friction forces (Grandfield 2001). In HDC casting there may be differences in forces acting on the top and bottom shell due to the difference in metal head pressure, but also if the mould is not correctly positioned vertically there is a misalignment between the emerging ingot held in place by the press-down roller on the conveyor and there may be difference forces on each shell. If the mould is positioned too high, the bottom shell would be forced against the mould base. If positioned too low, a pressure will be applied to the top shell. The metallostatic pressure will be applied to the shell, regardless of mould position. If the shell is hot and weak then it is pushed against the mould. In this case, friction might be expected to be affected by metal head pressure. Some measurements of friction loads between the shell and the mould have been published for DC casting (Donnerburg and Engler 2000). The load was 10–100 N on a 156 mm diameter VDC cast billet and found to decrease with increasing oil flow rate and to be dependent on the type of lubricant used.

Friction load data are in fact rare for all types of continuous casting. Yao and Fang (1996) calculated friction loads during continuous casting of steel by measuring the power on the motor oscillating the mould. Loads were around 4 kN; however, no friction coefficients were calculated. The coefficient of friction between hot copper and steel (to simulate the friction between a steel shell and mould in continuous casting of steel) was found to be 0.4 with rapeseed oil as the lubricant (Sorimachi and Nabeshima 1998). Friction loads for leaded gunmetal on a carbon mould have been measured at around 0.5–2 kN (Thomson *et al.* 1969). A value of 280–560 kPa is given for interfacial stresses for continuously casting of aluminium, in an unlubricated graphite mould, without a direct chill spray (Thomson and Ellwood 1972).

Friction and inadequate lubrication certainly play a role in surface tear formation and all DC casting is practised with some form of lubricant in order to prevent drag marks that may appear along the product in the casting direction, if there is insufficient lubrication where local welding takes place between the ingot and the mould. If casting continues with insufficient lubrication, then shell tearing eventually occurs (Ohm and Engler 1989; Laemmle and Bohaychick 1992).

Laemmle and Bohaychick (1992) examined the viscosity index and thermal stability of various lubricants. They argued that castor oil performs well because it maintains a high viscosity at high temperature, enabling the lubricant film to remain continuous. They also reported that oil vaporisation caused deposition of varnish-like deposits, which presumably increase friction, and may therefore result in tearing. This was also reported by Jacoby *et al.* (1986). The varnish-like deposits are attributed to fatty esters including triglycerides present in castor oil.

Weckman (1987) discussed the requirements for mould lubrication and compared a number of lubricants including castor oil, rapeseed oil and some synthetic oils. The main requirements are a high viscosity and high viscosity index (i.e. low sensitivity of viscosity to temperature). He also examined the defects that occur with either too little or too much lubricant. With insufficient lubricant, sticking, tears and heavy cold folding occur. With too much lubricant, a so-called 'Chevron' defect formed (i.e. a bent half-moon type shape) on the top surface and ripples formed on the bottom surface of HDC casting. The lubricant is presumed to act as a parting agent and prevent direct metal/mould contact. Direct contact would cause an increased heat transfer rate since heat flow would be directly through the metal rather than through the oil and oil vapour.

Waters (1952) examined the effect of alloy composition on shell tearing propensity, relating increased alloy freezing range to an increased probability of tearing. Waters suggested that both mechanical grabbing and poor lubrication contributed to tearing. Langerweger (1981) ascribed formation of 'transverse surface cracks', i.e. tears in aluminium VDC casting, to the presence of oxides combined with salt particles that stick to the mould wall. Nofal (1992) also suggested inclusions caused tearing. The importance of lubrication to prevent tearing is emphasised throughout the papers on HDC casting. Formation of lubricant tar deposits on the mould was reported to cause tearing and the mechanism was described as the lubricant deposits being 'detrimental to the thermal transfer' (Spear and Brondyke 1971). Presumably, a thinner, weaker shell was in the mind of these researchers. An alternative analysis is that sticking and tearing occurred as a result of high friction forces. In VDC casting, the cast times are short (<2 hours) and the moulds are cleaned between casts. In contrast, HDC casting can run for many days before a mould is changed. Tar build-up is therefore more likely to be a problem in HDC casting. Aitchison and Kondic (1953) discussed at length mould lubrication for continuous casting.

6.4.3 Cracking

The cooling of the ingot during DC casting generates thermal stresses due to the differential cooling rate for different parts of the casting. For example, the surface cools quickly due to the high heat fluxes from the water spray and thus is driven to contract rapidly. However, the interior of the ingot is cooled more slowly and at a later stage, and thus the different contraction rates in centre and surface result in tension in the interior and compression on the surface.

DC casting cracks can be divided into hot tears occurring above the solidus, i.e. in the semi-solid region and hot cracks occurring below the solidus. For most common aluminium alloys in the 1000, 3000, 5000 and 6000 series ranges, hot tearing is the usual type of cracking, while hot cracks tend to occur for the high strength, low thermal conductivity, low ductility 2000 and 7000 series aluminium alloys.

Hot tearing in DC casting is an area of ongoing research (Eskin *et al.* 2004), however the main mechanism is generally agreed upon. Hot cracks in DC casting form due to the semi-solid material in the solidification front being subjected to tensile strains generated by thermal contraction of the cooling solid. Different parts of the casting cool at different rates and consequently contract at different rates. The solid at the base of the liquid pool in the centre of the ingot is driven to contract after the solid at the surface has already cooled; consequently tension is generated in the mushy material at the base of the liquid pool in the centre of the ingot due to constrained thermal contraction. This tension causes dilation of the mushy material. If the liquid is unable to feed the imposed strain, crack-shaped voids appear between the grains. These are intergranular and show fracture surfaces that look like shrinkage porosity, i.e. evidently liquid was present during failure.

Eskin (2008) gives a full review of hot tearing in DC casting. Hot tearing at the start of billet casting is discussed in detail below, being one of the more important manifestations of the problem.

Schneider and Jensen (Jensen and Schneider 1990; Schneider and Jensen 1990) examined cracking at the start of extrusion billet casting and proposed the concept of using pool depth as a cracking criterion, i.e. above a critical pool depth cracks would form. This is consistent with recently developed hot cracking theories based around feeding models (Rappaz *et al.* 1999). As the pool gets deeper it is expected that so too does the width of the mushy zone. Consequently the feeding pressures increase and cracks are more likely to form. Also, the increase in pool depth reflects an increase in strain rate as the differential cooling and contraction rate between the centre and the surface of the billet increase. Schneider and Jensen also revealed the phenomenon of pool depth over-shoot. If one starts with a constant casting speed the pool depth goes through a maximum before reaching and maintaining a constant steady-state value. Flood *et al.* (1995) successfully modelled this phenomenon in non-dimensionalised heat flow models and this was also verified experimentally by Grandfield *et al.* (1997) (Fig. 6.8). This effect occurs due to the time delay before the cooling effect of the direct chill water spray is felt in the centre of the billet while the billet is being lowered.

Two types of approach are generally used in the industry to control start cast cracks:

- Slow start speed and ramp up.
- Dummy blocks with a centrally raised section (truncated cone or dome).

With a ramped cast start speed the cast starts at a slower speed than the usual run speed so that when the pool depth goes through its maximum it is at a lower level than that which causes cracking. The speed is then increased to the run speed. In the case of the dummy with a centrally raised section, the pool in the centre is higher at the start. When the effect of the water cooling has penetrated into the centre of the ingot the pool is at a higher position. The difficulty in practice is to know exactly what dummy height or ramp conditions to use for a given alloy and diameter.

Casters know well that certain alloys are more crack-prone than others (Jacoby 1995). Alloy composition and variation of minor elements are known to affect hot tearing tendency. For example, copper has been shown to increase cracking in aluminium alloy 6060 (Later 2001). Additionally, the grain size and morphology also affect hot tearing (Grandfield *et al.* 2001). This is the main reason grain refiner is added during DC casting to prevent hot tearing.

Hot cracks tend to be intra-granular rather than inter-granular as with hot tears. The fracture surface of hot tears tends to be similar to shrinkage porosity showing smooth surfaces and evidence of the presence of liquid. Trouser or J cracks can form (Fig. 6.17) during DC slab casting of these high-strength alloys. These cracks can occur after casting is completed upon cooling or when the product is sawn. There is a large amount of energy in the residual stresses and the failure can be quite dangerous; throwing large masses of material some distance.

Alloys that have high plasticity at high temperature such as dilute alloys tend not to cold crack as they yield plastically without cracking. However, the high strength aluminium alloys such as 7xxx series have low plasticity at high temperature and, combined with high strength, this allows the thermal stresses to accumulate during casting and then cracking failure occurs; sometimes during casting but sometimes after casting. The high alloy content of the high-strength alloys also results in a low thermal conductivity; this makes the temperature gradient difference and resulting thermal stresses greater when casting these alloys than more pure alloys.

Chang and Kang (1999) measured mechanical properties for aluminium alloy 7050 at high temperature and modelled stress evolution during DC casting. The predicted stresses increased as casting progressed. High-stress regions in the mid and quarter positions were predicted, as well as at the water quench point and the corners of the slab. Stress at the corner is a function of the increased water cooling in that region. Increasing casting speed was predicted to reduce butt curl stresses but made no difference to the internal high-stress regions.

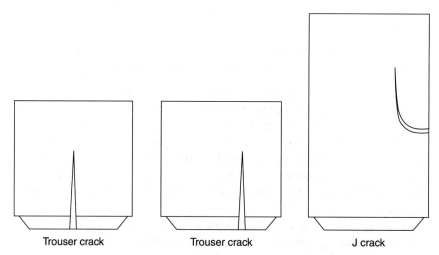

6.17 Types of cold crack that can form during DC casting of high strength low ductility alloys. (Source: Chang and Kang 1999.)

Direct chill casting of magnesium extrusion billet and rolling slab   259

A solution to cold cracking problems in high-strength aluminium alloys that has been widely adopted is the use of wipes to remove the falling water film at some point below the mould. This practice results in reheating of the surface while the centre cools, thus reducing the differential in contraction between the surface and the centre. The use of wipers or air jets to remove the falling film of water as a means of controlling cracking was first patented in 1952 (Zeigler 1952) (Fig. 6.18). The optimum position of the wipes and the degree of reheating to be achieved as a function of casting speed and alloy type need to be determined by modelling and experiment, and the exact implementation of wipe technology tends to be closely guarded within the industry. Both rubber wipes and air jets (Taylor *et al.* 1957) have been used. In US 3,763,921 (Behr *et al.* 1973) the idea of using a second spray at a point after the dry zone below the wipes is revealed. In US 3,891,024 (Gervais and Chollet 1975) further refinement of the wipe idea by controlling the position of the wipe to achieve higher casting speeds without

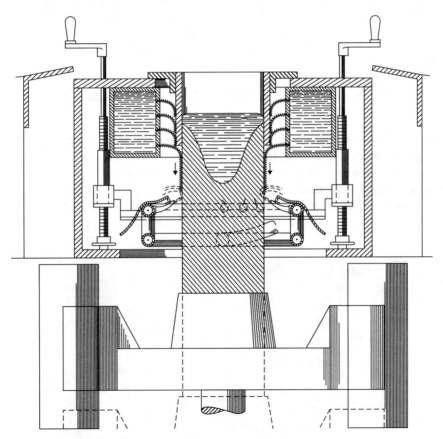

6.18 Application of wipers described in US patent 2,705,353. (Source: Ziegler 1955.)

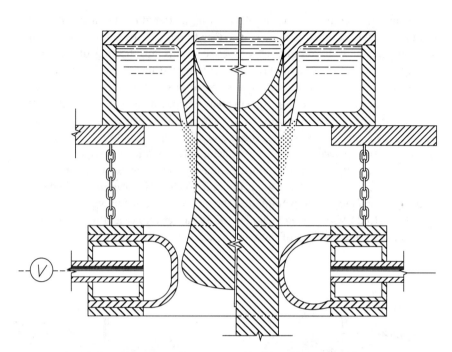

6.19 Inflatable wipes disclosed in US patent 4,237,961. (Source: Zinniger 1980.)

cracks is described. This patent is targeted at small diameter 5-inch zinc DC casting. Inflatable wipers may be used (Zinniger 1980) so that the butt of the ingot can easily pass through at cast start (Fig. 6.19). Measurements of the surface temperature showed that the surface does reheat when wipers are used (Fig. 6.20).

In order to predict hot crack occurrence a cracking criteria is needed. A simple criteria is that the strength has been exceeded. However, the difficulty is to

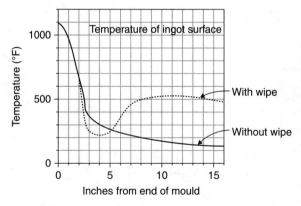

6.20 Reheating of ingot surface observed when wipes are used; from US patent 4,237,961. (Source: Zinniger 1980.)

compare uniaxial tension and compression strength data with the tri-axial stress state of the material during DC casting. Boender and Burghardt (2004) discuss failure criteria with respect to high-strength 2000 and 7000 series aluminium alloys and suggest von Mises stress is not appropriate, but that if the maximum principle stress exceeds the uniaxial tensile strength or uniaxial compressive strength then failure is likely.

In practice, cracking tends to occur at lower predicted stresses than the strength of the material. A fracture mechanics explanation that small defects can reduce the stress needed for failure can be invoked (Boender and Burghardt 2004; Ludwig *et al.* 2006). Even without a cracking criteria, mathematical models can be used to find optimum casting conditions, which minimise the internal stresses.

6.4.4 Effect of alloy type on heat flow, stresses and microstructures

The alloy composition has a profound effect on both its thermal properties, which determine the temperature field, and its mechanical properties, which influence the response of the material to the strains generated by the thermal field, i.e. the stresses generated and whether cracks form or not. In principle, any magnesium alloy can be DC cast; however, some alloys are more prone to defect formation than others.

As alloy content increases, the thermal conductivity drops. Specific heat and latent heat stay relatively unaffected although specific heat for some high-silicon aluminium alloys can be slightly higher than for pure aluminium. Density also varies with alloy, thus having a slight affect on heat flow. Liquidus temperature and freezing range are also affected by alloy composition and these in turn influence macrosegregation and other defects. In practice, a super heat of around 40–50 °C is used so that alloys with a lower liquidus temperature are cast at a lower casting temperature.

In general, as alloy content increases the resulting drop in thermal conductivity gives rise to deeper liquid pools and greater thermal stresses for the same casting speed. One finds that, as a consequence, casting speeds need to be lower the higher the alloy content is, to avoid cracking and, in practice, casting speed tends to vary inversely with thermal conductivity.

6.5   Conclusions

DC casting technology as applied to aluminium has developed significantly in terms of scientific understanding and commercial application. Due to the much smaller demand for and consequent low production volumes of magnesium DC cast wrought alloys, the drivers for development have not been as strong. Consequently the technology as applied to magnesium is generally rudimentary. However, when larger volumes of high-quality products are needed the science and technology of DC casting is available.

6.6 References

Adenis, D. J. P., Coats, K. H. and Ragonne, P. (1962–1963) 'An analysis of the direct chill casting process by numerical methods'. *Journal of the Institute of Metals* **91**, 395–403.

Aitchison, L. and Kondic, V. (1953) *The Casting of Non-Ferrous Ingots*. Macdonald & Evans Ltd., London.

Angleys, P. (1960) 'UgineVenthon process'. *Journal of Metals* **1960**, 42.

Angleys, P. (1964) 'Horizontal continuous casting process'. *Modern Metals* **20**, 46–48.

Apostolou, G. and Armaos S. (1988) Process and apparatus for top feed casting of metals, Aluminium Pechiney. US patent US4732209.

Bachowski, R. and Spear, R. E. (1975) 'Ingot shell formation'. *Light Metals* 147, 111–118

Bainbridge, I. F. and Grandfield J. F. (2007) 'Measurement of hot top direct chill casting mould ingot shell temperatures'. *Aluminium Cast House Technology 2007, 10th Australasian Conference and Exhibition*. CAST Consult Pty Ltd, Sydney, pp. 81–88.

Bainbridge, I. F., Grandfield, J. F. and Taylor, J. (2005) 'Meniscus shape and direct chill cast hot top mould design'. *Proceedings of the Australasian Conference and Exhibition – Aluminium Cast House Technology*. Melbourne, 2005. CASTconsult, Brisbane, pp. 93–100.

Baker, P. (1997) 'Issues in magnesium DC casting'. *International Symposium on Light Metals 1997*. Metaux Legers, Canadian Institute of Mining, Metallurgy and Petroleum, Montreal, CIM, pp. 355–367.

Baker, P. W. and Grandfield J. F. (1987) 'Mould wall heat transfer in air-assisted DC casting'. *Solidification Processing 1987*. The Institute of Metals, pp. 257–259.

Baker, P. W. and Grandfield J. F. (2001) 'The role of surface tension forces in gas pressurized VDC casting'. *Seventh Australian Asian Pacific Conference Aluminium Cast House Technology*. Hobart, Tasmania, TMS, pp. 195–204.

Baker, P. W. and McGlade P. T. (2001) 'Magnesium direct chill casting: A comparison with aluminium'. In *Light Metals 2001: Proceedings of Sessions, TMS Annual Meeting* (Warrendale, Pennsylvania). J. Anjier ed. New Orleans, LA, pp. 855–862.

Bakken, J. A. and Bergstrom T. (1986) 'Heat transfer measurements during DC casting of aluminum. I. Measurement technique'. *Light Metals 1986*, The Metallurgical Society/AIME, pp. 883–889.

Bamberger, M. and Jeschar, R., Prinz, B. (1979) *Z. Metallkunde* **70**, No 9, 553–560.

Behr, R. D., Couling, S. L. and Alfrey, T. (1973) 'Direct chill casting method'. Dow chemical corporation US 3763921, USA.

Benum, S., Hakonsen, A., Hafsås J. E. and Sivertsen J. (1999) 'Mechanisms of surface formation during direct chill (DC) casting of extrusion ingots'. *Light Metals 1999: Proceedings of Sessions*, TMS Annual Meeting (Warrendale, Pennsylvania). Minerals, Metals and Materials Society (TMS), San Diego, CA, pp. 737–742.

Bergmann, W. J. (1970) 'Solidification in continuous casting of aluminium'. *Metallurgical Transactions* **1**: 3361–3364.

Bergmann, W. J. (1973) 'Surface structures of continuously cast aluminium'. *J. Met.* **25** (2), 23–27.

Bergmann, W. J. (1975) 'Aluminium ingot surface improvement by advanced continuous casting technology'. *Aluminium* **51** (5), 336–339.

Boender, W. A. and Burghardt A. (2004) 'Numerical simulation of DC casting; interpreting the results of a thermo-mechanical model'. *Light Metals 2004*. A. T. Tabereaux ed., TMS, pp. 679–684.

Bohmer, J. R. and Jordan M. (1995) 'Verification of a mathematical model for continuous billet casting with a temperature and load history approach'. *Modeling of Casting, Welding and Advanced Solidification Processes VII*. TMS, pp. 809–816.

Brochu, C., Larouche, A. and Hark, R. (1993) 'Study of shell zone formation in lithographic and anodizing quality aluminum alloys: Experimental and numerical approach'. *Light Metals 1993*. The Minerals, Metals Materials Society, pp. 961–967.

Buxmann, K. (1982) 'Solidification conditions and microstucture in continuously cast aluminum'. *Journal of Metals* **34**, 28–34.

Caron, E. J. F. R., Wells, M. A., Sediako, D. and Hibbins, S. G. (2005) 'Evaluation of the heat flux in the secondary zone during the direct chill casting of magnesium alloy AZ31'. In *Magnesium Technology 2005*. N. R. Neelameggham, H. I. Kaplan and B. R. Powell eds, TMS, pp. 229–234.

Chang, K. M. and Kang, B. (1999) 'Cracking control in dc casting of high-strength aluminum alloys'. *Journal of the Chinese Institute of Engineers, Transactions of the Chinese Institute of Engineers*, Series A/Chung-kuo Kung Ch'engHsuchK'an **22** (1), 27–42.

Davidson, P. A. and Flood, S. C. (1994) 'Natural convection in an aluminum ingot: a mathematical model'. *Metallurgical Transactions B* **25B** (2), 293–302.

Devadas, C. and Grandfield, J. F. (1991) 'Experiences with modelling DC casting of aluminum'. *Light Metals 1991*. Minerals, Metals and Materials Society (TMS), New Orleans, LA, pp. 883–892.

Dobatkin, V. I. (1948) *Continuous Casting and Casting Properties of Alloys*. Oborongiz, Moscow.

Donnerburg, F. and Engler, S. (2000) In *Influence of Different Lubricants on the Friction Between the Solidifying Shell and the Mould During DC Casting of AlMgSi0.5*. K. Ehrke and W. Schneider eds. Wiley-VCH, Weinheim, p. 47.

Drezet, J. M., Ludwig, O. Jaquerod, C. and Waz, E. (2007) 'Fracture prediction during sawing of DC cast high strength aluminium alloy rolling slabs'. *International Journal of Cast Metals Research* **20** (3), 163–170.

Drezet, J. M. and Rappaz, M. (1996) 'Modeling of ingot distortions during direct chill casting of aluminum alloys'. *Metallurgical and Materials Transactions A: Physical Metallurgy and Materials Science* **27** (10), 3214–3225.

Drezet, J. M. and Rappaz, M. (1997) 'Direct chill casting of aluminum alloys: Ingot distortions and mold design optimization'. *Light Metals 1997: Proceedings of Sessions*, TMS Annual Meeting (Warrendale, Pennsylvania). Minerals, Metals and Materials Society (TMS), Orlando, FL, pp. 1071–1080.

Drezet, J. M., Rappaz, M., Ludwig, M. O. and Martin, C.-L. (1995) 'Experimental investigation of thermomechanical effects during direct chill and electromagnetic casting of aluminum alloys'. *Metallurgical and Materials Transactions B* **26** (4), 821–829.

Droste, W. and Schneider W. (1990) 'Laboratory investigations about the influence of starting conditions on butt curl and swell of DC cast sheet ingots'. *Light Metals 1991*. Minerals, Metals and Materials Society (TMS), New Orleans, LA, pp. 945–951.

Dunn, E. M. (1979) *Metallurgical Structure of Electromagnetically Cast Extrusion Billet*. HutnickeListy, Czechoslovakia.

Easton, M. and St John D. (1999a) 'Grain refinement of aluminum alloys: Part I. The nuclenat and solute paradigms – a review of the literature'. *Metallurgical and Materials Transactions A: Physical Metallurgy and Materials Science* **30** (6), 1613–1623.

Easton, M. and St John D. (1999b) 'Grain refinement of aluminum alloys: Part II. Confirmation of, and a mechanism for, the solute paradigm'. *Metallurgical and Materials Transactions A: Physical Metallurgy and Materials Science* **30** (6), 1625–1633.

Ekenes, J. and Peterson W. (1990) 'Visual observations inside an Airslip™ mold during casting'. *Light Metals 1990*. Minerals, Metals and Materials Society (TMS), Anaheim, CA, pp. 957–961.

Emley, E. F. (1976) 'Continuous casting of aluminium'. *International Metals Reviews* **21**, 75–115.

Ennor, W. T. (1942) Method of casting. US patent 974203 filed 1938, USA.

Eskin, D. (2008) *Physical Metallurgy of Direct Chill Casting of Aluminium Alloys*. CRC Press, London.

Eskin, D. G., Suyitno and Katgerman, L. (2004a) 'Mechanical properties in the semi-solid state and hot tearing of aluminium alloys'. *Progress in Materials Science* **49** (5), 629–711.

Etienne, J. F. R, Wells, M. A., Sediako, D. and Hibbins, S. G, (2005) 'Evaluation of the surface heat flux in the secondary zone during the direct chill casting of magnesium alloy AZ31'. In *Magnesium Technology 2005*. N. R. Neelameggham, H. I. Kaplan and B. Powell eds, TMS, pp. 229–234.

Fjaer, H. and Mo A. (1990) 'Mathematical modelling of thermal stresses during D.C. casting of aluminum billets'. *Light Metals 1990*. Minerals, Metals and Materials Society (TMS), Anaheim, CA, pp. 945–950.

Fjaer, H. G., Buchholz, A., Commet, B., Drezet, J.-M. and Mortensen, D. (2000) 'Investigations of the primary cooling in sheet ingot casting'. In *Continuous Casting*. K. Ehrke and W. Schneider eds. Wiley-VCH, Weinheim, pp. 131–137.

Fjaer, H. G., Jensen, E. K. and Mo. A. (1992) 'Mathematical modeling of heat transfer and thermal stresses in aluminium billet casting. Influence of the direct water cooling conditions'. *5th International Aluminum Extrusion Technology Seminar*, The Aluminum Association, pp. 113–120.

Flood, S. C., Davidson, P. A. and Rogers, S. (1995) 'A scaling analysis for the heat flow, solidification and convection in continuous casting of aluminium'. *Modeling of Casting, Welding and Advanced Solidification Processes VII*. Warrendale, Minerals, Metals and Materials Society/AIME, PA, pp. 801–808.

Furness, A. G. and Harvey J. D. (1975) *Mould Assembly and Method for Continuous or Semi-Continuous Casting*. British Aluminium Co Ltd, UK.

Gariepy, B., Mazerolle, D. and Weaver, C. (2005) 'Alcan HDC casting experience at Dubuc and Alma plants for the production of busbar, pure and foundry T ingots'. *Aluminium Cast House Technology 2005*. Cast Consult Pty Ltd, Melbourne, pp. 93–102.

Gervais, H. L. and Chollet, P. (1975) 'Method for the continuous casting of metal ingots or strips'. Noranda Mines Limited, USA.

Getselev, Z. N., Balakhontsev, G. A. and Cherepok, G. (1969) 'Method of continuous and semicontinuous casting of metals and a plant for same'. USSR, KMPO (Kuibyshev Metallurgical Works), US 3467166.

Goodrich, D. and Tarapore, E. (1991) 'Comparison of ingot structures for aluminum ingots produced by direct-chill and electromagnetic casting'. *Light Metals 1992*. Minerals, Metals and Materials Society (TMS), San Diego, CA, pp. 1371–1377.

Grandfield, J. F. (1997) 'DC casting of aluminium: a short review of process development'. *Proceedings of the Australasian Asian Pacific Conference on Aluminium Cast House Technology*. Minerals, Metals and Materials Society (TMS), Gold Coast, Australia, pp. 231–243.

Grandfield, J. F. (2000) 'Experimental methods for direct chill casting, including new techniques for magnesium horizontal direct chill casting'. *Aluminium Transactions* **3** (1), 41–51.

Grandfield, J. F. (2001) 'Hot tear defect formation during horizontal direct chill casting of magnesium'. PhD thesis. University of Queensland, Brisbane.

Grandfield, J. F. and Baker P. W. (1987) 'Variation of heat transfer rates in direct chill water spray of aluminium continuous casting'. *Solidification Processing 1987*. The Institute of Metals, pp. 260–263.

Grandfield, J. F. and Dahle A. (2000) 'Modelling and measurement of mould heat transfer during horizontal direct chill casting of magnesium'. In *4th Pacific Rim International Conference on Modeling of Casting and Solidification Processes*. Yonsei, C. P. Korea, J. K. Hong Choi and D. H. Kim eds. Centre for Computer-Aided Materials Processing, Yonsei, Korea, pp. 299–307.

Grandfield, J. F. and McGlade P. T. (1996) 'DC casting of aluminium: Process behaviour and technology'. *Materials Forum* **20**, 29–51.

Grandfield, J. F. and Wang, L. (2004) 'Application of mathematical models to optimization of cast start practice for DC cast extrusion billets'. In *Light Metals 2004*. A. T. Tabereauxed ed. TMS, pp. 685–690.

Grandfield, J. F. Bainbridge, I., Murray, J. and Oswald, K. (2007) 'Development of a gas pressurised hot top direct chill casting extrusion billet mould for aluminium and magnesium alloys'. *Light Metals Technology 2007*. K. Sadayappan and M. Sahoo eds. NRCC, Ontario, pp. 69–74.

Grandfield, J. F., Bainbridge I. F. and Taylor, J. A. (2005) 'Meniscus shape and direct chill cast hot top mould design'. In *Aluminium Cast House Technology 2005*. J. A. Taylor ed. CAST consult, pp. 85–92.

Grandfield, J. F., Davidson, C. J. and Taylor, J. A. (2001) 'Application of a new hot tearing analysis to horizontal direct chill cast magnesium alloy AZ91'. *Light Metals 2001*. J. Anjier ed. TMS, pp. 911–917.

Grandfield, J. F., Goodall, K., Misic, P. and Zhang, X. (1997a) 'Water cooling in direct chill casting: Part 2, effect on billet heat flow and solidification'. *Light Metals 1997: Proceedings of Sessions*. TMS Annual Meeting (Warrendale, Pennsylvania). Minerals, Metals and Materials Society (TMS), Orlando, FL, pp. 1081–1090.

Grandfield, J. F., Hoadley, A. and Instone, S. (1997b) 'Water cooling in direct chill casting: Part 1, boiling theory and control'. *Light Metals: Proceedings of Sessions*. TMS Annual Meeting (Warrendale, Pennsylvania). Minerals, Metals and Materials Society (TMS), Orlando, FL, pp. 691–699.

Grandfield, J. F., Nguyen, V. and I. F. Bainbridge (2009) 'Stresses and cracking during direct chill casting of az31 alloy billet'. In *Magnesium Technology 2009*. E. A. Nyberg, S. R. Agnew, N. R. Neelameggham and M. O. Pekguleryuz eds. TMS, pp. 129–133.

Grandfield, J. F., Young, C-C, Oswald, K. and Baker, P. (2003) 'Commissioning of a vertical direct chill caster for production of magnesium extrusion billet and slab'. In *Aluminium Cast House Technology – 8th Australasian conference 2003*. P. R. Whitely ed. pp. 215–221, Wiley, London.

Granger, D. A. (1989) 'Ingot casting in the aluminium industry'. *Treatise on Material Science and Technology* **31**, 109–135.

Gruen, G. U., Buchholz, A. and Mortensen, D. (2000) '3–D modeling of fluid flow and heat transfer during the DC casting process – Influence of flow modeling approach'. *Light Metals: Proceedings of Sessions*. TMS Annual Meeting (Warrendale, Pennsylvania), Nashville, TN, pp. 573–578.

Gruen, G. U. and Schneider W. (1996) '3–D modeling of the start-up phase of DC casting of sheet ingots'. *Light Metals: Proceedings of Sessions*. TMS Annual Meeting (Warrendale, Pennsylvania), Minerals, Metals and Materials Society (TMS), Anaheim, CA, pp. 971–978.

Hakonsen, A. and Mortensen D. (1995) FEM model for the calculation of heat and fluid flows in DC casting of aluminum slabs. In *Modeling of Casting, Welding and Advanced Solidification Processes*. M. Cross and J. Campbell eds. Minerals, Metals and Materials Society (TMS), London, UK, pp. 763–770.

Hakonsen, A. M. and Mihr, O. R. (1995) 'Dimensionless diagrams for the temperature distribution in direct-chill continuous casting'. *Cast Metals* **8** (3), 147–157.

Hamadaiid, A. (2004) 'The role of casting/mould interface on the heat transfer and solidification in aluminium die-casting.' PhD literature review, I. F. Bainbridge.

Hao, H., Maijer, D. M. and Wells, M. (2004) *Met. Trans. A* **35A**, 3843–3854.

Hao, H., Maijer, D. M. and Wells, M. (2005) 'Predictions and measurements of residual stresses/strains in a direct chill casting magnesium alloy billet'. In *Magnesium Technology 2005*. N. R. Neelameggham, H. I. Kaplan and B. R. Powell eds, TMS, pp. 223–228.

Hao, H., Maijer, D. M. and Wells, M. (2010) 'Modeling of stress-strain behaviour and hot tearing during direct chill casting of an AZ31 magnesium billet'. *Met. Trans. A* **41A**, 2067–2077.

Harrington, D. G. and Groce, T. E. (1971) Control of heat transfer in continuous casting of AL and Mg ingots, Kaiser, US patent 3,612,151.

Hasan, M. and Ragel K. R. (2009) Advanced CFD modeling of DC casting of aluminum alloys. *TMS Annual Meeting*. pp. 805–810.

Hibbins, S. G. (1998) 'Investigation of heat transfer in DC casting of magnesium alloys'. In *Met. Trans. A*. Métauéx Legers. M. Sahoo ed. CIM, pp. 265–280.

Ho, K. and Pehlke R. D. (1985) 'Metal/Mould Interfacial Heat Transfer'. *Metallurgical Transactions B* 16B (3), 585–594.

Hudault, G., Cheve, D., Meyer, J.-L. and Tartour, J. P. (1989) 'First experience of commercial operation with the CREM process'. *Light Metals: Proceedings of Sessions*. AIME Annual Meeting (Warrendale, Pennsylvania). Metallurgical Society of AIME, Las Vegas, NV, pp. 769–775.

Inoue, T. and Ju, D. Y. (1989) 'Temperature and viscoplastic stresses during vertical semi-continuous direct chill casting of aluminum alloy'. *International Conference on Residual Stress*. ICRS 2, Elsevier Applied Science, pp. 523–528.

Instone, S., Schneider, W. and Grun, G.-U. (2003) 'Improved VDC billet casting mould for Al-Sn Alloys'. *Light Metals*. TMS, San Diego.

Jacoby, J. (1995) 'Direct chill casting defects'. *Aluminium Cast House Technology, 5th Australian Asian Pacific conference*. TMS, Warrendale, pp. 245–254.

Jacoby, J. E., Laemmle, J. T. and Joseph, T. (1986) Lithium Alloy Casting. USA, Aluminum Company of America., US patent, USA 4628985.

Jensen, E. K. (1984) 'Mould temperatures during DC casting of 8 in. dia. extrusion ingots in alloy 6063.' *Light Metals 1984*. The Metallurgical Society/AIME, pp. 1159–1175.

Jensen, E. K. and Schneider W. (1990) 'Investigations about starting cracks in DC-casting of 6063–Type billets. II. Modelling results'. *Light Metals 1990*. The Minerals, Metals and Materials Society, pp. 937–943.

Jensen, E. K. J., Bergstrom, T. and Bakken, J. A. (1986) 'Heat transfer measurements during DC casting of aluminium. II. Results and verification for extrusion ingots'. *Light Metals 1986*. The Metallurgical Society/AIME, Warrendale, PA, pp. 891–896.

Jones Jr, W. K., Xu, D. and Evans, J. (1998) 'Physical modeling of the effects of non-symmetric placement of flow control bags used in semi-continuous casting of aluminum.' *Light Metals 1998*. Minerals, Metals and Materials Society/AIME, pp. 1051–1057.

Junghaus, S. (1938) Apparatus for the continuous casting of metal rods. USA Patent 2135184, filed 1933.

Katgerman, L. (1991) 'Developments in continuous casting of aluminium alloys'. *Cast Metals* **4** (3), 133–139.

Kettner, M., Pravdic F., Fragner W. and Kainer K. U., (2006) 'Vertical direct chill (VDC) casting of a novel magnesium wrought alloy with Zr and Re additions (Zk10): alloying issues'. In *Magnesium Technology 2006*. A. A. Luo, N. R. Neelameggham and R. S. Bealsed eds. TMS (The Minerals, Metals and Materials Society), pp. 133–138.

Kittilsen, B. and Oistad B. (1996) An apparatus, a mould and a stop procedure for horizontal direct chill casting of light alloys especially magnesium, and magnesium alloys. Australian Patent 1996062175, Norsk Hydro.

Kittilsen, B. and Pinfold, P. (1996) In *Light Metals 1996*. W. Hale ed. TMS, pp. 987–991.

Laemmle, J. T. and Bohaychick J. (1992) 'Mold lubricants for casting aluminum and its alloys'. *Lubrication Engineering* **48** (11), 858–863.

Langerweger, J. (1981) 'Nonmetallic particles as the cause of structural porosity, heterogeneous cell structure and surface cracks in DC cast aluminum products.' *Light Metals 1981*. Metallurgical Society AIME, pp. 685–705.

Later, D. (2001) 'Optimising pit recoveries on 6xxx extrusion billet'. *7th Aluminium Cast House Technology conference*. Minerals, Metals and Materials Society/AIME, pp. 213–222.

Lippert, T. W. (1944) 'Continuous casting'. *The Iron Age* **153** (8), 48–146.

Ludwig, O., Drezet, J. M. and Rappaz, M. (2006) 'Modelling of internal stresses in DC casting and sawing of high strength aluminum alloys slabs'. *Modeling of Casting, Welding and Advanced Solidification Processes – XI. Opio.* **1**, 185–192.

M'Hamdi, M., Benum, S. and Mo, A. (2003) 'The importance of viscoplastic strain rate in the formation of center cracks during the start-up phase of direct-chill cast aluminum extrusion ingots'. *Metallurgical and Materials Transactions A: Physical Metallurgy and Materials Science* **34 A** (9), 1941–1952.

M'Hamdi, M., Fjaer, H. G. and Mortensen, D. (2004) 'A new two-phase thermo-mechanical model and its application to the study of hot tearing formation during the start-up phase of DC cast ingots'. In *Solidification of Aluminum Alloys*. M. G. Chu, D. A. Granger and Q. Han eds. Charlotte, NC, pp. 191–200.

M'Hamdi, M. and Hakonsen, A. (2003) 'Experimental and numerical study of surface macrosegregation in DC casting of aluminium sheet ingots'. *Modeling of Casting, Welding and Advanced Solidification Processes*. M. Stefanescu, J. A. Warren, M. R. Jolly eds. Destin, FL, pp. 505–512.

M'Hamdi, M. and Mo, A. (2002) 'Microporosity and other mushy zone phenomena associated with hot tearing'. In *Light Metals 2002*. R. Peterson ed. TMS, pp. 709–716.

M'Hamdi, M. and Mo, A. (2005) 'On modelling the interplay between microporosity formation and hot tearing in aluminium direct-chill casting'. *Materials Science and Engineering A* **413–414**, 105–108.

M'Hamdi, M. and Mo, A. (2008) 'Modeling the mechanics of solidification defects in aluminum DC casting'. TMS *Light Metals*. New Orleans, LA, pp. 765–771.

M'Hamdi, M., Mo, A. and Mortensen, D. (2000) 'Mathematical modelling of surface segregation in DC casting of multicomponent aluminium alloys'. *Modelling of Casting, Welding and Advanced Solidification Processes – IX*. pp. 656–663, Wiley, London.

M'Hamdi, M., Mo, A. and Mortensen, D. (2002a) 'Two-phase modeling directed toward hot tearing formation in aluminum direct chill casting'. *Metallurgical and Materials Transactions A: Physical Metallurgy and Materials Science* **33** (7), 2081–2093.

M'Hamdi, M., Mo, A. and Mortensen, D. (2002b) 'Modelling of air gap development and associated surface macrosegregation in DC casting of aluminium sheet ingots'. *Light Metals 2002*, TMS. R. Peterson ed., 695–701.

M'Hamdi, M., Mo, A. and Mortensen, D. (2006) 'TearSim: A two-phase model addressing hot tearing formation during aluminum direct chill casting'. *Metallurgical and Materials Transactions A: Physical Metallurgy and Materials Science* **37** (10), 3069–3083.

Maenner, L., Magnin, B. and Caratini, Y. (1997) *Light Metals 1997*. TMS, Warrendale, pp. 701–707.

McGlade, P. (1999) 'Magnesium DC casting'. *6th Aluminium Asian Pacific Conference on Cast House Technology*. TMS, pp. 321–330.

McGlade, Baker P. (2001) 'Magnesium direct chill casting: A comparison with aluminium'. *Light Metals 2001*. TMS, pp. 855–864.

McGlade, P. and Ricketts N. (2001) 'Aluminium and magnesium: equipment and process comparison'. *Proceedings of the Australian Asian Pacific Conference on Aluminium Cast House Technology*. pp. 319–328.

Mitamura, R. and Itoh T. (1979) Process for direct chill casting of metals, Showa Denko Kabushiki Kaisha.

Mitamura, R. I., Takahashi, T. Y. and Hiraoka, T. (1978) 'New hot top continuous casting method featuring application of air pressure to mold'. *Light Metals 1978*. TMS, pp. 281–291.

Mo, A., Rusten, Thevik, T. H. J., Henriksen, B. R. and Jensen, E.K. (1997) 'Modelling of surface segregation development during DC casting of rolling slab ingots'. *Light Metals 1997: Proceedings of Sessions*. TMS Annual Meeting (Warrendale, Pennsylvania). Minerals, Metals and Materials Society (TMS), Orlando, FL, pp. 667–674.

Moriceau, J. (1975) 'Thermal stresses in continuous DC casting of Al alloys'. *Light Metals 1975*. TMS-AIME, p. 119.

Moritz, D. and Dietz W. (1979) 'Horizontal strip casting of aluminum using a new stationary mold (Horizontales bandgiessen von aluminium mit einer stationaeren kokille.)'. *Aluminium* **55** (6), 395–397.

Moritz, G. E. (1961) Metal casting system, Reynolds Metals Company.

Mortensen, D., Henriksen, B. R., M'Hamdi, M. and Fjaer, H. G. (2008) Coupled modelling of air-gap formation and surface exudation during extrusion ingot DC-casting. TMS *Light Metals*. New Orleans, LA, pp. 773–779.

Muojekwu, C. A., Samarasekera, I. V. and Brimacombe, J. K. (1995) 'Heat transfer and microstructure during the early stages of metal solidification'. *Metallurgical and Materials Transactions B* **26** (2), 361–382.

Tsunekawa, M., Nagae, K., Mutou, N., Hayashi, N. and Uno, T. (1996) 'Analysis of solidification process in aluminium semicontinuous casting slab by monitoring the mould temperature'. *Sumitomo Light Metals Technical Reports* **37** (3,4), 180–184.

Nagae, K., Hayashi, N. and Katgerman, L. (1988) 'Effect of some factors on the meniscus shape in electromagnetic Casting'. *Sumitomo KeikinzokuGiho/Sumitomo Light Metal Technical Reports* **29** (3), 1–8.

Niedermair, F. (2001) 'Horizontal direct chilled (HDC) casting technology for aluminium and requirements to metal cleanliness'. *Proceedings of the Australian Asian Pacific Conference on Aluminium Cast House Technology*. pp. 253–262.

Nishida, Y., Droste, W. and Schneider, W. (1986) 'The air-gap formation process at the casting/mould interface and the heat transfer mechanism through the gap'. *Metallurgical Transactions B* **17B** (4), 833–844.

Nofal, A. A. (1992) 'Surface defect formation in the DC-cast aluminium products'. *Advances in Continuous Casting: Research and Technology*. Woodhead Publishing Ltd, Abington Hall, Abington, Cambridge, pp. 205–218.

Ohm, L. and Engler S. (1989a) 'Driving forces of surface segregation of non-ferrous DC-casting'. *Metall* **43** (6), 520–524.

Ohm, L. and Engler S. (1989b) 'Mechanical properties of solidifying shells'. *Metall* **43** (6), 539–543.

Paddock, R. K. (1954) 'Continuous casting'. *The Iron Age* 149–151.

Porro, G. and Lombardi P. (1954) 'Studio sulla colata continua e calcoliempiricidellevelocita di abbassamentonellacolata di lingotti di legheleggere'. *Aluminio* **23**, 23–24.

Pritchett, T. R. (1973) 'Application of electromagnetism to aluminum metallurgy'. *Light Metal Age* **31** (11–12):,21–24.

Raffourt, C., Fautrelle, Y. and Rappaz, M. (1990) 'Liquid metal distribution in a slab DC casting: Experiments and modelling approach'. *Light Metals 1991*. Minerals, Metals and Materials Society (TMS), New Orleans, LA, pp. 877–882.

Rappaz, M., Drezet, J. M. and Gremaud, M. (1995) 'A new hot-tearing criterion'. *Metallurgical and Materials Transactions A: Physical Metallurgy and Materials Science* **30A** (2), 449–455.

Reese, J. M. (1997) 'Characterization of the flow in the molten metal sump during direct chill aluminum casting'. *Metallurgical and Materials Transactions B: Process Metallurgy and Materials Processing Science* **28** (3), 491–499.

Roberts, J. and A. Lowery (2009) 'Safety coatings to prevent molten aluminum-water explosions'. *Light Metals*. TMS, p. 667.

Roth, W. (1936) Verfarhenzum Giessen von metallblockenmitAusnahmesolcheraus Leichtmetall. Germany. Patent 974203.

Roth, W. (1943) 'Ueber die Abkuhlung des Strangesbeim Wasserguss'. *Aluminium* **25**, 283.

Roth, W. and Weisse, E. (1942) 'Strangeissen von magnesiumlegierungen, insbesonderenach dem wassergiessfahren'. *Aluminium* **26** (7/8), 134–136.

Roth, W. and Weisse, E. (1944) 'Strangeissen von magnesiumlegierungen, insbesonderenach dem wassergiessfahren'. *Aluminium* **26** (7/8), 134–136.

Roth, W., Welsch, M. and Röhrig, H. (1942) 'Uber die Eigenspannungen in Strangguss-Blockenauseinereutektischen Al-Si-Legierung', *Aluminium* **24**, 206.

Schneider, W. (2002) 'D. C. Casting of aluminium alloys – Past, present and future'. In *Light Metals: Proceedings of Sessions*. TMS Annual Meeting (Warrendale, Pennsylvania). R. Peterson ed. Seattle, WA, pp. 953–960.

Schneider, W. and Jensen E. K. (1990) 'Investigations about starting cracks in dc casting of 6063 type billets. Part I: Experimental results'. *Light Metals 1990*. Minerals, Metals and Materials Society (TMS), Anaheim, CA, pp. 931–936.

Schneider, W. and Kramer K. (1994) Method and apparatus for continuous casting, VAW AG (Vereinigte Aluminium-WerkeAktiengesellschaft): W. Schneider, K. Kramer 28 June 1993–05 July 1994 VAW AG; US5325910.

Siebel, G., Altenpohl, D. and Buxmann, K. (1953) 'PeriodischeSeigerungenBeiReinalumi nium-Strangguss'. *Zeitschrift fürMetallkunde* **44**, 173.

Sorimachi, K. and Nabeshima S. (1998) 'Lubrication and friction between mould and solidified shell in continuous casting of steel'. Tetsu to Hagane – *Journal of the Iron and Steel Institute Japan* **84** (2), 103–108.

Spear, R. E. and Brondyke K. J. (1971) 'Continuous casting of aluminium'. *Journal of Metals* **23**, 36–39.

Steen, I. K., Heggset, B., Torstein S. and Hanaset K. V. (1997) Casting equipment. USA, Norsk Hydro A. S. US patent US5678623.

Taylor, A. T., Thompson, D. H. *et al.* (1957) 'Direct chill casting of large aluminium ingots'. *Metal Progress* **72**, 70–74.

Thomson, R. and Ellwood E. C. (1972) 'Closed-head continuous casting: Part II – Mould billet interactions'. *The British Foundryman* **65**, 186–197.

Thomson, R., Mojab, F. and Ellwood, E. C. (1969) 'Experiments on continuous casting'. *The British Foundryman* **62**, 73–80.

Tremblay, S. P. and Lapointe, M. (2002) 'The manufacturing, design and use of a new reusable molten metal distributor for sheet ingot casting'. In *Light Metals: Proceedings of Sessions*. TMS Annual Meeting (Warrendale, Pennsylvania). R. Peterson ed. Seattle, WA, pp. 961–965.

Turski, M, Grandfield J. F., Wilks T., Davis B., DeLorme R. *et al.* (2010) 'Computer modeling of DC casting magnesium alloy WE43 rolling slabs'. *Magnesium Technology 2010*. S. R. Agnew *et al.* eds. TMS, pp. 333–338.

Wagstaff, F. E. and Wagstaff, W. G. C. and Collins, R. J. (1986) 'Direct chill metal casting apparatus and technique'. US, Wagstaff Engineering Inc. US Patent 4598763.

Wagstaff, R. B. and K. D. Bowles (1995) 'Practical low head casting (LHC) mold for aluminum ingot casting'. *Light Metals: Proceedings of Sessions*. TMS Annual Meeting (Warrendale, Pennsylvania). Minerals, Metals and Materials Society (TMS), Las Vegas, NV, pp. 1071–1075.

Watanabe, Y. and Hayashi, N. (1996) *Light Metals 1996*. W. Hale ed. TMS, pp. 979–984.

Waters, B. H. C. (1952) 'Continuous casting of non-ferrous metals: Part III Factors controlling permissible speed of casting'. *Metal Treatment and Drop Forging* **19**, 527.

Waters, B. H. C. (1954) 'Continuous casting of non-ferrous metals: Part V Surface appearance'. *Metal Treatment and Drop Forging* **20**, 79–84.

Weckman, D. C. (1987) 'Horizontal direct-chill continuous casting of non-ferrous alloy rods'. *Zeitschrift für Metallkunde/Materials Research and Advanced Techniques* **78** (12), 880–886.

Weckman, D. C. and Niessen P. (1982) 'A numerical simulation of the D. C. continuous casting process including nucleate boiling heat transfer'. *Metallurgical Transactions B* **13** (4), 593–602.

Weckman, D. C. and Niessen P. (1983) 'Optimum mould length design for the D. C. continuous casting of non-ferrous alloy rods'. *Zeitschrift für Metallkunde/Materials Research and Advanced Techniques* **74** (11), 709–715.

Weckman, D. C. and Niessen P. (1984) 'Heat-transfer conditions in hot-top mould and cold-shut formation mechanism on dc continuously cast non-ferrous alloy rods'. *Metals Technology* **11** (pt 11), 497–503.

Weckman, D. C. and Niessen P. (1984) 'Mathematical models of the DC continuous casting process'. *Canadian Metallurgical Quarterly* **23** (2), 209–216.

Weckman, D. C. and Niessen P. (1984) 'The mechanism of cold shut formation on D. C. direct-chill'. *Zeitschrift für Metallkunde* **75** (5), 332–340.

Weckman, D. C. and Niessen P. (1984) 'Mechanism of cold shut formation on DC continuously cast non-ferrous alloy products II The influences of oxidation, metallostatichead and mould insert geometry'. *Zeitschrift für Metallkunde/Materials Research and Advanced Techniques* **75** (6), 414–422.

Wilkinson, R. G. and Hirst S. B. (1952–1953) 'The control of quality in melting and casting magnesium alloys for hot working'. *Journal of the Institute of Metals* **81**, 393–400.

Yanagimoto, S. and Mitamura R. (1984) 'Application of new hot top process to production of extrusion and forging billet'. *ET'84: Extrusion Productivity Through Automation*. The Aluminum Assoc, pp. 247–256.

Yanagimoto, S. (1987) 'Application of Showa hot top process to casting of magnesium'. *Magnesium in the auto industry, Proceedings of the 44th Annual World Magnesium conference*. IMA, VA, pp. 58–63.

Yao, M. and Fang D. C. (1996) 'On line measuring method for mould friction in continuous casting'. *Ironmaking and Steelmaking* **23** (6), 522–527.

Yoh, I. (1989) 'HDC process for small diameter ingot'. *Light Metals: Proceedings of Sessions*. AIME Annual Meeting (Warrendale, Pennsylvania). Metallurgical Society of AIME, Las Vegas, NV, pp. 673–679.

Bong Sun You, Chang Dong Yim and Su Hyeon Kim (2005) 'Solidification of AZ31 magnesium alloy plate in a horizontal continuous casting process'. *Materials Science and Engineering A* **413–414**, 139–143.

Yu, H.(1985) *Light Metals 1985*. TMS, pp. 1331–1347.

Zeigler, P. P. (1952) 'Method of Continuous Casting'. Patent 2,708,297, USA.

Ziegler, P. P. (1955) 'Method of continuous casting'. USA Patent 2,705,353, Kaiser Aluminum and Chemical Corporation.

Zeillinger, H. and Beevis A. (1997) 'Universal continuous horizontal caster for ingot, billet or bar'. *Light Metal Age* **55** (5–6).

Zhang, Z., Qichi Le, Shijie Guo and Jianzhong Cui (2008) 'The effect of different direct chill casting process on microstructure macrosegregation and mechanical properties of 200 mm AZ31 billets'. *Magnesium Technology 2008*. TMS, pp. 233–237.

Zinniger, T. C. (1980) 'Direct chill casting method with coolant removal'. USA Patent 4237961, Kaiser Aluminium & Chemical Corporation, USA.

Zunkel, B. (1939) 'Giessvorrichtungzumununterbrochen Giessen von Blocken und ahnlichen Wekstuckenaus Leichtmetalloder Leichtmetallegierungen'. Patent 678534, Germany.

7
Twin roll casting of magnesium

E. ESSADIQI, CANMET Materials Technology Laboratory,
Canada, I.-H. JUNG, McGill University, Canada and
M. A. WELLS, University of Waterloo, Canada

Abstract: Although the twin roll casting (TRC) process offers a promising route for the economical production of Mg sheet, its application for commercial-scale magnesium strip production has proven difficult. This chapter explores developments in the TRC process and its use to produce magnesium sheet alloys, providing details of the technology, and its use historically, as well as the complex solidification process and resulting microstructures that occur. Thermodynamic calculations provide a deeper understanding of the complex chemistry of magnesium alloys that will assist in the design of new alloys for TRC and sheet production. Recent advances in TRC process modeling are outlined, providing a quantitative basis for optimizing important process variables to ensure high-quality sheet products.

Key words: twin roll casting process, magnesium strip, magnesium solidification microstructure, solidification, thermodynamic calculations, design of wrought magnesium alloys, twin roll casting process modeling, AZ31 magnesium alloy, wrought magnesium alloys.

7.1 Introduction

The high specific strength and density of magnesium makes it an attractive material for sheet applications where weight reduction is critical. In particular, the consumer electronics and transportation industries are prime candidates for increased use of magnesium. However, owing in large part to the high cost of production, the magnesium market share is a tiny fraction of the global sheet industry currently dominated by other more commonly used metals such as steel and aluminum.

Magnesium sheet has traditionally been produced via the direct chill (DC) ingot casting route, in which large rectangular ingots (400–500 mm thick) are produced, then scalped, homogenized, hot rolled, and annealed. The hexagonal close-packed (hcp) structure of magnesium limits the amount of deformation possible in each rolling step, thereby increasing the number of rolling passes and intervening heat treatments required to achieve the desired strip thickness. Alternatively, the twin roll casting (TRC) route combines both solidification and rolling in a single step, thereby producing a near net shape sheet which requires very little subsequent thermo-mechanical processing. As shown in Fig. 7.1, the TRC route eliminates the need for long homogenization times, scalping and, potentially, hot rolling; directly casting molten magnesium into thin metal strips.

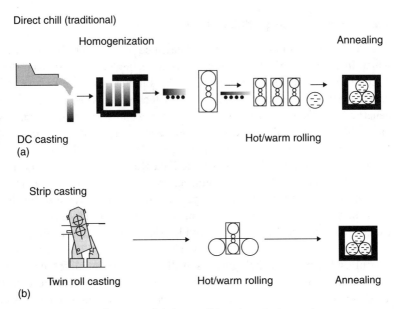

7.1 Comparison between (a) the traditional route to produce magnesium sheet and (b) the TRC process.

Thus, by comparison with the DC casting route, TRC can dramatically reduce the material losses and time, energy, and capital costs of the production cycle.

Twin roll casting has been successfully employed for the past sixty years to produce aluminum (Yun *et al.*, 2000), copper and, in the past ten years, steel sheet (Sosinsky *et al.*, 2008). Although the TRC process is relatively simple, its application for commercial-scale magnesium strip production has proven difficult. This is primarily due to inherent characteristics of magnesium alloys, such as high reactivity to oxygen, low specific heat, and large freezing ranges, which can induce formation of casting defects if various TRC processing parameters, such as metal delivery design, heat transfer in the roll gap, and casting speed, are not tightly controlled. Thus, research is underway worldwide to concurrently gain a better understanding of TRC processing variables in order to provide optimum casting conditions, which will reduce defects, and develop new magnesium alloys with properties tailored to the TRC process.

This chapter explores these developments, providing details of the technology, and its use historically, as well as the complex solidification process and resulting microstructures that occur during TRC for a range of magnesium alloys. Thermodynamic calculations are also explored to provide a deeper understanding of the complex chemistry of magnesium alloys that will assist in the design of new alloys for TRC and sheet production. Recent advances in TRC process modeling are also outlined, which will provide a quantitative basis for optimizing important process variables to ensure high-quality sheet products.

7.2 Industrial perspective

The TRC process is based on a concept patented by Sir Henry Bessemer in 1865, which integrated vertical casting and rolling technologies to produce steel strip directly from liquid steel (Bessemer, 1865). Limitations on refractory materials development, sensing and process control technologies prevented commercial penetration of the steel TRC process in the nineteenth century and it was not until the 1950s that the TRC process was successfully applied to commercial strip production for aluminum sheet. Aluminum TRC was developed by two major manufacturers: Hunter (later to become Fata Hunter) in the USA and Pechiney (later to become Novelis PAE) in France, both adopting horizontal placement of the roll casters, not the vertical configuration originally envisioned by Bessimer. Over the years, the casters have become larger in size to produce wider, thinner gauge, highly alloyed strip (Basson and Letzig, 2010), as seen in Table 7.1.

The commercial success of aluminum TRC, coupled with advances in process control and sensor technologies, as well as the potential for economic and environmental benefits cited earlier, sparked new and renewed interest in TRC for magnesium and steel sheet production. Japanese companies, Nippon steel and Mitsubishi Heavy industries, launched the world's first commercial-scale TRC technology for the production of stainless steel sheet (vertical configuration), 2–5 mm thick, 76–1300 mm wide early in the twenty-first century (Guthrie and Isac, 2004). Shortly afterward, American steel miller Castrip (Nucor/BHP/IHI) announced the first commercial TRC of low carbon steel (1.7–1.9 mm thick) (Wechsler and Campbell, 2002).

Magnesium alloy sheet production via TRC, on the other hand, is still in the early stages of commercialization. Experimental TRC of magnesium was carried out for the first time in the 1980s by Dow Chemical, who used a proprietary molten magnesium handling system, coupled with a Hunter caster originally designed for aluminum alloys (horizontal configuration) to produce 200 kg coiled magnesium strip, with a thickness of 6–7 mm and width of 610 mm, at a casting speed of 1.5 m/min (Hunter, 1958, Liang and Cowley, 2004, Park et al., 2009b). While the magnesium sheet market had not matured enough to warrant commercialization of the TRC process at that time (Park et al., 2009b, Liang and Cowley, 2004, Brown, 2002), interest in TRC of magnesium has advanced rapidly since 2000, with major research programs initiated in Australia, Germany, Korea, Japan, Turkey and Canada (Park et al., 2001, Allen et al., 2001, Park 2002 et al.,

Table 7.1 Historical evolution of aluminum strip casting technology

Roll diameter (mm)	Roll width (mm)	Strip thickness (mm)	Date
620	<1500	8–12	1960s
960, 840	Increased to 2000	6	Late 1970s, early 1980s
1150	Up to 2300	As low as 2–3	1990s

Kawalla *et al.*, 2003, Engl, 2005, Jung *et al.*, 2007, Park *et al.*, 2007b, Ding *et al.*, 2008, Park *et al.*, 2009b, Essadiqi and Lobo, 2010, Park *et al.*, 2011). These activities are summarized in Table 7.2.

To date, Korean steel producer, POSCO, is the only company currently capable of commercial magnesium sheet production using the TRC process. Figure 7.2

Table 7.2 Research and development activities related to TRC of magnesium

Country	Company/institute	Cast strip dimensions Width × thickness (mm)
Korea	POSCO (RIST)*	600 × 4–6
		2000 × 4–8
Japan	Mitsubishi Aluminium	250 × 5
	Gonda Metal	400 × 2–6
China	Luoyang Copper	600 × 7
	Yinguang Magnesium	600 × 2–8
Australia	CSIRO	600 × 3–5
Germany	MgF	700 × 4–7
	HZG†	600 × 4–7
Norway	Hydro Aluminium	700 × 4.5
Turkey	TUBITAK	1500 × 4.5–6.5
Canada	CANMET	250 × 4–6

* Commercialized in 2007. †Formerly known as GKSS.

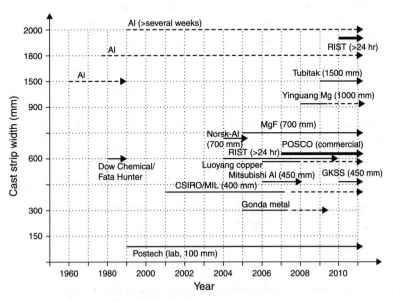

7.2 Strip width (mm) evolution with time.

7.3 As-cast strip AZ31 alloy coil of about 16 tons with 2000 mm width, 2500 mm diameter and 4–8 mm thickness. (Source: RIST, South Korea.)

highlights the current state of the art in the use of this technology to produce magnesium sheet. Although POSCO technically produces magnesium sheet at a commercial scale via the TRC process, this technology is not nearly as robust as that used to produce aluminum sheet. In the case of aluminum TRC, coils over 2000 mm wide can be produced at a continuous pace for over one month once the TRC process is started. In the case of magnesium TRC, POSCO have produced 600 mm wide coils for 24 h (Jung *et al.*, 2007). However, TRC of magnesium is expected to improve significantly with increased demand for magnesium alloy sheet. Recently, RIST (POSCO's subsidiary research center) successfully produced a 16 ton AZ31 coil of 2000 mm wide strip with a thickness of 4–8 mm as shown in Fig. 7.3.

A key component to successful commercialization of TRC produced magnesium sheet includes advances in knowledge of the TRC process and alloy development as well as the development of a viable commercial market for magnesium sheet. The following section will outline the current state of the art in magnesium TRC, and highlight areas of primary concern.

7.3 Twin roll casting (TRC) process

Twin roll caster configurations can be grouped into two main categories depending on the arrangement of the rolls and, hence, the orientation of the strip during

7.4 250 mm wide twin roll caster at CANMET.

production, namely vertical or horizontal. While a vertical layout is typically used for steel TRC, a horizontal configuration is adopted for aluminum and magnesium sheet production. The reasons for adopting the horizontal TRC process for magnesium include the ability to apply more force to the strip and hence more hot deformation, as well as safety issues around liquid metal break-outs during casting.

Figures 7.4 and 7.5 illustrate the horizontal TRC process. A pump transfers the molten metal from a holding furnace into a metal feeding system, which includes a headbox and feeding or nozzle tip. This system then feeds the molten metal between the rolls. At the exit of the nozzle tip magnesium starts to solidify against the surface of the rolls and the shell grows continuously from both sides of the rolls.

Conveniently, operating temperatures required during TRC of magnesium fall in the same range as aluminum TRC, suggesting the same TRC equipment can feasibly be used for both metals. However, there are some unique challenges to applying TRC technology to magnesium alloys as compared to aluminum (Liang and Cowley, 2004, Basson and Letzig, 2010, Park *et al.*, 2009b):

- Molten magnesium readily oxidizes and ignites; it is critical to provide a safe, protected environment for handling molten magnesium as it is being fed into the casting system.
- Liquid magnesium is very reactive to silicon (Si); direct contact between liquid magnesium and ceramic containing Si should be avoided.

- The volumetric heat capacity and latent heat of magnesium are lower than that of aluminum (Basson and Letzig, 2010, Park et al., 2009b), indicating that liquid magnesium freezes faster than liquid aluminum. This fast freezing will make it difficult to achieve uniform solidification across the strip.
- Magnesium wrought alloys have larger freezing ranges than typical aluminum wrought alloys produced by TRC, which increases the likelihood of both cracking and segregation.
- The surface tension of liquid magnesium (560 mJ/m^2) (Beck, 1940, Basson and Letzig, 2010) is lower than that of liquid aluminum (850–1100 mJ/m^2) (Kalazhokov et al., 2003), indicating that it is less resistant to an external force and has a higher tendency to wet and spread along the roll surface relative to aluminum. This property makes the formation of an adequate seal in the feeding equipment more challenging for magnesium (Basson and Letzig, 2010). This also implies that the meniscus between the lips of the tip and the surface of the rolls are less stable for liquid magnesium, which makes tip geometry consistency and tip positioning accuracy and stability of great importance.

7.3.1 Components of the TRC process (horizontal configuration)

In order to find the optimal casting conditions for industrial-quality magnesium sheet production, several processing variables must be taken into account. For example, in order to control solidification behavior and heat extraction through the caster rolls, processing parameters such as liquid metal temperature, liquid metal level in the headbox, setback, roll gap and casting speed can be varied. In addition, roll materials, roll roughness and cooling water temperature can influence heat extraction; nozzle tip design can directly influence laminar flow and temperature distribution across the nozzle length; and roll coating can vary the surface quality of cast strip.

Figure 7.5 shows a typical TRC equipment layout for magnesium sheet production based on the CANMET configuration:

- melting furnace
- pump and transfer tube
- headbox
- delivery nozzle (tip)
- twin roll caster stand
- pinch rolls
- moving shear unit
- stacking unit (or coiler).

The ingot preheating unit, data acquisition and the process control unit that are part of the TRC, are not shown in this figure. Magnesium ingots should be preheated prior to insertion into the melting furnace to remove surface moisture

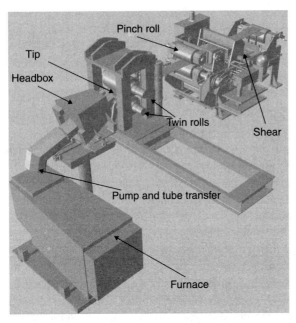

7.5 Typical TRC layout for magnesium based on the CANMET configuration. Magnesium ingots should be preheated prior to insertion into the melting furnace to remove surface moisture that can cause explosions during casting. Preheat temperatures typically range from 200–400 °C to prevent partial melting of the ingot.

that can cause explosions during casting. Preheat temperatures typically range from 200–400 °C to prevent partial melting of the ingot.

The melting furnace, shown in Fig. 7.5, is equipped with two chambers (one for melting and the other for pumping) and a steel crucible heated by electricity or natural gas. For casting stability, the melting furnace is equipped with liquid metal temperature and level sensors to control the liquid level, preventing overflow. A protective gas supply system is also in place to prevent contact between liquid magnesium and oxygen in both the furnace and the headbox, thereby avoiding oxidation of the magnesium melt and formation of oxide inclusions that may cause nozzle blockage and/or deteriorate the strip quality.

During the TRC process, the molten metal is pumped from a holding furnace into a metal feeding system, which includes a headbox and a feed nozzle. This nozzle is typically made of steel or refractory materials. Molten metal is introduced into the casting machine via the feeding tip, solidifies between the rotating water-cooled rolls, and leaves the casting machine in the form of a sheet. The design of the feeding tip is also crucial to the success of the TRC process. The feeding tip should provide an even flow as well as homogeneous temperature, minimizes turbulence and distributes the melt to the required strip width using refractory baffles. The internal baffling configuration affects the molten metal distribution and contributes to defect-free industrial sheet production. In particular, when casting the thinner gauges, or with higher speeds, it is necessary to modify the design to satisfy the larger feed requirements hydrodynamically and to promote a more uniform flow over the whole width of the tip.

The metal level in the headbox and its control during TRC is very important; if the metal level in the headbox is too high, the metal flow rate can exceed that required for TRC, causing molten magnesium to leak out between the nozzle and the rolls. However, if the metal level is too low, gaps will develop between the strip surface and the rolls resulting in surface defects and holes.

To prevent liquid magnesium from spreading laterally, edge dams are installed within the twin roll gap, extending from the two sides of the nozzle tip. Two main nozzle tip configurations have been developed for TRC of magnesium (Basson and Letzig, 2010). The first is a heated steel tip made of nickel-free, heat-resistant steel with embedded heating elements to maintain the tip temperature during casting, thereby ensuring no premature solidification occurs inside the nozzle. The second is composed of ceramic, based on the design developed for TRC of aluminum, where highly insulating materials limit heat losses through the tip. However, to prevent the liquid magnesium from reacting with the silica in the contained ceramic tip; the inner face of the tip plates are protected with a specialized coating such as boron nitride (Basson and Letzig, 2010).

To prevent metal from freezing and to remove residual moisture present inside the ceramic nozzle tip, the tip is preheated using an air blower. The location of the tip relative to the roll surface is well controlled during the casting operation since it influences the meniscus formation and subsequent solidification along the roll surface. The setback (the distance from the feeding tip to the kissing point of the rolls) directly affects the solidification front and deformation in the strip, and is therefore another important parameter that must be tightly controlled to ensure a high-quality strip is produced. The setback is a function of roll diameter, roll gap and nozzle tip height (Park, 2009b). Once the roll gap is fixed, the setback and casting speed mainly determine the extent of hot deformation during casting. In addition, a protective gas, typically in the form of SF_6 with CO_2 or N_2, is required between the nozzle tip end and rolls to prevent oxidation of the cast magnesium (Park et al., 2009b, Biedenkopf et al., 2005, Bach et al., 2006).

The water-cooled rolls shown in Fig. 7.5 are made of a core and a shell shrink-fitted on the core. Water circulates between the core and the shell to cool the latter, which is in direct contact with the magnesium. The rolls behave primarily as a heat exchanger. The uniformity of the heat extraction and hence cooling rates across the width and the circumference is affected by the design of the cooling channels (Basson and Letzig, 2010). To prevent metal sticking to the rolls, a lubricant typically consisting of a water suspension with graphite or other release agents can be applied to the rolls. This suspension also acts as a thermal barrier between the strip and the roll and hence controls the heat exchange at their interface. To further improve coating homogeneity, each roll is equipped with a rotating brush positioned on the caster exit side after the spray.

A pinch roll, shown in Fig. 7.5, guides the solidified strip to a moving shear that cuts the magnesium strip into prefixed lengths or to a coiler to be coiled into as-cast product for further thermo-mechanical processing.

7.3.2 Casting parameters important for TRC of magnesium

In addition to the above-mentioned TRC components, several other casting parameters strongly affect final strip quality, including melt cleanliness, superheat, casting speed, roll separating force and the profile of the roll surface.

One of the most important factor to achieving commercial-quality strip is consistent and homogeneous magnesium alloy chemistry across the sheet thickness. The chemical composition determines the solidification range, which in turn can affect segregation.

Liquid metal temperature

The liquid metal temperature has to be controlled and kept constant during casting operations. The superheat, which is the difference between the liquid pour temperature and alloy liquidus, influences the solidification microstructure features such as dendrite size (SDAS) and intermetallic size distribution.

Casting speed

For a given strip entry thickness, there is a corresponding casting speed range that produces good strip quality. Low casting speed may cause premature freezing of the metal in the nozzle tip and consequently increase rolling load, producing macrosegregation and hot cracking. A higher speed, on the other hand, increases productivity, but moves the liquid metal sump closer to the roll exit (Bae et al., 2007, Zeng et al., 2009, Lockyer et al., 1996), thus potentially producing a semi-solid zone at the center of the strip at exit. High superheat produces the same phenomenon (Cao et al., 2008).

Magnesium alloys cast using TRC

The most common wrought magnesium alloy cast using the TRC process is the commercially important AZ31. However, there continues to be a global push to develop new wrought magnesium alloys produced via TRC that maintain a random texture orientation during deformation and hence have improved formability. Several laboratories have tried casting a range of magnesium alloys (other than AZ31) at the pilot scale. At the commercial scale, POSCO/RIST (South Korea) produces AZ21, AZ31 and AZ61 alloy sheet (~500mm width), as well as sheet comprising the same alloys but containing small quantities of Ca.

MgF (Germany) and CSIRO (Australia) have also reported successful production of AZ alloys using semi-commercial scale twin roll casting lines. Recently, many wrought magnesium alloys have been twin roll cast at the laboratory-scale. To date, no magnesium alloys have been specifically designed for the TRC process but are variations of wrought magnesium alloys that have been successfully DC cast. Specifically, magnesium alloys cast using commercial and laboratory-scale TRC processes, as reported in the literature, include:

- AZ series: AZ21, AZ31, AZ61 and AZ91. Successful commercial twin roll castings of AZ21, AZ31 and AZ61 alloys with and without calcium have been carried out by CSIRO (Allen *et al.*, 2001), MgF (Engl, 2005) and POSCO/RIST (Jung *et al.*, 2007, Park *et al.*, 2011). The cast strips of these alloys were successfully rolled to produce Mg sheet with a minimum thickness of 0.12 mm. With increasing alloying elements, as-cast microstructure is changed from columnar dendrite structure (AZ31 or below) to equiaxed dendrite structure (AZ41 or higher) (Allen *et al.*, 2001, Chen *et al.*, 2008, Park *et al.*, 2011). Although AZ91 alloy has been cast using commercial TRC, the surface quality of AZ91 strip is still not suitable for sheet applications.
- AM series: AM31. AM31 alloy sheet has been successfully cast using lab-scale TRC and its as-cast microstructure has been found to be similar to that of AZ31 alloy sheet (Wang *et al.*, 2011).
- ZK, ZM and ZMA alloys series: ZK60, ZM61, ZMA611 and ZMA613. High Zn containing alloys have been cast using lab-scale TRC processes (Park *et al.*, 2007a, Park *et al.*, 2009a, Kim *et al.*, 2010, Liu *et al.*, 2009, Wang *et al.*, 2010). Although the yield strength and elongation of these twin roll cast alloys were reported to be higher than those of AZ31 alloys (Park *et al.*, 2007a), the surface quality control of these alloys in commercial production seems to be quite challenging due to the high Zn and Al contents.
- Rare earth (RE)-containing alloys: ZW41, ZW61 and Mg-4Zn-1Gd. Compared to previously mentioned alloys, yttrium (Y)- and gadolinium (Gd)-containing alloys show more random texture orientation after rolling (Kim *et al.*, 2010). These RE-containing alloys were cast using a lab-scale caster and demonstrated promising results for texture weakening. In addition, the high cooling rate of TRC enables increased super-saturation of Y and Gd in the magnesium matrix, as compared to the conventional DC casting route. The super-saturation of these elements seems to promote more random texture orientation after subsequent rolling.

7.4 Solidification and strip microstructure

7.4.1 Solidification microstructure

Defect-free strip is critical to successful implementation of the TRC process for commercial magnesium alloy sheet production. A typical as-cast microstructure

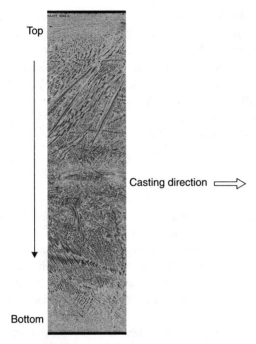

7.6 Mid-width solidification microstructure through the thickness of AZ31 strip cast using the twin roll caster at CANMET.

of twin roll cast AZ31 in the longitudinal direction through the sheet thickness is illustrated in Fig. 7.6. As shown, the microstructure can be divided into three zones: the chill, columnar, and central equiaxed zones.

A fine dendritic microstructure of α-Mg is observed in the chill zone, near the upper and lower surfaces of the strip. This is caused by the rapid cooling experienced when the metal is in direct contact with the roll surface. Columnar dendrites can be observed beneath the chill zones, upper and lower to the center, forming the columnar zone. These dendrites are not perpendicular to the casting direction, but are tilted roughly 45° to the casting direction due to the orientation of the heat flow into the roll and subsequent deformation during TRC. In the central area, equiaxed dendrites and a coarse eutectic structure (composed of α-Mg, $Mg_{17}Al_{12}$ and ternary Φ-phase) are frequently observed. This is due to the heat loss of the melt during solidification and the large freezing range of the last liquid enriched in alloying elements (Park *et al.*, 2009b, Jung *et al.*, 2007).

The measured secondary dendrite arm spacing (SDAS) through the strip thickness is in the range 4–12 μm (Essadiqi and Lobo, 2010, Aljarrah *et al.*, 2011). The grain size of twin roll cast AZ31 is finer than 200 μm, which is the typical grain size of ingot cast alloys (Park *et al.*, 2004). Fine Al_xMn_y intermetallic particles are also randomly distributed through the entire cast strip.

The morphology of the dendritic structure of an alloy depends on its freezing range, the temperature gradient in the liquid (Park et al., 2009b) and the heat extraction by the rolls. Alloys with wide freezing ranges and low temperature gradients (relatively slow solidification rate) produce equiaxed dendritic microstructures, while alloys with low freezing ranges and high temperature gradients (relatively high solidification rate) produce columnar dendritic microstructures (as in the case of AZ31) (Jung et al., 2007, Park et al., 2004, Kawalla et al., 2008, Nakaura and Ohori, 2005).

7.4.2 Strip defects

Various defects are possible in as-cast strips. For example, there are two common types of internal strip defects: non-metallic inclusions and segregation defects. The control of inclusions (typically oxide, fluoride and nitride) is very important to the production of high-quality commercial sheet. Enhanced melt cleanliness can contribute significantly to reducing the likelihood of inclusions during TRC. This requires the use of filtering processes and prevention of oxidation of liquid in the melting furnace and during liquid metal transfer to the headbox and to the inlet roll gap.

Inverse segregation is caused by the entrapment of highly alloyed liquid phase in the central region of the strip and the subsequent squeezing of this entrapped liquid outward to the strip surfaces by the pressure of the roll separating force (Park et al., 2011). This phenomenon is magnified for alloys with large solidification ranges, since they tend to have large mushy zone volumes in the central area.

Surface and edge cracks can develop in the strip for a number of reasons. For example, increased sump depth in combination with the non-uniformity of temperature, fluid flow, and heat transfer across the strip width are common causes of both defects (Li, 1995, Hunter, 1958). In addition, edge cracking can result from rapid cooling along the strip edge and/or from irregular melt flow in these zones. Other surface cracking may be caused by liquid flow turbulence and uneven solidification. Large grains produced in alloys with large freezing ranges are also susceptible to intergranular cracking, which can be reduced by adding Zr in the case of Al-free Mg alloys or Ca to produce finer grain structures.

7.5　Thermodynamic calculations

In order to understand the complex chemistry of magnesium alloys and assist in the design of new alloys for TRC, knowledge of their thermochemistry is invaluable. To keep pace with magnesium alloy developments and the expansion of the magnesium industry, magnesium alloy databases have been developed over the last 15 years by two major thermochemistry groups: FactSage-Center for Research in Computational Thermochemistry, Canada (FactSage-FTlite database: http://www.

factsage.com), and Pandat-Clausthal University, Germany (Pandat-Mg database: http://www.computherm.com). Values available through these thermodynamic databases are used to calculate multi-component phase diagrams, equilibrium phase distributions and Scheil cooling phase evolutions to gain a deeper understanding of the chemical reactions and phase transformations that occur during casting and annealing processes, and to design new magnesium alloys better suited for these processes.

7.5.1 Thermodynamic calculations

Alloy phase diagrams provide information on the stable phases at given temperatures and compositions. However, the phase diagram of a multi-component system does not provide the quantity of each phase formed at given temperatures. Thus, in order to understand phase transformations clearly, both phase diagrams and phase fraction calculations at given alloy compositions should be carried out. All calculations in this chapter were performed using FactSage thermochemical software with values available in the FTlite database. Similar calculations can be performed using Pandat software with their Mg database.

For instance, Fig. 7.7 provides the phase diagram of Mg-Al-1%Zn-0.3%Mn alloy, calculated to reflect a variation in aluminum content. According to the diagram, the solidus temperature decreases almost linearly with Al concentration, while the solidification range (temperature difference between solidus and liquidus) increases with Al concentration. The transformation temperature of the

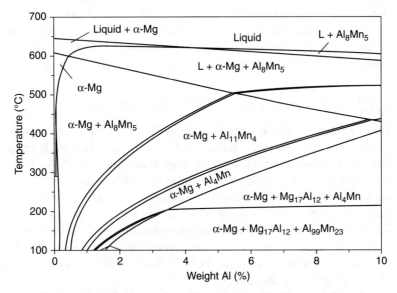

7.7 Calculated phase diagram of Mg-Al-1%Zn-0.3%Mn alloy showing a variation of Al content.

Al-Mn phases changes significantly with Al concentration as well. More importantly, the stability range of the $Mg_{17}Al_{12}$ phase increases significantly with an increase in aluminum concentration. Theoretically, the $Mg_{17}Al_{12}$ phase free temperature range for AZ31 is between 200–550 °C. This becomes narrower for AZ61 alloy (between 300–500 °C) and even narrower for AZ91 alloy (between 380–450 °C). Although equilibrium phase transformation temperatures are easily read from phase diagrams, the phase distribution cannot be interpreted from the diagram.

The most well-known wrought magnesium alloy is AZ31 (Mg-3%Al-1%Zn-0.3%Mn). The calculated phase distribution of AZ31 alloys across temperatures is shown in Fig. 7.8a. According to this equilibrium calculation, α-Mg solid can form at 630 °C and liquid can disappear completely at 550 °C. Furthermore, an Al_8Mn_5 phase can also form just below the liquidus temperature because manganese (Mn) has almost no solubility in α-Mg, and the strong affinity between Al and Mn results in the formation of intermetallic phases. The Al_8Mn_5 phase can transform into various Al_xMn_y phases with cooling, but the amount of Al_xMn_y phase is less than 1.2 wt% through the entire temperature range. $Mg_{17}Al_{12}$ phase can form below 200 °C and its amount can reach roughly 3 wt% at 100 °C.

Solidification of alloys can be simulated using the Scheil cooling calculation. The Scheil cooling calculation assumes that there is complete diffusion of all alloying elements in the liquid phase and that no diffusion occurs in the solid phase during solidification. The solidification behavior of AZ31 alloy is simulated using the Scheil cooling calculation in Fig. 7.8b. In comparison with the equilibrium phase fraction calculated in Fig. 7.8a, the Scheil cooling calculations show that primary α-Mg solid can form at 630 °C and Al_8Mn_5 phase can form subsequently. The amount of liquid phase can decrease with temperature and eventually disappear at 340 °C with the formation of the eutectic microstructure ($Mg_{17}Al_{12}$ + ternary Φ + ternary τ) phases.

7.5.2 Calculations for as-TRC microstructure

Understanding the TRC process is necessary to simulate the as-cast microstructure. A schematic solidification structure of Mg alloys during TRC is presented in Fig. 7.9. Although there is a small equiaxed layer formed on the surface of strip, it is possible to consider, for the sake of simplicity, that the solidification begins from the roll surface with columnar dendrite shape. This directional solidification behavior rejects a small amount of solute toward the center of the strip, thus accumulating solute in the central area (schematically sketched in the figure). This increase in solute content in the central area of the strip may eventually induce centerline segregation, which may be pushed out by the deformation near the roll nip to induce inverse segregation.

Although it is difficult to thoroughly understand the solute profile of twin roll cast strip just by using thermodynamic calculations, thermodynamic calculations

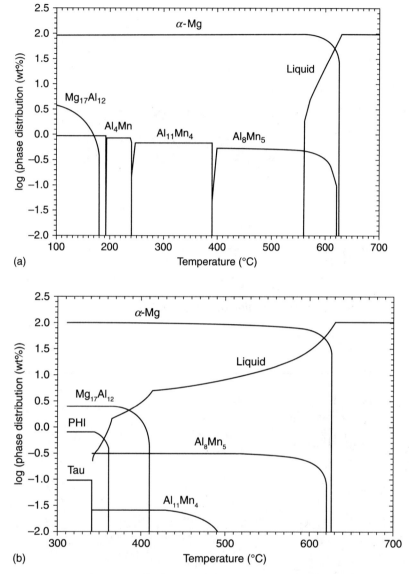

7.8 Calculated phase distribution of AZ31 alloy (Mg-3%Al-1%Zn-0.3%Mn). (a) Equilibrium phase distribution. (b) Scheil cooling phase distribution.

with the Scheil cooling assumption can provide guidance on what happens during solidification. Fig. 7.10 shows the solute profile and solidification temperature across the strip thickness, calculated using the Scheil cooling scheme. The solute profile in Fig. 7.10a is the same as that provided schematically in Fig. 7.9.

288   Advances in wrought magnesium alloys

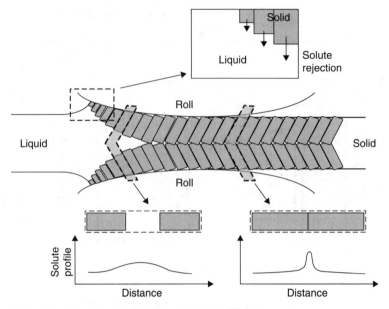

7.9 Schematic of solidification during the TRC process.

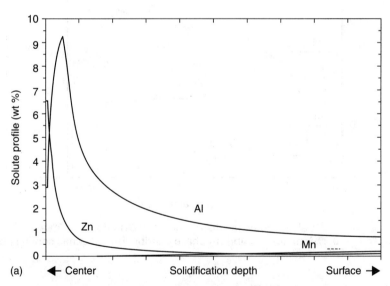

7.10 The calculated profile of solute and solidification temperature during the TRC process for AZ31 (Mg-3%Al-1%Zn-0.3%Mn) alloy using the Scheil cooling assumption. (a) Solute profile and (b) solidification temperature through the thickness of the strip.

(*Continued*)

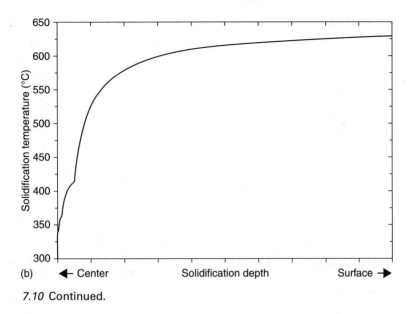

(b) ← Center   Solidification depth   Surface →

7.10 Continued.

The real solute profile measurement shows that the Al content from surface to ~45% of the thickness of the strip is almost constant, at about 2.7~2.8 wt% (0.2~0.3 wt% lower than the original AZ31 composition), while the Al content in the central region (only ~20% of strip thickness) reaches about 6 wt% (Aljarrah *et al.*, 2011). Although the calculated solute profile across the thickness does not exactly correspond to the real TRC strip profile, the results are qualitatively similar. From the surface to 80% of the central region, the Al concentration slowly increases but remains less than 3%. Then, a very steep increase in Al content is calculated in the very small area near the center of the strip. The decrease in Al concentration in the central region liquid phase is the result of a large amount of secondary phase formation, such as $Mg_{17}Al_{12}$ and ternary Φ phases, caused by eutectic reactions.

The solidification temperature through the thickness of TRC strip is calculated in Fig. 7.10b. Solidification of the strip surface begins at 630 °C. As cooling continues, 80% of solidification from the surface toward the center of the strip occurs above 580 °C. However, the central region remains in the mushy zone until the temperature decreases below 330 °C.

Once the overall chemical composition of the local area is known, Scheil cooling calculations can also be applied to predict the spatial as-cast microstructure of the twin roll cast strip. The typical overall compositions of the columnar dendrite and center line segregation region in central equiaxed zones are roughly Mg-2.7%Al-0.9%Zn-0.3%Mn and Mg-8%Al-4%Zn-0.2%Mn, respectively (Aljarrah, 2011). Figure 7.11 shows the phase distribution in the columnar and central zones, as calculated using the Scheil cooling scheme. As can be seen, the

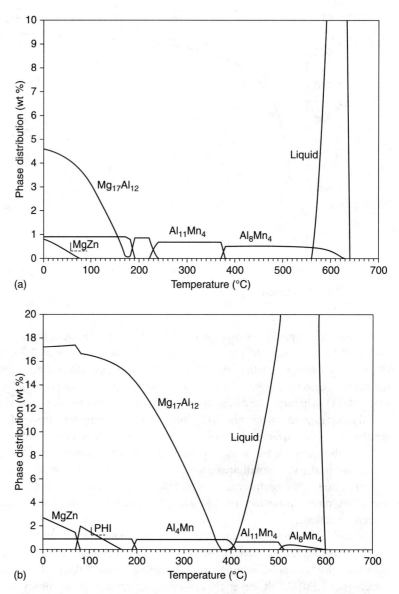

7.11 (a) Phase distribution in the columnar dendrite and central equiaxed regions, calculated using the Scheil cooling scheme. (a) Columnar region (Mg-2.7%Al-0.9%Zn-0.3%Mn). (b) Central region (Mg-8%Al-4%Zn-0.2%Mn).

calculated amounts of $Mg_{17}Al_{12}$ are 2.0% and 10.2% for columnar and central zone, respectively; suggesting that local composition changes (solute enrichment) can significantly influence the amount of $Mg_{17}Al_{12}$ phase formed. The Scheil calculation can even predict about 2% of ternary Φ phase in the central zone. In fact, the real as-cast microstructure of twin roll cast magnesium strip shows a very small amount of $Mg_{17}Al_{12}$ in the columnar zone and a large amount of $Mg_{17}Al_{12}$ concentration in centerline segregation, which can be considered as a large eutectic pocket in the central zone.

The difference between the local overall chemical composition and as-cast microstructure features raises issues during heat treatment. For example, although AZ31 alloy is typically homogenized at 400 °C, the homogenization of as-cast AZ31 strip using the TRC process can require a two-step heat treatment. As can be seen in Fig. 7.11b, the centerline segregation in the central zone is a eutectic mixture of α-Mg, $Mg_{17}Al_{12}$ and Φ phase, which has a melting temperature of 340 °C. This means the eutectic mixture may partially melt if the as-cast strip is directly heat-treated to 400 °C. If it is assumed that the diffusion lengths of Al and Zn are limited and the solute in the central area are homogenized (in reality, a certain amount of solute can diffuse out toward the adjacent columnar zone), the nominal composition of Al and Zn contents are 8% and 4%, respectively, which are still considerably higher than the normal AZ31 composition. Under this assumption, the equilibrium calculation just shows that partial melting can occur above 400 °C at the central zone. On the other hand, the equilibrium calculation shows that the melting of the columnar zone can occur at 550 °C.

A two-step heat treatment can be implemented to prevent this partial melting problem:

1 First heat treatment below 340 °C to dissolve a large amount of $Mg_{17}Al_{12}$ phase in the central segregation area.
2 Secondary heat treatment at about 400 °C to homogenize the chemical composition between the central and columnar zones.

The upper temperature limit for the heat treatment can be estimated from the solidus determined using the equilibrium cooling calculations.

7.5.3 Development of new wrought Mg alloys for TRC

In order to design wrought Mg alloys for TRC applications, several characteristics and constraints of the TRC process should be taken into account:

- Surface quality. Since TRC produces strip with thicknesses of approximately 5–8 mm, surface buffing is very limited. Thus, sound surface quality is very important in the as-cast state. It is believed that alloys with large solidification ranges are hard to produce via the TRC process because it is difficult to control uniform solidification during the rapid cooling process. Moreover, more severe segregation is expected in the central zone of the strip (centerline segregation), which increases the probability of inverse segregation (surface defect).
- Centerline segregation. Due to the solidification history during TRC, it is hard to avoid centerline segregation. If the segregation is composed of very thermally

stable secondary phases and α-Mg, it is difficult to remove them using a short heat treatment. Thus, formation of secondary phases that are thermally too stable should be avoided. If the alloying elements induce the formation of a eutectic microstructure with a very low melting temperature, partial melting may occur during heat treatment after the TRC process, which modifies the as-cast microstructure and induces the formation of internal voids.

- Limited heat treatment time. The TRC strip thickness is already less than 10 mm. If the grain size in the strip is too large, it would be difficult to obtain a fine grain structure after rolling. Thus, the heat treatment for homogenization should be shorter than that used following the typical DC casting process route. It is ideal to homogenize the solute distribution across the TRC strip as much as possible within a rather short homogenization processing time.
- Rolling and forming. High amounts of secondary phase alloys make sound rolling and forming processing difficult.
- Supersaturation of solutes in α-Mg. Since TRC incorporates a fast cooling process, a certain degree of supersaturation of the solute elements in α-Mg can be expected. Once the heat treatment is carried out, the supersaturated solute can precipitate back into the matrix. Of course, the severe deformation in TRC may reduce the super-saturation after the completion of solidification to produce deformation induced precipitations.
- Economy. Of course, the cost of the alloying element should be considered. For example, although RE elements are promising alloying elements for TRC alloys because they promote texture weakening of the final product, the amount of RE used has to be tightly controlled to low concentrations to produce cost-competitive sheet.

With knowledge of these constraints on the TRC process, the alloying elements suitable for TRC alloys can be narrowed down. In particular, the selection of the major alloying elements is very important since they can control the solidification range. Considering the success of the AZ31 alloys for TRC, for instance, new alloys with a solidification range lower than that of AZ31 can be relatively easily produced using the current AZ31 TRC technology. In addition, the major alloying elements should be soluble to a certain extent and not form secondary phases during solidification that are thermally stable. In other words, the secondary phases should be designed so that they can be removed during heat treatment. Secondary and ternary alloying elements produce precipitates after the rolling process (about 300~450 °C) and during the final annealing treatment (about 200~350 °C), and therefore can be added to improve the mechanical properties of the final product.

7.6 Process modeling and simulation

Knowledge-based process models of the TRC process for magnesium alloys are crucial to achieving a quantitative understanding of the process and conditions that lead to high-quality magnesium sheet production. Various models for the TRC process have recently been developed that consider transport phenomena coupled with solidification effects; however, in only very few cases has mechanical deformation been taken into account. Once selected, the governing equations and boundary conditions are solved using numerical solution methods. To date, very few attempts have been made to model the TRC process for magnesium alloys; however there has been significant work done to model the TRC process for other alloy systems such as steel and aluminum.

7.6.1 Basic phenomena and modeling

The twin roll casting process involves a complex series of phenomena at both the macroscopic and microscopic levels, including:

- Laminar to turbulent, transient fluid motion in a complex geometry (inlet nozzle geometry and liquid pool), thermal and solutal convection.
- Flow and heat transport within the liquid and mushy zone of the magnesium.
- Thermal, fluid, and mechanical interactions in the meniscus region (or first contact of the melt with the roll surface) between the solidifying meniscus, release/parting agent and roll surface.
- Heat transport through the solidifying magnesium solid shell at the interface between the shell and the roll (which may contain volatilized release/parting agents and growing air gaps).
- Distortion and wear of the roll surface.
- Nucleation of solid crystals, both in the melt and against the roll surface.
- Shrinkage of the solidifying magnesium shell, due to thermal contraction and solidification shrinkage.
- Stress/strain generation within the solidifying magnesium strip, due to external forces, thermal strains (friction, increased pressure between the rolls), creep, and plasticity (which varies with temperature, alloy and deformation and cooling rate).

An essential aspect of successful model development is the selection of the key phenomena of interest to a particular modelling objective and reasonable assumptions. This section will examine modeling research to date, with particular emphasis on macroscopic phenomena such as fluid flow, heat transfer and stress/strain development in magnesium strip as it is being cast.

7.6.2 Fluid flow

A key aspect in the successful production of high-quality strip from TRC is knowledge of the metal distribution to the casting rolls. Emphasis is placed on reducing the turbulence in the metal as it enters the melt pool and the need for effective distribution of the metal along the roll length. A further requirement of the metal delivery system is to provide metal to the meniscus at the roll surface in a stable and robust manner. Any disturbance at the meniscus invariably manifests itself as a strip defect; the process must be stable and in control at all times to ensure excellent surface quality.

Fluid flow from the nozzle to the rolls is of interest because it influences many important phenomena, which have consequences on strip quality. These effects include the dissipation of superheat (and temperature at the meniscus) and homogeneity of the temperature across the width of the strip.

In order to model the transient phenomena of flow and heat transfer in the tip, the mathematical model should have the ability to deal with the fluid flow, heat transfer, solidification and free surface movement at the same time.

Different numerical models are presented by various authors to simulate the TRC process. To date, much of the focus of modeling has been on the fluid flow and metal delivery systems (Guthrie and Tavares, 1998, Buechner, 2004, Ohler et al., 2003, Ju et al., 2005 and Bae et al., 2007).

The most common technique used to simulate the heat transfer and fluid flow during TRC is to modify the continuum equations developed by Bennon and Incropera (1987). These are derived based on the classical mixture theory and incorporate all pertinent macroscopic transport phenomena in the governing conservation equations. Typical assumptions in these models include the following:

- The molten metal is a Newtonian fluid.
- The solid and liquid phases are homogeneous and isotropic.
- The solidification shrinkage is negligible.
- The solid and liquid phases in the mushy zone are in local thermodynamic equilibrium.

Under the above assumptions, the continuum conservation equations for mass, momentum and energy in the Cartesian coordinate system may be solved.

To account for solidification, two important effects of this phenomenon on the fluid flow and heat transfer must be included: damping fluid flow in the mushy zone and latent heat of fusion. Since alloy solidification occurs over a temperature interval, a mushy zone forms that can inhibit fluid flow. To model this effect, it is assumed this mushy region acts as a porous medium and obeys Darcy's law of momentum (Lin, 2004, Guthrie, 1998, Zeng et al., 2009). The momentum equation is typically modified to account for solidification and changes as the material transforms from a liquid to a mushy medium, and finally a solid.

The source term requires velocities within the mushy zone to gradually approach the corresponding component of the roll velocity, as the liquid fraction decreases (Zeng et al., 2009 and Guthrie, 1998).

The general form for this source term when modeling TRC for magnesium is:

$$C \frac{(1-f_l)^2}{f_l^2 + \varepsilon} (\mu - \mu_r) \qquad [7.1]$$

The value of the morphology constant, C, determines the speed of this approach. Numbers in the range of 10^4–10^7 have typically been chosen to model TRC of aluminum, while values in the range of 10^5 to 10^7 have typically been used for TRC of steel. f_l is the fraction of liquid, ε is a very small number (typically) to avoid division by zero when the fraction liquid approaches zero. U_r is the horizontal or vertical configuration of the roll velocity.

During solidification, a significant amount of heat is released due to the latent heat of fusion of the liquid to solid phase change. This effect is typically incorporated into models by considering an equivalent specific heat equation as shown below.

$$H = H_{ref} + \int_{f_{ref}}^{f} C_p dT \qquad [7.2]$$

where H is the enthalpy of the system, C_p is the specific heat and T is the temperature. The conservation equations are discretized and then solved using either commercial software such as ANSYS Fluent, ProCAST, or in-house software.

To date, very little has been published on the fluid flow during TRC for magnesium alloys, or the effects of the nozzle shape and design on fluid flow. The most comprehensive modeling work to date has been done by Zeng (2009), who has investigated fluid flow during casting and the effects of various casting parameters; however, no details of the nozzle design were provided.

7.6.3 Boundary conditions

Typically, symmetry planes are used to simplify simulation of TRC sheet. In the horizontal configuration, this usually involves considering only half the sheet thickness for 2D simulations or a quarter section for 3D simulations. In addition, non-slip boundary conditions are typically considered on all solid boundaries for thermal-fluid simulations.

In cases where coupled thermal-fluid stress analysis is done, a friction coefficient between the roll and the strip is typically used, with a value ranging from 0.2–0.4. Sticking friction conditions along the arc of contact have also been assumed in modeling aluminum TRC (Bradbury and Hunt, 1995). Knowledge of what the friction coefficient should be during TRC of magnesium alloys is very limited; however, preliminary information can be obtained from the hot rolling literature for magnesium and aluminum. It should be noted, however, that the friction coefficient may vary dramatically depending on the use of lubricants or release agents on the surface of the roll.

On the free surfaces, heat loss due to radiation and convection are usually considered. Heat loss from the strip surface to the roll surface is usually approached using an overall heat transfer coefficient along the arc of contact. The value for this heat transfer coefficient is usually based on overall energy balances and/or comparisons between predictions of exit surface strip temperatures and measurements.

7.6.4 Heat transfer coefficient between strip and roll

Efforts to validate the accuracy of model predictions requires correct knowledge of the heat transfer coefficient (HTC) between the strip and the roll, from first contact of the molten metal with the roll surface through to last contact of the solid strip as it exits the rolls (i.e., the spatial variation of the HTC along the arc of contact). In addition, the effects of a range of parameters on the HTC must be understood, including roll texture and roughness, pressure at the interface, the role parting/release agents play and the roll/magnesium thermo-physical properties on the magnitude of the HTC.

Solidification during TRC can be subdivided into four zones (Kopp *et al.*, 1998), where different parameters control heat transfer from the rolls to the strip:

1. Preliminary contact between the liquid melt and rolls.
2. The formation of a growing strip shell; solidification shrinkage of the strip away from the roll and a reduction in the heat transfer across the interface occurs due to decreasing thermal contact.
3. As the kissing point of the rolls (position of least separation between the rolls) is approached, thermal contact increases as the pressure at the interface between the sheet and the roll increases significantly.
4. Beyond the kissing point of the rolls thermal contact decreases as the pressure at the interface between the strip and the rolls decreases until the strip leaves the rolls completely.

Figure 7.12 shows a schematic of the expected variation of the HTC along the arc of contact during the TRC process.

In Zone I, the melt directly covers the rolls. Fig. 7.13 is a schematic diagram of a typical interface between the roll and the magnesium during the early stages of twin roll casting, showing the types of heat transfer that can occur across the interface. As can be seen, this interface consists of discrete areas where there is good contact between the magnesium and the roll surface as well as regions where there are gaps. Depending on the parting agent used, these gaps may be filled with volatilized parting agent, air or other gases. The magnitude of the interfacial HTC is determined by the temperature difference between the roll and magnesium and the resistance to heat flow across the interface. Heat flow across the interface will be determined by two modes of heat transfer, namely:

- Heat conduction from the metal to the roll in regions of good contact.
- Heat transfer from the metal to the gas region and then to the roll in regions where gaps exist.

During solidification, latent heat is also released at the solid-liquid interface. Overall, thermal contact in Zone I is of medium quality; however, the melt becomes partially super-cooled very quickly.

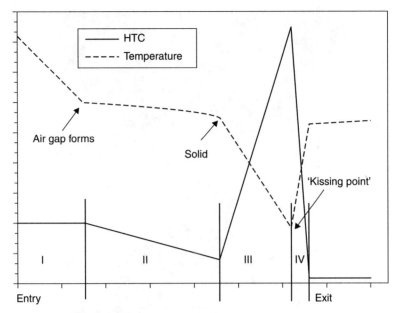

7.12 Schematic illustrating the expected change in heat transfer coefficient (HTC) and strip temperature along the arc of contact during TRC.

Zone II begins when the first layer of solidified strip covers the surface of the roll, forming an additional barrier through which the heat must pass. These parts of the just-solidified strip that stay in direct contact with the roll will cool more rapidly than the others with less contact. In fact, the gas niches that prevent good contact conditions cause the heat transfer to become heterogeneous. Further reduction of the heat flow is caused by the thermal shrinkage of the strip, which is counteracted by the increasing pressure between the rolls and the strip surface as the rolls come closer together.

Zone III begins as the kissing point of the rolls is reached and the contact pressure between the solid strip and roll increases, causing a rapid rise in the HTC, which quickly comes to a peak. After the kissing point, Zone IV starts where the HTC decreases in response to reduced pressure between the rolls and the strip. This is followed by a rebound in the surface temperature as heat from the center of the strip conducts to the surface.

Currently, HTC values used to model TRC of magnesium alloys vary from 3.3–20 kW/m^2 °C (Zeng 2009, and Hadadzadeh *et al.*, 2009).

In most models, the HTC between the strip and the roll surface is treated very simply and a constant value for the HTC is used along the arc of contact. The attempt to measure the variation in heat transfer along the arc of contact during TRC was undertaken by Guthrie *et al.* (2000), who used instrumented rolls to measure the temperature-time history in the roll during vertical TRC

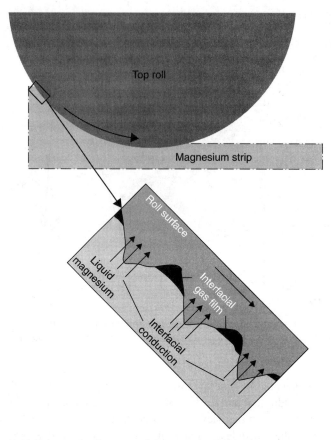

7.13 Schematic diagram of the metal/roll interface during the early stages of TRC. (Source: adapted from Loulou *et al.*, 1999.)

of steel in conjunction with inverse analysis to predict the interfacial HTC. Based on these measurements it was found that the HTC reaches a peak very rapidly in the arc of contact, consistent with the change in pressure between the strip surface and the roll. They also found that the HTC varied as a function of TRC casting speed.

7.6.5 Stress analysis

No published work has been reported on the development of stress in the strip during TRC or the influence of key variables on this development. Using ALSIM

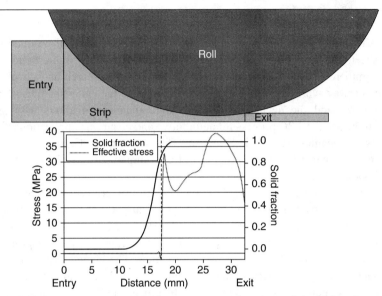

7.14 Model of predicted effective stress and fraction solid at the strip centerline during TRC of AZ31.

coupled thermal, fluid and mechanical software, analysis of the stress state in the strip was done based on the CANMET TRC configuration.

As can be seen in Fig. 7.14, as the coherency point is reached at the strip centerline, the stress quickly increases to a peak followed by a small decrease due to a lowering of the strain rate locally. The stress then continues to increase as the material solidifies completely and experiences plastic deformation.

7.6.6 Validation

Validation of TRC model predictions has been very limited owing to the difficulty in taking temperature measurements *in situ* in the roll bite region. As a result, very few experimental measurements have been done to gather details of the fluid flow and thermal-mechanical history during solidification and subsequent deformation for magnesium TRC. Measurements have instead consisted exclusively of strip surface temperatures after exiting the TRC apparatus. Although this will provide some data on the overall heat extracted, details of how this occurs along the roll bite cannot be elucidated.

Significant further work is needed to understand and quantify these phenomena and to apply the results to optimize the TRC process. In striving towards these goals, the importance of combining modeling and experimental analysis together cannot be overemphasized. In fact, a significant limitation of the current work is the lack of experimental data for detailed validation of model predictions.

7.7 Conclusions

Although TRC offers a promising route for the economical production of Mg sheet, the technology is still in its infancy with only limited success in its commercial ability to produce magnesium sheet. Research and development is underway worldwide to concurrently gain a better understanding of the TRC process and the development of wrought magnesium alloys with properties tailored to the TRC process for automotive and electronic applications. Currently, two companies, POSCO in Korea and Thyssen Krupp in Germany are leading the international effort to making the TRC commercially available for producing 2000 mm magnesium sheet.

The key elements of the TRC technology that are critical to its commercial success have been discussed in the chapter. Currently, the most common wrought magnesium alloy cast using TRC is AZ31. However, significant work is underway to identify new alloys that are castable, have good mechanical properties and maintain a weak texture during deformation. The chapter has also included a section related to thermodynamic calculations and their importance in alloying design of new alloys suitable for the TRC process and recent advances in the TRC process modeling, providing a quantitative basis for process optimization to produce high-quality sheet products.

Future research on the TRC process continues to include surface quality control and minimization of centerline segregation to obtain high-quality cast strip. To achieve these, roll texturing for homogeneous solidification with controlled cooling rate and nozzle and melt delivery system design for uniform melt distribution need to be optimized for each Mg alloy system being cast.

7.8 References

Aljarrah, M., Essadiqi, E., Kang, D. H. and Jung, In-Ho (2011) 'Solidification microstructure and mechanical properties of hot rolled sand annealed Mg sheet produced through twin roll casting route', *Proc. of the Fifth International Light Metals Technology Conf. 2011, LMT2011*, Lüneberg, Germany, pp. 331–334.

Allen, R. V., East D. R., Johnson, T. J., Borbidge, W. E. and Liang, D. (2001) 'Magnesium alloy sheet produced by twin roll casting', *Magnesium Technology 2001*. TMS, pp. 75–79.

Bach Fr. W., Schaper M., Hepke, M., Karger, A. and Werner, J. (2006) 'Reduction and avoidance of ecologically harmful cover gases', *Proceedings of the 7th International Conference on Magnesium Alloys and their Applications*, Weinheim: Wiley-VCH, Germany, pp. 215–220.

Bae, J. W., Kang, C. G. and Kang, S. B. (2007) 'Mathematical model for the twin roll type strip continuous casting of magnesium alloy considering thermal flow phenomena', *Journal of Materials Processing Technology*, **191**, 251–255.

Basson, F. and Letzig, D. (2010) 'Aluminum twin roll casting transfers benefits to magnesium', *Aluminium International Today*, November/December, pp. 19–21.

Beck Adolf (1940) In *The Technology of Magnesium and its Alloys*, Edited by E. H. Adolf Beck, translated from German by Magnesium Elektron Limited, published by F.A. Hughes & Co. Limited, p. 118.

Bennon, W. D. and Incropera, F. P. (1987) 'A continuum model for momentum, heat, and species transport in binary solid–liquid phase change systems. I: Model formulation,' *International Journal of Heat and Mass Transfer*, **30**, 2161–2170.

Bessemer, H. (1865),'Manufacture of Iron and Steel', No. 49,053 patented July 25, 1865.

Biedenkopf, P., Karger, A., Laukotter, M. and Schneider, W. (2005) 'Protecting liquid Mg by solid CO_2: New ways to avoid SF_6 and SO_2', TMS, *Magnesium Technology*, **2005**, 39–42.

Bradbury, P. J. and Hunt, J. D. (1995) 'A coupled fluid flow, deformation and heat transfer model for a twin roll caster', M. Cross and J. Campbell Editors, *Modeling of Casting and Advanced Solidification Processes VII Minerals*, Metals and Materials Society, Warrendale, PA, pp. 739–746.

Brown, R. E. (2002) 'Magnesium wrought and fabricated products yesterday, today and tomorrow', *TMS, Magnesium Technology*, **2002**, 155–163.

Buechner, A. R. (2004) 'Thin strip casting of steel with a twin-roll caster—correlations between feeding system and strip quality', *Steel Research*, **75**, 5–12.

Cao, G. M., Li, C. G., Liu, Z. Y., Wu, D., Wong, G. D. *et al.* (2008) 'Numerical simulation of molten pool and control strategy of kiss point in a twin-roll strip casting process', *Acta Metall. Sin. (Engl. Lett)*, **21** (2008), 459–468.

Chen Hongmei, Kang Suk Bong, Yu Huashun, Kim Hyoung Wook and Min Guanghui (2008) 'Microstructure and mechanical properties of Mg-4.5Al-1.0Zn alloys sheets produced by twin roll casting and sequential warm rolling', *Materials Science and Engineering: A*, **492**, 1–2, 25, 317–326.

Ding, P. D., Pan, F.-S., Jiang, B., Wang, J., Li, H. L. *et al.* (2008) 'Twin-roll strip casting of magnesium alloys in China', *Trans nonferrous Met Soc. China*, **18**, s7–s11.

Engl, B. (2005) 'Future aspects of magnesium sheet materials using a new production technology', *62nd Annual World Magnesium Conference*, Washington, D.C.: International Magnesium Association, pp. 27–34.

Essadiqi, E. and Lobo, N. (2010) 'Twin roll casting of AZ31', CANMET, 2010–02(CF).

Guthrie, R. I. L. and Tavares, R. P. (1998) 'Mathematical and physical modeling of steel flow and solidification in twin-roll/horizontal belt thin-strip casting', *Applied Mathematical Modeling*, **22**, 851–872.

Guthrie, R. I. L., Isac, M., Kim, J. S. and Tavares, R. P. (2000) 'Measurements, simulation, and analyses of instantaneous heat fluxes from solidifying steels to the surfaces of twin roll casters and of aluminum to plasma-coated metal substrates', *Metallurgical and Materials Transactions B*, **31B**, 1031–1047.

Guthrie, R. I. L. and Isac, M. (2004) 'The design of continuous casting process for steel', *Handbook of Metallurgical Process Design*, edited by G. E Totten, K. Funatani and L. Xie, New York: Marcel Dekker, pp. 251–293.

Hadadzadeh, A. Wells, M. A. and Essadiqi, E. (2009) 'Mathematical modeling of the twin roll casting for magnesium alloys – Effect of heat transfer coefficient between the roll and the strip', *Proceeding of 8th International Conference on Magnesium Alloys and their Applications*, Weimar, Germany, 26–29 October 2009, Weinheim: Wiley-VCH, Germany, pp. 138–144.

Hunter, J. L. (1958) 'Roll constructions for continuous casting machines', USA Patent, 2,850,776.

Ju, D-Y, Zhao, H. Y., Hu, H-D, Ohori, K. and Tougo, M. (2005) 'Thermal flow simulation on twin roll casting process for thin strip production of magnesium alloy', *Materials Science Forum*, **488–489**, 439–444.

Jung, I.-H., Bang, W., Kim, I. J., Sung, H.-J., Park, W.-J. et al. (2007) 'Mg coil production via strip casting and coil rolling technologies', *Proceedings of Magnesium Technology 2007*, TMS, pp. 85–88.

Kalazhokov, Kh., Kalazhokov, Z. and Khokonov, Kh. (2003) 'Surface tension of pure aluminum melt', *Technical Physics*, **48**, 2, 272–273. (MAIK Nauka/Interperiodica distributed exclusively by Springer Science+Business Media LLC.)

Kawalla, R., Vorobyov, K. Stolnikov, A. Pircher, H., Engl, B. et al. (2003) 'Magnesium ein wichtiger Werkstoff der Zukunft in Tagungsband', MEFORM 2003: 75 Jahre Metallformung in Freiberg, S, pp. 129–141.

Kawalla, R., Oswald, M., Schmidt, C., Ullmann, M., Vogt, H.-P. et al. (2008) 'Development of a strip-rolling technology for MG alloys based on the twin-roll-casting process', *Magnesium Technology 2008*, TMS, pp. 177–182.

Kim Kyung-Hun, Byeong-Chan Suh, Jun Ho Bae, Myeong-Shik Shim, Kim, S. et al. (2010) 'Microstructure and texture evolution of Mg alloys during twin roll casting and subsequent hot rolling', *Scripta Materialia*, **63**, 7, 716–720.

Kopp, R., Hagemann, F., Hentschel, L., Schmitz, J. W. and Senk, D. (1998) 'Thin-strip casting – Modeling of the combined casting: metal-forming process', *Journal of Materials Processing Technology* **80–81**, 458–462.

Li, B. Q. (1995) 'Producing thin strips by twin-roll casting – Part I: process aspects and quality issues', *JOM*, **47**, 5, 29–33.

Liang, D. and Cowley, C. B. (2004) 'The twin-roll strip casting of magnesium', *Journal of the Minerals, Metals and Materials Society*, **56**, 5, 26–28.

Lin, H-J. (2004) 'Modeling of flow and heat transfer in metal feeding system used in twin roll casting', *Modeling and Simulation in Materials Science and Engineering*, **12**, 255–272.

Liu Zhiyi, Song Bai and Kang Suk Bong (2009) 'Low-temperature dynamic recystallization occurring at a high deformation temperature during hot compression of twin roll cast Mg-5.51Zn-0.49Zr alloy', *Scripta Materialia*, **60**, 6, 403–406.

Lockyer, S. A., Ming Yun, Hunt, J. D. and Edmonds, D. V. (1996) 'Micro and macro defects in thin sheet twin roll cast aluminum alloys', *Mater. Charact.*, **37**, 301–310.

Loulou, T., Artyukhin, E. A. and Bardon, J. P. (1999) 'Estimation of thermal contract resistance during the first stages of metal solidification process II: Experimental setup and results', *International Journal of Heat and Mass Transfer*, **42**, 2129–2142.

Nakaura, Y. and Ohori, K. (2005) 'Properties of AZ31 magnesium alloy sheet produced by twin roll casting', *Mater. Sci. Forum*, **488–489**, 419.

Ohler, C., Odenthal, H-J. and Pfeifer, H. (2003) 'Physical and numerical simulation of fluid flow and solidification at the twin-roll strip casting process', *Steel Research International*, **74**, 11–12, 739–747.

Park, S. S, Park, Y. S. and Kim, N. J. (2002) 'Microstructure and properties of strip cast AZ91 Mg alloy', *Met. Mater.-Int.*, **8**, 551–554.

Park, S. S., Bae, G. T., Kang, D. H., You, B. H. and Kim, N. J. (2009a) 'Superplastic deformation behaviour of twin-roll cast Mg–6Zn–1Mn–1Al alloy', *Scripta Materialia*, **61**, 2, 223–226.

Park, S. S., Bae, G. T., Kang, D. H., In-Ho Jung, Shin, K. S. et al. (2007a) 'Microstructure and tensile properties of twin-roll cast Mg-Zn-Mn-Al alloys', *Scripta Materialia*, **57**, 9, 793–796.

Park, S. S., Jung G. Lee, Hak C. Lee and Nack J. Kim (2004) 'Development of wrought Mg alloys via strip casting', TMS, *Magnesium Technology*, **2004**, 107–112.

Park, S. S., Park, Y. S. and Kim, N. J. (2001) 'Strip casting of Mg alloys', *The Second International Conference on Light Materials for Transportation Systems (LiMAT-2001)*, Pusan, Korea, 6–10 May, Pohang, Pohang Kongkwa Taehak, Korea, pp. 225–232.

Park, S., Park, W-J., Kim, C.H., You, B. S. and Kim, N. J. (2009b) 'The twin-roll casting of Magnesium alloys', S, *JOM*, **61**, 8, 14–18.

Park, W. J., Jung, In-Ho, Bang, W., Kim I. J., Sung, H. J. *et al.* (2007b) 'Strip casting and coil rolling of magnesium alloy', *Materials Forum Korea*, **20**, 5, 9–18.

Park, W-J., Kim, J. J., Kim, I. J. and Choo, D. (2011) 'Wide strip casting technology of magnesium alloys', *Magnesium Technology 2011*, Edited by W. H. Sillekens, S. R. Agnew and R. N. Neelameggham,TMS 2011, Hoboken, NJ: John Wiley & Sons, Inc., pp. 143–146.

Sosinsky, D. J., Campbell, P., Mahapatra, R., Blejde, W. and Fisher, F. (2008) 'The CASTRIP® process – Recent developments at Nucor Steel's commercial strip casting plant', *Metallurg*, **52**, 11–12, 691–699.

Wang Yinong, Kang Suk Bong and Cho Jae-hyung (2011) 'Microstructure and mechanical properties of Mg–Al–Mn–Ca alloy sheet produced by twin roll casting and sequential warm rolling', *Journal of Alloys and Compounds*, **509**, 3, 704–711 (AM30-0.2Ca).

Wang Shou-ren, Wang Min, Kang Suk Bong and Cho Jae-hyung (2010) 'Microstructure comparison of ZK60 alloy under casting, twin roll casting and hot compression', *Transactions of Nonferrous Metals Society of China*, **20**, 5, 763–768.

Wechsler, R. and Campbell, P. (2002) 'The first commercial plant for carbon steel strip casting at Crawfordsville', *Dr. Manfred Wolf Symposium, 2002*, Zürich. Forch: MAIN GmbH, pp. 70–79.

Yun, M., Lockyer, S. and Hunt, J. D. (2000) 'Twin roll casting of aluminium alloys', *Materials Science and Engineering*, **A280**, 116–123.

Zeng, J., Koitzsch, R., Pfeifer, H. and Friedrich, B. (2009) 'Numerical simulation of the twin-roll casting process of magnesium alloy strip', *Journal of Materials Processing Technology*, **209**, 2321–2328.

8
Enhancing the extrudability of wrought magnesium alloys

A. G. BEER, Deakin University, Australia

Abstract: Whilst direct extrusion is an effective hot working process used for the manufacture of wrought magnesium sections, the cost of extrusion for existing magnesium alloys is high when compared to aluminium alloys. This can be directly ascribed to the comparatively slow extrusion speed of magnesium alloys. The chapter first discusses some of the metallurgical principles that govern the extrudability of wrought magnesium alloys. The chapter then examines the microstructural development during extrusion and details recent advances in alloy development specifically aimed at simultaneously improving the extrudability and mechanical properties of wrought magnesium.

Key words: extrusion, wrought magnesium, alloy development, microstructural development, recrystallisation.

8.1 Introduction

The consumption of magnesium alloys for structural applications has seen considerable growth over the past decade, particularly in the form of die castings. This has been driven primarily by the automotive industry whereby the low density of magnesium serves to reduce the weight of vehicles for increased fuel efficiency and reduced greenhouse gas emissions. Wrought magnesium products, however, only represent a small fraction of total magnesium consumption (1.5% of 2004 total consumption, as compared to 45% for die casting (Brown, 2004)) despite having the advantages of a higher strength and ductility than die-cast products.

Whilst direct extrusion is an effective hot working process used for the manufacture of wrought magnesium sections, the cost of extrusion for existing magnesium alloys is high when compared to aluminium alloys. A theoretical comparison of the production cost of equivalent magnesium and aluminium extruded automotive profiles (Schumann and Friedrich, 2003) suggests that the cost of magnesium extrusion is more than three times that of aluminium. The high extrusion cost of magnesium alloys can be directly ascribed to their slow extrusion speed. Compared to the extrusion of common aluminium alloys, the extrusion speeds of common wrought alloys such as ZM21 and AZ31 are typically two to five times slower. The problem is exacerbated for higher strength alloys, such as AZ61 and ZK60, which are required to be extruded up to ten times slower. It is thus apparent that increasing the rate of extrusion, through process optimisation

Enhancing the extrudability of wrought magnesium alloys 305

and/or alloy development, has the potential to significantly reduce the final cost of extruded profiles and in turn expand the market for wrought magnesium.

Whilst it has been shown that it is possible to use alternative extrusion processes, such as hydrostatic (Bohlen et al., 2005) and indirect (Savage and King, 2000) extrusion, to increase the allowable extrusion speed for magnesium alloys, direct extrusion is utilised most commonly by industry and is thus the focus for the current chapter. The chapter first introduces extrusion limit diagrams and discusses some of the metallurgical principles that govern the extrudability of wrought magnesium alloys. The operative deformation mechanisms, both during and after extrusion, are then examined and their influence on extrudability and microstructural control is described. Finally, some recent advances in alloy development aimed at simultaneously improving the extrudability and mechanical properties of wrought magnesium are detailed.

8.2 Extrudability of magnesium alloys

8.2.1 Extrusion limits

Extrusion limit diagrams, an example of which is given in Fig. 8.1, provide an effective means by which to assess the relative extrudability of metals and have been explored extensively for aluminium alloys by Sheppard (1999). In general, there are two factors that limit the maximum speed at which a metal can be extruded: the available extrusion pressure, and the avoidance of surface defects. These limits create a window within which the process variables need to be controlled such that successful extrusion is achieved.

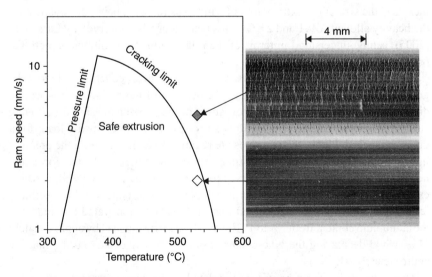

8.1 Extrusion limits for as-cast AZ31 including surfaces developed on billets extruded either side of the cracking limit.

At low billet temperatures, a limit arises as the stress required to make the material plastically flow reaches that of the pressure capacity of the press (pressure limit). At high billet temperatures, a limit is reached when the surface quality deteriorates. This limit is typically associated with the onset of cracking, whereby local temperatures in the die land area exceed the solidus temperature of the alloy, leading to incipient melting of the extrudate surface and subsequent cracking (cracking limit). The transition from an acceptable surface finish to one exhibiting minor cracking can be seen in Fig. 8.1 (for AZ31 billets extruded at 530 °C). It is worth noting, however, that other surface defects may occur prior to reaching the cracking limit. Atwell and Barnett (2007) found that in magnesium alloy M1, an orange peel surface defect reduced the limit at high temperatures by approximately 60 °C. Surface discolouration at higher extrusion speeds, due to oxidation of the material as it exits the die, has been reported for AZ31 (Sillekens, 2005) as well as M1 and ZM21 (Atwell and Barnett, 2007). For AZ alloys, the degree of surface oxidation following extrusion has been found to be more severe with higher aluminium and zinc levels (Murai et al., 2003).

8.2.2 The role of composition on extrudability

The extrudability of magnesium alloys is strongly dependent upon the alloying composition. This can be seen in Fig. 8.2, in which the extrusion limits for a range of commercially available magnesium alloys are compared to a relatively fast extruding aluminium alloy, AA6063 (adapted from Atwell and Barnett, 2007). It is evident that the size of the extrusion limit window for the leanest magnesium alloy, M1, is similar to that of AA6063. As the level of alloying additions is increased the size of the extrusion limit window is reduced, being the smallest for the heavily alloyed AZ61 and ZK60. This behaviour, also observed by Lass et al. (2005), can be understood in terms of the role of alloying additions on both the pressure and cracking limits.

For constant extrusion parameters, such as billet size, reduction ratio and die geometry, the extrusion load, and thus the position of the pressure limit, can be directly related to the hot working flow stress of the alloy. Increasing the alloying content increases the hot working flow stress and thus the extrusion load; for a given ram speed, higher temperatures are required to avoid reaching the pressure limit. Higher extrusion loads also translate to a higher degree of deformation heating during extrusion, which serves to reduce the maximum speed at which extrusion can be performed without developing surface cracking. In examining leaner versions of AZ31, Davies and Barnett (2004) demonstrated that reducing the aluminium content from 3% to 1% lowered the extrusion load by approximately 11%, while decreasing the zinc content from 1% to 0% decreased the load by approximately 4%.

The cracking limit can be generally related to the solidus temperature of the alloy; the higher the solidus temperature, the higher the ram speed and billet

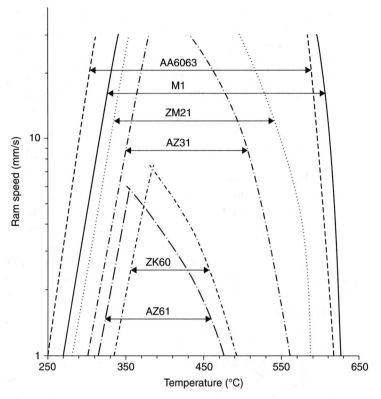

8.2 Extrusion limits for a range of commercially available magnesium alloys and aluminium alloy AA6063. (Source: Atwell and Barnett, 2007.)

temperature can be before incipient melting and subsequent cracking of the extrudate surface occurs. Both aluminium and zinc lower the solidus temperature of magnesium and so it is of no surprise that the cracking limit of alloys from the AZ series is observed to extend to higher speeds and temperatures when the aluminium and zinc level is reduced (Davies and Barnett, 2004). Removing the zinc from AZ31, to produce the alloy AM30, has been shown to increase the maximum extrusion speed by 20% (Luo and Sachdev, 2007). Manganese does not reduce the solidus temperature and thus has been shown to have little influence on the maximum extrusion speed prior to cracking (Murai *et al.*, 2003).

For alloys in the ZK series, reducing the Zn content from 6.8% to 3.8% increased the speed at which hot cracking occurred by a factor of three (Doan and Ansell, 1947). Doan and Ansell (1947) also revealed that the extrudability of Mg-Zn alloys increased with the addition of zirconium; for alloys containing 6% zinc, the maximum extrusion speed was approximately ten times higher with the addition of 0.8% zirconium. This was reported to be due to zirconium and, to a lesser extent, manganese increasing the solidus temperature of the Mg-Zn alloy.

It is worth noting that incipient melting, and thus cracking, can occur during extrusion at temperatures well below the equilibrium solidus temperature of an alloy due to the presence of local low melting point constituents in the as-cast billet (Sillekens, 2005). A homogenisation heat treatment is typically employed to overcome this issue; homogenising AZ31 billets was found to increase the maximum extrusion speed by over 20% as compared to as-cast billets (Atwell and Barnett, 2007).

8.2.3 The extrudability/mechanical property trade off

It is clear that a key factor in increasing the extrudability of magnesium alloys is to reduce the level of alloy additions, particularly of those that reduce the solidus temperature. However, a major downside of this approach is that certain mechanical properties, in particular yield strength, are adversely affected. This is evident from Fig. 8.3, where the average tension/compression yield strength for a range of magnesium alloys is plotted against their extrudability (taken as the width of the corresponding extrusion limit window in Fig. 8.2 at a ram speed of 1 mm/s).

Whilst the contribution to strength provided by solutes and precipitates is expected to be diminished in leaner alloys, the reduced strength of leaner alloys is

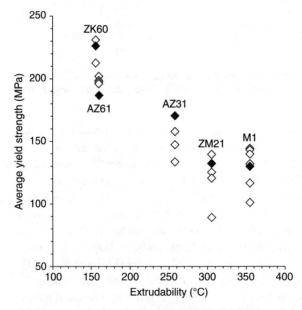

8.3 The average tension/compression yield strength for a range of commercially available magnesium alloys as a function of their extrudability (taken as the width of the extrusion limit window at a ram speed of 1 mm/s). Data is provided for alloys extruded over a wide range of temperatures and ram speeds. Filled symbols represent a ram speed of 5 mm/s and temperature of 375 °C.

predominantly related to the larger extruded grain sizes that they develop. It is well established that grain size has a strong influence on the strength of wrought magnesium; reducing the extruded grain size of AZ31 from 22 μm down to 3 μm has been shown to double the compressive yield strength (Barnett *et al.*, 2004). Grain refinement can also improve ductility and reduce the tension–compression yield anisotropy of wrought magnesium.

8.3 Microstructural development during extrusion

As the extruded microstructure, in particular grain size, has such a strong influence on the final mechanical properties of wrought magnesium, it is important to understand how the microstructure develops in the extrusion process, both during the deformation step and after deformation (during slow cooling, as the material exits the extrusion die, or during subsequent annealing).

8.3.1 Microstructural development during hot deformation

During the extrusion of aluminium alloys, a fibrous microstructure (consisting of original as-cast grains elongated in the extrusion direction) is commonly developed due to the operation of dynamic recovery (DRV). In magnesium alloys, however, dynamic recrystallisation (DRX) operates during hot deformation and a fine-grained microstructure is progressively developed. This process can be seen in Fig. 8.4, which shows the microstructures developed in a partially extruded AZ31 billet (a corresponding finite element method (FEM) simulation, indicating the material flow pattern and the associated strains, is included). In this case, extrusion was conducted using a relatively low temperature and slow ram speed, of 300 °C and 0.01 mm/s respectively, and a reduction ratio of 30. The deformed microstructure was maintained by immediately water-quenching both the billet and container.

In the low strain region of the billet (Fig. 8.4a), DRX grains have nucleated at the boundaries of the large as-cast grains, forming a 'necklace' type structure. Twins are also present and in some regions they have acted as nucleation sites for further DRX. In the middle of the billet (Fig. 8.4b), where the deformation strain is slightly higher, more of the material has undergone DRX and the necklaces of DRX grains decorating the pre-existing grains, and the DRX grains associated with twins, have continued to thicken. In the main deformation zone (Fig. 8.4c), where the strain is the highest, the degree of DRX has increased significantly and the overall grain size of the alloy has been significantly reduced; the dynamically recrystallised grain size was determined to be 2.5 μm in this region.

Figure 8.4c also reveals that, for the imposed extrusion conditions, the microstructure is not completely dynamically recrystallised; the remnants of original grains that did not undergo DRX can be seen elongated in the extrusion direction (this is despite a relatively high imposed deformation strain of over four). This behaviour is most likely a result of the deformation becoming localised

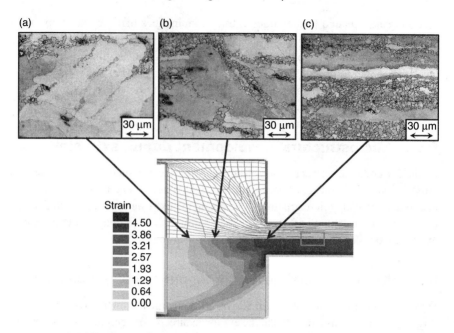

8.4 Microstructures developed in a partially extruded AZ31 billet and corresponding FEM simulation indicating the flow pattern and strains attained. Extrusion was conducted using a temperature of 300 °C, a ram speed of 0.01 mm/s, a reduction ratio of 30 and immediately water-quenching the billet and container.

in the dynamically recrystallised fraction of the material, brought about by the large difference between the initial and DRX grain sizes, or a lack of further nucleation sites once the original grain boundaries have been completely decorated with DRX grains (Beer and Barnett, 2007a). The retention of these elongated non-recrystallised grains is undesirable due to their detrimental influence on mechanical properties. A more homogeneous microstructure can be developed by enhancing the operation of DRX, which also serves to reduce the loads attained during deformation. This can be achieved by reducing the grain size of the billet prior to extrusion; the higher density of grain boundaries in a finer-grained structure enhances the kinetics of DRX by providing more nucleation sites.

A common approach to refining the grain size of billets is to pre-extrude them. Atwell and Barnett (2007) demonstrated that for AZ31 the maximum extrusion speed of wrought, or pre-extruded, billets was 35% higher than that of as-cast and homogenised billets. The microstructure of billets can also be refined during solidification by the use of alloying additions and/or optimised casting methods. Barnett *et al.* (2005) showed that small zirconium additions to ZM20 markedly reduced the initial billet grain size from 2 mm down to 60μm. Whilst the extrudability of the alloy was slightly reduced with a finer billet grain size, possibly due to softening from enhanced DRX being outweighed by hardening from the zirconium addition, the grain refinement did result in smaller extruded grain sizes being developed.

8.3.2 Microstructural development after deformation

The grain sizes developed by DRX are much smaller than those typically observed in commercially extruded alloys. This is due to the fact that a significant degree of grain coarsening occurs during annealing (considered as either slow cooling, as the material exits the extrusion die, or during a subsequent heat treatment of the extrudate). This restoration of the deformed microstructure can be seen in Fig. 8.5,

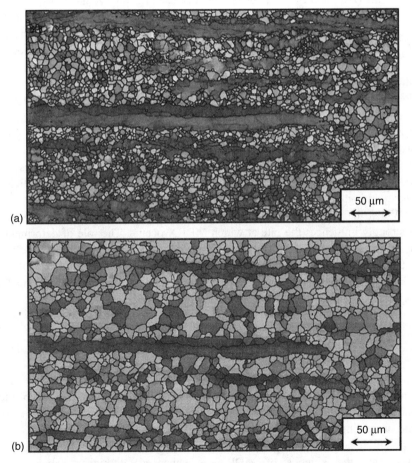

8.5 Orientation map of partially extruded AZ31 (generated on material after the die exit, as indicated by the grey box in Fig. 8.4) (thin black line >5° misorientation, thick black line >15° misorientation); (a) as extruded, and (b) after *in situ* annealing at 300 °C for 300 s.

whereby extruded AZ31 (from the region as indicated by the grey box in Fig. 8.4) has been annealed *in situ* in a scanning electron microscope at 300 °C for 300 s. In the as-extruded condition (Fig. 8.5a), the microstructure consists of elongated unrecrystallised grains and small DRX grains with an average grain size of 2.5 μm. After annealing, (Fig. 8.5b), larger grains can be seen to have developed in the region of the prior DRX grains; the average grain size in this area has doubled to 5 μm and several grains over 12 μm in diameter are evident. In some regions, the growing grains have started to consume the unrecrystallised regions.

The dominant mechanism by which the deformed microstructure is restored during annealing is that of metadynamic recrystallisation (MDRX), which involves the continued growth of nuclei, formed during DRX, during subsequent annealing (Beer and Barnett, 2008). The increase in grain size due to MDRX can be seen in Fig. 8.6 (for magnesium alloy AZ31) where the dynamically recrystallised grain size, d_{DRX}, and the metadynamically recrystallised grain size, d_{MDRX}, are plotted as a function of the Zener–Hollomon parameter, Z ($\dot{\varepsilon}\exp(135000/RT)$) (Beer and Barnett, 2008). Both d_{DRX} and d_{MDRX} obey a power law relationship with Z; for MDRX, this indicates that the final grain size developed is governed solely by the deformation conditions employed and not the temperature at which the material is annealed. For the conditions examined in Fig. 8.6, the MDRX grain size ranges from two to five times that of the DRX grain size, with finer grain sizes (for both d_{DRX} and d_{MDRX}) attained at higher values of Z. This indicates that, for extrusion, higher speeds and lower temperatures will yield smaller MDRX grain sizes (provided that that DRX has provided sufficient nuclei during the deformation stage).

Whilst the annealing temperature has no bearing on the MDRX grain size, it strongly influences the rate at which MDRX occurs. The role of deformation and annealing conditions on the kinetics of MDRX in AZ31 has been established by examining the increase in flow stress observed during double-hit compression testing (Beer and Barnett, 2009) and can be described using the expression:

$$t_{0.5} = 1\times 10^{-7}\left[\dot{\varepsilon}\exp\left(\frac{135000}{RT_{def}}\right)\right]^{-0.83} \exp\left(\frac{200000}{RT_{ann}}\right) \quad [8.1]$$

where $t_{0.5}$ is the time to reach 50% recrystallisation, $\dot{\varepsilon}$ is the strain rate and R is the ideal gas constant (= 8.314 J/mol K). T_{def} and T_{ann} are the deformation and annealing temperatures (K) respectively. Predictions of $t_{0.5}$ for three different annealing temperatures are also included in Fig. 8.6, where it is clear that the kinetics of MDRX is enhanced at higher values of Z (high strain rates and low temperatures) and at higher annealing temperatures.

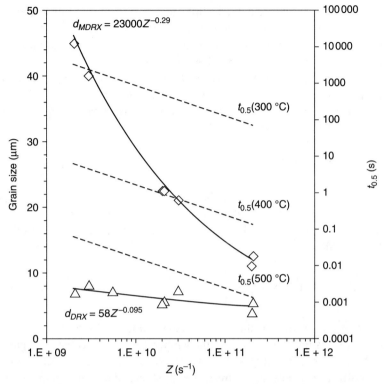

8.6 The influence of the Zener–Hollomon parameter, Z ($\dot{\varepsilon}\exp(135000/RT)$) on the dynamically recrystallised grain size, d_{DRX}, and the metadynamically recrystallised grain size, d_{MDRX}, after the hot deformation of magnesium alloy AZ31 to a strain of 1.5 (annealing temperature equal to the deformation temperature). Predictions of $t_{0.5}$, the time to reach 50% recrystallisation, are also included for annealing temperatures of 300 °C, 400 °C and 500 °C.

8.3.3 Predicting the grain size of extrudates

The relationship between grain size (both d_{DRX} and d_{MDRX}) and the imposed deformation conditions (see Fig. 8.6) can be used to predict the grain sizes and mechanical properties that may be expected to develop during the direct extrusion of AZ31. To do this, the process variables of ram speed and billet temperature are required to be converted to strain rate and deformation temperature.

The mean strain rate in extrusion, $\bar{\varepsilon}$, can be described by Chandra and Jonas (1970):

$$\bar{\varepsilon} = 4V_0 D_0^2 \tan\theta \left\{ \frac{2}{3m \ln r} \left[\left(\frac{1}{D_L}\right)^{3m} - \left(\frac{1}{D_0}\right)^{3m} \right] \right\}^{1/m} \quad [8.2]$$

where V_0 is the ram speed, D_0 the billet diameter, D_L the extrudate diameter, r the extrusion ratio and m the strain rate sensitivity (determined to be 0.12). The die semiangle, θ, is taken as the angle of the dead metal zone and was measured (from the partial extrusion used in Fig. 8.4) to be 78°.

The deformation temperature is determined by adjusting the initial billet temperature to account for deformation heating. The degree of deformation heating in the main deformation zone during extrusion can be approximated using the expression (after Sheppard, 1999):

$$\Delta T = \frac{\sigma \ln r}{\sqrt{3}(\rho C_p)} \qquad [8.3]$$

where r the extrusion ratio, ρ is the density and C_p is the specific heat capacity of the metal. The stress, σ, is taken as the steady state stress during hot working and, for AZ31, can be described using a power law of the form (adapted from Beer and Barnett, 2006):

$$\sigma_{ss} = A\left[\dot{\varepsilon}\exp\left(\frac{Q}{RT}\right)\right]^m \qquad [8.4]$$

where A is constant (equal to 2.5), m is the average rate sensitivity (determined to be 0.12), $\dot{\varepsilon}$ is the strain rate, T is the temperature (K), R is the ideal gas constant (= 8.314 J/mol K), and Q is taken as the activation energy for self diffusion (135 kJ/mol). The steady-state stress was first calculated using the initial billet temperature, and the degree of deformation heating was determined. The steady-state stress was then re-calculated using the initial billet temperature adjusted for deformation heating. The final amount of deformation heating was determined using an average value of the two steady-state stresses.

The predicted DRX and MDRX grain sizes are plotted as contour lines over the extrusion limit diagram in Figs 8.7a and 8.7b respectively. It is evident that at higher extrusion speeds and lower extrusion temperatures, smaller DRX and MDRX grains sizes are predicted. If the dynamically recrystallised microstructure were maintained after extrusion, then extrudates with grain sizes ranging from 4–7.5 μm could be expected (Fig. 8.7a). However, if MDRX were to occur, a much wider range of extruded grain sizes, between 7.5–45 μm, could be realised for processing conditions over that of the extrusion limit window (Fig. 8.7b). The measured grain sizes of extruded AZ31 profiles (using a reduction ratio of 30 and air cooling) are also included in Fig. 8.7b, where it can be seen that a fairly close correlation between the actual grain sizes and predicted MDRX grain sizes exists. This confirms that MDRX is determining the final microstructure for the extrusion conditions employed and that the current approach can adequately predict the extruded grain size over a wide range of extrusion conditions.

It is interesting to note that the MDRX grain size is much more sensitive to changes in billet temperature than changes of ram speed, i.e. for a given billet

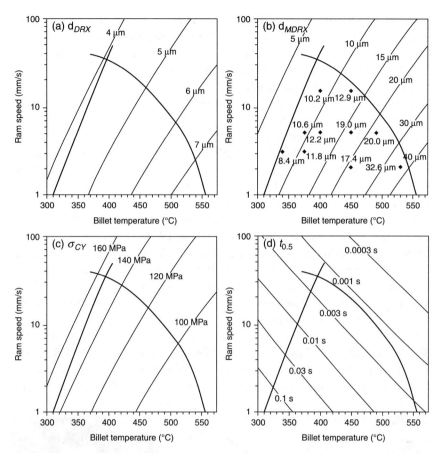

8.7 The incorporation of property/process predictions over the extrusion limit diagram. (a) The dynamically recrystallised grain size. (b) The metadynamically recrystallised grain size (including actual extruded grain sizes obtained over a range of process conditions). (c) The 0.2% compressive yield strength. (d) The time for metadynamic recrystallisation to be 50% complete.

temperature very little change in grain size is observed as the ram speed is increased. This behaviour can be attributed to the increased amount of deformation heating generated at higher ram speeds; the increase in the MDRX grain size due to the higher deformation temperature offsets the reduction in the MDRX grain size expected from the higher applied strain rate. A similar observation was made by Barnett et al. (2005) for the extrusion of magnesium alloy ZM20 whereby, for a given billet temperature, very little difference in the extruded grain size was seen when the ram speed was increased from 5 to 15 mm/s.

As the final extruded grain size has a marked influence on the strength of wrought magnesium, through the well known Hall–Petch relationship, the predicted MDRX grain size can also be used to predict the yield strength that would be expected to develop. This is shown in Fig. 8.7c, where the compressive yield strength is plotted as contour lines on the extrusion limit diagram (using Hall–Petch parameters from Barnett *et al.*, 2004). Compressive yield strengths in the range of 80–150 MPa are predicted, with higher yield strengths at lower billet temperatures and higher ram speeds.

An important consideration in predicting the microstructural development during the extrusion of magnesium alloys relates to the rate at which the material undergoes MDRX. The time that it takes for the DRX microstructure to become 50% recrystallised (using Eq. 8.1 and assuming that the annealing temperature is equivalent to the deformation temperature) is plotted over the extrusion limit diagram in Fig. 8.7d. It is clear that, under the strain rates and temperatures imposed during direct extrusion, the time for the deformed microstructure to undergo MDRX is extremely fast for AZ31. At the bottom left of the extrusion limit window, where the temperature and ram speed are at their lowest, the time for 50% recrystallisation is still 0.1 s. It is thus highly likely that during the commercial direct extrusion of AZ31, MDRX is complete before any quenching can be applied at the die exit (however quenching is still required to avoid any normal grain growth that may occur once MDRX is complete and to reduce the extrudate temperature for handling).

In lean alloys (which can be extruded at relatively low temperatures), or in alloys containing alloying additions that reduce the kinetics of MDRX, it is possible to maintain a dynamically recrystallised microstructure after extrusion. This can be seen for the commercial extrusion of M1 rods at 300 °C (see Fig. 8.8).

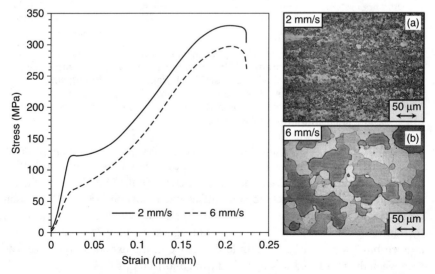

8.8 The developed microstructures, and corresponding compressive flow curves, of commercially extruded M1 billets at a temperature of 300 °C and at ram speeds of (a) 2 mm/s and (b) 6 mm/s (with a reduction ratio of 50).

When the alloy was extruded with a ram speed of 2 mm/s, a fine DRX microstructure, with an average grain size of 3.5 µm, prevailed in the centre of the extrudate (some large MDRX grains were observed on the periphery of the rod profile). When the extrusion speed was increased to 6 mm/s, MDRX occurred and the average grain size increased to 36 µm. This increase in grain size drastically reduced the compressive strength of the alloy by approximately 40%. Whilst 'locking-in' a dynamically recrystallised microstructure is possible, it is an impractical approach for enhancing the mechanical properties of commercially extruded magnesium alloys as very low extrusion speeds are required, it is difficult to control the microstructure in complex profiles and the microstructure is likely to undergo MDRX during subsequent annealing or warm forming operations.

8.3.4 The role of composition

As well as the deformation conditions employed during extrusion, alloy composition also has a marked influence on the grain size developed during MDRX. The grain sizes for a range of magnesium alloys that were hot deformed and subsequently quenched within 1 s or annealed for 1000 s, are compared in Fig. 8.9. For alloys quenched within 1 s, the microstructures were dynamically recrystallised with average grain sizes of between 6–8 µm (the exception to this being the lean Mg-Zr and Mg-Zn-Zr alloys that had already begun to undergo MDRX prior to quenching). All samples annealed for 1000 s were completely recrystallised apart from M1, which remained unchanged, and Mg-1Zn-1Zr and Mg-6Zn-0.5Zr, which were partially recrystallised.

In general, increased alloy additions serve to reduce the MDRX grain size (see Fig. 8.9). For the AZ alloys, with between 1–6 wt% Al, the annealed grain size is seen to decrease as the aluminium level is increased. For the deformation and annealing conditions employed, the MDRX grain size for Mg-6Al-1Zn is 60% smaller than that of Mg-1Al-1Zn. For the ZK alloys, with between 0.3–1 wt% Zr and up to 6 wt% Zn, there is a significant difference between the annealing behaviour of the various alloys. For the leanest alloy investigated, Mg-0.3Zr, the grain size increases rapidly and develops a MDRX grain size of over 80 µm. Conversely, the heavily alloyed Mg-6Zn-0.5Zr develops a grain size of only 10 µm after annealing for 1000 s. Increasing the zirconium level from 0.3 wt% to 0.5 wt% reduced the MDRX grain size from 89 µm to 30 µm, which was reduced down to 17 µm as the zirconium level was further increased to 1 wt%. The addition of 1 wt% Zn to the Mg-Zr alloys reduced the annealed grain size further, apart from the alloy containing 0.5 wt% Zr. The addition of 1 wt% Mn to Mg was sufficient to retard MDRX under the deformation and annealing conditions imposed; however, the addition of 2 wt% Zn to the Mg-1Mn alloy allowed MDRX to occur and a 31 µm grain size was developed (it should be noted that although MDRX is retarded with the addition of 1 wt% Mn, when MDRX does occur in this alloy then significantly larger grain sizes are developed).

318 Advances in wrought magnesium alloys

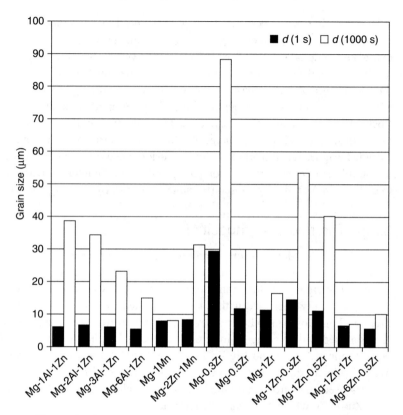

8.9 The grain sizes for a range of magnesium alloys that were hot deformed in compression to a strain of 1.5 at a temperature of 350 °C and a strain rate of 0.1 s^{-1}, and subsequently quenched within 1 s or annealed at 350 °C for 1000 s.

Differences between the degrees of grain coarsening in these magnesium alloys can be related to two main factors. Firstly, increasing the level of alloying additions is expected to increase the stored energy in the deformed matrix (when deformed under the same conditions, increasing the alloying level increases the steady-state flow stress); this would favour a higher nucleation rate that would subsequently lead to a smaller final grain size (Beer and Barnett, 2007b). Secondly, the presence of fine, closely spaced intermetallic particles, particularly in alloys containing higher levels of alloys additions, can pin grain boundaries during MDRX and thus limit the grain size developed. This mechanism has been observed in magnesium alloy AZ61, whereby boundaries were pinned by fine particles of the intermetallic phase $Mg_{17}Al_{12}$, which had been fragmented during prior extrusion (Beer and Barnett, 2007b).

Of the alloy additions examined here, zirconium appears to be the most effective in reducing the MDRX grain size. Despite being significantly leaner, the alloy containing only 1.0 wt% Zr develops a similar annealed grain size to that of Mg-6Al-1Zn. This observation highlights the fact that specific alloying elements can be added in small amounts to magnesium to obtain small MDRX grain sizes, and yet extrudability is maintained due to the alloy remaining comparatively lean (deformation loads and the solidus temperature are not greatly affected).

8.4 Recent extrusion alloy development

The strategy of using small alloy additions, or 'microalloying', to maintain fine extruded grain sizes in lean magnesium alloys has been highlighted by several researchers (e.g. Easton et. al., 2008, Hänzi et al., 2009, Barnett et al., 2010). Most attention has been directed towards the use of rare earth (RE) additions, such as cerium, yttrium, neodymium, or cerium rich misch-metal, which have been shown to successfully maintain fine grain sizes in extruded alloys of the Mg-Zn (Hänzi et al., 2009, Luo et al., 2010) and Mg-Mn (Barnett et al., 2010, Bohlen et al., 2010) alloy systems.

Overall, it appears that the effect of RE additions on grain refinement is related to the retardation of recrystallisation (Barnett et al., 2010). It is not entirely clear by which mechanisms smaller grains are maintained, with solute drag, particle pinning and enhanced nucleation during annealing due to particle stimulated nucleation all being suggested. In any case, the effect is quite potent, being observed in a range of magnesium alloys with RE additions less than 0.4 wt% (Luo et al., 2010, Barnett et al., 2010). An RE addition of up to 0.4 wt% has been shown to reduce the extruded grain size of Mg-1Mn from over 20 μm down to 6 μm (Barnett et al., 2010). Interestingly, the same study revealed that the addition of 0.5 wt% aluminium to the Mg-Mn-RE alloy 'poisoned' any grain refining effect from the RE addition, and large extruded grain sizes of approximately 50 μm were attained.

RE additions are also of interest due to the weakened extruded texture that they develop (e.g. Stanford et al., 2008). This texture modification can be seen in Fig. 8.10, in which the extruded textures of AZ31, M1 and ME10 (M1 with an 0.4 wt% Ce-rich Misch metal addition) are shown as inverse pole figures relative to the extrusion direction. Both AZ31 and M1 display typical fibre textures with the majority of grains having ⟨1010⟩ aligned with the extrusion direction. The texture of ME10, however, displays a slightly weaker overall texture strength (2.17 times random, as compared to 4.87 times random for AZ31) and the majority of grains are rotated such that ⟨1121⟩ is aligned with the extrusion direction.

This texture component, referred to as the 'rare earth' texture component (Stanford and Barnett, 2008), provides grains that are more favourably aligned for basal slip and extension twinning, leading to an increased tensile ductility and improved yield asymmetry in the alloys. Whilst further research is required to understand the mechanisms by which RE additions modify the microstructure, recent studies have indicated that the weakened texture develops as the DRX

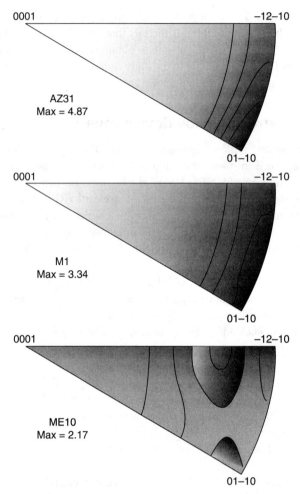

8.10 Texture of magnesium alloys AZ31, M1 and ME10 (with an 0.4% Misch Metal addition), measured using EBSD and shown as inverse pole figures referring to the extrusion direction (billets extruded at 400 °C, with a ram speed of 5 mm/s and a reduction ratio of 30).

microstructure recrystallises statically (Bohlen *et al.*, 2010) and may originate from shear bands in the deformed structure (Stanford and Barnett, 2008). The development of the RE texture component has also been observed to revert back to a traditional fibre texture as the extrusion temperature is increased (Stanford and Barnett, 2008) or as extrusion speeds are decreased (Luo *et al.*, 2010) and thus a more complete understanding of the role of deformation conditions on the microstructural development in RE-containing magnesium alloys is still required for these alloys to be used confidently in commercial extrusion.

8.5 Conclusions

The comparatively slow extrusion speed of existing wrought magnesium alloys, when compared to aluminium alloys, is a major contributor to the high cost of extruded magnesium products. It has been shown that reducing the level of alloy additions, particularly of those that reduce the solidus temperature, can greatly increase the extrudability; however, this comes at the expense of the final mechanical properties that are diminished due to the larger extruded grains sizes that develop. It has also been shown that the final microstructure can be controlled through understanding the operative mechanisms during deformation and annealing and by employing the appropriate choice of deformation conditions and alloying additions. These observations underlie recent advances in alloy development aimed at simultaneously improving the extrudability and mechanical properties of wrought magnesium, whereby micro-alloying is employed to maintain fine grain sizes during the extrusion of lean magnesium alloys. Whilst this aim appears to have been met by the use of small rare earth additions, further research is required to understand the mechanisms by which RE additions modify the microstructure and how they are influenced by the imposed extrusion conditions.

8.6 References

Atwell, D. and Barnett, M. R. (2007) 'Extrusion limits of magnesium alloys', *Metallurgical and Materials Transactions A*, **38A**, 3032–3041.

Barnett, M., Keshavarz, Z., Beer, A. and Atwell, D. (2004) 'Influence of grain size on the compressive deformation of wrought Mg-3A1-1Zn', *Acta Materiala*, **52**, 5093–5103.

Barnett, M. R., Atwell, D., Davies, C. and Schmidt, R. (2005) 'Grain refinement of magnesium alloy ZM20 prior to and during extrusion', in *Light Metals Technology*, edited by H. Kaufmann. St. Wolfgang, Austria, 8–10 June 2005, pp. 161–166.

Barnett, M. R., Beer, A., Atwell, D., Davies, C. H. J. and Abbott, T. (2010) 'Exploiting low levels of rare earth addition in Mg extrusion alloys', in *Magnesium Technology 2010*, edited by Agnew, S. R., Neelameggham, N. R., Nyberg, E. A. and Sillekens, W. H. The Minerals, Metals & Materials Society (TMS), Warrendale, PA, pp. 353–357.

Beer, A. and Barnett, M. (2006) 'Influence of initial microstructure on the hot working flow stress of Mg-3Al-1Zn', *Materials Science & Engineering A* **423**(1–2), 292–299.

Beer, A. and Barnett, M. (2007a) 'Microstructural development during hot working of Mg-3Al-1Zn', *Metallurgical and Materials Transactions A*, **38A**, 8, 1856–1867.

Beer, A. and Barnett, M. (2007b) 'The microstructural evolution in hot worked and annealed magnesium alloys', in *Magnesium Technology 2007*, edited by Beals, R., Luo, A., Neelameggham, N. and Pekguleryuz, M. The Minerals, Metals and Materials Society (TMS), Warrendale, PA, pp. 427–432.

Beer, A. and Barnett, M. (2008) 'Microstructure evolution in hot worked and annealed magnesium alloy AZ31', *Materials Science and Engineering A*, **485**, 1–2, 318–324.

Beer, A. G. and Barnett, M. R. (2009) 'The post-deformation recrystallization behaviour of magnesium alloy Mg–3Al–1Zn', *Scripta Materialia*, **61**, 12, 1097–1100.

Bohlen, J., Swiostek, J., Sillekens, W. H., Vet, P.-J., Letzig, D. *et al.* (2005) 'Process and alloy development for hydrostatic extrusion of magnesium: The European community research project MAGNEXTRUSCO', in *Magnesium Technology* 2005, edited by Neelameggham, N. R., Kaplan, H. I. and Powell, B. R. The Minerals, Metals and Materials Society (TMS), Warrendale, PA, pp. 241–246.

Bohlen, J., Yi, S., Letzig, D. and Kainer, K. U. (2010) 'Effect of rare earth elements on the microstructure and texture development in magnesium-manganese alloys during extrusion', *Materials Science and Engineering A*, **527**, 7092–7098.

Brown, R. E. (2004) 'Magnesium', *Mining Journal Annual Review for 2004*, 1–14.

Chandra, T., Jonas, J. (1970) 'The extrusion force and the mean strain rate during the extrusion of strain rate sensitive materials', *Metallurgical and Materials Transactions*, **1**, 2079–2082.

Davies, C. and Barnett, M. (2004) 'Expanding the extrusion limits of wrought magnesium alloys', *Journal of Metals*, **56**, 22–24.

Doan, J. P. and Ansell, G. (1947) 'Some effects of zirconium on extrusion properties of magnesium-base alloys containing zinc', *AIME*, **171**, 286–305.

Easton, M., Beer, A., Barnett, M., Davies, C., Dunlop, G. *et al.* (2008) 'Magnesium alloy applications in automotive structures', *Journal of Metals*, **60**, 11, 57–62.

Hänzi, A., Ebeling, T., Bormann, R. and Uggowitzer, P. (2009) 'New microalloyed magnesium with exceptional mechanical properties', in *Magnesium Technology* 2009, edited by Nyberg, E. A., Agnew, S. R., Neelameggham, N. R. and Pekguleryuz, M. The Minerals, Metals and Materials Society (TMS), Warrendale, PA, pp. 521–526.

Lass, J.-F., Bach, F.-W. and Schaper, M. (2005) 'Adapted extrusion technology for magnesium alloys', in *Magnesium Technology*, San Francisco, CA, 13–17 February 2005, edited by Neelameggham, N. R., Kaplan, H. I. and Powell, B. R. TMS, Warrendale, PA, pp. 159–64.

Luo, A. and Sachdev, A. (2007) 'Development of a new wrought magnesium-aluminum-manganese alloy AM30', *Metallurgical and Materials Transactions A*, **38A**, 1184–1192.

Luo, A., Mishra, R. and Sachdev, A. (2010) 'Development of high ductility magnesium-zinc-cerium extrusion alloys', in *Magnesium Technology 2010*, edited by Agnew, S. R., Neelameggham, N. R., Nyberg, E. A. and Sillekens, W. H. The Minerals, Metals and Materials Society (TMS), Warrendale, PA, pp. 313–318.

Murai, T., Matsuoka, S., Miyamoto, S. and Oki, Y. (2003) 'Extrudability of Mg-Al-Zn alloys', *Materials Science Forum*, **419–422**, 349–354.

Savage, K. and King, J. K. (2000) 'Hydrostatic extrusion of magnesium', in *Magnesium alloys and their Applications*, edited by Kainer, K. U. Munich, pp. 609–614.

Schumann, S. and Friedrich, H. (2003) 'The route from the potential of magnesium to increased application in cars', *Proc. 60th Annual IMA World Magnesium Conference*, Stuttgart, Germany, International Magnesium Association, pp. 35–41.

Sheppard, T. (1999) *Extrusion of Aluminium Alloys*, Kluwer Academic, Dordrecht, p. 420.

Sillekens, W. H. (2005) 'Extrusion technology for magnesium: avenues for improving performance', in *Light Metals Technology*, edited by Kaufmann, H. St. Wolfgang, Austria: LKR, pp. 137–142.

Stanford, N., Atwell, D., Beer, A., Davies, C. and Barnett, M. (2008) 'Effect of microalloying with rare-earth elements on the texture of extruded magnesium-based alloys', *Scripta Materialia*, **59**, 7, 772–775.

Stanford, N. and Barnett, M. (2008) 'The origin of 'rare earth' texture development in extruded Mg-based alloys and it effect on tensile ductility', *Materials Science and Engineering A*, **496**, 399–408.

9
Hydrostatic extrusion of magnesium alloys

W. H. SILLEKENS, TNO, The Netherlands and
J. BOHLEN, Helmholtz-Zentrum Geesthacht, Germany

Abstract: This chapter deals with the capabilities and limitations of the hydrostatic extrusion process for the manufacturing of magnesium alloy sections. Firstly, the process basics for the hydrostatic extrusion of materials in general and of magnesium in particular are introduced. Next, some recent research and development issues are discussed, demonstrating the technical assets of hydrostatic extrusion over conventional extrusion methods and with particular reference to the extension of the process limits (extrusion speed and temperature) and the associated microstructures and mechanical properties (notably strength and its anisotropy). Finally, prospects for further development and anticipated application areas are considered.

Key words: hydrostatic extrusion, capabilities, limitations, microstructures, mechanical properties.

9.1 Introduction

Extrusion is a means to process metals such as aluminium and magnesium alloys into long semi-finished products of a wide variety of cross-sections. These extrusion products are commonly called sections, extrusions or profiles and are used for door and window frames, railings, tubing, structural and architectural shapes, and so on. Different methods of extrusion are direct, indirect, and hydrostatic extrusion. As for industrial use, direct extrusion is the mainstream process, while indirect extrusion and hydrostatic extrusion are currently rather niche processes, mainly because of their more elaborate equipment and handling.

Direct extrusion of magnesium alloys is typically carried out at temperatures between 300 and 400 °C. An impediment to a more widespread use is that the extrusion speeds are quite low. Actually, extrusion exit speeds of 20–30 m/min are reported for low-strength magnesium alloys and basic shapes, but these values drop by a factor of ten or more for high-strength alloys or shapes with higher complexity such as multi-hollow sections (Becker *et al.*, 1998). Effectively, extrusion speeds are typically five to ten times lower than for comparable aluminium sections. Along with the substantial costs of the feedstock (billets), this makes magnesium alloy sections expensive.

In Table 9.1, indications for the extrudability of magnesium alloys are given, together with some other relevant data. The principal limitation is the occurrence of hot cracking – also known as hot shortness – which is initiated by incipient melting of the material in the plastic zone due to excessive temperature rise. This

Table 9.1 Material data for selected magnesium wrought alloys and their workability by direct extrusion

Alloy	Nominal composition (wt%)				Physical properties			Extrudability: maximum exit speed (m/min)
	Al	Mn	Zn	Zr	ρ (kg/m^3)	C (J/kgK)	T_i (°C)	
M1	1.2				1770	1050	648[b]	6–20
ZM21		0.5	2.0		1780	1040		6–20
AZ31	3.0	0.20	1.0		1780	1040	532	3–6
AZ61	6.5	0.15	1.0		1800	1050	418	1–3
AZ80	8.5	0.12	0.5		1806	975	427	1–3
ZK60			5.5	0.45[a]	1830		518	1–3

[a] Minimum value. [b] Solidus temperature.
ρ = density, C = specific heat, T_i = incipient melting temperature.
Sources: Avedesian and Baker, 1999; Anon., 2001.

phenomenon is obviously alloy-dependent, as it is affected by such properties as flow stress and incipient (or non-equilibrium) melting temperature. Another restriction for high-strength alloys such as AZ80 is that complicated cross-sections cannot be realised. Further, a more general issue with magnesium wrought products is that the directionality of mechanical properties can be substantial due to the inherent anisotropic behaviour of the hexagonal close-packed (hcp) crystal lattice in conjunction with the crystallographic textures resulting from plastic working. For instance, compressive yield stress can be as low as half the tensile yield stress for specific situations and alloys (Busk, 1987).

In this chapter it will be demonstrated that hydrostatic extrusion can relieve some of the current limitations, notably on productivity (extrusion speed) and product quality (mechanical performance), but also that it has distinct restrictions regarding complex and large cross-sections. In Section 9.2, the process basics for the hydrostatic extrusion of materials in general and of magnesium alloys in particular are introduced. Section 9.3 discusses some recent research and development issues related to the differences with conventional extrusion methods and with a particular notion for the mentioned assets. In Section 9.4, prospects for further development and anticipated application areas are considered. Section 9.5 gives some guidance to further information, while Section 9.6 concludes the chapter.

9.2 Process basics

Hydrostatic extrusion differs from direct and indirect extrusion in several ways. In this section its principle will be outlined and its distinguishing features versus the other extrusion methods clarified, on the basis of a simple analytical process model. This should facilitate the understanding of the phenomena and underlying mechanisms as presented in the remainder of this chapter.

9.2.1 Principle of hydrostatic extrusion

In hydrostatic extrusion, a billet (of the product material) is forced through a die opening by pressurised fluid (the pressure medium), as is shown in Fig. 9.1, for the manufacture of solid sections. Unlike direct and indirect extrusion, there is no metallic contact between ram/container and billet. Characteristic of the process is that plastic deformation takes place under high hydrostatic pressure, and that there is nearly no friction. Working procedure is that a billet is loaded into the container, the container sealed by the ram and filled with the fluid, and the ram then advanced into the container. This pressurises the chamber and forces the billet through the

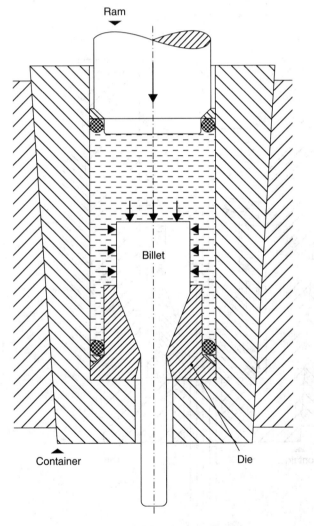

9.1 Hydrostatic extrusion of solid sections.

die. The die has a conical entry, imposing a converging material flow and simultaneously preventing leakage of the fluid between billet and die. To seal the pressure chamber, further specially designed fixed and sliding high-pressure seals are provided between die and container and between ram and container, respectively. Due to the conical die entry, plastic shearing of the material that is being formed is generally less than for conventional extrusion in which so-called flat or square dies are used. Single-hollow sections such as tubes can be made by forming the billet around a mandrel, requiring billets with a bore. A schematic of this is given in Fig. 9.2. The mandrel is kept in place with a back plate and a

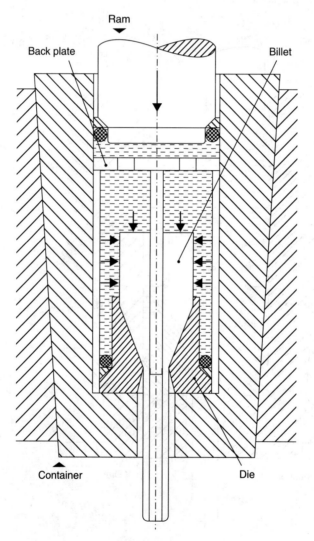

9.2 Hydrostatic extrusion of hollow sections.

supporting structure, and for symmetrical shapes is, to a certain extent, self-centring due to the converging flow of the product material around it. For the pressure medium, vegetable oils such as castor oil are commonly used in cold and warm hydrostatic extrusion because they have lower pressure coefficient of viscosity than mineral oils.

The process was originally developed in the 1960s. Initial focus was on enhancing cold workability of brittle materials, based on the observation that high hydrostatic pressure has a distinctly beneficial effect on ductility (Pugh, 1970), but it is now also used for warm and hot working. Elevated-temperature hydrostatic extrusion is usually implemented by heating the billets prior to loading rather than by heating the fluid. The terms extrusion temperature and billet temperature are thus used interchangeably. Current applications include copper tubing, copper-clad aluminium wire, high-strength aluminium alloys and (low-temperature) superconducting materials. Due to the virtual absence of friction, the flow of the material needs to be balanced in order to prevent deflection and warpage of the sections upon exiting the die, so cross-sectional symmetry is highly beneficial. Figure 9.3 shows some examples of magnesium alloy sections that have been manufactured with hydrostatic extrusion.

Advantages of hydrostatic extrusion over conventional extrusion methods are that difficult-to-work (such as compound) materials can be processed, that high extrusion ratios and geometrical accuracy of the produced sections can be achieved, that long billets and billets in a variety of diameters can be processed with the same set-up, and that high extrusion speeds are possible. In that sense, flexibility and productivity of the process are high. Disadvantages are that the handling and equipment is more complicated (implying longer dead-cycle times), that billet

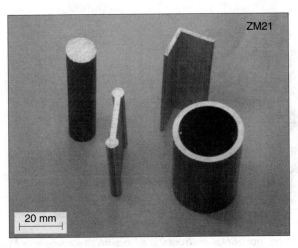

9.3 Hydrostatically extruded magnesium alloy sections (clockwise from top left: bar, angle, tube, dumb-bell).

preparation is more elaborate (requiring prior machining of a nose to fit the conical die entry and possibly a bore for hollow sections) and material efficiency generally lower, and that the range of possible shapes is much more limited.

Since its inception, ample research has been done on hydrostatic extrusion (Inoue and Nishihara, 1985). The processing principle is meanwhile widely known, but industrial use nevertheless remains limited. As compared to the thousands of conventional extrusion presses, only a small number of industrial hydrostatic extrusion presses have been manufactured worldwide and have been or are operational on a commercial basis. ASEA (Sweden, currently merged into ABB) has delivered two types of horizontal hydrostatic extrusion presses: a smaller one with a maximum ram force of 12 MN (1200 tonf) and a larger one with a maximum ram force of 40 MN (4000 tonf), both featuring an associated hydrostatic extrusion pressure of 1400 MPa (14 000 bar). A picture of the second type of press is given in Fig. 9.4. Typical for these presses is the wire-winding frame to meet the required stiffness. The container of this press has an inner diameter of 0.19 m (7.5 inch) and length of 1.6 m; regular pressure medium is castor oil. Considering that industrial direct extrusion presses range in container diameter between roughly 0.10 m (4 inch) and 0.36 m (14 inch), this press is to be categorised as a small/medium-size press. Effectively this also limits the size (circumscribed diameter) of the sections that can be produced.

Although the emphasis in this chapter is on microstructural development and resulting mechanical properties, surface condition of the sections is an important quality aspect as well as it affects post-extrusion operations (forming, coating) and functional performance. Table 9.2 summarises some distinct surface imperfections that can be encountered in hydrostatically extruded sections, including the measures that may be taken to avoid them. Where hot shortness and

9.4 Industrial horizontal hydrostatic extrusion press (Hydrex Materials, Waalwijk, Netherlands). Make, type: ASEA Quintus QEH40; ram force 40 MN, hydrostatic pressure 1400 MPa.

Table 9.2 Surface imperfections observed in the hydrostatic extrusion of magnesium alloys

Defect	Appearance	Countermeasures
Hot shortness (or hot cracking): cracks perpendicular to the extrusion direction due to incipient melting of the material	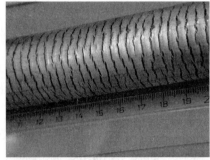	Reduce billet temperature and/or extrusion speed, so as to achieve a lower temperature of the material at the die exit
Scoring (or die lines): scratches longitudinally to the extrusion direction due to scoring in the cone and die		Alter tooling geometry and/or finish to improve metal flow through and reduce friction in the die
Contamination: remnants of the pressure medium, difficult to remove when burned onto the surface or appearing on the inside of hollow shapes		Ensure correct sealing of press channel by billet and/or avoid complete extrusion of the billet
Roughening: irregular, coarse topography, occasionally accompanied by breaking out of particles		Use billets with more homogeneous and finer microstructure

scoring are the more generic surface defects in extrusion, contamination with the pressure medium – which is not even an actual surface defect, but evidently a problem – is typical for hydrostatic extrusion. Roughening is a particular issue that may appear when using feedstock with a very pronounced and coarse casting morphology, as in the case of the M1 alloy. Regular feedstock, with a finer cast or pre-extruded microstructure, is not likely to yield this particular defect.

9.2.2 Process analysis

As a basic means of analysis, an upper-bound model (Avitzur, 1983) will be used here to reveal how process limits are influenced by the particularities of the extrusion method.

The rotationally symmetrical process model is summarised in Fig. 9.5. It essentially provides a set of equations for the steady-state extrusion pressure, following from an assumed spherical pattern of the (plastic) material flow. This pressure p consists of some distinct terms. These are attributed to the internal deformation in zone II (p_i), the shear deformation along the discontinuity surfaces Γ_1 and Γ_2 (p_s), the shear deformation or friction along the surface Γ_3 ($p_{s/f}$), and the friction along the interface surface Γ_4 (p_f). Direct, indirect, as well as hydrostatic extrusion can be described by choosing appropriate values for the friction factors m_3 and m_4 (Von Mises law). The overall deformation that is imposed by the extrusion process is characterised by the extrusion ratio (or area-reduction ratio) R. The flow stress σ_0 of the material is assumed to be constant. For direct and indirect extrusion, the semi-cone angle α is an optimisation parameter: it adopts a value that minimises the required power/pressure. For hydrostatic extrusion, the semi-cone angle matches the angle α_D of the die entry. Further, there is a

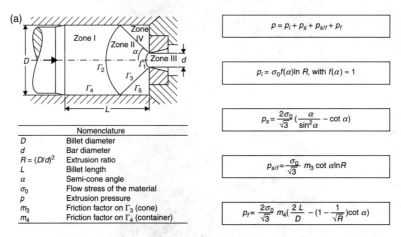

9.5 Analytical upper-bound model of steady-state extrusion ((a) schematic for direct extrusion with a flat die).

geometrical function $f(\alpha)$ involved, which represents redundancy of internal deformation.

Bearing this in mind, the influence of the extrusion method on the process window can be clarified as follows. The nearly ideal lubrication conditions in hydrostatic extrusion reduce the friction-related terms in the extrusion pressure to almost zero; only a small component from friction at the conical die entry remains. For direct extrusion, however, these friction-related terms are typically of the same magnitude as the deformation-related terms. A lower extrusion pressure now acts twofold: firstly, the imposed deformation – or extrusion ratio – can be higher for the same press capacity; secondly, the heat originating from the dissipation of the mechanical work (plastic deformation and friction) and therefore the temperature rise is lower so that hot shortness is impeded (Seido *et al.*, 1977). This is visualised in Fig. 9.6. Here the process limits are represented in plots of the billet temperature T_b versus the extrusion ratio R, the latter on a logarithmic scale. For hydrostatic extrusion, the press-capacity limit is shifted upward and the hot-shortness limit is tilted upward as compared to direct extrusion. A third process limit – low-temperature cracking – will not be considered here.

To extend on the previous, the temperature T_e at which the extruded material leaves the die can be estimated from the original billet temperature T_b and the temperature rise during extrusion ΔT as follows:

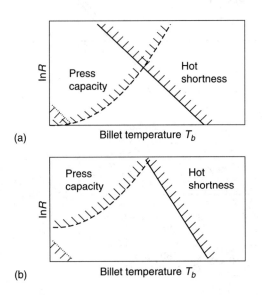

9.6 Schematic representation of extrusion process limits (temperature-deformation plots). (a) Direct extrusion; (b) hydrostatic extrusion.

$$T_e = T_b + \Delta T = T_b + K \frac{p}{\rho C} \qquad [9.1]$$

Here, it is supposed that all mechanical work is converted into heat. K denotes a factor that accounts for heat conduction to the tooling and indirectly represents the effect of extrusion speed. It ranges from quasi-static/isothermal ($K = 0$) to adiabatic processing ($K = 1$). Further, ρ and C are the density and the specific heat of the material, respectively. It should be noted that this is a first-order approximation: for instance, thermal effects are averaged over the billet volume, whereas hot shortness is clearly a local phenomenon.

Combining Avitzur's solution for the extrusion pressure p with Eq. 9.1 enables quantification of the temperature rise that is associated with each of the considered extrusion methods. This is shown in Fig. 9.7 for a particular set of extrusion parameters and magnesium alloy property data. Evidently the temperature rise is highest for direct extrusion and lowest for hydrostatic extrusion, with indirect extrusion in between. For hydrostatic extrusion with a fixed die semi-cone angle α_D, this thermal effect is logarithmically dependent on the extrusion ratio, which appears in the graph as a straight line. For direct and indirect extrusion, the optimal semi-cone angle increases as the extrusion ratio increases, giving deflecting lines.

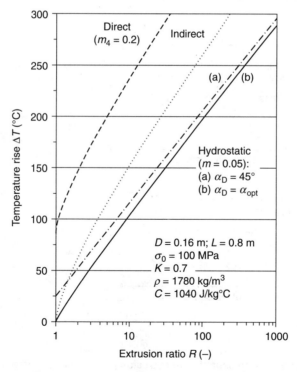

9.7 Estimated thermal effect for extrusion of magnesium alloy bar.

The same applies if the die angle in hydrostatic extrusion is adapted to minimise pressure ($a_D = a_{opt}$). Although it is recognised that the presented results are merely indicative, it is clear that temperature rise for direct extrusion is much more severe than for indirect and for hydrostatic extrusion (under similar processing conditions). Direct extrusion is therefore the most susceptible option regarding hot shortness. That indirect extrusion offers an advantage in this respect has also been experimentally demonstrated (Müller, 2002). With typical values ranging between 10–100, it is also obvious that the extrusion ratio is of major significance in connection with these thermal effects.

9.3 Research and development issues

As outlined in the previous section, hydrostatic extrusion holds the potential of an enlarged process window, which will be especially beneficial for the processing of magnesium alloys. This section addresses some aspects that have recently been subject of dedicated research efforts concerning the distinctions between the conventional and hydrostatic extrusion, the feasibility of high extrusion speeds, and the feasibility of low extrusion temperatures. These topics will be considered in relation to resulting product quality.

9.3.1 Indirect extrusion versus hydrostatic extrusion

To assess the influence of the extrusion method on microstructural development and the resulting mechanical properties, the following trials were conducted using an 8 MN indirect extrusion press at the Extrusion Research and Development Centre TU Berlin (Berlin, Germany) and a 12 MN ASEA hydrostatic extrusion press at Compound Extrusion Products GmbH (Freiberg, Germany). The alloys AZ31, AZ61 and AZ80 were investigated, representing a selection of magnesium-aluminium wrought alloys with different aluminium contents (ranging from nominally 3.0–8.5 wt%). Commercial direct-chill (DC) cast feedstock was homogenised (for 12 h at 350 °C for AZ31 and AZ61, and for 12 h at 385 °C for AZ80) and machined to billets (Ø95 mm for indirect extrusion and Ø80 mm for hydrostatic extrusion).

The selected test section was a round bar (Ø20 mm for indirect extrusion and Ø15 mm for hydrostatic extrusion, corresponding to the fairly similar extrusion ratios of $R = 23$ and 28, respectively). Billets were heated in a furnace to a pre-set billet (or extrusion) temperature of $T_b = 300$ °C. This temperature was chosen as it is a regular setting for conventional magnesium alloy extrusion. For indirect extrusion the highest extrusion exit speed avoiding hot shortness was used, being 8 m/min for AZ31, 6 m/min for AZ61 and 4 m/min for AZ80. For hydrostatic extrusion the lowest setting of the ram's hydraulic system was used, matching an extrusion exit speed of 8 m/min and yielding sound products for all alloys. The sections were left to cool in ambient air after leaving the die exit.

Figure 9.8 shows light-optical microscopy (LOM) images of the indirectly and hydrostatically extruded bars along with the matching average grain sizes. In all micrographs, the horizontal direction represents the extrusion direction. For AZ31, an inhomogeneous microstructure appears for both extrusion methods with smaller equi-axed grains and larger elongated grains, the latter representing non-recrystallised remains of the original cast structure. As such, these microstructures are typical for a partially recrystallised material. For hydrostatic extrusion, however, the grains in general are much finer. For AZ61 and AZ80, the microstructures are more homogeneous and the grains generally equi-axed, suggesting a more complete dynamic recrystallisation (DRX) during thermomechanical treatment. Again, hydrostatic extrusion yields finer grains than indirect extrusion although the effect is less pronounced than for AZ31.

As stated, the speed for indirect extrusion was set just below the hot-shortness limit, while for hydrostatic extrusion it was below this limit for any of the alloys. Effectively this means that the temperature rise for indirect extrusion has been higher, which in turn affects the DRX including any associated secondary grain growth. This follows the general notion that a higher (peak) temperature during extrusion leads to a coarser microstructure (Uematsu, 2006). As for the differences between individual alloys, it is to be noted that for indirect extrusion the resulting grain size tends to decrease with aluminium content while for hydrostatic extrusion it increases. The background to that may be complicated. The aluminium content in the alloys affects the homologous temperature (or ratio T/T_m of absolute processing temperature and melting temperature) as well as the presence of the

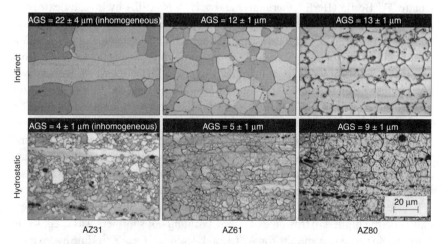

9.8 Micrographs of magnesium alloy bar obtained by different extrusion methods; headings denote average grain size AGS (linear intercept values; for AZ31: DRX areas only). Longitudinal cross-sections, etchant picric acid, size bar as indicated for all images.

intermetallic compound $Mg_{17}Al_{12}$, both having a bearing on DRX and grain growth. Furthermore, the different speeds that were implemented for indirect extrusion affect the thermomechanical cycle and as such the kinetics of the microstructural development.

Figure 9.9 shows the yield stress of the extruded bars as obtained from tension and compression tests. For each alloy, hydrostatic extrusion results in a higher strength than indirect extrusion, which is consistent with the observed smaller grains in connection with the Hall–Petch relationship (Bohlen et al., 2007). Further, compressive yield stress (*CYS*) is lower than tensile yield stress (*TYS*) for any alloy and extrusion method. This yield stress anisotropy is related to a different activation of deformation mechanisms in tension and compression for the hcp lattice structure, namely expansion twinning of a {10.2} type. As such it is associated to grain size as well as to texture. For the indirectly extruded bars the yield stress ratio ranges from *CYS*/*TYS* = 0.52 for AZ31 to 0.75 for AZ61 and AZ80, and for the hydrostatically extruded bars it ranges from 0.78 for AZ80 to 0.92 for AZ61. The latter values are higher due to the increased compressive yield stresses. Effectively the hydrostatically extruded bars are both stronger and less anisotropic. Regarding ductility (or toughness), the tension tests yielded elongations at fracture that ranged between 21–23% with no significant trends between the investigated alloys and extrusion methods.

9.3.2 Extrusion speed

To explore the viability of high-speed hydrostatic extrusion of magnesium alloys and assess its influence on the microstructural development and the resulting

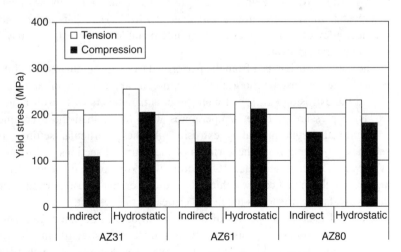

9.9 Mechanical properties of magnesium alloy bar obtained by different extrusion methods. Room-temperature testing, strain rate 10^{-3} 1/s.

mechanical properties, the following trials were conducted using a 40 MN ASEA hydrostatic extrusion press at Hydrex Materials, formerly HME (Waalwijk, Netherlands). The alloys M1, ZM21 and AZ31 were investigated, representing a selection of medium-strength magnesium wrought alloys with different alloying elements. Commercial DC-cast feedstock (Ø76 mm) was homogenised (4 h at 385 °C and then 8 h at 420 °C) and machined to billets (Ø73 mm × 500 mm, including nose and bore for the hollow sections). The billets for these experiments were relatively small, being about half the diameter and length of those that are usual for this industrial press.

The selected test sections were a round bar (Ø15 mm), a tube (Ø35 mm × 2.5 mm), an angle (20 mm × 20 mm × 2 mm), and a so-called dumb-bell (2Ø6.5 mm + 25 × 3 mm); extrusion ratios for these sections varied accordingly between $R = 12$ and 55. Billets were heated in a furnace to a pre-set temperature in the range of $T_b = 220$–290 °C and then loaded into the container. For all trials, the same setting of the ram's hydraulic system was used, scaling the extrusion exit speed with the extrusion ratio. The sections were left to cool in ambient air after leaving the die exit.

Images of longitudinally split billets are reproduced in the upper part of Fig. 9.10; the vertical direction is the casting direction. M1 shows a pronounced and coarse casting morphology. ZM21 and AZ31 have a much finer and equi-axed grain structure, where the latter is not fully homogeneous. LOM images of the extruded bars are shown in the lower part of the figure; horizontal direction is the extrusion direction. M1 and AZ31 turn out to be partially recrystallised. AZ31 again shows elongated grains in the extrusion direction embedded in smaller equi-axed grains. ZM21 appears to be fully recrystallised. In AZ31 the refinement of the microstructure (decrease in average grain size) due to the thermomechanical treatment is much more pronounced than in ZM21. Here it is to be noted that the latter alloy essentially does not form intermetallic compounds that may inhibit secondary grain growth.

The main extrusion data from the performed trials are summarised in Table 9.3. Basically, the extrusion ratio R, the die semi-cone angle α_D, the billet temperature T_b and the extrusion exit speed v are process parameters that result in a certain extrusion pressure p. Billet temperatures were lower than those common for conventional magnesium alloy extrusion. At these particular settings, no hot shortness – judging from the surfaces of the extruded sections – was observed. With values even greater than 100 m/min, the extrusion exit speeds were much higher than those that can be achieved for direct extrusion, supporting the notion that hydrostatic extrusion is much less critical in this respect.

Table 9.4 lists the properties of the extruded bars as obtained from hardness, tension and compression tests. The measured (tensile) strength values are quite similar to the handbook reference data, whereas ductility is much better for ZM21 and AZ31. The data also show the known difference between tensile and compressive yield stress, although this anisotropy is less prominent for

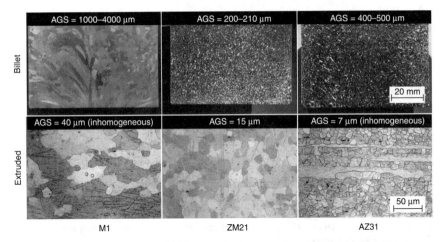

9.10 Macrographs of billets and micrographs of magnesium alloy bar obtained by hydrostatic extrusion; headings denote average grain size AGS (linear intercept values; for M1 and AZ31: DRX areas only). Longitudinal cross-sections, etchant picric acid, size bar as indicated for each row.

Table 9.3 Experimental data for hydrostatic extrusion of magnesium alloy test sections

Alloy	R	α_D (°)	T_b (°C)	v (m/min)	p (MPa)	R	α_D (°)	T_b (°C)	v (m/min)	p (MPa)
Bar:				●		Tube:			○	
M1	24	40	285	50	490	12	55	285	25	590
ZM21	24	40	280	50	590	12	55	220	25	850
AZ31	24	40	260	50	720	12	55	285	25	850
Angle:				L		Dumb-bell:			●—●	
ZM21	55	30	280	114	720	30	30	290	62	620
AZ31	55	30	285	114	780	30	30	290	62	690

R = extrusion ratio, α_D = die semi-cone angle, T_b = set-point billet temperature, v = nominal extrusion exit speed, p = steady-state extrusion pressure.

AZ31 with a yield stress ratio CYS/TYS = 0.78 than for M1 with 0.45 and ZM21 with 0.66.

9.3.3 Extrusion temperature

To explore the viability of low-temperature hydrostatic extrusion of magnesium alloys and assess the influence of billet/extrusion temperature on the microstructural development and the resulting mechanical properties, the following trials were conducted using a 12 MN ASEA hydrostatic extrusion press at Compound

Table 9.4 Mechanical properties of magnesium alloy bar obtained by hydrostatic extrusion

Alloy	Hardness	Tension				Compression
	HV 200/30	E (GPa)	TYS (MPa)	UTS (MPa)	Elong. (%)	CYS (MPa)
M1	36	40 (45)	192 (180)	268 (255)	12 (12)	86 (83)
ZM21	48	43	175 (155)	258 (235)	23 (8)	116
AZ31	53	42 (45)	198 (200)	278 (255)	23 (12)	155 (97)

Room-temperature testing: HV = Vickers hardness, E = modulus of elasticity, TYS = tensile yield stress, UTS = ultimate tensile strength, Elong. = tensile elongation, CYS = compressive yield stress.
Note: the table represents some distinct classes of properties
Hardness: HV 200/30
Tension: E, TYS, UTS, Elong.
Compression: CYS
Reference data in brackets. Source: Avedesian and Baker, 1999.

Extrusion Products GmbH (Freiberg, Germany). The alloys AZ31, AZ61 and AZ80 were investigated, representing a selection of magnesium-aluminium wrought alloys with different aluminium contents. Commercial DC-cast feedstock was homogenised (12 h at 350 °C for AZ31 and AZ61, and 12 h at 380 °C for AZ80) and machined to billets (⌀80 mm).

The selected test section was a round bar (⌀15 mm, corresponding to an extrusion ratio of $R = 28$). Billets were heated in a furnace to a pre-set temperature of either $T_b = 300$ °C or 100 °C. The high temperature was chosen as a regular setting for magnesium alloy extrusion. As for the low temperature, preliminary trials at room temperature showed cold cracking in all alloys, so that some heating was necessary to obtain visually sound products (actually, the cold-cracking issue persisted in AZ80 at 100 °C so that an increase to 110 °C was implemented in this case). For all trials, the lowest setting of the ram's hydraulic system was used, matching an extrusion exit speed of 8 m/min. Directly behind the die exit, the sections were cooled with water in order to suppress microstructural changes after extrusion (Bohlen *et al.*, 2005).

Figure 9.11 shows LOM images of the extruded bars along with the accompanying average grain sizes. In all micrographs, the horizontal direction represents the extrusion direction. Microstructures are at least partly if not completely recrystallised. For AZ31, microstructure is again rather inhomogeneous and, although grains are generally fine, the typical larger elongated grains appear as before. For AZ61, a more homogeneous microstructure appears with some stringers in the extrusion direction that consist of precipitates, presumably $Mg_{17}Al_{12}$ intermetallic compounds. For AZ80, homogeneous microstructures appear and while the precipitates are mainly concentrated on the grain boundaries,

Hydrostatic extrusion of magnesium alloys 339

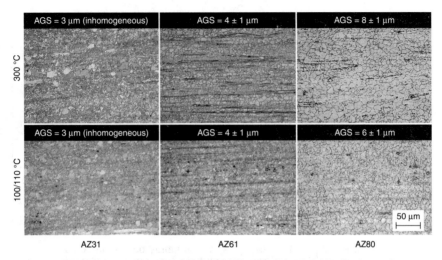

9.11 Micrographs of magnesium alloy bar obtained by hydrostatic extrusion at different billet temperatures; headings denote average grain size AGS (linear intercept values; for AZ31: DRX areas only). Longitudinal cross-sections, etchant picric acid, size bar as indicated for all images.

single particles within the grains exist as well. Based on these micrographs, differences between the two extrusion temperatures are not apparent except that for the low-temperature setting the microstructure for AZ31 seems to be somewhat more inhomogeneous and the microstructure for AZ80 somewhat finer.

A notable result from these experiments is that a recrystallised microstructure is obtained even at the low extrusion temperature. In this temperature range, plastic deformation is governed by twinning and dislocation glide, where dislocations are mainly of the <a> type in the basal plane (Ion *et al.*, 1982) and twinning is primarily active at low strains. These deformation mechanisms are thus suggested to initiate DRX, which will also affect the resulting textures for these low extrusion temperatures.

Figure 9.12 shows the matching inverse pole figures, obtained from texture measurements on cylindrical samples (Ø15 mm × 25 mm) taken from the extruded bars. Inverse pole figures were used because of the symmetry of the applied deformation. For each sample four complete pole figures of (10.0), (00.2), (10.1) and (11.0) were measured using neutron diffraction to derive 'global' texture information. This allows the orientation distribution to be determined and the inverse pole figures to be calculated. For AZ31 and AZ61 extruded at the high temperature, a typical <10.0> fibre texture is found as often observed for round bars from hcp metals after unilateral deformation such as extrusion. In AZ80, a second fibre component appears around the <11.0> pole; thus a double fibre texture exists with two prismatic-type poles. For AZ31, AZ61 and AZ80 extruded at the low temperature, also a double fibre texture is found with the relative significance of the concerned poles depending somewhat on the

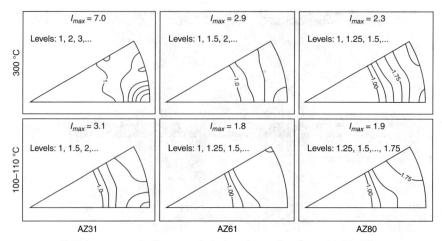

9.12 Inverse pole figures of magnesium alloy bar obtained by hydrostatic extrusion at different billet temperatures; headings denote maximum intensity$_{max}$.

alloy. In general a lower pole intensity – and thus weaker texture – is found as the aluminium content in the alloy is higher. If the results for both settings are compared, the low temperature yields lower pole intensities and thus weaker textures, but also a higher significance of the <11.0> pole versus the <10.0> pole in all three alloys.

In newer works it is shown that the significant <10.0> fibre can also be understood as a remains of those grains that did deform but not recrystallise during extrusion, while the development of grains with <11.0> orientation is seen as a result of recrystallisation (Yi et al., 2010). Notably only in AZ31 strong <10.0> fibre textures are found, corresponding to partly recrystallised microstructures.

Figure 9.13 shows the yield stress of the extruded bars as obtained from tension and compression tests. For each alloy, tensile strength is quite similar for both investigated temperatures but compressive strength is higher for the low-temperature setting. While the yield stress ratio for the high extrusion temperature ranges from $CYS/TYS = 0.79$ for AZ31 to 0.90 for AZ80, it increases for the low extrusion temperature from 0.89 for AZ31 to 0.99 for AZ80. Effectively the strength anisotropy has almost disappeared for the bars that were extruded at the low temperature. Regarding ductility, the tension tests yielded elongations at fracture that ranged between 17–21% with no significant trends between the investigated alloys and extrusion temperatures.

Upon relating the metallographical results to the corresponding mechanical properties, it is noted that for any alloy the microstructures (e.g. average grain sizes) are quite similar for both extrusion temperatures, so that the differences in textures must be mainly accountable for the differences in strength. From

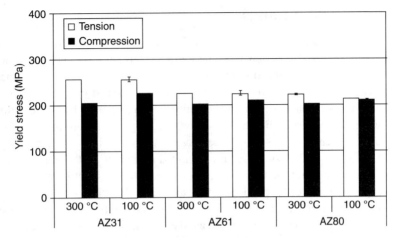

9.13 Mechanical properties of magnesium alloy bar obtained by hydrostatic extrusion at different billet temperatures. Room temperature testing, strain rate 10^{-3} 1/s.

this perspective it appears that a weaker or more 'randomised' texture (e.g. lower I_{max}) correlates with a higher compressive strength and associated lower strength anisotropy. The effect is more pronounced in AZ31 than in AZ61 and AZ80 as the differences in textures are the largest for this first-mentioned alloy.

As observed above, the extruded bars show a typical single <10.0> fibre or <10.0> <11.0> double fibre texture and both are of prismatic type. This means that the basal planes of the crystals are mainly oriented parallel to the extrusion direction. As the easiest twinning mode, {10.2} twinning contributes to the macroscopic strain only if stress is applied in tension along the c-axis of the hcp unit cell or in compression perpendicular to it. Thus, in the case of the extruded bars, {10.2} twinning is preferred in compression rather than in tension, with a correspondingly lower yield stress (Kleiner and Uggowitzer, 2004). From this perspective, a weaker texture will also imply an increase in compressive strength as fewer grains have a favourable orientation for twinning.

A further observation, which also applies to some extent to the results of Section 9.3.1, is that the mechanical strength tends to decrease with increasing aluminium contents in the alloy. This seems to contradict the common knowledge that higher aluminium content is associated with higher strength due to solid-solution and precipitation hardening (Avedesian and Baker, 1999). For this particular case, however, it should also be noted that the average grain sizes increase with increasing aluminium content in the concerned alloys for these extrusion conditions and that the associated textures become weaker. The latter effect appears to be dominating overall mechanical response (Bohlen *et al.*, 2007).

9.4 Future trends

The previous section has shown that hydrostatic extrusion offers distinct advantages over conventional extrusion of magnesium alloys. Its industrial application, however, is still pending. This section identifies some directions for further advancement.

The asset of high extrusion speed should be translated into higher productivity and eventually into lower cost price of magnesium sections. This suggests the use of hydrostatic extrusion for situations in which the pre-treatment (of the billets) and post-treatment (of the semi-finished products) are relatively easy, pointing in the direction of solid sections, requiring limited machining of the billets, that can be coiled rather than have to be stretched and cut-off in distinct lengths. This further suggests the use of long billets (which are especially difficult to extrude by direct extrusion), which effectively translates into a favourable proportion of extrusion time versus dead-cycle time. From another perspective, it puts forward the use of high-strength alloys for which conventional extrusion speeds are particularly low, and the leverage by using hydrostatic extrusion consequently high. Appropriate product choices thus would be welding wire/rod and feedstock (slugs) for forging. Common alloys for the former application are AZ31 and AZ61 for wrought products and AM50 for cast products, coming in standardised diameters ranging from 1.2–3.0 mm or larger. Common alloys for the latter application are the high-strength alloys AZ80 and ZK60, coming in custom-made diameters within a wide range. A further choice could be tubing and other symmetrical single-hollow sections for serving in a variety of structural applications (e.g. for bicycle frames). Here, medium-strength weldable alloys such as ZM21 could be an appropriate choice. Thin-walled sections would be especially interesting as their shaping capability by conventional extrusion is quite limited. For this aim, the manufacturability limits for the different magnesium alloys (e.g. minimum wall thickness) need to be further explored.

Enhanced compressive yield stress of the produced sections is especially beneficial for parts under bending stresses and thus would add to the weight-saving potential of magnesium components in automotive applications, amongst others. Here, a further development of process capabilities in the direction of more complicated hollow sections would be in order, noting that the possibilities of extending this are intrinsically limited. Further, the reduced directionality due to the finer microstructures and weaker textures will reduce anisotropic flow during subsequent forming operations. This again makes it an interesting prospect for forging stock, but also for such processes as tube hydroforming. For the latter example, the absence of (possibly weaker) longitudinal weld seams in the extruded tube could be an additional advantage.

Finally, a major drawback that currently seems to prevent the upheaval of hydrostatic extrusion of magnesium alloys is the limited availability of appropriate infrastructure. This concerns not only the hydrostatic extrusion press, but also the auxiliary equipment that needs to be geared to the processing of magnesium (e.g. machining operations and storage).

9.5 Sources of further information

While this chapter merely gives a concise introduction to hydrostatic extrusion and its use for magnesium alloys, additional information on the subject may be gained from the sources below.

For a general treatment of the hydrostatic extrusion process, the reader is referred to Inoue and Nishihara (1985). This reference text provides an introduction to the subject for engineers who work in industries that plan to employ this technique, and gives comprehensive information on the capabilities of the process, the properties of the extruded products, as well as on the performance and durability of production facilities.

Current players with hydrostatic extrusion facilities are, at the time of writing:

- Hitachi Cable Ltd (Tsuchiura, Japan).
- Hydrostatic Extrusions Ltd (Perth, Scotland).
- Compound Extrusion Products GmbH (Freiberg, Germany).
- Extrusion Research and Development Centre TU Berlin (Berlin, Germany).

The first two companies apply hydrostatic extrusion for the industrial manufacturing of superconducting and other compound materials, the others are research and development centres.

The results regarding hydrostatic extrusion of magnesium alloys as presented in this text were originally gained in the context of the European FP5 Growth project 'Hydrostatic extrusion process for efficient production of magnesium structural components' (MAGNEXTRUSCO), contract number G5RD-2000-00424, and within the framework of the strategic research programme of GKSS Forschungszentrum Geesthacht (currently Helmholtz-Zentrum Geesthacht). While the former was application-driven – including for instance the development of a number of demonstration components – the latter efforts were basically directed towards a more fundamental understanding. This chapter summarises a more extensive collection of experimental data as published in a number of conference and journal papers (e.g. Swiostek *et al.*, 2003; Sillekens *et al.*, 2003; Bohlen *et al.*, 2006). Further, the MAGNEXTRUSCO project has yielded a patent on new magnesium alloy compositions in conjunction with the hydrostatic extrusion process (Kainer *et al.*, 2005).

9.6 Conclusions

Based on the information in the previous sections, the main advantages and disadvantages of hydrostatic extrusion versus conventional extrusion of magnesium alloys are summarized in Table 9.5. From this it follows that hydrostatic extrusion and conventional (direct) extrusion are to be regarded as complementary rather than as competing processes.

Table 9.5 Characteristic features of hydrostatic extrusion of magnesium alloys

Assets	Drawbacks
• Extrusion speed can be 5–10 times higher than for direct extrusion • Operation temperature can be lower than for conventional extrusion (100–300 °C) • Enhanced compressive yield stress and low strength anisotropy of the sections • Hollow sections are produced without longitudinal weld seams (possible weak spots) • High extrusion ratios, long billets and a variety of billet diameters can be processed with the same set-up	• Limited to symmetrical solid and single-hollow sections; no complex cross-sections • Elaborate billet preparation (machining of nose and possibly bore) • Complicated handling and equipment, especially for hollow sections • Few operational industrial presses world-wide and with a limited product range

The current state of the art can also be assessed in terms of its so-called technology readiness level (TRL). Within this methodology, the maturity of evolving technologies prior to incorporating them into (sub)-systems or regular practice is rated from TRL1 (basic principles observed and reported) to TRL9 (system adequacy proven in practice). In this view, hydrostatic extrusion technology for magnesium sections would have to be rated TRL7–8, meaning that the technology itself is demonstrated but that (sub)systems – or in this case the process operation – requires some further development and testing before eventual launch and regular industrial use.

The largest application potential for hydrostatic extrusion of magnesium alloys is in basic solid and single-hollow sections (such as for welding wire/rod, forging feedstock and thin-walled tubes) and, notably, for high-strength alloys. Despite the considerable basis for efficient and cost-effective manufacturing of these (semi-finished) products as well as their superior mechanical strength, hydrostatic extrusion will at least for the foreseeable future remain a niche process as compared to the mainstream direct extrusion process that caters for more complicated and multi-hollow sections.

9.7 References

Avedesian MM and Baker H (1999) *Magnesium and Magnesium Alloys – ASM Specialty Handbook*, Materials Park, OH: ASM International.

Avitzur B (1983) *Handbook of Metal-Forming Processes*, New York: John Wiley.

Becker J, Fischer G and Schemme K (1998) 'Light weight construction using extruded and forged semi-finished products made of magnesium alloys', *Proc 4th Int Conf Magnesium Alloys and their Applications*, pp. 15–28, Matt Infor, Frankfurt.

Bohlen J, Yi SB, Swiostek J, Letzig D, Brokmeier H-G and Kainer KU (2005) 'Microstructure and texture development during hydrostatic extrusion of magnesium alloy AZ31', *Scripta Mat*, **53** (2), 259–264.

Bohlen J, Swiostek J, Brokmeier H-G, Letzig D and Kainer KU (2006) 'Low temperature hydrostatic extrusion of magnesium alloys', *Magnesium Technology*, **2006**, 213–217.

Bohlen J, Dobrex P, Swiostek J, Letzig D, Chmelík F *et al.* (2007) 'On the influence of the grain size and solute content on the AE response of magnesium alloys tested in tension and compression', *Mater Sci Eng A*, **462** (1–2), 302–306.

Busk RS (1987) *Magnesium Products Design*, New York: Marcel Dekker.

Inoue N and Nishihara M (1985) *Hydrostatic Extrusion – Theory and Applications*, London: Elsevier.

Ion SE, Humphreys FJ and White SH (1982) 'Dynamic recrystallisation and the development of microstructure during the high temperature deformation of magnesium', *Acta Metall*, **30**, 1909–1919.

Kainer KU, Bohlen J, Vet P-J, Hoogendam P, Meijer L *et al.* (2005) Method for the production of profiles of a light metal material by means of extrusion, international patent publication WO 2005/087962 A1.

Kleiner S and Uggowitzer PJ (2004) 'Mechanical anisotropy of extruded Mg-6%Al-1%Zn alloy', *Mater Sci Eng A*, **379** (1–2), 258–263.

Müller KB (2002) 'Direct and indirect extrusion of AZ31', *Magnesium Technology*, **2002**, 187–192.

Pugh HLID (1970) *The Mechanical Behaviour of Materials Under Pressure*, Amsterdam: Elsevier.

Seido M, Mitsugi S and Oelschlägel D (1977) 'Warm hydrostatic extrusion of high strength aluminum alloys', *Proc 2nd Int Aluminum Extrusion Tech Sem*, pp. 133–141.

Sillekens WH, Schade van Westrum JAFM, Bakker AJ van and Vet P-J (2003) 'Hydrostatic extrusion of magnesium: Process mechanics and performance', *Mater Sci Forum*, **426–432**, 629–636.

Swiostek J, Bohlen J, Letzig D and Kainer KU (2003) 'Comparison of microstructures and mechanical properties of indirect and hydrostatic extruded magnesium alloys', *Proc 6th Int Conf Magnesium Alloys and their Applications*, 278–284.

Uematsu Y, Tokaji K, Kamakura K, Uchida K, Shibata H *et al.* (2006) 'Effect of extrusion conditions on grain refinement and fatigue behaviour in magnesium alloys', *Mater Sci Eng A*, **434**, 131–140.

Yi S, Brokmeier H-G and Letzig D (2010), 'Microstructural evolution during the annealing of an extruded AZ31 magnesium alloy', *J Alloys Compd*, **506**, 364–371.

10
Rolling of magnesium alloys

J. BOHLEN, G. KURZ, S. YI and D. LETZIG,
Helmholtz-Zentrum Geesthacht, Germany

Abstract: Attempts to expand the existing industrial applications of magnesium alloys are currently focused on the utilisation of semi-finished products such as sheets. This chapter on the rolling of magnesium alloys will consider all aspects along the processing chain for producing magnesium-based sheet materials. A short historical review and an outline of the challenges in the modern industrial environment will also be given. The effects of processing parameters and special aspects of the rolling process on the mechanical properties and sheet formability will be examined. Recent developments in magnesium sheet alloys will be discussed and future trends will be outlined.

Key words: magnesium sheet, rolling, process parameters, rare earth elements, formability, mechanical properties, crystallographic texture.

10.1 Introduction

Magnesium alloys find widespread interest as the lightest available metallic construction materials for weight-saving purposes. An expansion of existing applications, e.g. of cast products, will require the utilisation of semi-finished products such as sheets for the fabrication of a large variety of components and structures.

Sheet metal is a fundamental form of material that is used in numerous industrial applications. Almost every metal is also available in the form of sheets, which allows the production of formed parts based on this well standardised semi-finished product. Sheets are produced using the rolling process, which takes conventional cast products and transforms them using a massive deformation schedule. The material is deformed between rolls for the purpose of producing sheet to a homogeneous thickness and large area. The resulting sheet is then available for transfer and further processing in sheet metal forming procedures, which are viable for a large variety of applications. Thus, sheets are important products if the full potential of a class of materials is to be realised.

Unfortunately, magnesium and its alloys belong to those materials with a hexagonal close-packed lattice structure, which limits the number of active deformation mechanisms, at least in comparison to cubic metals. Therefore, ductility and formability, which are basic requirements for the production process, the forming procedures and the properties of parts in application, are also limited. The feasibility of the fabrication of magnesium sheets was shown decades ago, but there are still no extensive applications at present.

Newer studies deal with the influence of processing parameters, especially the temperature, on the feasibility of conventional forming procedures. Furthermore, it was only recently discovered that alloys containing certain alloying elements, namely rare earth elements, tend to exhibit different microstructural, and especially textural, development during rolling, which has a significant influence on resulting sheet properties. Still, a major obstacle to the widespread use of magnesium sheets appears to be poor economic efficiency, if compared to other lightweight construction materials. Since there is no significant use of magnesium sheets at present it is difficult to improve the economics of manufacture. The rolling procedures remain complex and are still not well understood for a large variety of magnesium alloys. Technologies such as twin roll casting may contribute to more efficient production of sheets with a positive impact on the costs of these semi-finished products.

This chapter on the rolling of magnesium will cover all aspects along the processing chain of the production of magnesium sheets. In the first section, the historical and present situation and importance of magnesium sheets will be reviewed with special focus on current and competitive challenges in industrial environments today. The following two sections will focus on special aspects of microstructure and texture development and their impact on mechanical properties and formability based on the processing itself, i.e. the influence of processing routes and parameters, as well as the influence of the alloy composition, which may change fundamental mechanisms important for sheet properties. The impact on sheet formability will be reviewed in a subsequent section. Enabling technologies will be discussed briefly to present ways to progress from the semi-finished products to the requirements for industrial applications. The possible industrial impact of the implementation of new technologies and new concepts of alloy development will complete this chapter.

10.2 Potential of magnesium sheets

An important goal of different branches of industry is to reduce the weight of their products. Each industrial sector has its own motivation for reducing the weight of products. The motivation of the tool and electronics industry is easier handling, the motivation of the transport and aviation industry is to increase load capacity, and the motivation of the automotive industry is the reduction of CO_2 emissions to meet legal requirements. The greatest potential for savings offered by magnesium alloys can be achieved with components such as sheet metal, body shell panels or the housings of electronic products that are exposed to bending stresses (Schumann and Friedrich, 1998). In the past there have been some ambitious structural applications of magnesium sheets in the aeronautics and aerospace industry, in satellite and rocket technology as well as in ground vehicles (Emley, 1966; Beck, 1939, Juchmann et al., 2002). One example of a complete car body is the Allard sports car from the year 1952, which showed the possible potential of magnesium sheets in design and formability.

Sheet metal-formed parts are characterised by advantageous mechanical properties and good surface quality without pores, in comparison to die cast components. Therefore, the substitution of conventional sheet materials by magnesium sheets should lead to essential weight savings. It can be assumed that aluminium and steel sheets will be replaced by magnesium sheets in some applications. The use of magnesium alloys in lightweight structures offers significant weight saving potential. Compared with aluminium, which can be seen as the competing material, a weight reduction of 37% could be realised by the direct substitution of an Al alloy component by a Mg alloy component. In the future, magnesium sheet components should succeed if the sheets fulfil all challenging requirements concerning quality, geometry and quantity. Most importantly, however, magnesium sheets will have to compete with conventional aluminium sheets, both technically and economically.

Currently, the development of magnesium sheets is focusing on advanced technology for the mass production of wide magnesium sheets. For typical automotive sheet applications, semi-finished products with a width of up to 2 m are required to realise inner and outer panels. The key issue with magnesium sheets is to ensure the high reliability of supply. In parallel, companies must gain experience in the material and processing by prototyping and developing magnesium sheet components.

Further research is necessary to design new wrought alloys with increased formability, because at the time of writing the choice of sheet alloys is very limited; mainly sheets of the alloy AZ31 being available (Dröder and Janssen, 1999; Sebastian *et al.*, 2000; Enß *et al.*, 2000).

10.3 The sheet rolling process and its influence on sheet properties

Conventional rolling is the main established production method to produce thin, flat sheets as well as profiled products. Rolling of magnesium alloys is carried out following the same principles as for other metals and alloys.

10.3.1 Sheet rolling

The production of sheets starts with the feedstock used for rolling. Conventional feedstock for the rolling of magnesium alloys are slabs, e.g. produced by direct chill (DC) casting (Magnesium Elektron, 2011b). The continuity of this process promises optimum microstructural homogeneity throughout the casting. For smaller quantities, and especially in laboratory rolling studies, gravity casting is also used. Slabs are subject to homogenisation prior to processing and are then machined to suitable dimensions for the respective rolling procedure and the size of the mill. Appropriate annealing does not necessarily have a significant influence on the grain structure if temperatures are carefully selected for the respective alloys, but is used to homogenise the element distribution and dissolve second-phase particles that result from the casting process. The main issue appears to be the homogeneity of the microstructure throughout the slab. Typical annealing temperatures are collected in Table 10.1 as presented in Chabbi and Lehnert (2000) where the annealing temperature is related to the solidus temperature of the respective alloy.

The metal slab is then passed through a pair of rolls, where the roll gap is smaller than the thickness of the original material and therefore plastic deformation is applied to the material. Table 10.1 also includes a number of intermediate heating temperatures typically applied during sheet rolling (Beck, 1939). For other metals and alloys there are two types of processing scheme: hot rolling, if the temperature of the material is higher than the recrystallisation temperature, and cold rolling if the temperature is below the recrystallisation temperature. Examples of the processing of magnesium alloys during rolling have been published by Emley (1966). Different types of mills are designed for different functions along the processing route. Typically, a two-high mill is the most basic setup with two rolls through which the slab is passed during processing. Four-high mills are also described in the literature (e.g. Brown, 2002). The introduction of further rolls, characterised by smaller diameters, is aimed at fulfilling the requirements of thickness reduction, avoiding roll bending and improving rolling speed.

One requirement of the rolling schedule itself is the breakdown of the original cast microstructure by recrystallisation of the material. This cast microstructure is changed to finer grained microstructure during the first passes of the rolling schedule. It is well known that for aluminium alloys such breakdown rolling is carried out using hot rolling, with high degrees of deformation per pass. As a

Table 10.1 Heat treatment of cast magnesium alloys prior to and during rolling

Alloy	Temperature (°C)	Time (h)	Remarks
AZ31	< 520	Up to 8	Annealing and homogenisation of cast slab
AZ61 (AZM)	< 490	Up to 12	Annealing and homogenisation of cast slab
ZK10	< 480	Up to 12	Annealing and homogenisation of cast slab
M2 (AM503)	< 470	Up to 8	Annealing and homogenisation of cast slab
AZ21	320–500		Intermediate annealing during rolling
AZ61 (AZM)	280–320		Intermediate annealing during rolling
M2 (AM503)	320–500		Intermediate annealing during rolling
ME20	320–460		Intermediate annealing during rolling

Sources: Chabbi and Lehnert, 2000; Beck, 1939.

result of the limited formability of magnesium, such a processing step does not appear appropriate for this metal and its alloys. Rather, a more complex schedule of rolling is used with moderate degrees of deformation per pass (Emley, 1966) and a matching of the rolling temperature to the specific rolling reduction. In between passes, re-heating of the rolled product is needed in order to maintain the appropriate rolling temperature. Temperatures are highest in the breakdown stage of the rolling process and may decrease afterwards. Typical ranges of the rolling temperature are shown in the literature and are presented in Table 10.2. It is reported that some alloys are finish rolled at final gauge using cold rolling schedules. Subsequent levelling and annealing might be required. Sheets designed for adjusted strength, e.g. in a strain-hardened and stress-relieved H24 temper condition (Avedesian and Baker, 1999), are commercially available products, e.g. DS482 (Magnesium Elektron, 2011a).

10.3.2 Relation between process parameters and sheet properties

The sheet properties, i.e. the mechanical properties, can be related to the applied rolling schedule. Further important aspects of the sheet quality depend on additional features of the mills that are used. The homogeneity of the final gauge and shape are directly related to the rolls used. The surface quality is also dependent on the lubrication used during the rolling process. The influence of processing parameters on the sheet properties is addressed below.

The hot processing during sheet rolling of magnesium alloys is directly accompanied by dynamic recrystallisation (DRX), whereby microstructural

Table 10.2 Rolling of magnesium alloys: hot rolling temperature

Alloy	Temperature (°C)	Remarks
AZ21	260–370	Balanced range of processability
AZ31	250–450	Limit due to recrystallisation temperature
		Temperature range of rolling schedules
AZ61	320–420	Limits due to recrystallisation temperature
	280–320	Balanced range of processability
AZ80	300–420	Limit due to recrystallisation temperature
M2 (AM503)	250–440	Limit due to recrystallisation temperature
	250–500	Balanced range of processability
	Cold–500	Cold finish rolling, rolling schedules
ZK10	250–400	Limit due to recrystallisation temperature
ZK30 (ZW3)	320–430	Limit due to recrystallisation temperature
	380–480	Temperature range of rolling schedules
ZK60	300–400	Limit due to recrystallisation temperature
ME20 (AM537)	250–400	Balanced range of processability

Sources: Beck, 1939; Chabbi and Lehnert, 2000; Emley, 1966.

features related to the deformation history are easily lost. As a result of this, it is typically observed that the spectrum of microstructures obtainable from rolling experiments is quite narrow, i.e. grain sizes and texture intensities do not vary significantly. This has resulted in limited research activities regarding the influence of rolling parameters on the properties of flat magnesium products compared to cubic structured materials such as steels and Al sheets. Crystallographic texture control has been highlighted as a key to obtaining improved sheet formability, which is crucial for the widening of the number of industrial applications. Therefore, recent studies have focused on understanding the relationship between the initial microstructures, the rolling temperature, the reduction per pass, the total rolling reduction and the resulting texture.

The rolling temperature is the determining process parameter, because certain deformation mechanisms in Mg alloys, such as non-basal dislocation slip, as well as the occurrence of DRX, are activated by increasing temperature, and the subjected deformation can be accommodated without failure of the sheets. As a result, a variety of other process parameters can be taken into account for their influences on the resulting sheet properties. Although a study (Jeong and Ha, 2007) demonstrated that AZ31 sheet can be successfully rolled at 200 °C, a rolling temperature higher than 300 °C is selected to prevent cracking. As summarised in Fig. 10.1, a coarser grain structure resulted at higher temperature

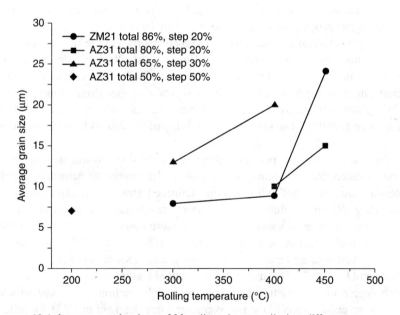

10.1 Average grain sizes of Mg alloy sheets rolled at different temperatures. ZM21 (Hantzsche et al., 2007); AZ31 total 80%, step 20% (Chino et al., 2009); AZ31 total 65%, step 30% (Jin et al., 2009); AZ31 total 50%, step 50% (Jeong et al., 2009);

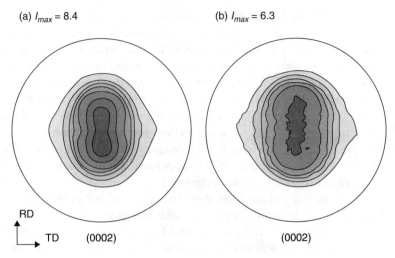

10.2 (0002) pole figures of laboratory rolled ZM21 sheets. Contour levels = 0.5, 1, 1.5, 2, 3, 5, 7. (a) 300 °C. (b) 450 °C. Source: Hantzsche et al., 2007.

as a result of DRX as well as grain growth during subsequent slow cooling of the rolled sheets. Furthermore, the texture sharpness of the rolled sheet decreases with increasing rolling temperature (Fig. 10.2). This texture weakening is related to the higher propensity for the occurrence of DRX and the activation of non-basal slip mechanisms, which result in a broader spread of the basal poles into the sheet rolling direction (Jin *et al.*, 2010; Bohlen *et al.*, 2003; Nestler *et al.*, 2007). A recent study (Chino and Mabuchi, 2009) examined the influence of the rolling temperature on the formability of AZ31 sheets. A higher stretch formability and lower plastic anisotropy (r value) are achieved in the sheet rolled at 450 °C than that rolled at 390 °C. The improved formability of the sheet is based on the weaker texture.

The rolling reduction per pass influences the final microstructure and texture, since it determines the strain rate as well as the amount of deformation before recovery and recrystallisation of the deformed structure during intermediate annealing. A bimodal distribution in grain size results from employing a small thickness reduction of about 10%, while a more homogeneous and fine-grained structure is obtained by using larger reductions (Sun *et al.*, 2009). Moreover, it is reported that weaker texture results from greater reductions (Sun *et al.*, 2009; Jeong and Ha, 2007). The above microstructural features are due to the fact that a high degree of deformation gives rise to a high fraction of DRXed structure. Moreover, examination of the microstructural development in AZ31 for different rolling reductions per pass showed that rolling reductions larger than 50% at 400 °C result in a high fraction of shear banding (Yim *et al.*, 2009). These shear bands are highly localised deformed regions in which more random grain orientations and thus weaker texture are developed. In ZM21 sheets rolled with reductions per pass from 10–30%, a decreasing tendency in the plastic anisotropy was observed on increasing the reduction per pass, based on the low texture sharpness and homogeneous grain structures (Hantzsche et al., 2007). However, grain growth can occur with an increase in process temperature as a result of deformation heating when the reduction per pass is set to a high value. An example can be found in Liang et al. (2009), in which a larger grain size results in AZ31 sheets after rolling with 40% thickness reduction than after rolling with 20% or 30% reductions.

The influence of the total rolling reduction on the final microstructure and texture can be interpreted in a similar way to the reduction per pass. Although the grain size and texture sharpness decrease in general with increasing the total reduction (Jin et al., 2010, Yim et al., 2009), as shown also by comparing the results of Chino and Mabuchi (2009) and Jin et al. (2010) presented in Fig. 10.1, the resulting microstructures are more dependent on the rolling temperature, duration of the intermediate annealing and reduction per pass. Because sheets experience static recrystallisation during intermediate annealing and the accumulated deformation is determined mainly by the degree of deformation during the last rolling pass, the total reduction plays a minor role in the final microstructure. The importance of the total reduction can be found in relation to the final thickness. Upon increasing the total reduction, the sheet thickness decreases, which results in fast cooling of the sheet during rolling and lowers the process temperature, especially in the case of non-heated rolling mills.

Increasing the rolling speed leads to a decrease in the average grain size of magnesium sheets as shown by Essadiqi et al. (2006) for the rolling of AZ31. It is noted that the effect of the rolling speed is similar to the effects of other rolling parameters as described above, and differences in typical ranges of the speed appear to be small. Notably, it has been reported that very high rolling speeds in the range of 1000 m/min and greater can be used during the manufacture of magnesium sheets (as described in a number of laboratory studies using AZ31 or AZ80 (Sakai et al., 2007; Sakai et al., 2009)). The short contact time between work piece and the roll surface gives rise to better temperature stability as the single rolling passes are short.

Since the rolling process in general involves both massive deformation and intermediate annealing, the initial microstructures are usually destroyed by deformation, DRX and static recrystallisation. It was reported that the rolling of differently prepared AZ31 alloys, e.g. gravity cast, squeeze-cast blocks and rolled sheet, leads to comparable microstructures and textures after about 80% thickness reduction (Bohlen et al., 2003). That is, the influence of the initial material state on the final microstructures becomes weak on increasing the rolling reduction. However, if the initial material already has a strong texture, this can influence the final microstructure. An example can be found in Huang et al. (2010) in which the microstructural evolution during the rolling of an extruded plate containing two strong texture components was examined. One component corresponds to the basal poles in the sheet normal direction (ND) and the other to the basal pole in the sheet transverse direction (TD). Whereas rolling of a cast ingot typically leads to the development of a texture component corresponding to the basal pole parallel to the ND, the latter texture component is retained during rolling up to 70% thickness reduction at 550 °C. The retention of this (TD) texture component after rolling results in a significant improvement in the sheet formability.

Beside these processing parameters, other aspects are important for the sheet properties, such as the rolling method itself, e.g. uni-directional rolling, or reversing the sheet direction after each rolling pass or cross rolling, as well as the lubrication method used. It was reported (Chen et al., 2010) that an AZ31 sheet rolled in the reverse-wise manner shows a higher strength and more isotropic mechanical properties than uni-directionally rolled sheets. This was related to the homogeneous grain structure and the stronger texture resulting from the reverse-rolling. In addition, cross rolling may be used to adjust the dimensions of the sheets. The effect of cross rolling implies that a change of strain path during rolling affects the microstructure and texture developed. If cross rolling is applied instead of uni-directional rolling, weaker textures are achieved for alloy AZ31, although no distinct change in the type of texture could be found (Al-Samman and Gottstein, 2008). The same effect is described for some experimental sheet alloys by Lim et al. (2009) and improved mechanical properties and stretch formability were obtained. Employment of a proper lubricant is also important, especially during hot rolling, to prevent adhesion of the sheet to the rolling mill. The above short review on the process parameters suggests that high rolling temperatures and large reductions per pass will result in improved sheet formability. However, the probability of adherence of the sheet to the rolling mill will also increase at higher process temperatures if the wrong lubrication concept is employed. Consequently, to obtain improved sheet formability through a tailored microstructure and texture it is necessary to optimise the process parameters with regard to their inter-relationships.

10.4 Alloying effects on sheet properties

The previous section showed the influence of the processing schedule on the sheet properties. This influence is related to the thermomechanical behaviour of the alloy used during sheet rolling. The role of the alloy composition itself and its influence on the sheet properties is reviewed in this section.

Alloy AZ31 has been used as a well-balanced magnesium sheet alloy for decades, although it was not supposed to be the easiest alloy for the rolling process. Table 10.3 collects mechanical properties as described in both earlier and recent literature. If strength is an issue, modifications to higher aluminium contents, e.g. alloy AZ61 (old name: AZM), can be used. Aluminium-free alloys would be those of a magnesium-zinc composition modified using zirconium, such

Table 10.3 Mechanical properties of conventional magnesium sheets

Sheet alloy	TYS (MPa)	UTS (MPa)	Elongation (%)
AZ31 (0-temper)	145–150	250–255	17.0–21.0
AZ31 (H24)	145–220	255–290	14.0–19.0
AZ31 (H26)	170–205	260–275	10.0–16.0
AZ61	154–216*	263–340	8.0–18.0
AZ80	154–216*	278–340	8.0–14.0
M1	100	232	6
ZK10	178	263	10
ZK30	185	270	8
ZK60	201–216*	293–309	8.0–10.0
ZM21 (0-temper)	131	232	13
ZM21 (H24)	Min. 180	Min. 275	Min. 8

*Value at maximum uniform strain 5%. TYS: tensile yield stress. UTS: ultimate tensile stress. Sources: Kammer, 2000; Emley, 1966; Busk, 1987.

as ZK30 or ZK10 in the modern ASTM nomenclature, or binary magnesium-manganese alloys denoted M1 or M2, respectively. Other alloys used contained thorium as an alloying element (HK31, HM21) if high-temperature usage was a goal (Emley, 1966), but these are no longer used.

One of the major issues in assessing the properties of magnesium sheets is the anisotropy of the mechanical behaviour. Table 10.4 collects a number of mechanical properties for some magnesium sheets tested in the rolling (RD) and transverse (TD) directions. A typical result is a high tensile yield stress (TYS) measured along the TD and a low one along the RD if conventional alloys are addressed, as shown for AZ31 or ZM21 in Table 10.4. The elongation to failure is only slightly higher in the RD than in the TD. The strain hardening is rather low and similar along the different directions if the difference between ultimate tensile stress (UTS) and TYS is taken as a rough measure for this. The planar anisotropy (r value), which is the ratio of plastic strain in width vs the thickness of the sheet sample, exhibits a distinct increase from the RD to the TD. This type of behaviour has been considered typical for those magnesium sheet alloys that exhibit a distinct texture with their basal planes parallel to the sheet plane and a moderate difference in the angular spread of the basal planes to the RD and the TD. Mechanical results obtained on magnesium sheets that contain a certain amount of rare earth elements exhibit a somewhat different behaviour. In these sheets, the yield strength is highest in the RD and decreases towards the TD. This effect is much more pronounced than in the AZ31 sheets. However, the elongation to failure is also very dependent on the orientation and increases significantly from the RD to the TD. There is also a difference in the strain-hardening behaviour as a function of the orientation. It is rather low along the RD and much higher along the TD. Finally, the r values for the RD and TD are close to 1. Despite the significant differences in the yield stress, these results were seen as a potential improvement for the sheet formability under selected forming conditions, a topic that will be addressed below. The reason for the different mechanical behaviour of these alloys is seen in the significant differences in the textures of the sheet. Typical magnesium sheets such as AZ31, ZM21 or other older alloys collected in Table 10.3 develop strong textures during rolling with a preferred orientation of basal planes parallel to the sheet plane as described above (see e.g. Fig. 10.2). The result is a grain structure with orientations difficult for deformation, due to the intrinsic plastic anisotropy based on the hexagonal close-packed lattice structure and, thus, higher flow stresses. Furthermore, the ability to work harden is limited under these conditions and this therefore affects the uniform elongation, fracture strain and, more generally, the formability of such magnesium sheets (Yukutake *et al.*, 2003). Therefore applications are limited to forming processes at elevated temperature where the material is easier to form (Doege and Dröder, 2001).

Only recently, new magnesium alloys containing rare earth elements were described that do not tend to form such distinct textures during rolling and therefore have different mechanical behaviour (Table 10.4). These sheets typically form microstructures with grain sizes between 10–20 μm, but have textures that are significantly weaker compared to those of the conventional sheets. Examples are given in Fig. 10.3. Furthermore, the character of these textures is different with a distinct preferential tilt of basal planes towards the TD. Moreover, these sheets do not have a high concentration of basal planes parallel to the sheet plane, but exhibit a preferential tilt of the basal planes of 10–20° to the RD (Bohlen *et al.*, 2007) or the TD (Mackenzie and Pekguleryuz, 2008; Wendt *et al.*, 2009).

Table 10.4 Orientation dependence of mechanical properties

Alloy	Orientation	TYS (MPa)	UTS (MPa)	Uniform elongation (%)	Elongation (%)	r_8
ZM21	RD	127	236	18.4	24.3	0.9
	TD	144	235	14.6	18.1	2.1
ZK10	RD	194	254	10.0	15.6	0.9
	TD	226	272	11.2	22.5	1.3
ZE10	RD	191	216	7.7	19.8	0.9
	TD	138	226	18.9	29.7	0.9
ZEK100	RD	203	234	7.1	23.7	0.9
	TD	154	241	15.7	31.9	1.2
ZEK410	RD	258	291	4.9	8.8	1.3*
	TD	182	280	17.5	23.7	0.9
ZW41	RD	209	258	10.8	17.4	0.9
	TD	130	248	22.1	30.1	0.7

*Value at maximum uniform strain 5%. TYS: tensile yield stress. UTS: ultimate tensile stress. r_8: r-value at 8% strain. Source: Bohlen *et al.*, 2007.

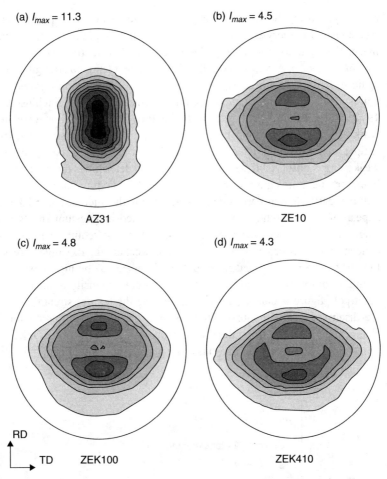

10.3 Recalculated pole figures from texture measurements on magnesium sheets. Contour levels = 0.5, 1, 1.5, 2, 3, 5, 7. (a) AZ31 (Bohlen *et al.*, 2007, 2009). (b) ZE10. (c) ZEK100. (d) ZEK410.

Generally, a wider spread of the basal planes towards the TD of the sheet is observed. It is not only rare earth-containing Mg-Zn-RE alloys that form such different textures. A similar tendency is also found in experimental studies if other elements are used, e.g. an addition of Ca to AZ31 (Ebeling *et al.*, 2006).

Understanding the factors that lead to the development of the microstructure and texture during rolling of magnesium sheets requires a combination of mechanisms. The main deformation mechanisms in magnesium are known to be basal slip and extension twinning if critical resolved shear stresses are reviewed (Ion *et al.*, 1982; Agnew, 2002). Activation of these mechanisms is likely to orient the basal planes parallel to the sheet plane. As dynamic and static recrystallisation are active during hot processing, it is believed that the important nucleation mechanism for recrystallisation is a discontinuous grain boundary nucleation mechanism that does not change the underlying deformation texture significantly (Ion et al., 1982). A well recrystallised AZ31 sheet tends to strengthen its texture again if exposed to a heat treatment that leads to overall grain growth, a feature that is also very often observed in cubic metals and alloys (Hutchinson and Nes, 1992). In the example in Fig. 10.4, the texture intensity is shown (related to the work of Kaiser et al. (2003)) for a magnesium sheet in the strain-hardened H24 temper condition and its change with heat treatment. First, a slight texture weakening is observed if static recrystallisation determines the microstructure. At higher temperatures grain growth is found and the texture intensity increases.

If rare earth elements are added to magnesium, the microstructural development appears to be different. It was shown that rolled magnesium sheets containing single rare earth elements as alloying additions require certain minimum concentrations in order to achieve texture weakening. Examples are shown in Fig. 10.5. The texture weakening potential appears to be higher when the solid solubility of the element is lower, if the maximum intensity of the corresponding basal pole figure is considered as representing the texture strength. Two aspects are discussed in this context. There is evidence that deformation mechanisms change their balance during the rolling process, e.g. contraction or other twins are found more extensively, or, non-basal slip plays a more significant role (Hantzsche

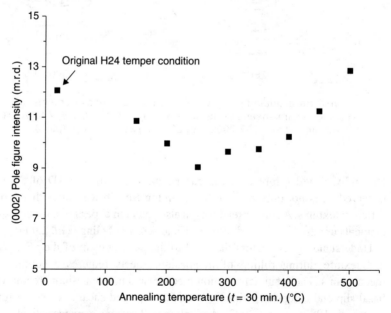

10.4 Pole figure intensity after annealing of a H24-temper AZ31 sheet. Source: Kaiser et al., 2003.

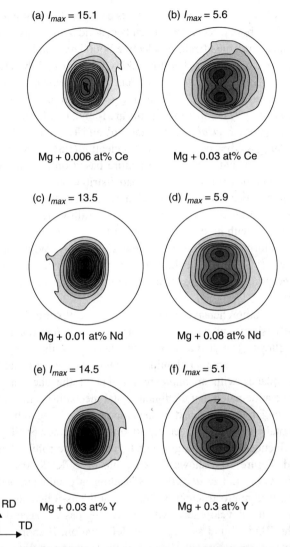

10.5 (0002) Pole figures and maximum intensities for as-rolled binary magnesium sheets with varying content of rare earth elements. Contour levels = 0.5, 1, 1.5, 2, 3, 5, 7.

et al., 2010; Yi et al., 2010). This gives support to the assumption that new orientations formed during deformation contribute to the texture development of the sheet (Hadorn et al., 2011). Furthermore, it is noted that the grain size of the sheets after the same processing and subsequent heat treatment decreases with increasing content of the particular rare earth element. All the annealed sheets are completely recrystallised and changes in the microstructure can therefore be related to grain growth. As the alloying elements are very different with regard to their solute solubility, both effects, the elements in solid solution as well as in the form of precipitates, are considered. A selective boundary pinning due to solute segregation or particle pinning is therefore likely to occur and therefore restrict the growth of the dominant texture component if the element content is high enough (Hantzsche et al., 2010). This finding is not limited to the rolling of magnesium sheets but very comparable results are found, e.g. for the extrusion of magnesium alloys (Bohlen et al., 2010; Stanford and Barnett, 2008). Ball and Prangnell (1994) suggested particle stimulated nucleation (PSN) of recrystallisation as the main mechanism that weakens the texture based on the high content of alloying elements and the appearance of a broad distribution of particles. In more recent studies, it has been reported (Mackenzie et al., 2007) that PSN is a mechanism that has to be considered in magnesium alloys if they contain rare earth elements. Alloys with moderate contents of RE or Y exhibit shear bands as a result of plastic deformation (Senn and Agnew, 2006, 2008). Within shear bands, grain nuclei growing into the band exhibit a more random texture compared to the surrounding unrecrystallised regions (Mackenzie and Pekguleryuz, 2008; Stanford and Barnett, 2008).

There is still no comprehensive explanation for the texture development in ternary Mg-Zn-RE sheets, which again is somewhat different compared to the binary Mg-RE alloys. The textures of such sheets are also very weak but show a tendency to form another dominant texture component that favours a tilt of the basal planes with an angle to the TD of the sheet as shown in Fig. 10.3. It can be anticipated that the fundamental mechanisms that lead to such changes in the microstructure and texture development are comparable to those in binary experimental alloys. It has been reported that in the newer alloys an even more distinct influence of the rolling schedule, namely the rolling temperature, can be observed. Figure 10.6 shows the example of ZEK100 sheets rolled at 300 °C and 450 °C (Wendt et al., 2009). If rolling is conducted at 450 °C, a partially recrystallised microstructure is obtained in the as-rolled condition. This has an effect on texture development. After the rolling pass, a basal double peak tilted towards the RD is found as a result of deformation. If short annealing is applied at the same temperature, weakening of the texture and the formation of another texture component tilted towards the TD is observed. If the same rolling experiment is repeated at the lower temperature of 300 °C, the microstructure does not significantly recrystallise although the massive deformation initiates DRX. The corresponding texture shows an off-basal character due to the two peaks in the (00.2) plane being aligned with the RD again. Still, a broad angular distribution of the basal planes between the ND and the TD is found. Thus, the observed texture differs from the typical, strong basal type as seen in AZ31 sheets but is also comparably strong. During annealing at 300 °C or 400 °C, static recrystallisation occurs and the texture weakens. Only in the case of annealing at higher temperature does the texture lose its basal peak split character in the RD.

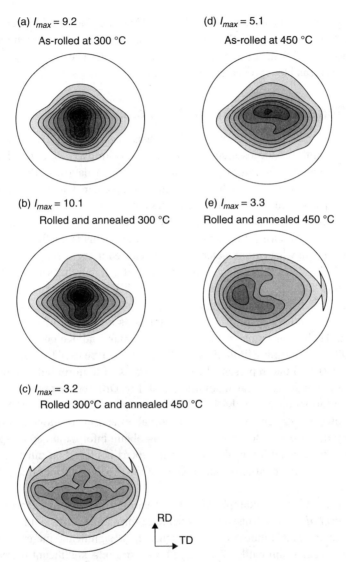

10.6 Texture development during the rolling and annealing of ZEK100.

A very comparable result was found by Mackenzie and Pekguleryuz (2008) after rolling of the ZE10 alloy at 150 °C, which produces a texture with basal poles tilted towards the RD, whereas after annealing at 400 °C this component vanishes and a preferred orientation with a tilt to the TD is found. Based on the recrystallisation kinetics that are modified by the addition of the alloying elements considered, the results for the sheets after conventional rolling will be influenced. Vice versa, the processing route, e.g. characterised by the rolling temperature or the rolling speed, will have the same influence. Thus, the objective for magnesium sheet developers will be to outline alloy compositions that meet the requirements for microstructural features, such as a weak texture, and their dependence on the applied – and feasible – processing parameters.

10.5 Formability of magnesium sheets

Sheet performance is often assessed on the basis of uniaxial mechanical testing in which strength properties, ductility, in-plane anisotropy (r value) and strain-hardening behaviour (n value) are characterised. Such data are often used to optimise the subsequent step in the product processing chain, which is the forming of structural parts. Generally, forming involves procedures such as stretch forming and deep drawing in which the sheet is subjected to a variety of strain paths. Technically, the main aspect of the different forming paths is the flow behaviour of the sheet, i.e. to what extent material flow is realised in the sheet plane or in the thickness of the sheet. A convenient way to present a comprehensive assessment of the overall formability of a sheet with these different aspects is based on a description by Keeler (1965). In this basic presentation it was suggested to use different strain paths with limiting major strains as a function of minor strains (Goodwin, 1968; Nakazima *et al.*, 1968; Marciniak and Kolodziejski, 1972; Hasek, 1978). The forming limit diagram (FLD) can then be used to show forming limit curves (FLC) that represent the critical strains of uniform deformation for different strain paths as a measure of the formability. Different test setups are used to reveal information about the FLC. An overview is presented in Hsu *et al.* (2008). However, stretch forming (Erichsen value) and deep drawing (limiting drawing ratio) are often used in the literature to obtain information about specific features of the formability. In this section the focus is placed on intrinsic sheet properties, i.e. the microstructure and texture of the sheets and their effects during forming.

Figure 10.7a gives an example of FLCs at various temperatures for an AZ31 sheet (Stutz *et al.*, 2011). The room temperature formability turns out to be poor. An increase in temperature to 150 °C leads to a significant increase in the formability under strain paths of negative minor strain. A significant increase in stretch formability is found when increasing the test temperature to 200 °C and beyond. The maximum major strain is reached with a specimen tested at 200 °C under negative minor strain. The minimum major strain is found for a specimen with minor strain close to zero, regardless of the temperature. This is a typical finding for sheet metals as this describes a plane strain condition that appears to be critical for sheets and is also described, e.g. for aluminium alloys (Hsu *et al.*, 2008). The corresponding major strain value is denoted FLD_0 and is sometimes used to give first hand information on the formability of sheets as the minimum value achievable (Wagoner *et al.*, 1989). Besides the effect of temperature there are few publications dealing with the influence of testing parameters during

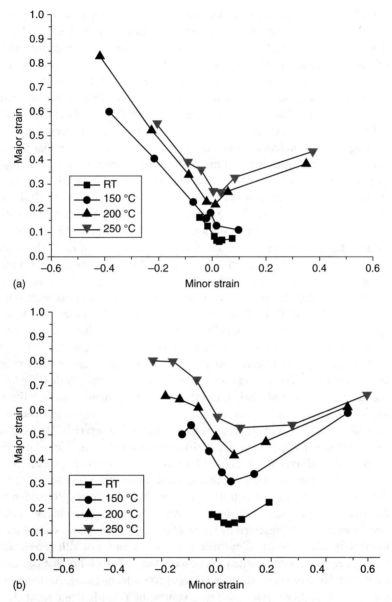

10.7 Forming limit curves of magnesium sheet at various temperatures. (a) AZ31. (b) ZE10. Source: Stutz *et al.*, 2011.

forming (Hsu *et al.*, 2008). It was shown by Bruni *et al.* (2010) that the strain rate dependency is more pronounced on the right hand side of the FLD than on the left hand side. A comprehensive analysis of the effect of rolling parameters on sheet formability is not yet available. Some aspects can be derived for the influence of the rolling parameters on sheet formability. Typically these results are related to changes in the microstructure and texture as described in Section 10.3. It was shown by Chino and Mabuchi (2009) that higher stretch formability and lower r values are obtained if the material is rolled at higher temperatures. Interestingly, lower r values appear to correlate with increased stretch formability.

The alloy composition can have a more significant effect on the microstructure and texture. Lower limiting drawing ratios are found for the sheet alloys AZ61 and M1 (Doege and Dröder, 2001), but the general result of an increase in the limiting drawing ratio is similar in these alloys. It is worth mentioning that during deep drawing, the rate dependency of magnesium sheets appears to be more significant than for aluminium sheets. Apparently this also holds for stretch tests (Lee *et al.*, 2008). In Fig. 10.7b an example of an FLD is shown for alloy ZE10 sheet (Stutz *et al.*, 2011). This sheet exhibits a different texture compared to AZ31 as shown in Fig. 10.3. In this case, lower r values and higher formability are found compared to AZ31 at the same temperatures. The forming limit curves of ZE10 show generally higher values of major strain compared to AZ31. At room temperature, the formability under biaxial stretching conditions is already higher than for AZ31 at 150 °C. FLD_0 depends strongly on temperature and shows an increase from room temperature to 250 °C. ZE10 generally outperforms AZ31 in terms of formability. At 150 °C, comparable formability can be achieved; however the degree of deformation under plane strain conditions and biaxial stretching is unmatched by AZ31 even at 250 °C. Dreyer *et al.* (2010) have confirmed this correlation between texture and FLC_0. For the same alloy, a higher deep drawability than for AZ31 was also described (Yi *et al.*, 2010). Despite this improvement in formability and the lower r value with less in-plane anisotropy, ZE10 shows a higher tendency to earing than AZ31. Results from electron backscattering diffraction (EBSD) using sections from deep-drawn cups (Yi *et al.*, 2010) also show higher activity of compression twins in the RE-containing alloy. Moreover, a TEM-based analysis (Sandlöbes *et al.*, 2011) clarified that the addition of Y encourages the activation of <c + a> slip and compression twins, which leads to more homogeneous deformation without the occurrence of macro-shear banding and, consequently, to a higher formability. Similar results were presented by Chino *et al.* (2009, 2010) who studied the stretch formability of Mg-Zn-RE alloys where RE stands for Ce or Y. It is concluded that the weaker textures contribute to improved stretch formability that can match the results for 5000 or 6000 Al series alloys at room temperature. An increase in the fraction of precipitates with increasing content of Y in Mg-Zn-Y sheets leads to a decrease in stretch formability. Again, low r values are found in these studies with average values below 1 as obtained from uniaxial tensile tests. In binary Mg-Ce (Chino *et al.*, 2008) alloys, where textures are weaker than in pure magnesium sheets, higher stretch formability is also observed.

Other reports in the literature indicate improved formability in Mg-Li alloys, e.g. as shown by Takuda *et al.* (2000). They show significantly improved room temperature formability in Mg-Li-Zn alloy, which also exhibits significant earing.

10.6 Enabling technologies and future trends

10.6.1 Post-processing of magnesium sheets and parts

Industrial acceptance of magnesium sheets depends on the integration of parts into assemblies and their successful protection. This lays emphasis on the full process chain and requires solutions for corrosion protection and joining. Suitable processes are available today and are also recommended by sheet manufacturers (Magnesium Elektron, 2011c, 2011d).

Optimisation criteria for magnesium sheets also include those connected with subsequent processing steps such as deep drawing, joining and coating. The formability in the deep drawing process is closely linked to the rolling process and the alloy composition as described above. Deep drawing of magnesium can be performed but the forming operation must be carried out at elevated temperatures (approximately 225 °C) (Doege et al., 2001). At this temperature very good forming results can be achieved; however the aim is the development of magnesium sheet materials with better formability at lower temperatures.

The next step in the process chain is joining the magnesium part to the assembly. This can be done by welding or by forming. The welding characteristics can be compared to those of aluminium, but the tendency to show cracks or pores is marked, and depends on the alloy composition chosen (Haferkamp et al., 1997). Suitable methods for welding of magnesium alloys are arc welding processes, such as TIG, MIG, and plasma arc welding, or beam processes, such as laser and electron beam welding, as well as special processes such as friction welding. An essential problem in the fusion welding process is the tendency towards thermal fissures, which especially occur when dealing with materials of high zinc content. Further problems during the joining process for magnesium alloys are caused by oxide layers containing amounts of $Mg(OH)_2$, because of porosity and oxides that are included in the welding material (Ferjutz and Davis, 1990; Stolbov et al., 1991; Ryazantsev, 1991, Chen et al., 1993).

Mechanical joining processes such as clinching or punch riveting are also suitable for magnesium sheet components, especially if the magnesium is to be joined to other materials such as steel or aluminium. However, as with deep drawing, clinching operations need elevated temperatures of 250–275 °C (Hübner, 2005). Another aspect of the mechanical joining process is galvanic corrosion. Galvanic corrosion does not only appear between the joining element and the sheet material, but also between magnesium and other materials. In order to avoid galvanic corrosion between the magnesium sheet and the joining elements, suitable layers of the joining elements should be developed. Galvanic corrosion between the magnesium sheet and other materials can be avoided by adding an adhesive layer, as is usually applied during these types of joining processes.

Components formed from magnesium sheets as well as assemblies require a cleaning step prior to further treatment in order to avoid surface contamination by lubricants and other residues (Juchmann and Wolff, 2010). Conversion coatings

366   Advances in wrought magnesium alloys

and anodic oxidation methods are available, as well as various organic coatings. Further general requirements for surface protection have been described in Haferkamp *et al.* (2000) or by Grey and Luan (2002) and find application to formed products from magnesium sheets.

10.6.2 Twin roll casting

Twin roll cast strips are currently the focus of interest as an alternative to DC cast feedstock for the production of magnesium sheets. The main difference is the shape, which is already comparable to a rolled plate.

Twin-roll casting is the only process that combines both solidification and rolling in a single step. Fig. 10.8 shows the general principle of strip casting. In this process, liquid metal is pumped from a furnace or cast over a pipe into a tundish with a protective gas coverage. The melt is then dragged into the roll gap of a pair of counter-rotating, internally cooled rolls. The metal solidifies upon contact with the cooled rolls and is rolled to a strip. These strips can be rolled using conventional rolls in a continuous process. In the next step the strip is rolled to final gauge. The overall process includes different heat and surface treatments, with some technical and economic advantages. The reduction in the number of processing steps to final gauge leads to shorter production times and a decrease in production costs. It is reported that twin roll cast strips are significantly more fine-grained compared to conventional castings as a result of the combined casting-rolling process and

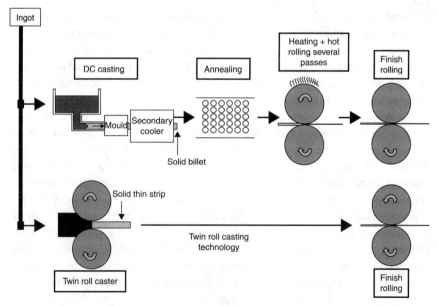

10.8 Comparison of the technical procedures for conventional rolling and twin roll casting followed by rolling.

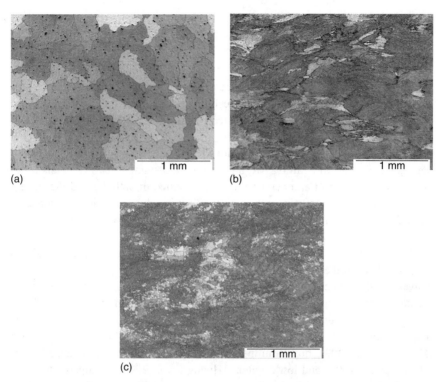

10.9 Microstructure of different materials used for subsequent rolling of alloy AZ31. (a) Conventional direct chill cast magnesium slab. (b) Example of twin roll cast strip. (c) Example of an annealed strip.

subsequent heat treatment of the material. A key difference between twin roll casting and DC casting is the solidification rate. While in DC casting it is limited to 1–50 °C/s, it reaches 500–1000 °C/s in a twin roll caster, leading to greatly refined structures with small and finely distributed intermetallic phases, extended solid solutions and fine grain sizes. (Basson and Letzig, 2010). Micrographs of alloy AZ31 demonstrate this difference in Fig. 10.9. This type of material is very similar to a plate rolled from a slab and is therefore seen to have the potential to overcome economic limitations in the production of magnesium sheets. Although the first preparation of strip goes back to the early 1980s (Brown, 2002), there are still issues to overcome that have to do with the homogeneity of the material. Thus, twin roll casting is believed to be an excellent method for the generation of fine-grained feedstock materials that can subsequently be warm rolled to thin sheets.

10.6.3 Asymmetrical sheet processing

Other types of process development are based on the rolling process and offer ways to alter the strong basal-type texture of sheets and, consequently, improve their formability. One way is to impose a large shear deformation during processing using asymmetrical rolling, which especially finds application in the research and development field. Two variants of this method are differential speed rolling (DSR) (Watanabe et al., 2004) and equal channel angular rolling (ECAR) (Cheng et al., 2007a), which are presented as prospective techniques for special property development. These techniques were originally used to obtain ultra-fine grain sizes in Al alloys and steels (Lee et al., 2002; Park et al., 2004; Raab et al., 2004).

In DSR, the shear deformation is produced by the difference in velocity between the top and bottom rolls. In ECAR, a forming die containing two channels (inlet and exit) that intersect at a certain angle, is installed directly behind the rolling mill. As the rolled sheet passes through the forming die the sheet experiences a shear deformation in the same way as in equal channel angular pressing (ECAP) (Furukawa et al., 1996).

The microstructural changes taking place during DSR and ECAR of Mg alloys generally lead to the formation of a weak texture and finer grains compared to conventional symmetrical rolling. Furthermore, the textures exhibit a tilt of the basal poles from the ND to the RD (Huang et al., 2008; Kim et al., 2007; Cheng et al., 2007b; Kim et al., 2009). This texture change leads to strong activation of basal slip and thus to high fracture strains (Kim et al., 2007), high Erichsen values (Huang et al., 2009), high limiting deep drawing ratios at room temperature (Huang et al., 2009) and low r values (Huang et al., 2008). Compared to other technologies based on severe shear deformation, such as ECAP, shear rolling provides a technical method to achieve a fine grain structure without additional contamination during processing. A recent report (Kim et al., 2009) showed that a combination of ECAR and DSR results in higher ductility without decreasing strength. By conducting DSR on an ECARed sheet, even finer grain sizes and weaker basal type textures were obtained compared to both conventional rolling or solely asymmetrically rolled sheet. The sheet fabricated using the optimised process combination has a tensile elongation of 30% and a tensile yield strength of about 210 MPa.

An example combining plate extrusion with sheet rolling was shown by Huang et al. (2009), which demonstrated the influence of the initial microstructure on the properties of a rolled sheet. The sheets produced using the combined technology retain the texture component corresponding to the basal poles tilted into the TD, as generally observed in RE-containing Mg alloy sheets.

10.7 Sources of further information

The number of research groups at universities having a strong impact on the topic of this chapter is increasing. The selective list presented here is limited to larger research groups in government funded institutions.

10.7.1 Sheet manufacturers

- Beijing Meros Magnesium Rolling Sheet Co., Ltd. (http://www.cnmeros.com/english/index.asp). Joint venture company set up by Beijing Guangling Jinghua ST Co., Ltd. and the University of Science and Technology, Beijing; AZ31 sheet production, maximum width 350 mm.
- Luoyang Copper (http://www.lycopper.cn). Chinese producer of magnesium plates and sheets. Used to be a special state-owned enterprise for the military industry. China Aluminium Corporation is the largest shareholder. Production programme includes sheets up to 1400 mm width and ZK series alloys.
- Gonda Metal Industry Co., LTD (http://www.gondametal.co.jp). Nonferrous metal manufacturer in Japan. Uses twin roll casting and rolling; AZ31 and AZ61 sheets available.
- Magnesium Elektron MEL (http://www.magnesium-elektron.com). Division of the Luxfer Group in Manchester (UK). Processing magnesium since 1936 with manufacturing sites in the UK, USA, Canada and the Czech Republic; commercial production of AZ31 sheet, plate and coil in various conditions.
- Magnesium Flachprodukte MgF (http://www.thyssenkrupp-mgf.com). Subsidiary of the German steel manufacturer Thyssen-Krupp in Freiberg (Germany), established in 2001. Strong cooperation with Freiberg University of Mining and Technology (Germany); production of AZ31 sheet and strip.
- POSCO-Mg (http://www.poscomg.co.kr). Division of the South Korean steel manufacturer POSCO, established in 2007; commercial production of AZ31 sheet using twin roll casting technology and subsequent rolling to sheet thickness of 0.4 mm.
- Salzgitter Magnesiumtechnologie (SZMT) (http://www.szmt.de). Division of the German steel manufacturer Salzgitter AG; production of AZ31 sheets and other tailored solutions, out of business December 2010.
- Stolfig Group (http://www.stolfig.de). Solutions in the construction of technical systems and components of magnesium, aluminium and polymers; manufacturer of magnesium sheet products based on an alloy named MnE21, subsidiaries in Germany and China.

10.7.2 Research groups

- CANMET – MLT (http://www.nrcan-rncan.gc.ca/mms-smm). Division of Natural Resources Canada, research centre for metals and materials fabrication, processing and evaluation. Twin roll casting of magnesium and sheet rolling.
- CSIRO Process Science and Engineering (http://www.csiro.au/science/Magnesium.html). Division of Australia's national science agency, research platform for magnesium and its alloys. Twin roll casting of magnesium strips.
- Magnesium Innovation Centre (http://www.hzg.de/institute/materials_research/structure/magic). Division of the Helmholtz-Zentrum Geesthacht in Germany,

research platform for process technology, wrought alloys, corrosion and surface protection as well as biomedical application. Uses twin roll casting technology and sheet rolling.
- MagNet (http://www.magnet.ubc.ca). Strategic network funded by the Canadian Government, research platform for Canadian industry and university partners, fundamental research project with a large variety of topics with respect to magnesium, including magnesium sheets and alloy development.
- Research Institute of Industrial Science and Technology (RIST) (http://www.rist.re.kr). South Korea, Materials Research Division, research on magnesium sheet production processes.
- The Science and Technology Research Council of Turkey (TUBITAK) MRC Materials Institute. Division of the Marmara Research Centre of TUBITAK. Industrial scale magnesium twin roll casting plant providing 1500 mm wide sheets of the alloys AZ31, AZ61, AZ91, AM50 and AM60.

10.8 References

Agnew SR (2002) 'Plastic anisotropy of magnesium alloy AZ31B sheet', in *Magnesium Technology 2002*, edited by Kaplan HI. Warrendale: The Minerals, Metals and Materials Society (TMS), pp. 169–174.

Al-Samman T and Gottstein G (2008) 'Influence of strain path change on the rolling behaviour of twin roll cast magnesium alloy', *Scripta Materialia*, **59**, 760–763.

Avedesian MM and Baker H (1999) *ASM Specialty Handbook – Magnesium and Magnesium Alloys*, Materials Park: ASM International.

Ball EA and Prangnell PB (1994) 'Tensile-compressive yield asymmetries in high strength wrought magnesium alloys', *Scripta Metallurgica et Materialia*, **31**, 111–116.

Basson F and Letzig D (2010) 'Twin roll casting of Magnesium – Differences and challenges compared to Aluminium', *Aluminium International Today*, in press.

Beck A (1939) *Magnesium und seine Legierungen*, Berlin: Springer-Verlag.

Bohlen J, Dobron P, Hantzsche K, Letzig D, Chmelik F and Kainer KU (2009) 'Acoustic emission study of the deformation behaviour of magnesium sheets', *International Journal of Materials Research*, **100**, 790–795.

Bohlen J, Horstmann A, Kaiser F, Styczynski A, Letzig D and Kainer KU (2003) 'Influence of the thermomechanical treatment on the microstructure of magnesium alloy AZ31', in *Magnesium Technology 2003*, edited by Kaplan HI. Warrendale: The Minerals, Metals and Materials Society TMS, pp. 253–258.

Bohlen J, Nürnberg MR, Senn JW, Letzig D and Agnew SR (2007) 'The texture and anisotropy of magnesium-zinc-rare earth alloy sheets', *Acta Materialia*, **55**, 2101–2112.

Bohlen J, Yi S, Letzig D and Kainer KU (2010) 'Effect of rare earth elements on the microstructure and texture development in magnesium-manganese alloys during extrusion', *Materials Science and Engineering A*, **527**, 7092–7098.

Brown RE (2002) 'Magnesium wrought and fabricated products yesterday, today, and tomorrow', in *Magnesium Technology 2002*, edited by Kaplan HI. Warrendale: The Minerals, Metals and Materials Society TMS, pp. 155–164.

Bruni C, Forcellese A, Gabrielli F and Simoncini M (2010) 'Effect of temperature, strain rate and fibre orientation on the plastic flow behaviour and formability of AZ31 magnesium alloy', *Journal of Materials Processing Technology*, **210**, 1354–1363.

Busk RS (1987) *Magnesium Products Design*, New York: Marcel Dekker Inc.

Chabbi L and Lehnert W (2000) 'Walzen von Magnesiumwerkstoffen' in Aluminium-Zentrale Düsseldorf, *Magnesium Taschenbuch*, Düsseldorf, Aluminium Verlag, pp. 415–433.

Chen G, Rothz G and Maisenhälder F (1993) 'Laserstrahlschneiden und-schweißen von Gußmagnesium', *Laser und Optoelektronik*, **25**, 43–47.

Chen X, Shang D, Xiao R, Huang G and Liu Q (2010) 'Influence of rolling ways on microstructure and anisotropy of AZ31 alloy sheet', *Transactions of Nonferrous Metals Society of China*, **20**, 589–593.

Cheng YQ, Chen ZH and Xia WJ (2007a) 'Drawability of AZ31 magnesium alloy sheet produced by equal channel angular rolling at room temperature', *Materials Characterization*, **58**, 617–622.

Cheng YQ, Chen ZH and Xia WJ (2007b) 'Effect of crystal orientation on the ductility in AZ31 Mg alloy sheets produced by equal channel angular rolling', *Journal of Materials Science*, **42**, 3552–3556.

Chino Y and Mabuchi M (2009) 'Enhanced stretch formability of Mg-Al-Zn alloy sheets rolled at high temperature (723K)', *Scripta Materialia*, **60**, 447–450.

Chino Y, Huang X, Suzuki K, Saassa K and Mabuchi M (2010) 'Influence of Zn concentration on stretch formability at room temperature of Mg-Zn-Ce alloy', *Materials Science and Engineering A*, **528**, 566–572.

Chino Y, Kado M and Mabuchi M (2008) 'Enhancement of tensile ductility and stretch formability of magnesium by addition of 0.02wt% (0.035at%) Ce', *Materials Science and Engineering A*, **494**, 343–349.

Chino Y, Sassa K and Mabuchi M (2009) 'Texture and stretch formability of a rolled Mg-Zn alloy containing dilute content of Y', *Materials Science and Engineering A*, **513–514**, 394–400.

Doege E and Dröder K (2001) 'Sheet metal forming of magnesium wrought alloys – Formability and process technology' *Journal of Materials Processing Technology*, **115**, 14–19.

Doege E, Sebastian W, Dröder K and Kurz G (2001) 'Increased formability of Mg-sheets using temperature controlled deep drawing tools', in *Innovations in Processing and Manufacturing of Sheet Materials*, edited by Demeri M. The Minerals, Metals and Materials Society (TMS), pp. 53–60.

Dreyer CE, Chiu WV, Wagoner RH and Agnew SR (2010) 'Formability of a more randomly textured magnesium alloy sheet: Application of an improved warm sheet formability test', *Journal of Materials Processing Technology*, **210**, 37–47.

Dröder K and Janssen S (1999) 'Forming of Magnesium Alloys – A Solution for Light Weight Construction', Warrendale: SAE International, SAE-Paper 1999-01-3172.

Ebeling T, Hartig C, Bohlen J, Letzig D and Bormann R (2006) 'Effects of calcium on texture evolution and plastic anisotropy of the magnesium alloy AZ31' in Kainer KU, *Magnesium Alloys and their Applications*, Weinheim: VCH Wiley, pp. 158–164.

Emley EF (1966) *Principles of Magnesium Technology*, Oxford: Pergamon Press.

Enß J, Evertz T, Reier T and Juchmann P (2000) 'Properties and Perspectives of Magnesium Rolled Products', in *Magnesium Alloys and their Applications*, edited by Kainer KU. Weinheim: VCH Wiley, pp. 590–595.

Essadiqi E, Galvani C, Amjad J, Shen G and Spencer K (2006) 'Hot rolling of AZ31 magnesium alloy to sheet gauge', Warrendale: SAE International, SAE-Paper 2006-01-0295.

Ferjutz K and Davis JR (1990) *Metals Handbook Vol.6*, Metals Park: ASM, pp. 772–782.

Furukawa M, Horita Z, Nemoto M, Valiev RZ and Langdon TG (1996) 'Microstructural characteristics of an ultra-fine grain metal processed with equal-channel angular pressing', *Materials Characterization*, **37**, 277–283.

Goodwin GM (1968) 'Application of strain analysis to sheet metal forming problems in the press shop', Warrendale: SAE International, SAE paper 680093.

Grey JE and Luan B (2002) 'Protective coatings on magnesium and its alloys – A critical review', *Journal of Alloys and Compounds*, **336**, 88–113.

Hadorn JP, Hantzsche K, Yi S, Bohlen J, Letzig D and Agnew SR (2011) 'Role of solute in the texture modification during hot deformation of Mg-Rare earth alloys', *Metallurgical Transactions A, submitted for publication*.

Haferkamp H, Bach FW, Kaese V, Niemeyer M, Tai P and Wilk P (2000), 'Chemisches Verhalten von Magnesium' in Aluminium-Zentrale Düsseldorf, *Magnesium Taschenbuch*, Düsseldorf, Aluminium Verlag, pp.285–343.

Haferkamp H, Burmester I, Niemeyer M, Doege E and Dröder K (1997) 'Innovative production technologies for magnesium light-weight constructions – laser beam welding and sheet metal forming', 30th ISATA, Florence, Italy 16–19 June 1997, Paper No. 97NM072.

Hantzsche K, Bohlen J, Wendt, J, Kainer KU, Yi S and Letzig D (2010) 'Effect of rare earth additions on microstructure and texture development of magnesium alloy sheets', *Scripta Materialia*, **63**, 725–730.

Hantzsche K, Kurz G, Bohlen J, Kainer KU and Letzig D (2007) 'Magnesium sheet alloys for structural applications', in *Proceedings of Third International Conference on Light Metals Technology*, Ottawa, edited by Sadayappan K and Sahoo M. Canada: Public Works and Government Services, pp. 189–200.

Hasek V (1978) 'Untersuchung und theoretische Beschreibung wichtiger Einflußgrößen auf das Grenzformänderungsschaubild', *Blech Rohre Profile*, **25** (10), 493–499, 213–220, 285–292, 493–499, 613–627.

Hsu E, Carsley JE and Verma R (2008) 'Development of forming limit diagrams of aluminium and magnesium sheet alloys at elevated temperatures', *Journal of Materials Engineering and Performance*, **17**, 288–296.

Huang X, Suzuki K and Chino Y (2010), 'Influences of initial texture on microstructure and stretch formability of Mg-3Al-1Zn alloy sheet obtained by a combination of high temperature and subsequent warm rolling', *Scripta Materialia*, **63**, 395–398.

Huang X, Suzuki K, Watazu A, Shigematsu I and Saito N (2008) 'Mechanical properties of Mg-Al-Zn alloy with a tilted basal texture obtained by differential speed rolling', *Materials Science Engineering A*, **488**, 214–220.

Huang X, Suzuki K, Watazu A, Shigematsu I and Saito N (2009) 'Improvement of formability of Mg-Al-Zn alloy sheet at low temperature using differential speed rolling', *Journal of Alloys and Compounds*, **470**, 263–268.

Hübner S (2005) *Clinchen moderner Blechwerkstoffe*, Garbsen, PZH Produktionstechnisches Zentrum GmbH.

Hutchinson B and Nes E (1992) *Materials Science Forum*, **94–96**, 385–390.

Ion SE, Humphreys FJ and White SH (1982) 'Dynamic recrystallisation and the development of microstructure during the high temperature deformation of magnesium' *Acta Metallurgica*, **30**, 1909–1919.

Jeong HT and Ha TK (2007) 'Texture development in a warm rolled AZ31 magnesium alloy', *Journal of Materials Processing Technology*, **187–188**, 229–561.

Jin L, Dong J, Wang R and Peng LM (2010) 'Effects of hot rolling processing on microstructures and mechanical properties of Mg-3%Al-1%Zn alloy sheet', *Materials Science and Engineering A*, **527**, 1970–1974.

Juchmann P, Wolff S and Kurz G (2002) 'Magnesiumblech – Werkstoffalternative für den Ultraleichtbau', in *Umformtechnik – Erschließung wirtschaftlicher und technologischer Potenziale*, edited by Doege E. Hannover, Hannoversches Forschungsinstitut für Fertigungsfragen, Institut für Umformtechnik und Umformmaschinen Universität Hannover, Wissenschaftliche Gesellschaft für Produktionstechnik, pp. 179–192.

Juchmann P and Wolff S (2010) 'A closed process chain for magnesium sheet parts – Requirements and present possibilities', in *Magnesium: Proceedings of the 8th International Conference on magnesium alloys and their applications*, edited by Kainer KU. Weinheim: VCH Wiley, pp. 1059–1065.

Kaiser F, Letzig D, Bohlen J, Styczynski A, Hartig C and Kainer KU (2003) 'Anisotropic properties of magnesium sheet AZ31', *Materials Science Forum* **426–432**, 315–320.

Kammer C (2000) 'Mechanische Eigenschaften' in Aluminium-Zentrale Düsseldorf, *Magnesium Taschenbuch*, Düsseldorf, Aluminium Verlag, pp. 207–236.

Keeler, SP (1965) 'Plastic instability and fracture in sheet stretched over rigid punches', *ASM Transactions*, **56**, 25–48.

Kim WJ, Lee JB, Kim WY, Jeong HT and Jeong HG (2007), 'Microstructure and mechanical properties of Mg-Al-Zn alloy sheets severely deformed by asymmetrical rolling', *Scripta Materialia*, **56**, 309–312.

Kim WJ, Yoo SJ, Chen ZH and Jeong HT (2009) 'Grain size and texture control of Mg-3Al-1Zn alloy sheet using a combination of equal channel angular rolling and high speed ratio differential speed rolling processes', *Scripta Materialia*, **60**, 897–900.

Lee YS, Kwon YN, Kang SH, Kim SW and Lee JH (2008) 'Forming limit of AZ31 alloy sheet and strain rate on warm sheet metal forming', *Journal of Materials Processing Technology*, **201**, 431–435.

Lee C, Seok HK and Suh JY (2002) 'Microstructural evolutions of the Al strip prepared by cold rolling and continuous equal channel angular pressing', *Acta Materialia*, **50**, 4005–4019.

Liang S, Sun H, Liu Z and Wang E (2009) 'Mechanical properties and texture evolution during rolling process of an AZ31 Mg alloy', *Journal of Alloys and Compounds*, **472**, 127–132.

Lim HK, Lee JY, Kim DH, Kim WT, Lee JS and Kim DH (2009) 'Enhancement of mechanical properties and formability of Mg-MM-Sn-Al-Zn alloy sheets fabricated by cross-rolling method', *Materials Science and Engineering A*, **506**, 63–70.

Mackenzie LWF, Davis B, Humphreys FJ and Lorimer GW (2007) 'The deformation, recrystallisation and texture of three magnesium alloy extrusions', *Material Science and Technology*, **23**, 1173–1180.

Mackenzie LWF and Pekguleryuz MO (2008) 'The recrystallization and texture of magnesium-zinc-cerium alloys', *Scripta Materialia*, **59**, 665–668.

Magnesium Elektron (2011a) DS482, Magnesium Elektron Datasheet 482, 'Elektron AZ31B Sheet, Plate & Coil', available online: http://www.magnesium-elektron.com

Magnesium Elektron (2011b) DS490, Magnesium Elektron Datasheet 490, 'Elektron Cast Billet', available online: http://www.magnesium-elektron.com

Magnesium Elektron (2011c) DS250, Magnesium Elektron Datasheet 250, 'Joining magnesium alloys', available online: http://www.magnesium-elektron.com

Magnesium Elektron (2011d) DS256, Magnesium Elektron Datasheet 256, 'Surface treatments for magnesium alloys in aerospace and defense', available online: http://www.magnesium-elektron.com

Marciniak Z and Kolodziejski J (1972) 'Assessment of sheet metal failure sensitivity by method or torsioning the rings' in *Proceedings of the 7th Biennial Congress*, Amsterdam: International Deep Drawing Research Group (IDDRG), pp. 61–64.

Nakazima K, Kikuma T and Hasuka K (1968) 'Study of the formability of steel sheet', *Yawata Technical Report*, **294**, 140–141.

Nestler K, Bohlen J, Letzig D and Kainer KU (2007) 'Influence of process parameters on the mechanical properties of rolled magnesium ZM21-sheets', in *Magnesium Technology 2007*, edited by Beals RS, Luo AA, Neelameggham NR and Pekguleryuz MO. Warrendale: The Minerals, Metals and Materials Society, TMS, pp. 95–100.

Park JW, Kim JW and Chung YH (2004) 'Grain refinement of steel plate by continuous equal channel angular process', *Scripta Materialia*, **51**, 181–184.

Raab GJ, Valiev RZ, Lowe TG and Zhu YT (2004) 'Continuous processing of ultrafine grained Al by ECAP-Conform', *Materials Science and Engineering A*, **382**, 30–34.

Ryazantsev VI (1991) 'Weldability of new aluminium and magnesium alloys', *Paton Welding Journal*, **3**, 521–523.

Sakai T, Utsunomiya H, Koh H and Minamiguchi S (2007) *Materials Science Forum* **539–543**, 3359–3364.

Sakai T, Watanabe Y and Utsunomiya H (2009) *Materials Science Forum* **618–619**, 483–486.

Sandlöbes S, Zaefferer S, Schestakow I, Yi S and Gonzalez-Martinez R (2011) 'On the role of non-basal deformation mechanisms for the ductility of Mg and Mg–Y alloys', *Acta Materialia*, **59**, 429–439.

Schumann S and Friedrich H (1998) 'The use of Magnesium in cars – today and in future', in *Magnesium Alloys and their Application*, edited by Kainer KU. Wolfsburg: VCH Wiley, pp. 3–13.

Sebastian W, Dröder K and Schumann S (2000) 'Properties and Processing of Magnesium Wrought Products for Automotive Applications', in *Magnesium Alloys and their Applications*, edited by Kainer KU. Weinheim: VCH Wiley, pp. 602–607.

Senn JW and Agnew SR (2006) 'Texture randomization during thermomechanical processing of a magnesium-yttrium-neodymium alloy', in *Proceedings of Magnesium Technology in the Global Age*, edited by Pekguleryuz MO and Mackenzie LWF. Montreal: The Canadian Institute of Mining, Metallurgy and Petroleum, pp. 115–130.

Senn JW and Agnew SR (2008) 'Texture randomization of magnesium alloys containing rare earth elements', in *Magnesium Technology 2008*, edited by Pekguleryuz MO, Neelameggham NR, Beals RS and Nyberg EA. Warrendale: The Minerals, Metals and Materials Society, TMS, pp. 153–158.

Stanford N and Barnett MR (2008) 'The origin of "rare earth" texture development in extruded Mg-based alloys and its effect on tensile ductility', *Materials Science and Engineering A*, **496**, 399–408.

Stolbov VI, Eltsov, VV, Oleinik IA and Matyagin VF (1991) 'Effect of the nature of thermal processes on cracking of repair welding components of magnesium alloys', *Welding International*, **5**, 799–802.

Stutz L, Bohlen J, Letzig D and Kainer KU (2011) 'Formability of magnesium sheet ZE10 and AZ31 with respect to initial texture', in *Magnesium Technology 2011*, edited by Sillekens W, Agnew SR, Mathaudhu SN and Neelameggham NR. Warrendale: The Minerals, Metals and Materials Society (TMS), pp. 373–378.

Sun H, Liang S and Wang E (2009) 'Mechanical properties and texture evolution during hot rolling of AZ31 magnesium alloy', *Transactions of Nonferrous Metals Society of China*, **19**, 349–354.

Takuda H, Enami T, Kubota K and Hatta N (2000) 'The formability of a thin sheet of Mg-8.5Li-1Zn alloy', *Journal of Materials Processing Technology*, **101**, 281–286.

Wagoner RH, Chan KS and Keeler SP (1989) *Forming Limit Diagrams: Concepts, Methods and Applications*, Oxford: The Minerals, Metals and Materials Society (TMS).

Watanabe H, Mukai T and Ishikawa K (2004) 'Differential speed rolling of an AZ31 magnesium alloy and the resulting mechanical properties', *Journal of Materials Science*, **39**, 1477–1480.

Wendt J, Kainer KU, Arruebarrena G, Hantzsche K, Bohlen J and Letzig D (2009) 'On the microstructure and texture development of magnesium alloy ZEK100 during rolling', in *Magnesium Technology 2009*, edited by Nyberg EA, Agnew SR, Neelameggham NR and Pekguleryuz MO. Warrendale: The Minerals, Metals and Materials Society (TMS), pp. 289–293.

Yi S, Bohlen J, Heinemann F and Letzig D (2010) 'Mechanical anisotropy and deep drawing behaviour of AZ31 and ZE10 magnesium alloy sheets', *Acta Materialia*, **58**, 592–605.

Yim CD, Seo YM and You BS (2009) 'Effect of the reduction ratio per pass on the microstructure of a hot-rolled AZ31 magnesium alloy sheet', *Metals and Materials International*, **15**, 683–688.

Yukutake E, Kaneko J and Sugamata M (2003) 'Anisotropy and non-uniformity in plastic behaviour of AZ31 magnesium alloy plates', *Materials Transactions*, The Japan Institute of Metals (JIM), **44**, 452–457.

11
Forging technology for magnesium alloys

B.-A. BEHRENS, I. PFEIFFER and J. KNIGGE, Leibniz Universität Hannover, Germany

Abstract: This chapter discusses forging technology, and how forged components may be superior to die cast. Forming behaviour is described, and the most important magnesium alloys are introduced. The forging process, including heat treatment and lubrication, is reviewed, with a description of finite element analysis and simulation of the process, and a sample forged component.

Key words: magnesium, alloy, forging technology, grain structure, stress, deformation twin, finite element analysis, near net shape forming.

11.1 Introduction

Compared to all technically relevant light metals, magnesium and its alloys exhibit a good combination of low specific weight and high specific strength. These advantages promote ambitious efforts to extend their use into different fields of application. Today most magnesium applications are based on casting processes. However, the mechanical properties of castings are unsatisfactory for their use in safety applications, for example. To achieve improved material properties forging is a promising alternative production technology. With their high requirements concerning weight reduction and safety, the aerospace and automotive industries offer a wide range of applications for magnesium forgings.

11.2 Forging technology

Compared to alternative manufacturing methods, forgings have outstanding material and component characteristics. Forged parts possess a high mechanical and dynamic strength due to process-related grain refinement and the unbroken grain flow. Furthermore, they are almost free of pores and shrink holes. This results in a higher mechanical load capacity of the parts and hence advances the trend for lightweight construction (Fig. 11.1).

Due to their material characteristics and their load-adapted scope for design, forgings are excellently applicable for components for force transmission such as gears, chassis parts, wheels and suspension arms, as well as steering and brake parts, for example.

The manufacturing method of forging can basically be divided into open die forging and drop forging. In order to increase the forming capacity of the material

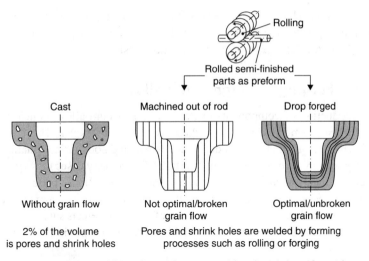

11.1 Comparison of fibre flow of cast, machined and drop forged parts. (Source: Behrens and Doege 2010.)

and to reduce stresses and strengths, forging is usually carried out at elevated temperatures.

Open die or hammer forging is the oldest forging method. Today it is mainly used for the fabrication of preforms for drop forging processes or for the production of large forgings. In contrast to drop forging, the forged parts are manufactured without limiting dies. The forging geometry evolves from a suitable guiding of the forging between the dies.

The relevant process for magnesium forging is drop forging. It can be subcategorized into drop forging with and without flash (Fig. 11.2). During drop

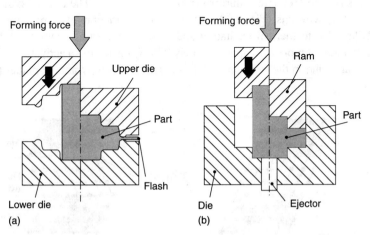

11.2 Drop forging. (a) With flash. (b) Without flash. (Source: Behrens and Doege 2010.)

forging with flash the workpiece is enclosed by the die, and the excessive material can run off at the so-called flash gap. Flashless drop forging is performed in a closed die. A discharge of the material is not possible.

11.3 Forging of magnesium alloys

The production of components made from magnesium alloys is nearly entirely done by smelt-metallurgical procedures, such as die casting. These procedures yield complex geometries, but due to brittleness and poor surface quality, inferior mechanical properties may occur (Mertz 1999). In many cases these parts cannot meet the mechanical requirements. For these applications, parts produced by forming processes represent an alternative. Due to their load-adapted fiber flow, the fine-grained structure and the absence of structure defects such as segregations or pores, forged parts exhibit considerably improved properties.

11.3.1 Principles of the forming behaviour of magnesium and its alloys

Magnesium crystallizes in an hcp lattice structure. Below a temperature of approximately 225 °C sliding occurs solely in the basal plane of the hcp structure during a forming process. The exact temperature lies between 200–225 °C and depends on the alloy composition. Consequently, magnesium and its alloys behave in a brittle manner at room temperature and have a poor formability up to a temperature of approximately 225 °C.

At room temperature and with compressive stress orthogonal to the basal plane there is another forming mechanism apart from slip on the basal plane, called deformation twinning. This means that a shear strain causes a transformation of a microstructure area into a mirror-symmetric condition. The axis of symmetry is called the twin axis (Fig. 11.3). At increasing temperatures the atoms start to oscillate due to thermal excitation and show an increased mobility. A forming temperature of more than approximately 225 °C results in an activation of further slip planes and in the creation of deformation twins on pyramidal planes (Bauser

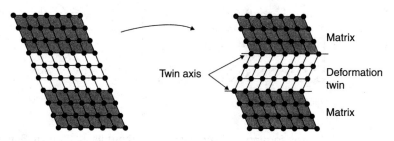

11.3 Principle of deformation twins.

et al. 2001, Reed-Hill and Robinson 1957). This causes the dislocations to cross slip so that obstacles may not only be climbed over but also circumvented. This results in a significant improvement of the material's formability.

11.3.2 Important magnesium alloys for magnesium forging technology

Magnesium alloys are basically categorized into casting and wrought alloys. As alloying elements, aluminium (A), manganese (M), rare earths (E), yttrium (W), zirconium (K) and zinc (Z) are used. These elements increase the strength, heat resistance and corrosion resistance of magnesium alloys. Regarding cast alloys, the most important characteristics are related to mould filling and castability whereas with wrought alloys the focus is on good plastic formability (Becker and Fischer 1999, Kammer 2000). An overview of the most relevant wrought alloys and their properties is given in Table 11.1.

The most relevant alloys for the production of magnesium forgings are AZ80, ZK30, ZK60 and WE alloys. AZ alloys show a good combination of strength and ductility as well as a good corrosion resistance as high purity (hp) alloys. They are particularly suitable for the production of extrusions as well as for simple forgings. In forming technology AZ alloys are used in preference to other alloys. The reasons for this are their favourable material properties as well as their relatively low material costs. The production costs of WE alloys are significantly higher than those of AZ alloys. WE alloys can be used for complex forgings or for applications requiring high-temperature strength and ductility. Compared to AZ alloys, ZK alloys show a better formability at high temperatures. Despite the limited number of available wrought alloys, a wide range of properties can be covered by adequate heat treatment of the material (Becker and Fischer 1999, Kammer 2000, Boehm 2005).

11.3.3 Forging process design

During forging process design the adjustment of forging parameters has to be carried out with reference to material characteristics and procedural demands.

Table 11.1 Magnesium wrought alloys for extrusions and forgings

Alloy	Properties
AZ31, AZ61, AZ80	Extrusion and forging alloys with good combination of strength and ductility and with excellent corrosion resistance as high-purity alloy
ZK30, ZK60	Excellent forgeability
WE43, WE54	Extrusion and forging alloys with good high-temperature strength

Source: Becker and Fischer 1999.

Simulation and calculation processes based on finite element analysis (FEA) are commonly used for the design of forming processes. The most important basis for using these FEA tools is the knowledge of material characteristics and process parameters.

Material characterization

High stresses lead to a elastic and plastic deformation of metallic materials. A specific stress, the flow stress, is needed to initiate and maintain the forming process. Flow stress, as the fundamental factor, is given a major significance within the field of forging technology, since it also serves as a basis for the calculation of forces as well as the amount of energy required.

For drop forging processes the cylinder compression test is a useful laboratory-scale test procedure to determine the material characteristics concerning the formability of metallic material at different forming temperatures, forming velocities and different microstructure. These characteristics provide a basis for numerical simulation processes by FEA, for example. At the Institute of Forming Technology and Machines, Hanover, Germany, the required material data are determined with a servo-hydraulic forming simulator (Fig. 11.4). This has a maximum load of 400 kN and is able to realise a stamp velocity between quasi-static and 2.2 m/s. It is possible to realise a constant strain rate that is required for the material data for finite element method by reducing the tappet speed in a logarithmic manner during the upsetting tests.

The procedure is an open die upsetting test with a material flow perpendicular to the stamp movement. The standardized specimens have a diameter of 11 mm and a height of 18 mm in most cases, and are placed between two upsetting tools within a thermal container. The container guides the tools in a parallel manner and guarantees a constant testing temperature during the forming procedure. The friction between the tools and the specimen has to be reduced as far as possible with a suitable lubricant. With this system, compression tests with a constant forming temperature above and below room temperature can be carried out.

Simulation-based forging process design by finite element analysis (FEA)

The increasing demands on forged products, where increasing complexity of the work piece geometry, combined with the increased cost of higher pressure, make it necessary to pay special attention to process optimization concerning the forging sequences and material flow in the tool.

The application of process simulation can contribute significantly to this optimization. Numerical simulation especially has a high potential to support the design and optimization of forming processes. The finite element method (FEM) has established itself for the modelling of numerous forming processes. This is a mathematical method for the numerical computation of partial differential

Forging technology for magnesium alloys 381

11.4 Servo-hydraulic forming simulator.

equations, and is ideal for the calculation of numerous physical-technological tasks. In forming technology it is mainly applied to the simulation of the material flow and for the determination of stress, temperature and strain distributions (Behrens and Doege 2010). By means of this method it is thus possible to gain deeper knowledge of the later stages and already in the development stage and thus to avoid mistakes in the planning and design stage. FE simulation also helps to decrease the amount of expensive and complex 'trial and error'. Apart from purely economic aspects, numerical simulation is the only scientific option to analyse a process at reasonable cost and to identify the influence of several forming parameters at the same time.

For the simulation of processes with extreme plastic workpiece deformation, such as forging, a number of commercial FEM programs are available. Among others, the software systems MSC.MARC™, Simufact.forming™, Transvalor Forge™ and Deform™ have been applied successfully. However, the simulation results can only match physical reality if all input parameters have been accurately defined in advance. Thus, the biggest challenge for numerical simulation is to ensure a realistic process description. Therefore the following aspects are currently the main foci of research into applications of FEM in metal forming (Alasti 2008, Kurz 2004):

- Contour-close description of the geometries by means of very fine discretization and application of new elements and improved element formulation,
- Implementation of temperature, stress and strain dependent material parameters,
- Complex material modelling for considering all effects occurring within the material during the forming process, such as strain hardening,
- Application of modern friction models, which take into account, *inter alia*, the surface properties of the contact partners, the current local pressures, temperatures, and the current relative velocity between the work piece and the tool surface,
- Considering the machine influences and dynamic effects during forming in the simulation model.

The highest potential for improving simulation results for forging processes is to be found in material modelling. It is a big challenge to obtain material parameters for implementation in the simulation, especially for a material such as magnesium, and there is a great need for research (Behrens and Doege 2010).

Process chain

In this section the process steps for magnesium forging such as cutting, heat treatment, heating, the forging process itself and lubrication are presented.

Raw materials for the forging of magnesium alloys are usually extruded, or sometimes rolled. Continuous cast material can be used for the production of heavy and compact parts. In this case a sufficient kneading of the basic raw material has to be guaranteed in order to weld the pores and casting defects.

The raw magnesium parts for forging are usually cut from the rod by high-speed cutting. Due to the brittleness of magnesium and a large fracture surface, cold shearing is not an alternative for the production of raw parts. A defect-free surface has high priority, since surface defects may cause part failure (Kammer 2000).

By means of heat treatment before or after the forging process the part properties can be improved for special applications or further processing without having a negative impact on other properties such as corrosion resistance or fracture resistance. The aims of heat treatment are the generation of a homogeneous microstructure, the reduction of concentration differences between grain boundaries and the inner part of the grain, and the dissolution of eutectic microstructure ingredients at the grain boundaries. Additionally, supersaturated dissolved elements, which could influence the recrystallization behaviour and the formability at high temperatures, should be eliminated (Boehm 2005, Kammer 2000).

An example of an important heat treatment procedure is the soft annealing of raw parts. During this process a work-hardened microstructure is transformed into a microstructure with good formability by means of annealing. This procedure is a recrystallization annealing process, which for magnesium is carried out at 300 °C.

For cast raw materials, homogenization can be required. During solidification of cast parts, phases of different chemical compositions occur in the material. Forging processes require a homogeneous microstructure that can be obtained by a homogenization annealing process. Sometimes the forging process and the thermomechanical treatment are accompanied by a homogenization so that a separate homogenization step is not required (Kammer 2000).

Temperature control is of vital importance for forging of magnesium alloys. If the temperature is too low, the formability of the material is decreased. The lower limit of the forging temperature corresponds to the recrystallization temperature. If the temperature is too high, initial fusing at grain boundaries, coarse grain development or hot cracking at the forged part can occur. The resulting forging temperature depends on the heat lost due to cooling and the energy input during deformation.

Due to the high heat conductivity of magnesium, handling and contact time, which lead to a fast temperature decrease after the heating process, should necessarily be as short as possible. All tool components coming into contact with the material during the forging process have to be heatable. This is especially important during the forging of delicate magnesium parts, as the high heat conductivity leads to a heat loss during material flow and thus to a decrease of formability (Kammer 2000).

Due to their high-temperature strength the following materials are established for magnesium forging tools:

- 1.2344 (X40CrMoV5-1)
- 1.2365 (32CrMoV12-28)
- 1.2367 (X38CrMoV5-3)
- 1.2714 (56NiCrMoV7).

The forging temperature range for different magnesium alloys is strictly limited. It depends on the respective alloy and is $T = 290$–420 °C for AZ and ZK and $T = 420$–480 °C for WE alloys (Avedesian and Baker 1999, Becker 2002, Kleiner 2002).

In general, in view of the economic aspects and also for metallurgical reasons, the process temperature during the forming of magnesium parts should be as low as possible. During formation of the hcp magnesium crystals the basal planes align with the flow direction. This causes anisotropy of mechanical properties, which is typical for magnesium forming processes. This anisotropy gradually decreases with lower forming temperatures (Beck 2001).

The heating of the raw parts is carried out in gas or electric furnaces. Due to the reasons mentioned above it is essential that the temperature of the raw parts does not exceed the specified forging temperature, in spite of the benefits from the additional forming energy. At usual forging temperatures an inert gas atmosphere in the furnace is not necessary. The raw parts have to be heated up as homogeneously as possible and therefore for magnesium forging processes, circulating air furnaces are currently preferred (Kammer 2000).

Generally for the production of magnesium forgings, hydraulic presses are used. With these presses it is possible to achieve low forming velocities. This supports process-accompanying regeneration and recrystallization of the microstructure and thus the formability of the material. A high forming velocity may lead to localised temperature fluctuation above the part temperature, due to the applied energy of deformation. This happens particularly in areas of high relative movement between tool and part. This may cause phase boundary melting or grain growth, which leads to a weakening of the part. Furthermore, hydraulic presses provide complete process control, in terms of forming velocity and forming force, as well as the ability to keep the maximum load constant for a specified time. This reduces the danger of crack initiation. For that reason, for the forging of parts with complex geometries, superior cavity filling can be provided (Becker and Fischer 1999, Doege *et al.* 2000).

The tribological system is of great importance for magnesium forging processes. Common lubricants for magnesium forging processes are water-based or oil-based graphite suspensions, or oil-based lubricants without graphite. After the application of the lubricant the carrier medium vaporizes, leaving an even lubricating film. On very hot forging tools a graphite–water suspension often fails to produce an even lubrication film so that a graphite–oil suspension is required. Another possibility is a direct application of the lubricant to the raw parts prior to their insertion in the die. For complex and dedicated tool cavities, wax can be used (Kammer 2000).

11.4 Near-net-shape forming of magnesium alloys

Precision forging is an innovative manufacturing process for highly stressable near-net-shape components. It belongs to the category of drop forging and is performed flashless in a closed die. In exceptional cases fabrication tolerances up to IT 7 can be achieved by precision forging (Behrens and Doege 2010). Due to the fact that precision forging is a flashless process, machining of the part usually required after forging can be avoided.

11.4.1 Examples of near-net-shape forming

Pulley wheels

Within the scope of this chapter the design of a forging process for the production of pulley wheels is described. These parts exhibit a complex shape with thin- as well as thick-walled segments, which is challenging for the forging process design (Fig. 11.5). In Fig. 11.6 the tool concept for the production of the pulley wheels is depicted.

11.5 Pulley wheels made of AZ31 and AZ61. (Source: Behrens and Pfeiffer 2007.)

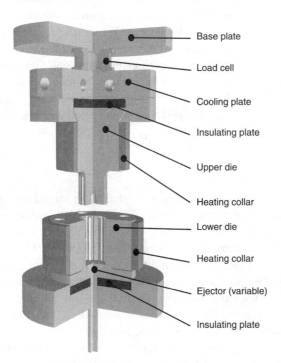

11.6 Tool concept for production of pulley wheels. (Source: Behrens and Pfeiffer 2007.)

The pulley wheels were forged in a closed die in a one-step process. By using different ejectors this single tool offers the possibility of achieving different cross-sections of the forged components (Fig. 11.5). To heat both the upper and the lower die pieces respectively to a process temperature of 350 °C, heating sleeves are adjusted. The heating coils are embedded in a ceramic base and can be flexibly arranged around the die. The constant temperature control is carried out using thermocouples. Insulating boards are used to avoid heat transfer between the die, the bed plate and the ram. Cylindrically extruded raw parts are placed into the heated lower forging die and are formed through the closing of the press. After the forging process the component is ejected from the lower die and immediately cooled with water.

During the forging process of the pulley wheels, long flow paths and high frictional forces occur. In order to determine optimized process parameters such as die temperature, bulk temperature, forming speed or choice of lubricant, a macroscopic evaluation of the cavity filling, filling accuracy and surface quality, material analysis and metallographic analysis of the flow characteristics were performed.

The pulley wheels are manufactured with graphite-free molybdenum sulphite spray (MoS_2). Due to the deep die cavity the lubricant is applied from an aerosol can. This process guarantees a complete and consistent surface of lubricant throughout the forging die, at die temperatures of 350 °C. During manufacture of the pulley wheels the ejector has to be well lubricated, since the components cool down very fast after the forging process and may shrink on the ejector.

Forging temperatures for AZ31 have been varied between 380–400 °C and for AZ61 between 330–400 °C. The raw parts were heated in small lots in a circulating air furnace for 15 minutes. This heating procedure prevents long exposure times and structural changes to the material. The temperature of the forging dies was constant $T_{WZ} = 350$ °C. Manufacture of the pulley wheels was carried out by a hydraulic press (press capacity $F_{St} = 12\,500$ kN). The ram speed was approximately $v_{St} = 30$ mm/s (Behrens and Pfeiffer 2007).

Door stop fitting

The forging process for the door stop fitting, presented as the second example for magnesium forging processes, was developed through the MagForming project, which was funded by the European Commission within the sixth framework programme, priority four, aeronautics and space. The objective of MagForming was to advance the state of the art in seven different technologies of plastic and super-plastic forming of magnesium wrought alloys for aeronautical applications by developing methodologies and tools for industrial applications, and showing their feasibility in aeronautics. The MagForming consortium consisted of ten companies and two universities from Europe and Israel.

In the scope of MagForming, design of the forging processes and the production of the door stop fittings made of AZ80 were tasks for the Institute of Forming Technology and Machines in Hanover, Germany.

In order to build the required material data files for the FEA of the forging process, the first step was the determination of flow curves for AZ80 magnesium alloy. Upsetting tests at temperatures between 280–450 °C were carried out on a servo-hydraulic forming simulator as described above. Specimens with a diameter of 11 mm and a height of 18 mm were tested, varying the parameters of strain rate and temperature. By means of the detected load and path of the ram the required flow curves were measured and the true stress, true strain dependence on temperature and strain rate were determined (Behrens and Knigge 2010).

CAD models of the forging dies were then developed using 3D models of the forging parts. Due to the elevated temperature of billets and tools (400 °C) the temperature-related shrinking of the part following the forging process was considered; therefore the dies were scaled up. Geometries of the forging billets were developed. The initial models of dies and billets were transferred to FEA software.

The first result of the FE analysis of the door stop fitting process was that a one-step forging process with a single die set was possible. With subsequent FE analysis the design process for both tools was supported in order to receive information about optimal die and billet temperatures, the required forming force and the forming speed. Geometry of the raw part was adapted to the magnesium forging process. Based on these investigations the die geometry was determined.

The forging die was constructed in a tool guiding frame to ensure an optimal part quality within close manufacturing tolerance. Therefore, adaptation parts and insulating plates were designed for the connection of dies and tool guiding frame. The required heating sleeves were determined for a maximum die temperature of 400 °C. Figure 11.7 shows the forging tool for the door stop fitting within a 4000 kN hydraulic press.

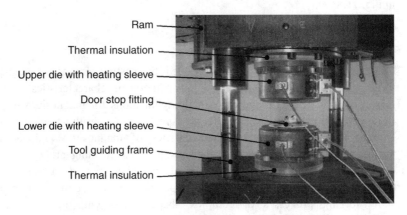

11.7 Door stop fitting forging tool. (Source: Behrens and Knigge 2010.)

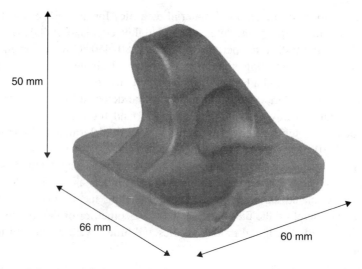

11.8 Door stop fitting made of AZ80.

For the heating of the billets a circulating air furnace was used. To determine the billet temperature a reference billet equipped with thermocouples on the surface and in the centre was heated at the same time in the furnace.

To investigate the influence of temperature and ram speed the parts were forged with different parameters in a first step. Die and billet temperatures were varied in a range from 300–400 °C. The ram speed was varied between 10–30 mm/s. Due to the complex shape, the optimization of the forging process of the door stop fitting was challenging. Low temperatures led to incomplete form filling. At a higher temperature the forging part stuck to the tool. Finally, the part was forged at a temperature of 300 °C. The forging part of the door stop fitting is depicted in Fig. 11.8.

11.5 Conclusions

Compared to magnesium parts produced by casting processes magnesium forgings show improved material and component characteristics. For that reason there is a wide range of possible applications for magnesium forgings. In this chapter an overview of different aspects of magnesium forging is given. Due to the special material characteristics of magnesium and its alloys the heating of the raw parts and the design of the forming process itself are of vital importance for the component characteristics. Two examples of pulley wheels and door stop fittings respectively made of magnesium alloys show forging processes of complex shaped parts with good mechanical properties.

11.6 Acknowledgements

The authors would like to thank the German Research Foundation (DFG) and the European Commission for the financial support of the presented studies.

11.7 References

Alasti M (2008) 'Modellierung von Reibung und Waermeuebergang in der FEM-Simulation von Warmmassivumformprozessen', Dissertation, Hannover.
Avedesian M and Baker H (1999) *ASM Specialty Handbook Magnesium and Magnesium Alloys*. ASM International, USA.
Bauser M, Sauer G and Siegert K (2001) *Strangpressen*, Duesseldorf, Aluminium-Verlag.
Beck A (2001) *Magnesium und seine Legierungen*, Berlin, Springer.
Becker J and Fischer G (1999) Strangpress- und Schmiedeerzeugnisse aus Magnesium – sicheres und leistungsfaehiges Halbzeug für den Leichtbau; proceedings 'Umformtechnik an der Schwelle zum naechsten Jahrtausend', 16: Umformtechnisches Kolloquium Hannover.
Becker J (2002) 'Stranggepresste und geschmiedete Magnesiumblech-Bauteile: Legierung, Herstellung, Eigenschaften und Anwendungen', *Zukunftsorientierter Einsatz von Magnesium im Automobilbau*, Muenchen.
Behrens B-A and Pfeiffer I (2007) 'Geschmiedete Komponenten aus Magnesium – optimierte Eigenschaften durch Anpassung der Prozessparameter'. *Metall – Internationale Fachzeitschrift für Metallurgie*, Band 61.
Behrens B-A and Doege E (2010) *Handbuch der Umformtechnik*, Berlin, Springer.
Behrens B-A and Knigge J (2010) 'EU project Magforming: Development of a magnesium forging process for aeronautical applications', *Magnesium, International Conference on Magnesium Alloys and Their Applications*, Weinheim, Wiley-VCH.
Boehm R (2005) *Entwicklung anwendungsoptimierter Magnesiumlegierungen und ihre Verarbeitung*, Berichte aus dem IW, Hannover.
Doege E, Droeder K and Elend LE (2000) *Umformverhalten von Magnesiumfeinblechen*, proceedings Neuere Entwicklungen in der Blechumformung, MAT-INFO Werkstoff-Informationsgesellschaft mbH, Frankfurt.
Kammer C (2000) *Magnesium Taschenbuch*, Duesseldorf, Aluminium Verlag.
Kleiner S (2002) 'Magnesium und seine Legierungen', *6. internationales IWF-Kolloquium Feinstbearbeitung technischer Oberflaechen*, Egerkingen, Schweiz, 18–19 April 2002.
Kurz G (2004) 'Temperiertes hydromechanisches Tiefziehen von Magnesiumblechen', Dissertation, Hannover.
Mertz A (1999) 'Karosseriestrukturen aus Strangpressprofilen und Mg-Druckguss', *proceedings Fortschritte mit Magnesium im Automobilbau*; Bad Nauheim.
Reed-Hill RE and Robertson WD (1957) *Trans. AIME*, **209**, 496.

Part III
Applications of magnesium alloys

12
Applications of magnesium alloys in automotive engineering

A. A. LUO and A. K. SACHDEV, General Motors Global Research and Development, USA

Abstract: This chapter discusses the material properties and mass-saving potential of magnesium alloys in comparison with major automotive materials: mild steel, advanced high strength steel (AHSS), aluminium, polymers and polymer composites. The alloy development and manufacturing processes of magnesium extrusion and sheet products are summarized. The opportunities and challenges of wrought magnesium alloys for automotive applications are discussed.

Key words: magnesium alloy development, magnesium extrusion, magnesium sheet, forming processes, automotive applications.

12.1 Introduction

Magnesium components are increasingly being used by major automotive companies including General Motors (GM), Ford, Volkswagen and Toyota.[1–8] Current major automotive magnesium applications include instrument panel beams, transfer cases, steering components, and radiator supports. However, the magnesium content in a typical family sedan built in North America is only about 0.3% of the total vehicle weight.[1] While high-pressure die casting is the dominant manufacturing process for current automotive magnesium applications, wrought magnesium alloys and their manufacturing processes are receiving increasing attention from academia and industry. As magnesium is expanding into more critical automotive structural applications, there is a great need for developing wrought magnesium products to provide improved mechanical and physical properties, crash performance and corrosion resistance.

Most critical is the improvement of the plasticity of wrought magnesium, which becomes dramatically reduced as the various working methods used to create the primary fabrications, such as sheet and extrusions, rotate the grains to less favorable orientations for isotropic plasticity. As a result, secondary forming processes such as stamping and tube hydroforming need to be executed at warm temperatures to promote non-basal slip to occur so that the larger strains targeted can be accommodated. This need increases manufacturing costs and puts magnesium at a disadvantage compared with traditional metals. Further, since final performance occurs at ambient temperatures, the plasticity of magnesium at these temperatures remains paramount to their performance in structural applications. Finally, the high electrochemical activity of magnesium has limited its application due to concerns over galvanic corrosion. Mitigating solutions are either still not robust enough or too expensive for high-volume manufacturing. Similar to magnesium plasticity, the use of magnesium for exterior exposed panels, requiring a high degree of finish for their appearance, remains an equally challenging situation.

In this chapter, the mechanical properties, structural performance and mass-saving potential of cast and wrought magnesium alloys are compared with those of several major automotive materials: mild steel, advanced high strength steel (AHSS), aluminium, polymers and polymer composites. The formability and manufacturing processes including welding and joining of magnesium wrought products are critically reviewed, and current and potential applications of extrusions and sheet products in the automotive interior, body, chassis and powertrain areas are identified and technical challenges discussed.

12.2 Materials properties

12.2.1 Property comparison with other automotive materials

Table 12.1 summarizes the mechanical and physical properties of typical cast and wrought magnesium alloys in comparison with other materials for automotive applications.[9-12] Being the lightest structural metal, magnesium has a density less than one-fourth that of ferrous alloys (cast iron, mild steel and AHSS) and offers similar mechanical and physical properties as aluminium alloys, but with about one-third mass saving. Extrusion and sheet alloys, for example alloy AZ80 and AZ31, respectively, provide comparable tensile strength to aluminium extrusion alloy 6061 and commonly used 5XXX and 6XXX sheet alloys, but are less ductile. The wrought magnesium alloys are also much less formable than steel or aluminium at room temperature due to their hcp (hexagonal close-packed) crystal structure, although their tensile ductilities often appear reasonable.

While magnesium components are slightly heavier than polymers or polymer composites, they are generally stiffer due to their higher elastic modulus. A plastic blend of polycarbonate and acrylonitrile-butadiene-styrene (PC/ABS), for example, has a modulus of about 1/20 that of magnesium and has been used in a limited way in instrument panel beams where the design requirements are met through the geometry of the closed sections. Fiber reinforcements can be used to increase the modulus of polymers, as seen in many polymer-based composites such as the conventional glass-fiber reinforced polymer (GFRP) and advanced carbon fiber reinforced polymers (CFRP). Depending on applications, polymer composites can be classified into two groups:

- Structural composites (structural-GFRP and structural-CFRP) in which glass or carbon fibers are uniaxially orientated to provide unidirectional strength/ stiffness for structural applications.

Table 12.1 Comparison of mechanical and physical properties of various automotive materials[9–12]

Material	Cast Mg		Wrought Mg		Cast iron	Steel		Cast Al		Wrought Al		Polymers (PC/ABS)	GFRP[b] (glass/polyester)			CFRP[c] (carbon/epoxy)		
Alloy/grade	AZ91	AM50	AZ80-T5	AZ31-H24	Class 40	Mild steel Grade 4	AHSS[a] DP340/600	380	A356-T6	6061-T6	5182-H24/6111-T8X[e]	Dow Pulse 2000	Structural (50% uniaxial)	Exterior (27%)		Structural (58% uniaxial)	Exterior (60%)	
Process/product	Die cast	Die cast	Extrusion	Sheet	Sand cast	Sheet	Sheet	Die cast	P/M[d] cast	Extrusion	Sheet	Injection molding	Liquid moulding	Compression molding		Liquid/compression molding	Autoclave moulding	
Density (d, g/cm³)	1.81	1.77	1.80	1.77	7.15	7.80	7.80	2.68	2.76	2.70	2.70	1.13	2.0	1.6		1.5	1.5	
Elastic modulus (E, GPa)	45	45	45	45	100	210	210	71	72	69	70	2.3	48	9		189	56	
Yield strength (YS, MPa)	160	125	275	220	N/A	180	340	159	186	275	235/230	53	1240	160		1050	712	
Ultimate tensile strength (S_t, MPa)	240	210	380	290	293	320	600	324	262	310	310/320	55						
Elongation (e_f, %)	3	10	7	15	0	45	23	3	5	12	8/20	5 at yield and 125 at break	<1	2		<1	<1	
Fatigue strength (S_f, MPa)	85	85	180	120	128	125	228	138	90	95	120/186		N/A	N/A		N/A	N/A	

(Continued)

Table 12.1 Continued

Material	Cast Mg		Wrought Mg		Cast iron	Steel		Cast Al		Wrought Al		Polymers (PC/ABS)	GFRP[b] (glass/polyester)		CFRP[c] (carbon/epoxy)	
													Structural	Exterior	Structural	Exterior
Alloy/grade	AZ91	AM50	AZ80-T5	AZ31-H24	Class 40	Mild steel Grade 4	AHSS[a] DP340/600	380	A356-T6	6061-T6	5182-H24/ 6111-T8X5	Dow Pulse 2000	(50% uniaxial)	(27%)	(58% uniaxial)	(60%)
Process/ product	Die cast	Die cast	Extrusion	Sheet	Sand cast	Sheet	Sheet	Die cast	P/M[d] cast	Extrusion	Sheet	Injection molding	Liquid moulding	Compression molding	Liquid/ compression molding	Autoclave moulding
Thermal cond. (l, W/m.K)	51	65	78	77	41	46		96	159	167	123		0.6	0.3	0.5	0.5
Melting temperature (T_m, °C)	598	620	610	630	1175	1515		595	615	652	638/585	143 (softening temp.)	130–160 (molding temp.)		175 (max. service temp.)	

[a] AHSS: advanced high strength steel. [b] GFRP: glass fiber reinforced polymer. [c] CFRP: carbon fiber reinforced polymer. [d] P/M: permanent mould. [e] T8X: simulated paint-bake (2% strain plus 30 min. at 177°C).

- Exterior composites (exterior-GFRP and exterior-CFRP) in which glass or carbon fibers are appropriately aligned to provide 'quasi-isotropic' strength/stiffness in planar directions but not in the thickness direction, for exterior panel applications.

Exterior-GFRP is used in many low/medium-volume body panel applications (e.g., Corvette). CFRP, on the other hand, is much more expensive and is today generally used in aerospace and performance cars; a recent GM application of exterior CFRP is the hood outer for the Commemorative Corvette Z06. Recently, however, there has been a greater push to use CFRP in automotive body and closures, but an extensive discussion of this is outside the scope of this chapter. Polymers and composites are prone to creep, and are thus not suitable for elevated temperature applications due to their low service temperatures (e.g., 143 °C for PC/ABS and 175 °C for CFRP). Compared with aluminium, magnesium has a higher thermal expansion coefficient and lower thermal conductivity, which needs to be considered when substituting for aluminium in elevated temperature applications.

12.2.2 Structural performance and mass-saving potential

Materials selection for automotive structural applications is an extremely complex process in which component geometries, loading conditions, material properties, manufacturing processes and cost provide conflicting requirements.[3,12,13] As the bending mode is often the primary loading condition in many automotive structures such as instrument panel beams and frame rails, the following analyses on structural performance and mass-saving potential of magnesium over mild steel (presently the dominant automotive material) are based on bending stiffness and strength calculations.

For a panel (plate) under bending loads, the minimum thickness (t) and mass (m) can be calculated using the 'materials performance index' concept.[13] Designating steel and magnesium properties with subscripts S and Mg, the thickness ratios and mass ratios of components made of the two materials for an equal stiffness design may be expressed as:

$$t_{Mg}/t_S = (E_S/E_{Mg})^{1/3} \qquad [12.1]$$

$$m_{Mg}/m_S = (d_{Mg}/d_S)(E_S/E_{Mg})^{1/3} \qquad [12.2]$$

where E and d are the elastic modulus and density of the materials, respectively. Using the property data as shown in Table 12.1, the thickness and mass ratios of magnesium (AZ91 alloy) compared with a mild steel beam can be calculated:

$$t_{Mg}/t_S = 1.67 \qquad [12.3]$$

$$m_{Mg}/m_S = 0.39 \qquad [12.4]$$

Therefore, in order to achieve the same bending stiffness, a magnesium panel will be required to have 1.67 times the thickness of a steel one, with a mass saving of 61%. For bending strength-limited design (same bending strength at minimum mass), such ratios for AZ91 magnesium alloy compared with mild steel become:

$$t_{Mg}/t_S = (YS_S/YS_{Mg})^{1/2} = 1.06 \qquad [12.5]$$

$$m_{Mg}/m_S = (d_{Mg}/d_S)(YS_S/YS_{Mg})^{1/2} = 0.25 \qquad [12.6]$$

where YS is the yield strength of the materials.

For a beam in an equal stiffness design, Eqs 12.1 and 12.2 became:

$$t_{Mg}/t_S = (E_S/E_{Mg})^{1/2} \qquad [12.7]$$

$$m_{Mg}/m_S = (d_{Mg}/d_S)(E_S/E_{Mg})^{1/2} \qquad [12.8]$$

Similarly, for a solid beam in an equal strength design, Eqs 12.5 and 12.6 became:

$$t_{Mg}/t_S = (YS_S/YS_{Mg})^{2/3} \qquad [12.9]$$

$$m_{Mg}/m_S = (d_{Mg}/d_S)(YS_S/YS_{Mg})^{2/3} \qquad [12.10]$$

Based on the above nomenclature and the property data in Table 12.1, Table 12.2 summarizes the thickness and mass ratios of various materials compared with mild steel for solid panel and beam designs, respectively. For a generic comparison, the uniaxial properties of structural composites are not included in the analysis. Instead, the 'quasi-isotropic' properties of exterior GFRP and CFRP composites are compared with the isotropic properties of other materials for mass-saving analysis. Figure 12.1 and Fig. 12.2 highlight the thickness ratios and the resultant percentage mass savings of the materials when replacing a mild steel component. These results show that magnesium alloys have higher mass-saving potential when compared with AHSS, aluminium and polymers, in substituting for mild steel structures for equal stiffness or strength. To overcome its much lower elastic modulus, a polymer part will have to be reinforced with fibers, or a metal-polymer hybrid structure has to be used. Compared to GFRP composites, magnesium alloys have higher mass-saving potential for equal stiffness and similar savings for equal strength. While CFRP has the highest mass-saving potential, wrought magnesium alloys provide slightly less mass savings, but at a lower cost.

12.3 Alloy development

12.3.1 Commercial extrusion alloys

Table 12.3 lists the nominal composition and typical room-temperature tensile properties of extruded magnesium alloy tubes.[9–11] Of the commercial extrusion

Table 12.2 Thickness and mass ratios of various materials compared with mild steel for equal bending stiffness- and strength-limited design

Material		AHSS	Al (cast)	Al (wrought)	Mg (cast)	Mg (wrought)	PC/ABS	GFRP (exterior)	CFRP (exterior)
Equal stiffness panel	Thickness ratio	1.00	1.44	1.45	1.67	1.67	4.50	2.86	1.55
	Mass ratio	1.00	0.49	0.50	0.39	0.39	0.65	0.59	0.30
Equal strength panel	Thickness ratio	0.73	1.06	0.81	1.06	0.81	1.84	1.06	0.50
	Mass ratio	0.73	0.37	0.28	0.25	0.19	0.27	0.22	0.10
Equal stiffness beam	Thickness ratio	1.00	1.72	1.74	2.16	2.16	9.56	4.83	1.94
	Mass ratio	1.00	0.59	0.60	0.50	0.50	1.38	0.99	0.37
Equal strength beam	Thickness ratio	0.65	1.09	0.75	1.08	0.75	2.26	1.08	0.40
	Mass ratio	0.65	0.37	0.26	0.25	0.17	0.33	0.22	0.08

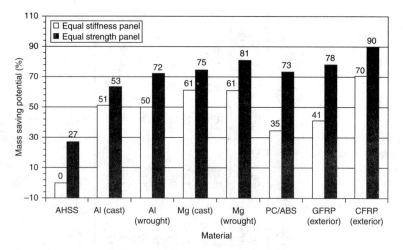

12.1 Percentage mass savings of various materials vs mild steel for designing a structural panel with equivalent bending stiffness or bending strength. AHSS: advanced high strength steel; PC/ABS: polycarbonate/acrylonitrile-butadiene-styrene; GFRP: glass-fiber reinforced polymer; CFRP: carbon fiber reinforced polymer.

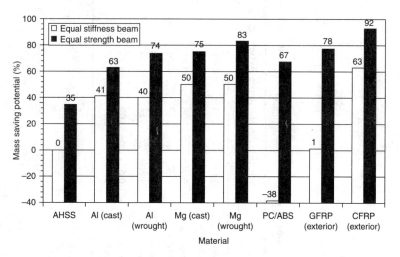

12.2 Percentage mass savings of various materials vs mild steel for designing a structural beam with equivalent bending stiffness or bending strength.

alloys, AZ31 is most widely used in non-automotive applications. With higher aluminium contents, AZ61 and AZ80 offer higher strength than AZ31 alloy, but with much lower extrudability. The high-strength Zr-containing ZK60 was designed for applications in racing cars and bicycles, such as wheels and stems.[10] The extrusion speed for making ZK60 tubes is extremely low, rendering it

Table 12.3 Nominal composition and typical room-temperature tensile properties of extruded magnesium alloys[9–12]

Alloy	Temper	Composition (wt%)				Tensile properties		
		Al	Zn	Mn	Zr	Yield strength (MPa)	Ultimate tensile strength (MPa)	Elongation (%)
AZ31	F*	3.0	1.0	0.20		165	245	12
AZ61	F	6.5	1.0	0.15		165	280	14
AZ80	T5	8.0	0.6	0.30		275	380	7
ZK60	F		5.5		0.45	240	325	13
ZK60	T5†		5.5	–	0.45	268	330	12
AM50	F	5.0		0.30	–	168	268	18

* F signifies as extruded. † T5 signifies artificially aged after extrusion.

uneconomical for automotive applications. The maximum extrusion speed of AZ31 alloy, the most extrudable commercial magnesium alloy, is only about half that of aluminium extrusion alloy 6063, which makes magnesium extrusions much more expensive due to the higher material and processing costs. New magnesium alloys are being developed to improve the extrusion speed while maintaining good mechanical properties.

12.3.2 New extrusion alloys

This section introduces two experimental extrusion alloys developed at General Motors for automotive structural application: AM30 (Mg-3%Al-0.3%Mn) alloy for high-strength applications and ZE20 (Mg-2%Zn-0.2%Ce) alloy for high-ductility applications.

AM30 alloy

A new experimental extrusion alloy, AM30 (Mg-3%Al-0.3%Mn), was recently developed with improved extrudability and formability.[11] Aluminium also improves strength, hardness and corrosion resistance among the alloying elements considered[9] but reduces ductility. An aluminium content of about 5–6% yields the optimum combination of strength and ductility for structural applications. Increasing aluminium content widens the freezing range and makes the alloy easier to cast, but more difficult to extrude due to increased hardness. For example, alloys containing less than 3% Al can be extruded at higher extrusion speeds compared with high-strength alloys such as AZ61 (Mg-6%Al-1%Zn) and ZK60 (Mg-6%Zn-0.5%Zr).[10] To maximize the ductility and extrudability, while maintaining reasonable strength and castability (for billet casting prior to extrusion), an aluminium content of 3% was selected for the new alloy.

Zinc is next to aluminium in effectiveness as an alloying ingredient to strengthen magnesium.[9] However, it reduces ductility and increases hot-shortness of Mg-Al-based alloys. Zinc-containing magnesium alloys are prone to microporosity.[14] Zinc was also reported to have mild to moderate accelerating effects on corrosion rates of magnesium as determined by alternate immersion in 3% NaCl solution.[9] Therefore, unlike most commercial magnesium alloys, Zn was not selected for this experimental alloy. Manganese does not have much effect on tensile strength, but it does slightly increase the yield strength of magnesium alloys. Its most important function is to improve the corrosion resistance of Mg-Al-based alloys by removing iron and other heavy-metal elements into relatively harmless intermetallic compounds, some of which separate out during melting. For this purpose, Mn is added at about 0.4% as recommended by the ASTM Specification B93–94a.

Based on these analyses and experiments, a new magnesium alloy, AM30 (Mg-3%Al-0.4%Mn), was formulated. The extrudability of an alloy billet is defined by its maximum extrusion speed at which the billet can be pushed through an extrusion die at a given temperature without causing visible surface defects, such as cracking or fracture.[11] Compared with the current workhorse commercial magnesium wrought alloy AZ31 (Mg-3%Al-1%Zn), the new AM30 alloy can be extruded 20% faster (Fig. 12.3), has a 50% increase in room-temperature ductility

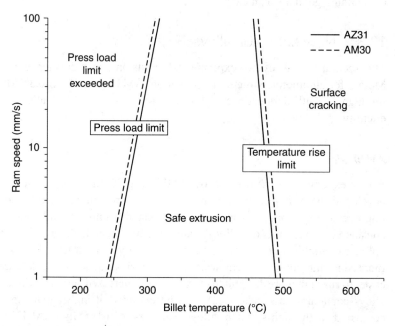

12.3 Extrusion limit diagram for AM30 and AZ31 alloys.[11]

with similar strength, and has up to 30% improvement in ductility at elevated temperatures up to 200 °C (Fig. 12.4).

Mg-Zn-Ce (ZE) alloys

It was recently found that a small addition of only 0.2% Ce to pure Mg significantly improved the ductility of extruded bars (Fig. 12.5) due to a change in texture that favoured basal slip. However, the strength of the Mg-0.2%Ce alloy remained too low for automotive structural applications. A follow-on investigation[16] examined the addition of aluminium (Al) to Mg-0.2%Ce binary alloy, which improved its strength but decreased its ductility considerably. This is due to the fact that Ce has

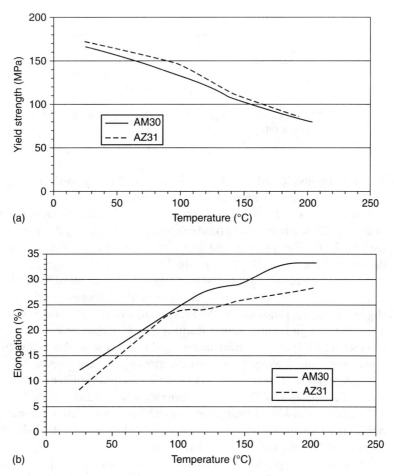

12.4 Effect of temperature on tensile properties of extruded tubes of AM30 and AZ31 alloys.[11] (a) Tensile yield strength. (b) Tensile elongation.

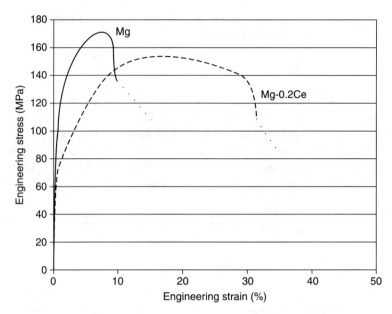

12.5 Tensile curves for pure Mg and Mg-0.2%Ce alloy, after extruding bars with a ratio of 25:1.[15]

a higher affinity for Al and forms $Al_{11}Ce_3$ in the Mg-Al-Ce ternary alloys, offsetting the beneficial effect of the Ce addition.

Zinc additions of 2–6%, on the other hand, significantly improved the strength of the Mg-0.2%Ce alloy, due to solid solution strengthening by Zn, while retaining the beneficial effect of Ce on randomizing the texture and the associated high ductility.[17] Clearly, the ZE20 alloy (Mg-2%Zn-0.2%Ce) has significantly higher strength compared with Mg-0.2%Ce alloy (135 MPa vs 69 MPa yield strength and 225 MPa vs 170 MPa ultimate tensile strength). Although the ZE20 alloy has slightly lower elongation (27.4% vs 31%), compared with the binary Mg-0.2%Ce alloy, it has significantly higher ductility (27.4%) compared with 16.9% for commercial AZ31 alloy; a 62% increase. This increase is obtained with only a minor reduction of about 16% in tensile strength. As also shown in Fig. 12.6, increasing the Zn content from 2% to 8% increased the ultimate tensile strength of the Mg-Zn-Ce (ZE) alloy, but the elongation was reduced considerably. The ZE20 alloy is considered most promising and has a somewhat lower strength but much higher ductility compared with AM30 (170 MPa yield strength, 240 MPa ultimate tensile strength and 12% elongation).

The extrusion experiments on ZE alloys show that ZE20 alloy has excellent extrudability, about 25% higher maximum extrusion speed compared with the AZ31 alloy (0.08 m/s) at an extrusion temperature of 400 °C, and is similar to that of AM30, which also showed about 20% better extrudability than AZ31.[11] When

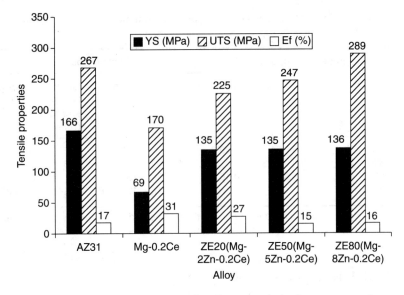

12.6 Tensile properties of Mg-Zn-Ce alloy extruded tubes compared with AZ31 and Mg-0.2%Ce alloy tubes processed in similar conditions.[8]

the Zn content is increased to 5% (ZE50 alloy), the extrudability is reduced to the level for AZ31. A further increase in Zn content to 8% (ZE80 alloy) results in a dramatic decrease in extrudability (0.02 m/s). Higher Zn contents have been reported to increase surface cracking and oxidation during extrusion and, hence, to lower the extrusion speed limits.[18]

12.3.3 Commercial sheet alloys

There are three types of commercial sheet magnesium alloys: Mg-Al-Zn (AZ), Mg-Th (HK and HM), and Mg-Li-Al (LA) alloys, which are summarized in Table 12.4. Alloy AZ31 is the most widely used alloy for sheet and plate and is available in several grades and tempers. It can be used at temperatures up to 100 °C. The HK31 and HM21 alloys are suitable for use at temperatures up to 315 °C and 345 °C, respectively. However, HM21 has superior strength and creep resistance.

Alloys with very low density and good formability have been developed by adding lithium to magnesium. For example, LA141A, which contains 14% Li, has a density of 1.35 g/cm^3, or only 78% that of pure magnesium. This alloy has found limited use in missile and aerospace applications because of its ultra-low density and excellent formability at room temperature or slightly elevated temperatures.[19]

Table 12.4 Nominal composition and typical room-temperature tensile properties of magnesium sheet alloys[9,19]

Alloy	Temper	Composition, wt%					Tensile properties		
		Al	Zn	Mn	Zr	Other	Yield strength (MPa)	Ultimate tensile strength (MPa)	Elongation (%)
AZ31	H24*	3.0	1.0	0.20			180	325	15
HK31	H24			3.0	0.3	0.7	3.25 Th	160	285	9
HM21	T8†			0.45		2.0 Th	130	270	11
LA141	T7‡	1				14 Li	105–160	132–165	11–24

* H24 signifies strain hardened and partially annealed. †T8 signifies solution heat treated, cold worked, and artificially aged. ‡T7 signifies solution heat treated and stabilized.

12.3.4 New sheet alloys

As demonstrated in the commercial LA141A alloy, the Mg-Li alloy system is very promising for high formability, and can potentially be formed at room temperature. Similar to the extrusion alloys developed, the Mg-Zn-Ce alloy system has also been shown to provide significant improvement in tensile ductility and sheet formability. Microalloying has been found to be effective in refining Mg-Al-based sheet alloys for formability enhancement and property improvement.

Mg-Li-based alloys

According to the Mg-Li phase diagram (Fig. 12.7), the alloys exhibit two phase structures of α (hexagonal close packed (hcp)) Mg-rich and β (body-centered cubic (bcc)) Li-rich phases at room temperature when 5–11% Li is added to magnesium. Further additions of Li (more than 11%) can transform the hcp α-Mg solid solution into highly workable, body-centered cubic alloys.[20] It has been shown that the rolling textures of Mg-14% Li alloy sheet with a bcc structure are very similar to other bcc metals such as interstitial-free (IF) steel, ferritic steel and Fe-3%Si when they are similarly processed.[21] However, the binary Mg-Li alloys offer poor corrosion resistance, and limited mechanical properties (especially low fatigue strength) at room and elevated temperatures. Alloy development efforts have been on the effect of third alloying elements (Al, Zn, Si, RE, etc.) on the mechanical properties of Mg-Li alloys.[22–26] While aluminium improves the creep resistance and mechanical properties at elevated temperatures, Mg-Li-Al alloys have poor corrosion resistance due to the existence of the chemically active face-centered cubic (fcc) LiAl phase in the microstructure, in addition to the β (bcc) phase solid solution. Rare earth additions can improve the formability and chemical and thermal stability of Mg-Li-Al alloys, especially when the Li content

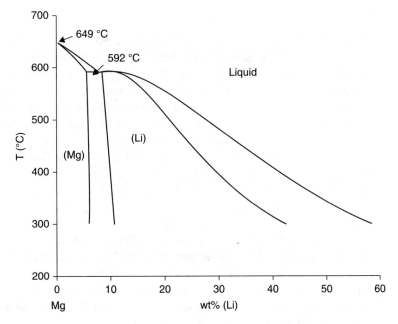

12.7 Mg-Li phase diagram.[9]

is less than 6% and the alloys are in the α-Mg single phase region, due to the formation of Al_2RE and Al_4RE phases in the microstructure.

The microstructure and formability of Mg-Li-Zn alloy sheets were studied and the results are summarized in Table 12.5.[27] The Mg-6%Li-1%Zn alloy with dominant α-Mg (hcp) phase has limited ductility and formability, while the Mg-12%Li-1%Zn alloy with dominant β-Li (bcc) microstructure has high ductility at comparatively low strain rates. On the other hand, the Mg-9.5%Li-1%Zn alloy sheet with an (α + β) two-phase microstructure has a better combination of tensile properties and formability compared with the alloys with higher or lower Li contents.

Table 12.5 Microstructure and tensile properties of Mg-Li-Zn alloys[27] obtained from uniaxial tension tests for an initial stain rate of 8.3×10^{-4} s^{-1}

Alloy	Mg-6%Li-1%Zn	Mg-9.5%Li-1%Zn	Mg-12%Li-1%Zn
Microstructure	α-Mg (hcp)	α-Mg + β-Li	β-Li (bcc)
Yield strength (MPa)	112	121	124
Ultimate tensile strength (MPa)	155	134	125
Elongation (%)	32.2	71.4	56.0
Work-hardening exponent, n	0.15	0.06	0.00
Normal anisotropy parameter, r	5.98	0.87	0.70

Mg-Zn-Ce (ZE) alloys

Similar to ZE20 extrusion alloy (Section 12.3.2.), a Mg-1.5%Zn-0.2%Ce sheet alloy was reported[28] to provide high-tensile elongation (around 30%) and improved formability, attributed to the TD (transverse direction)-split texture produced when the alloy was rolled at 450 °C and annealed at 350 °C. However, the strength of this alloy is still considered low (yield strength <120 MPa and ultimate tensile strength <210 MPa) for many automotive structural applications.

The conventional Mg-Al-based alloy sheets generally show strong basal texture, in which the basal plane is parallel to the rolled-sheet surface. The basal and prismatic slip can operate parallel to the rolling direction and width direction, but not in the thickness direction. Therefore, the rolled sheets hardly deform in the thickness direction, resulting in premature fracture at an initial stage of forming at room temperature, thus the poor formability.

A Mg-Zn alloy with a small addition of rare earth (such as Ce) processed by hot rolling has significantly different texture, with the basal poles at 35° from the thickness direction as shown in Fig. 12.4.[19] As a result, the alloy sheet can readily deform in the thickness direction, thus significant improvement in room temperature formability. The unique texture in the Mg-Zn-Ce alloy is attributed to the activation of prismatic slip by the Ce addition.

Microalloying of Mg-Al-based alloys

Various alloying elements, such as Sr, Ti, Ce, Ca, Sb, Sn and Y, have been studied aimed at improving the mechanical properties and formability of Mg-Al-Zn alloys. These alloying elements, either individually or in combination, form additional phases and/or refine the grain structure of the alloys.

Small additions of Sr (0.01–0.1%) and Ti (0.01–0.03%) can effectively decrease the grain size of the AZ31 alloy billets,[18] promising improved formability and mechanical properties in the resultant sheet or extrusion products. Additions of up to 1% Ce or about 0.1% Ca can refine the grains of AZ31 alloy, and small additions of Sb and Sn can improve the strength of AZ alloys due to the formation of precipitation phases Mg_3Sb_2 and Mg_2Sn.[29] An addition of 0.7% Y to AZ31 was reported to modify the $Mg_{17}Al_{12}$ intermetallic phase to more rounded particles, leading to better properties in the sheet samples.[29]

12.4 Manufacturing process development

12.4.1 Extrusion and forging processes

Conventional extrusion processes

Magnesium alloys can be warm- or hot-extruded in hydraulic presses to form bars, tubes, and a wide variety of profiles.[9] It is well known that a tubular section is significantly stiffer than a solid beam of the same mass. While hollow magnesium extrusions can be made with a mandrel and a drilled or pierced billet, it is generally preferable to use a bridge die where the metal stream is split into several branches which recombine before the die exit.

Hydrostatic extrusion process

Hydrostatic extrusion processing, typically used for copper tubing fabrication, is a much faster extrusion process compared with the conventional direct extrusion. It was reported that seamless magnesium tubes were extruded using the hydrostatic process at speeds up to 100 m/min, due to the absence of friction between the billet and container since the billet is suspended in hydraulic oil.[30] Although the process is capable of extrusion ratios up to 700, the outer diameter of tubes produced by this process is limited to about 45 mm, even with a large 4000-ton press.[30]

Forging processes

Commercial extrusion alloys listed in Table 12.3 (AZ31, AZ61, AZ80 and ZK60), can also be forged into high-integrity components using hydraulic presses or slow-action mechanical presses. Forging is normally done within 55 °C of the solidus temperature of the alloy. Corner radii of 1.6 mm, fillet radii of 4.8 mm, and panels or webs 3.2 mm thick can be achieved by forging. The draft angles required for extraction of the forgings from the dies can be held to 3° or less.[9]

12.4.2 Tube bending and hydroforming processes

Tube hydroforming

Tube hydroforming is a metal forming process that uses pressurized fluids such as water to make various perimeter shapes from tubes. Compared with stampings and castings, hydroformed tubular sections provide further mass savings for structural components. While hydroformed steel and aluminium tubes are currently used in many structural applications including frame rail, engine cradle, radiator support and instrument panel beam,[31] hydroforming of magnesium alloy tubes is not yet developed.

Tube bending

Tube bending is generally needed as a pre-form step for hydroforming automotive parts, but the bendability of magnesium extrusions at room temperature is very limited.[32] A moderate temperature (150–200 °C) bending process has been developed at GM for magnesium alloy extrusions.[33] A bend radius twice the tube outer diameter (2D) was achieved on magnesium alloy extrusions at 150 °C (Fig. 12.8). The mechanical properties and microstructure of magnesium alloys at

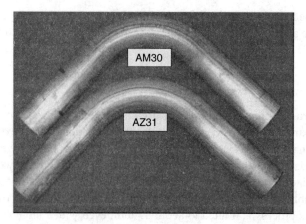

12.8 Magnesium AM30 and AZ31 tubes bent (2D/90°) at 150 °C.[33]

elevated temperatures indicate that moderate temperature hydroforming and other forming processes at this temperature range (150–200 °C) should be explored.[34]

12.4.3 Sheet production and forming processes

Sheet production processes

The direct-chill (DC) casting process is generally used to cast magnesium slabs (about 50 mm thick), which are then hot-rolled at 315–370 °C to produce magnesium sheet and plates.[9] Unlike aluminium, for which a cold-roll is usually the final step in sheet production, a warm finish roll is applied to magnesium sheet products. Large grains of 200–300 µm in the slab can be reduced by warm rolling to fine recrystallized grains between 7–22 µm for a sheet gauge of 1.3–2.6 mm.[35,36] Recently, the twin-roll continuous casting (CC) process has been investigated for the production of low-cost magnesium sheet. Pilot plants have been established in Germany, Austria and Australia,[37–39] and production plants are being built in China and Korea. In this process, molten magnesium is poured into a gap between two rolls to produce a continuous strip about 2.5–10 mm thick, which is then warm-rolled to a final gauge. Currently, magnesium sheet using this process is available from Salzgitter (Germany) with a maximum width of 1850 mm and a minimum gauge of 1.0 mm.[40]

Sheet forming processes

The majority of the processes used to convert sheet metal into automobile components occur at room temperature, including stamping, flanging, bending, hemming, and trimming. The processes are very robust for high-formability materials such as mild steel but, with some concessions on draw depth and corner radii, have been successfully used with less formable materials such as aluminium and high-strength steel. Unfortunately, the limited formability of magnesium due to its hcp structure (discussed above) makes the use of these processes very difficult. An example of the problem is shown in Fig. 12.9, where the results of a forming trial using a simple rectangular pan for a mild steel and AZ31B magnesium sheet are compared. The pan could be formed to a depth of 125 mm with the steel (Fig. 12.9a), but split after only about 12 mm with the magnesium (Fig. 12.9b).[41]

Warm stamping (200–400 °C) of magnesium sheet has been used to make complex products for aerospace and luggage components, and the optimum temperature for warm stamping was reported as approximately 350 °C.[36] Fig. 12.9(c) shows that magnesium alloy sheet can be stamped at reasonably moderate temperatures of 150–175 °C,[42,43] similar to these for tube bending temperatures.[33] Sheet hydroforming differs from the conventional stamping process in that the solid punch (upper die) or female (bottom) die is replaced with a forming medium. When the female die is replaced with a fluid or a pad of flexible polyurethane, it is also called hydro-mechanical drawing.[44,45] Another variant, called active sheet hydroforming, is to use a forming medium instead of a solid punch to directly press the blank against a die contour.[46] The optimum conditions for hydro-mechanical drawing magnesium sheet were reported to be temperatures of 180–220 °C and pressures of 600–800 bar.[43] Obviously, forming at lower temperatures would be more economical and yield much better

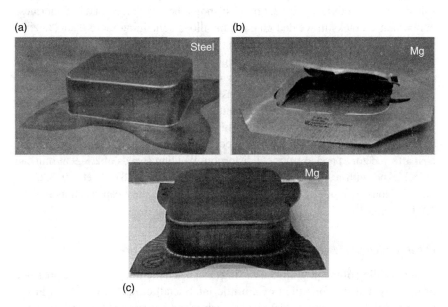

12.9 Comparison of a forming trial on a 125 mm deep pan with: (a) mild steel formed at room temperature; (b) AZ31B magnesium formed at room temperature; (c) AZ31B magnesium formed at 350 °C.

dimensional accuracy due to less thermal contraction and distortion during cooling to room temperature.

Superplastic forming (SPF) is a gas-forming process using one-sided tools (such as sheet hydroforming) based on the superplasticity of many materials (aluminium, titanium, magnesium, etc.) at elevated temperatures and under controlled strain rates. SPF is often used to fabricate large and complex aluminium parts in the aerospace industry. Extensive research and development at GM on aluminium SPF has shortened the process cycle time to an acceptable level for automotive panels, leading to the development of a quick plastic forming (QPF) process.[47] QPF has been implemented to produce aluminium decklids for the Cadillac STS. Recently, the QPF process has been used to produce many prototype magnesium inner closure panels at a forming temperature of about 475 °C.[48] Examples of QPF prototype products are shown in Fig. 12.12.

12.4.4 Welding and joining techniques

Welding processes

Various welding processes can be used for joining magnesium to magnesium. Most magnesium alloys are readily fusion-welded with higher speeds than aluminium due to the lower thermal conductivity and latent heat of magnesium.[9,49–51] While gas metal arc welding (MIG or GMAW) is used for joining magnesium sections ranging from 0.6–25 mm, gas tungsten arc welding (TIG or GTAW) is more suited for thin sections up to 12.7 mm.[9] It should be recognized that hot-shortness may produce cracks in welded magnesium alloys containing more than 1% zinc, which can often be overcome by using proper filler wires: ER AZ61A is the preferred filler wire for welding wrought alloys containing aluminium, while ER AZ91A has been found to lower crack sensitivity in AZ and AM cast alloys.[51] Magnesium sheet and extrusions ranging from 0.5–3.3 mm can be joined by all types of resistance welding, including seam, projection, and flash, but the most common type is spot welding.[9] A higher welding speed for thin magnesium sheet can be achieved with laser or electron beam welding, due to the narrow heat affected zones and lower weld distortion. Welding speeds of 2.5–9 m/min can be achieved with a 4 kW solid-state laser for welding AZ31 sheet 1.0–3.2 mm thick.[51] Non-vacuum electron beam welding of AZ31 alloy can reach as high as 12–15 m/min.[51]

Other joining processes

Fusion welding of magnesium die castings can be challenging due to the presence of porosity and the formation of a brittle intermetallic phase ($Mg_{17}Al_{12}$) in the welds.[49] Solid-state welding techniques, such as friction-stir welding and magnetic pulse welding can be used to improve the weld quality of magnesium die castings.[52–54] While these solid-state welding techniques can potentially be used to join magnesium to dissimilar materials such as aluminium,[52,55] mechanical joining (self-piercing rivets, clinching, and hemming) and adhesive bonding are preferred for dissimilar material joining involving magnesium to aluminium or steel. Adhesive bonding of magnesium parts (with or without dissimilar materials) requires proper pre-treatment of the joint surfaces, which includes cleaning, etching and wet chemical passivation. Many adhesives including epoxy and polyurethane can be used as long as they are chemically stable and have aging stability.[56]

12.5 Automotive applications of magnesium alloys

12.5.1 Historical applications

Wrought magnesium has been used as a structural material in the transportation industry since World War II. The most famous application was the B-36 bomber, which contained 5500 kg (12 200 lbs) of magnesium sheet and 680 kg (1500 lbs) of magnesium forgings, along with 300 kg (660 lbs) of magnesium castings.[57] The first commercial ground transportation applications were developed in the early 1950s. Metro-lite trucks were manufactured between 1955 and 1965 and featured magnesium sheet panels as well as structures made with magnesium plate and extrusions.[58] These trucks had increased payload capacity and were excellent applications for magnesium because they did not require extreme formability. However, wrought magnesium has not been used in high-volume production in the mainstream automobile industry.

12.5.2 Current and future applications

Wrought magnesium alloys are used in aerospace, nuclear, luggage, hand tools, bicycle and motorcycle applications, but there have been no reported applications of magnesium extrusions in the automotive industry. The only current production application of magnesium sheet is the centre console in the low-volume Porsche Carrera as shown in Fig. 12.10. Table 12.6 summarizes the potential applications of magnesium extrusions in automotive interior, and body and chassis areas, while powertrain magnesium applications will remain castings.

Interior

Since corrosion is of less concern in the interior, this area has seen the most applications of magnesium, with the biggest growth in the instrument panels (IP) and steering structures. The first magnesium IP beam was die cast by GM in 1961 with a mass saving of 4 kg over the same part cast in zinc. The design and die casting of magnesium IP beams have advanced dramatically in the last decade. For example, current IPs normally have a thickness of 2–2.5 mm (compared with

12.10 Sheet magnesium centre console cover in Porsche Carrera GT automobile.[41]

Table 12.6 Potential automotive applications of wrought magnesium components

System	Component	Note
Interior	Instrument panel	Extrusion/sheet
	Seat components	Extrusion/sheet
	Trim plate	Sheet
Body	Door inner	Sheet
	Tailgate/liftgate inner	Sheet
	Roof frame	Extrusion
	Sunroof panel	Sheet
	Bumper beam	Extrusion
	Radiator support	Extrusion
	Shotgun	Sheet/extrusion
	A and B pillar	Sheet
	Decklid/hood inner	Sheet
	Hood outer/fender	Sheet
	Decklid/door outer	Sheet
	Dash panel	Sheet
	Frame rail	Extrusion
Chassis	Wheel	Forging
	Engine cradle	Extrusion
	Subframe	Extrusion
	Control arm	Forging

4–5 mm for the earlier IP beam application) with more part consolidation and mass savings. However, the use of cast magnesium IP beams has recently been facing strong competition. IP beams made of aluminium extrusions are used by Mercedes in Europe. IP designs using bent steel tubes (with or without hydroforming) are slightly heavier than magnesium die casting, but significantly less expensive. To maintain and grow its use in IP production, magnesium design and thin-wall casting technology must continue to improve, further reducing weight and cost. Tubular designs using magnesium extrusions and sheet components should also be explored.

The use of magnesium seat structures began in Germany, where Mercedes used magnesium die castings in its integrated seat structure with a three-point safety belt in the SL Roadster.[59] The latest example in North America is the 'Stow-n-Go' seating and storage system for the Chrysler and Dodge minivans, where the folding mechanisms require light weight for easy operation; thus, some aluminium is used in the second-row seats, and the back frame of the third-row folding seats is a magnesium casting.[60] Compared to magnesium castings, wrought magnesium provides further mass-saving opportunities for seat applications. Research is needed to reduce the cost of wrought magnesium and its forming processes.

Body

The use of magnesium in automotive body applications is limited but has expanded recently. GM has been using a one-piece die cast roof frame since the introduction of the C-5 Corvette in 1997. Magnesium is also used in the Cadillac XLR roadster's retractable hard-top convertible roof and the roof-top frame. The Ford F-150 has coated magnesium castings for its radiator support. In Europe, Volkswagen and Mercedes have pioneered the use of thin-wall magnesium die castings in body panel applications. The one-piece die cast door inner for the Mercedes S-Class Coupe is only 4.56 kg.[61] The key to manufacturing these thin-wall castings (approximately 1.5 mm) lies in casting design using proper radii and ribs for smooth die filling and to stiffen the parts. These thin-wall die castings, such as door inners, A and B pillars, can often offset the material cost penalty of magnesium over steel sheet metal construction due to part consolidation.

While production applications are limited, numerous prototype components have been made using sheet magnesium. General Motors made prototype hoods for the Buick LeSabre in 1951, various body panels for the Chevrolet Corvette SS Race Car in 1957, and hoods for the Chevrolet Corvette in 1961 (Fig. 12.11). Recently, Volkswagen made a prototype hood for the Lupo a few years ago.[39] GM has made numerous panels including a hood, door inner panel, decklid inner, liftgate, and various reinforcements, some of which are shown in Fig. 12.12.[37,48,62] Chrysler LLC has performed a number of studies using magnesium sheet, including the inner panel and a magnesium-intensive body structure.[63]

12.11 (a) 1951 Buick LeSabre concept car with magnesium and aluminium body panels; (b) 1961 Chevrolet Corvette with prototype hood made from magnesium sheet; (c) 1957 Chevrolet Corvette SS race car with 'featherweight magnesium body'.[41]

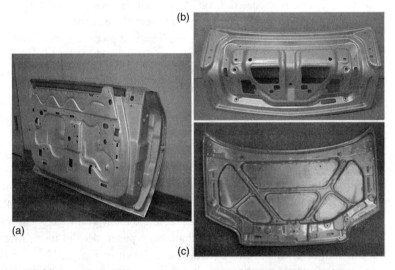

12.12 Magnesium sheet panels formed recently by General Motors. (a) Door inner panel;[37] (b) decklid inner panel;[62] (c) hood.[48]

Applications of magnesium alloys in automotive engineering 417

In body panel applications where bending stiffness is frequently the design limit, magnesium sheet metal can offer as much as 61% mass saving (see Fig. 12.1). The majority of the applications discussed above were 'inner' panels, which create the structure of the vehicle closures but are not visible on the outside of the vehicle. This is due to two factors. First, the surface quality of the currently available magnesium sheet requires significant finishing compared with aluminium or steel, and second, the limited formability at room temperature makes assembly processes for outer panels, such as hemming, difficult. This means it is likely that the first commercial applications for magnesium sheet will be inner panels.[41] However, elevated-temperature forming of magnesium and corrosion protection coatings further impose a cost penalty for magnesium sheet applications. The development of new sheet alloys for near-room temperature forming is needed, along with low cost and robust coatings for corrosion resistance.

Similar to magnesium sheet, magnesium extrusions for automotive applications are still in the development stage, with many prototype parts, such as bumper beams and frame rail for the Volkswagen 1-Liter Car.[42] However, the use of magnesium tubes/extrusions in body applications would require more development to meet all structural and cost targets, as well as a supply base for high-volume automotive production.

Chassis

Cast or forged magnesium wheels have been used in many high-priced race cars or high-performance roadsters including GM's Corvette. However, the relatively high cost and potential corrosion problems of magnesium wheels prevented their use in high-volume vehicle production. The first-in-industry one-piece magnesium die cast cradle for the Chevrolet Corvette Z06 weighs only 10.5 kg, and demonstrates a 35% mass savings over the aluminium cradle it replaced.[64]

Aluminium tubes are presently used in high-volume cradle production such as the welded extrusion design for GM's mid-size cars and the hydroformed tubular subframe for the BMW 5 and 7 series. Tubular designs are generally more mass-efficient than solid castings. Cradles using magnesium tubes would offer significant mass savings. The development of low-cost, corrosion-resistant coatings and new magnesium alloys with improved fatigue and impact strength will also accelerate the further penetration of magnesium in chassis applications.

12.6 Future trends

While magnesium is the lightest structural metal and the third most commonly used metallic material in automobiles following steel and aluminium, many challenges remain in various aspects of alloy development and manufacturing processes to exploit its high strength-to-mass ratio for widespread lightweight applications in the automotive industry.

12.6.1 Material challenges

Compared with the numerous aluminium alloys and steel grades, there are only a limited number of low-cost wrought magnesium alloys available for automotive applications. The conventional Mg-Al-based alloys offer moderate mechanical properties due to the limited age-hardening response of this alloy system. New alloy systems with a potential for precipitation hardening such as Mg-Sn[65,66] and Mg-RE[67] should be developed with improved mechanical properties. Also, the strong plastic anisotropy and tension-compression asymmetry due to texture remain obstacles for many structural applications. Alloys systems such as Mg-Li and Mg-RE have shown more 'isotropic' mechanical properties. Computational thermodynamics and kinetics will be used to design and optimize these new alloys.

12.6.2 Process challenges

Various forming processes need to be optimized for magnesium alloys. Elevated temperature forming is needed for most extrusion and sheet components. Research efforts have been directed to lowering the forming temperatures and reducing the cycle times. New forming processes should be developed to utilize the dramatically improved formability of magnesium alloys at certain ranges of temperature and strain rate. Room temperature (RT) or near-RT forming techniques are also being explored for new magnesium alloys such as Mg-Zn-Ce alloys.[68]

12.6.3 Performance challenges

There are several performance-related challenges that need significant research efforts. Some of these are highlighted in the current Canada-China-USA Magnesium Front End Research & Development project.[69]

Crashworthiness

Magnesium castings have been used in many automotive components such as the instrument panel beams and radiator support structures. High-ductility AM50 or AM60 alloys are used in these applications and performed well in crash simulation and tests, and many vehicles with these magnesium components achieved five-star crash ratings. However, there is limited material performance data available for component design and crash simulation. A recent study showed that magnesium alloys can absorb significantly more energy than either aluminium or steel on an equivalent mass basis.[70] While steel and aluminium tubes fail by progressive folding in crash loading (a more desirable situation), magnesium alloy (AZ31 and AM30) tubes tend to fail by sharding or segment fracture.[70,71] However, the precise fracture mechanisms for magnesium under crash loading are still not clear,

Applications of magnesium alloys in automotive engineering

and material models for magnesium fracture are needed for crash simulation involving magnesium components. Additionally, new magnesium alloys need to be developed to have progressive folding deformation in crash loading.

Noise, vibration and harshness (NVH)

It is well known that magnesium has a high damping capability, but this can be translated into better NVH performance only for mid-range sound frequencies; 100–1000 Hz.[63] The low-frequency (<100 Hz) structure-borne noise can be controlled by the component stiffness between the source and receiver of the sound. The lower modulus of magnesium, compared with steel, is often compensated by thicker gauges and/or ribbing designs. For high-frequency (>1000 Hz) airborne noise, a lightweight panel, regardless of material, would transmit significantly more road and engine noise into the occupants' compartment unless the acoustic frequencies could be broken up and damped. Magnesium, with its low density, is disadvantaged for this type of application unless new materials with laminated structures are developed for sound isolation.

Fatigue and durability

Fatigue and durability are critical in magnesium structural applications and there is limited data in the literature, especially for wrought alloys. The effect of alloy chemistry, processing and microstructure on the fatigue characteristics of magnesium alloys need to be studied. Extrusion and sheet products need to be characterized sufficiently to establish links between microstructural features and fatigue behavior. Multi-scale simulation tools can be used to predict the fatigue life of magnesium components and sub-systems, which can be validated for automotive applications.

Corrosion and surface finishing

Pure magnesium has the highest standard reduction potential of the structural automotive metals (Table 12.7). As noted earlier, while pure magnesium (at least with very low levels of iron, nickel, and copper) has atmospheric corrosion rates that are similar to that of aluminium, magnesium's high reduction potential makes it very susceptible to galvanic corrosion when it is in electrical contact with other metals below it in the reduction potential table. The impact of this susceptibility to galvanic corrosion on the application of magnesium in exposed environments is severe in both the macro-environment and the micro-environment. In the macro-environment, magnesium alloys must be electrically isolated from other metals to prevent the creation of galvanic couples: e.g. steel bolts cannot be in direct contact with magnesium. Isolation can be achieved by replacing the bolt with a less reactive metal, as has been done in the Mercedes automotive transmission case where steel bolts have been replaced with aluminium bolts.[72] Isolation can also be achieved by coating the 'other' metal. Finally, isolation can be achieved by the use of shims or spacers of compatible materials of sufficient geometry and size to prevent electrical contact in the presence of salt water, as shown, for example, for the Corvette cradle (Fig. 12.13).[64] While the component cost can be competitive with aluminium, the isolation strategies required can often make the application more expensive and thus restrictive in its use.

Table 12.7 Standard reduction potential of common metals[9]

Electrode	Reaction	Potential (V)
Li, Li$^+$	Li$^+$ + e$^-$ → Li	−3.02
K, K$^+$	K$^+$ + e$^-$ → K	−2.92
Na, Na$^+$	Na$^+$ + e$^-$ → Na	−2.71
Mg, Mg^{2+}	Mg^{2+} + 2e$^-$ → Mg	−2.37
Al, Al^{3+}	Al^{3+} + 3e$^-$ → Al	−1.71
Zn, Zn^{2+}	Zn^{2+} + 2e$^-$ → Zn	−0.76
Fe, Fe^{2+}	Fe^{2+} + 2e$^-$ → Fe	−0.44
Cd, Cd^{2+}	Cd^{2+} + 2e$^-$ → Cd	−0.40
Ni, Ni^{2+}	Ni^{2+} + 2e$^-$ → Ni	−0.24
Sn, Sn^{2+}	Sn^{2+} + 2e$^-$ → Sn	−0.14
Cu, Cu^{2+}	Cu^{2+} + 2e$^-$ → Cu	0.34
Ag, Ag$^+$	Ag$^+$ + e$^-$ → Ag	0.80

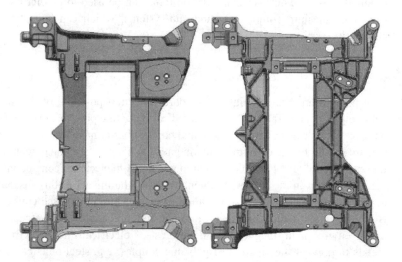

12.13 Aluminium isolator locations for Chevrolet Corvette Z06 magnesium cradle (bottom and top views).[64]

A major challenge in magnesium automotive applications is to establish the surface finishing and corrosion protection processes. The challenge is two-fold since surface treatments for magnesium play roles in both manufacturing processes (e.g. adhesive bonding) as well as the product life cycle that demands corrosion resistance. Furthermore, the current manufacturing paradigm for steel-intensive body structures employs chemistries in the paint shop that are corrosive to magnesium and additionally are aggravated by galvanic couples, primarily steel fasteners. Future research will explore novel coating and surface treatment technologies including pretreatments such as micro-arc anodizing, non-chromated conversion coatings, and 'cold' metal spraying of aluminium onto magnesium surfaces. Since most studies of corrosion protection and pre-treatment of magnesium have focused on die castings, the behavior of sheet, extrusion and high-integrity castings will be explored for process compatibility.

12.7 Conclusions

The future success of magnesium as a major automotive material will depend on how these technical challenges are addressed. These challenges are huge and global, and would require significant collaboration among industries, governments and academia from many counties. One example is the current Canada-China-USA Magnesium Front End Research & Development project funded by the three governments. This project has brought together a unique team of international scope, from the United States, China and Canada, and has developed some key enabling technologies and a knowledge base for automotive magnesium applications. Such technologies will be demonstrated and validated in a demonstration structure currently being designed. The technologies and knowledge base developed in this project not only benefit automotive magnesium applications using the front-end structure as a test bed (see Fig. 12.14), they also promote primary magnesium production, component manufacturing, fundamental research and advanced computational tools such as integrated computational materials engineering (ICME). It is very encouraging that many of these international and interdisciplinary collaborations are being nurtured for magnesium. It is expected that future developments exploiting the new characterization tools available will provide the much needed breakthroughs to design new wrought magnesium alloys and low-cost corrosion mitigating solutions to increase the use of magnesium, the lightest structural metal.

12.8 Acknowledgements

The authors gratefully acknowledge the collaboration and discussions with Dr Paul Krajewski, Dr Raj K. Mishra and Dr Bob R. Powell of General Motors Global Research & Development, Warren, MI, USA.

12.14 A schematic of the front end structure of a production sedan.[69]

12.9 References

1. A.I. Taub, P.E. Krajewski, A.A. Luo and J.N. Owens, 'The evolution of technology for materials processing over the last 50 years: The automotive example', *JOM (Journal of Metals)*, February 2007, **59**, (2), 48–57.
2. K.U. Kainer, *Magnesium – Alloys and Technologies*, Wiley-VCH, Weinheim, Germany, 2003.
3. A.A. Luo, 'Magnesium: Current and Potential Automotive Applications', *JOM (Journal of Metals)*, 2002, **54**, (2), 42–48.
4. H.-H. Becker, 'Status, potential & challenges for automotive magnesium applications from the point of view of an OEM,' presentation at 65th Annual World Magnesium Conference, May 18–20, 2008, Warsaw, Poland.
5. Y. Michiura, 'Current magnesium research and application in automotive industry in Japan', presentation at International Automotive Body Congress, 7–8 November 2007, Troy, MI, USA.
6. J.S. Balzer, P.K. Dellock, M.H. Maj, G.S. Cole, D. Reed *et al.*, 'Structural magnesium front end support', SAE Technical Paper 2003-01-0185, SAE International, Warrendale, PA, 2003.
7. M. Hoeschl, W. Wagener and J. Wolf, 'BMW's magnesium-aluminium composite crankcase, state-of-the-art light metal casting and manufacturing', SAE Technical Paper 2006-01-0069, SAE International, Warrendale, PA, 2006.
8. C.J. Duke and S. Logan, 'Lightweight magnesium spare tire carrier', *Proceedings of the 64th Annual World Magnesium Conference*, International Magnesium Association, 2007, Wauconda, IL 60084, USA, pp. 75–80.
9. M.M. Avedesian and H. Baker, *ASM Specialty Handbook, Magnesium and Magnesium Alloys*, ASM International, Materials Park, OH, 1999.

10. Timminco Corporation, *Timminco Magnesium Wrought Products*, Timminco Corporation Brochure, Aurora, CO, 1998.
11. A.A. Luo and A.K. Sachdev, 'Development of a new wrought magnesium-aluminium-manganese alloy AM30', *Metallurgical and Materials Transactions A*, 2007, **38A**, 1184–1192.
12. ASM, *Metals Handbook, Desk Edition*, ASM International, Materials Park, OH, 1998.
13. M.F. Ashby, 'Performance Indices', *Metals Handbook*, Vol. 20, ASM International, Materials Park, OH, 1998, pp. 281–290.
14. R.S. Busk, *Magnesium Product Design*, Marcel Dekker, Inc., New York and Basel, 1987.
15. R.K. Mishra, A.K. Gupta, P.R. Rao, A.K. Sachdev, A.M. Kumar *et al.*, 'Influence of cerium on the texture and ductility of magnesium extrusions', *Scripta Materialia*, 2008, **59**, 562–565.
16. A.A. Luo, W. Wu, L. Jin, R.K. Mishra, A.K. Sachdev *et al.* 'Microstructure and mechanical properties of extruded magnesium-aluminium-cerium alloy tubes', *Metallurgical and Materials Transactions A*, 2010, **41A**, 2662–2674.
17. A.A. Luo, R.K. Mishra and A.K. Sachdev, 'High-ductility magnesium-zinc-cerium extrusion alloy', *Scripta Materialia*, 2010, 10.1016/j.scriptamat.2010.10.045.
18. X. Zeng, Y. Wang, W. Ding, A.A. Luo and A.K. Sachdev, 'Effect of strontium on the microstructure, mechanical properties, and fracture behaviour of AZ31 magnesium alloy', *Metallurgical and Materials Transactions A*, 2006, **37A**, 1333–1341.
19. Alloy Digest, 'Magnesium – LA141A', Filing Code M-56 Magnesium Alloy, February 1964.
20. J. H. Jackson, P.D. Frost, A.C. Loonam, L.W. Eastwood and C. H. Lorig, *Trans. AIME*, 1949, **185**, 149.
21. G.J. Shen and B.J. Duggan, 'Texture development in a cold-rolled and annealed body-centered-cubic Mg-Li alloy', *Metallurgical and Materials Trans. A*, 2007, **38A**, 2593–2601.
22. G.V. Raynor and J.R. Kench, *J. Inst. Metals*, 1959–60, **88**, 209.
23. J.C. McDonald, *J. Inst. Metals*, 1969, **97**, 353.
24. Y. Kojima, M. Inoue and O. Tanno, *J. Japanese Inst. Metals*, 1990, **54**, 354.
25. S. Hori and W. Fujitani, *J. Japanese Inst. Light Metals*, 1990, **40**, 285.
26. K. Matsuzawa, T. Koshihara, S. Ochiai and Y. Kojima, *J. Japanese Inst. Light Metals*, 1990, **40**, 659.
27. H. Takuda, H. Matsusaka, S. Kikuchi and K. Kubota, 'Tensile properties of a few Mg-Li-Zn alloy thin sheets', *Journal of Materials Science*, 2002, **37**, 51–57.
28. Y. Chino, K. Sassa and M. Mabuchi, 'Texture and stretch formability of Mg-1.5mass%Zn-0.2mass%Ce alloy rolled at different rolling temperatures', *Materials Transactions*, 2008, **49**, 2916–2918.
29. F. Pan and E. Han, *High-Performance Wrought Magnesium Alloys and Processing Technologies*, Science Press, Beijing, 2007.
30. K. Savage and J.F. King, 'Hydrostatic extrusion of magnesium', in *Magnesium Alloys and Their Applications*, ed. K.U. Kainer, Wiley-VCH, Weinheim, Germany, 2000, pp. 609–614.
31. A.A. Luo and Anil K. Sachdev, 'Bending and hydroforming of aluminium and magnesium alloy tubes', *Hydroforming for Advanced Manufacturing*, Woodhead Publishing Ltd, Cambridge, UK, 2008, pp. 238–266.
32. J. Becker, G. Fischer and K. Schemme, 'Light weight construction using extruded and forged semi-finished products made of magnesium alloys', in *Magnesium Alloys and Their Applications*, eds. B.L. Mordike and K.U. Kainer, Werkstoff-Informationsgesellschaft Frankfurt, Germany, 1998, pp. 15–29.

33. A.A. Luo and A.K. Sachdev, 'Development of a moderate temperature bending process for magnesium alloy extrusions', *Materials Science Forum*, 2005, **488–489**, 447–482.
34. A.A. Luo and A.K. Sachdev, 'Mechanical properties and microstructure of AZ31 magnesium alloy tubes', in *Magnesium Technology 2004*, ed. A.A. Luo, TMS, Warrendale, PA, 2004, pp. 79–85.
35. J. Enss, T. Everetz, T. Reier and P. Juchmann, 'Properties of magnesium rolled products', in *Magnesium Alloys and Their Applications*, ed. K.U. Kainer, Wiley-VCH, Weinheim, Germany, 2000, pp. 591–595.
36. P.E. Krajewski, 'Elevated temperature forming of sheet magnesium alloys', in *Magnesium Technology 2002*, ed. H.I. Kaplan, TMS, Warrendale, PA, 2002, pp. 175–179.
37. P.E. Krajewski, 'Elevated temperature behaviour of sheet magnesium alloys', SAE Technical Paper 2001-01-3104, SAE International, Warrendale, PA, 2001.
38. H. Palkowski and L. Wondraczek, 'Thin slab casting as a new possibility for economic magnesium sheet production', in *Magnesium – Proceedings of the 6th International Conference Magnesium Alloys and Their Applications*, ed. K.U. Kainer, Wiley-VCH, Weinheim, Germany, 2003, pp. 774–782.
39. A. Moll, M. Mekkaoui, S. Schumann and H. Friedrich, 'Application of Mg sheets in car body structures', in *Magnesium – Proceedings of the 6th International Conference Magnesium Alloys and Their Applications*, ed. K.U. Kainer, Wiley-VCH, Weinheim, Germany, 2003, pp. 935–942.
40. S. Braunig, M. During, H. Hartmann and B. Viehweger, 'Magnesium sheets for industrial applications', in *Magnesium – Proceedings of the 6th International Conference Magnesium Alloys and Their Applications*, ed. K.U. Kainer, Wiley-VCH, Weinheim, Germany, 2003, pp. 955–961.
41. B.R. Powell, P.E. Krajewski and A.A. Luo, 'Magnesium alloys', in *Materials Design and Manufacturing for Lightweight Vehicles*, Woodhead Publishing Ltd, Cambridge, UK, 2010, pp. 114–168.
42. V. Kaese, L. Greve, S. Juttner, M. Goede, S. Schumann *et al.*, 'Approaches to use magnesium as structural material in car body', in *Magnesium – Proceedings of the 6th International Conference Magnesium Alloys and Their Applications*, ed. K.U. Kainer, Wiley-VCH, Weinheim, Germany, 2003, pp. 949–954.
43. P. Juchmann and S. Wolff, 'New perspectives with magnesium sheet', in *Magnesium – Proceedings of the 6th International Conference Magnesium Alloys and Their Applications*, ed. K.U. Kainer, Wiley-VCH, Weinheim, Germany, 2003, pp. 1006–1012.
44. R. Neugebauer and M. Seifert, 'Results of tempered hydroforming of magnesium sheets with hydroforming fluids', in *Magnesium – Proceedings of the 6th International Conference Magnesium Alloys and Their Applications*, ed. K.U. Kainer, Wiley-VCH, Weinheim, Germany, 2003, pp. 306–312.
45. G. Kurz, 'Heated hydro-mechanical deep drawing of magnesium sheet metal', in *Magnesium Technology 2004*, ed. A.A. Luo, TMS, Warrendale, PA, 2004, pp. 67–71.
46. K. Siegert and S. Jaeger, 'Pneumatic bulging of magnesium AZ31 sheet metal at elevated temperatures', in *Magnesium Technology 2004*, ed. A.A. Luo, TMS, Warrendale, PA, 2004, pp. 87–90.
47. P.E. Krajewski and J.G. Schroth, 'Overview of quick plastic forming technology', *Materials Science Forum*, 2007, **551/552**, 3–12.
48. J.T. Carter, P.E Krajewski and R. Verma, 'The hot blow forming of AZ31 Mg sheet: formability assessment and application development', *Journal of Minerals, Metals, and Materials*, 2008, **60** (11), 77–81.

49. A. Stern, A. Munitz and G. Kohn, 'Application of welding technologies for joining of Mg alloys: Microstructure and mechanical properties', in *Magnesium Technology 2003*, ed. H.I. Kaplan, TMS, Warrendale, PA, 2003, pp. 163–168.
50. S. Lathabai, K.J. Barton, D. Harris, P.G. Lloyd, D.M. Viano *et al.*, 'Welding and weldability of AZ31B by gas tungsten arc and laser beam welding processes', in *Magnesium Technology 2003*, ed. H.I. Kaplan, TMS, Warrendale, PA, 2003, pp. 157–162.
51. K.G. Watkins, 'Laser welding of magnesium alloys', in *Magnesium Technology 2003*, ed. H.I. Kaplan, TMS, Warrendale, PA, 2003, pp. 153–156.
52. R. Johnson and P. Threadgill, 'Friction stir welding of magnesium alloys', in *Magnesium Technology 2003*, ed. H.I. Kaplan, TMS, Warrendale, PA, 2003, pp. 147–152.
53. N. Li, T.-Y. Pan, R. Cooper and D.Q. Houston, 'Friction stir welding of magnesium AM60 alloy', in *Magnesium Technology 2004*, ed. A.A. Luo, TMS, Warrendale, PA, 2004, pp. 19–23.
54. J.I. Skar, H. Gjestland, L.D. Oosterkamp and D.L. Albright, 'Friction stir welding of magnesium die castings', in *Magnesium Technology 2004*, ed. A.A. Luo, TMS, Warrendale, PA, 2004, pp. 25–30.
55. A.C. Somasekharan and L.E. Murr, 'Fundamental studies of the friction stir welding of magnesium alloys to 6061-T6 aluminium', in *Magnesium Technology 2004*, ed. A.A. Luo, TMS, Warrendale, PA, 2004, pp. 31–36.
56. L. Budde, J. Bischoff and T. Widder, 'Low-temperature joining of magnesium-materials in vehicle constructions', in *Magnesium Alloys and Their Applications*, eds. B.L. Mordike and K.U. Kainer, Werkstoff-Informationsgesellschaft Frankfurt, Germany, 1998, pp. 613–618.
57. R.E. Brown, 'Future of magnesium developments in 21st century', presentation at *Materials Science and Technology Conference*, Pittsburgh, PA, USA, 5–9 October 2008.
58. L.T. Barnes, 'Rolled magnesium products, 'what goes around, comes around', *Proceedings of the International Magnesium Association*, Chicago, IL, 1992, pp. 29–43.
59. A. Hector and W. Heiss, 'Magnesium die-castings as structural members in the integral seat of the new Mercedes-Benz roadster', SAE Technical Paper No. 900798, SAE, Warrendale, PA, 1990.
60. D. Alexander, 'Intier seats are customer driven', *Automotive Engineering International*, May 2004, 22–24.
61. L. Riopelle, 'Magnesium applications', *International Magnesium Association (IMA) Annual Magnesium in Automotive Seminar*, Livonia, MI, 20 April 2004.
62. R. Verma and J.T. Carter, 'Quick Plastic Forming of a Decklid Inner Panel with Commercial AZ31 Magnesium Sheet', SAE International Technical Paper No. 2006-01-0525, 2006, SAE International, Warrendale, PA.
63. S. Logan, A. Kizyma, C. Patterson and S. Rama, 'Lightweight Magnesium-Intensive Body Structure', SAE International Technical Paper No. 2006-01-0523, 2006, SAE International, Warrendale, PA.
64. J. Aragones, K. Goundan, S. Kolp, R. Osborne, L. Ouimet and W. Pinch, 'Development of the 2006 Corvette Z06 Structural Cast Magnesium Crossmember', SAE International Technical Paper No. 2005-01-0340, 2005, SAE International, Warrendale, PA.
65. C.L. Mendis, C.J. Bettles, M.A. Gibson and C.R. Hutchinson, 'An enhanced age hardening response in Mg–Sn based alloys containing Zn', *Materials Science and Engineering A* 2006, **435/436** 163–171.
66. A.A. Luo and A.K. Sachdev, 'Microstructure and mechanical properties of Mg-Al-Mn and Mg-Al-Sn alloys', in *Magnesium Technology 2009*, eds. E.A. Nyberg, S.R. Agnew, N.R. Neelameggham and M.O. Pekguleryuz, 2009, TMS, Warrendale, PA, pp. 437–443.

67. P. Fu, L. Peng, H. Jiang, J. Chang and C. Zhai, 'Effects of heat treatments on the microstructures and mechanical properties of Mg-3Nd-0.2Zn-0.4Zr (wt.%) alloy', *Materials Science and Engineering A*, 2008, **486**, 183–192.
68. Bohlen, M. Nuernberg, J.W. Senn, D. Letzig and S.R. Agnew, 'The texture and anisotropy of magnesium-zinc-rare earth alloy sheets', *Acta Materialia*, 2007, **55** (6), 2101–2112.
69. A.A. Luo, W. Shi, K. Sadayappan and E.A. Nyberg, 'Magnesium front end research and development: Phase I progress report of a Canada-China-USA collaboration', *Proceedings of IMA 67th Annual World Magnesium Conference*, International Magnesium Association (IMA), Wauconda, IL, USA.
70. M. Easton, A. Beer, M. Barnett, C. Davies, G. Dunlop *et al.*, 'Magnesium alloy applications in automotive structures', *JOM*, 2008, **60** (11), 57–62.
71. D.A. Wagner, S.D. Logan, K. Wang, T. Skszek and C.P. Salisbury, 'Test results and FEA predictions from magnesium AM30 extruded beams in bending and axial compression', in *Magnesium Technology 2009*, eds. E.A. Nyberg, S.R. Agnew, N.R. Neelameggham and M.O. Pekguleryuz, 2009, TMS, Warrendale, PA.
72. J. Greiner, C. Doerr, H. Nauerz and M. Graeve, 'The new "7G-TRONIC" of Mercedes-Benz: Innovative transmission technology for better driving performance, comfort, and fuel economy', SAE Technical Paper No. 2004-01-0649, 2004, SAE International, Warrendale, PA.
73. E.F. Emley, *Principles of Magnesium Technology*, Oxford, Pergamon Press, 1966.

13
Biomedical applications of magnesium alloys

W. H. SILLEKENS, TNO, The Netherlands and
D. BORMANN, Leibniz Universität Hannover, Germany

Abstract: This chapter deals with the emerging field of biomedical applications for magnesium-based materials, envisioning degradable implants that dissolve in the human body after having cured a particular medical condition. After outlining the background of this interest, some major aspects concerning degradable implants in general and magnesium implants in particular are addressed. These aspects relate to the requirements and testing (of mechanical and corrosion properties) as well as to the degradation characteristics and the tuning thereof (corrosion behaviour and major influences). Next, some categories of applications are discussed for which magnesium is currently being explored, namely cardiovascular and orthopaedic implants. Finally, prospective trends towards further developments are considered.

Key words: biomedical applications, biodegradation, corrosion, cardiovascular implants, orthopaedic implants.

13.1 Introduction

Life expectancy has increased considerably over the last fifty years. In line with this, the market for implantable medical devices is booming due to upward trends in medical conditions and patient activity, changing patient treatment approaches and technical advances. For instance, the demand for cardiac implants quadrupled, for orthopaedic implants doubled and for all other implants tripled between 1997 and 2007 (Apelian, 2007).

Current (metallic) materials for biomedical implants are essentially neutral in the body. Such non-degradable bio-passive implants remain either in place (with the risk of loosening, fracturing and tissue inflammation) or are explanted after healing (implying additional surgical risk, patient discomfort and costs). Degradable or absorbable materials that are under development partially meet these drawbacks as they do need less repeated invasive surgery, since once absorbed they leave behind only the healed natural tissue. Apart from magnesium, other materials that are being considered are certain types of polymers and iron. Further trends are towards implants and coatings that are also vehicles for drug delivery (bio-active devices), for instance to repress inflammation and aid the healing process. Biodegradable materials are of particular interest here as bio-active substances may be incorporated that gradually release upon dissolution of the implant in the body.

As for magnesium, its biocompatibility including its decomposition in the body (corrosion in the electrolytic environment), non-toxicity to the human body, as well as its functional role in the physiological system, render it an attractive candidate for degradable implants. Its mechanical properties – notably strength and modulus of elasticity – as well as its density are quite similar to natural bone material and hence of interest for hard-tissue engineering applications.

Pioneering work in this field goes back to the end of the nineteenth century and indicated a significant reduction in healing time and acceleration in mineralisation of bone fractures, while no toxic effects were observed (Witte, 2010). The rapid degradation of the magnesium implant and formation of hydrogen gas in the associated corrosion process, however, posed a problem and the interest shifted in favour of stainless steel and titanium. It was not before the late 1970s that the corrosion resistance of magnesium alloys was substantially improved by the use of specific alloying elements and high-purity alloys, so that hydrogen formation was suppressed as well. Most recently the use of magnesium for medical applications has seen a growing research interest, also because of its perceived osteo-conductive activity favouring bone apposition (Staiger et al., 2006). China appears to be especially active in this field (Zheng and Gu, 2011).

Figure 13.1 illustrates the steep increase in the number of published research works on the subject since 2005. Acknowledging that the outcome of such a basic literature search is merely indicative, the trend is undeniable. Yet, it is also obvious that the magnitude of the research in this particular field is still modest as compared to the overall magnesium and/or biomedical research efforts. The current state of the art for biomedical implants and the envisaged role for magnesium are outlined in Fig. 13.2.

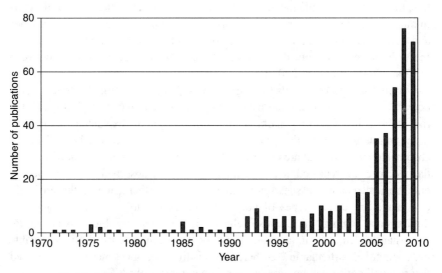

13.1 Publications count for biomedical applications of magnesium, SciVerse Scopus literature database (search terms 'magnesium' and 'implant' in key words).

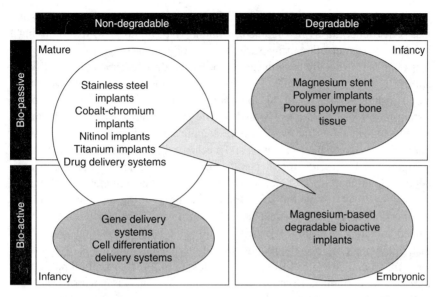

13.2 Trends in biomedical implants (represented by the arrow) and the position of magnesium.

In this chapter, recent advances in the development of biomedical applications for magnesium-based materials will be introduced. In Section 13.2 some major aspects concerning the functional behaviour of degradable implants in general and of magnesium implants in particular are addressed. Sections 13.3 and 13.4 discuss the main categories of applications for which magnesium is currently being explored, namely cardiovascular and orthopaedic implants. In Section 13.5 prospective trends towards further developments are considered. Section 13.6 gives some guidance to further information, while Section 13.7 concludes the chapter.

13.2 Functionality of magnesium implants

The essential qualities of a degradable implant are that it maintains its integrity in the body for a certain time in order to fulfil its supporting (or scaffolding) function, after which it gradually dissolves, e.g. by chemical reaction with constituents of the body fluid, and without leaving harmful residues behind. This section will deal with some general aspects associated with this degradation regarding the functional requirements and testing as well as regarding the actual decomposition of magnesium alloys and means to control this.

13.2.1 Requirements and testing

The development of magnesium implants aims at certain requirements that must be accordingly tested. A major distinction in this respect is whether the testing is done '*in vitro*' or '*in vivo*'. *In vitro* (literally 'within the glass') refers to any testing that is not done in a living organism, while *in vivo* (literally 'within the living') involves experimentation using a whole living organism.

As implants generally have to support or fix (hard) tissue, strength and stiffness are among the primary demands. Testing of the mechanical properties of magnesium-based materials for biomedical applications generally is based on well-known testing means such as the tensile and fatigue test. Nevertheless, the gathering of reliable and accurate data may be more complicated and costly, especially when interactions between biodegradation and mechanical properties are considered.

Figure 13.3 presents the generally desirable mechanical behaviour of a degradable implant (or artificial scaffolding function) in conjunction with that of the healing tissue (or natural scaffolding function). During its functional phase, the implant has to maintain mechanical integrity until the supported tissue has gained sufficient strength and/or stiffness to again carry the natural loads. After that, the implant has become redundant and mechanical integrity is of no further concern, so that strength/stiffness may be lost during the absorption phase until the remaining implant structure collapses and in time disappears. While the desired duration of this process does highly depend on the particular circumstances, typical times for healing would be between a few months and a year. Further, it is of concern that the mechanical properties of the implant on the one hand and those of the natural tissue on the other hand are alike. This last aspect is referred to as structural biocompatibility, a concept that also involves the actual implant design and inner structure (such as the orientation of fibres in anisotropic materials).

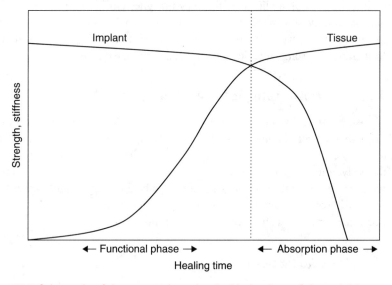

13.3 Schematic of the targeted mechanical behaviour of degradable implants (linear scales).

With the previous in mind, it is clear that the degradation behaviour should be controlled, which effectively poses the biggest challenge for magnesium implant development. The general view is that magnesium forms non-toxic reaction products in body fluid that harmlessly remain or are excreted. Biocompatibility, however, concerns also the effects of the alloying elements on the surrounding tissue and the body as a whole. Wrought alloys probably are to be favoured here over casting alloys as they generally contain lower alloying contents and feature higher strength and ductility. Table 13.1 lists some basic technological and biological aspects of selected alloying elements in magnesium. Current commercial magnesium alloys were primarily developed for use in structural applications (such as for automotive components) and alloying elements selected on the basis of their technological merits. For implants the biological assets and drawbacks are essential as well, calling for the development of a new category of alloys geared for biomedical use. Data on the biocompatibility of alloying elements in magnesium traditionally are scarce; yet, more thorough studies on the toxicity, carcinogenicity and general influence on cells of a variety of elements have meanwhile been conducted (e.g. Feyerabend *et al.*, 2009). Such biocompatibility evaluations are typically performed by either seeding cells directly on the implant material samples, or by exposing these samples for a certain time to a test solution and then using this solution with the degradation products for cell cultures.

The corrosion testing of magnesium-based materials for biomedical applications is a particular concern. For screening purposes the common laboratory potentiodynamic polarisation test is widely used, in which the physiological environment of the body is mimicked by means of some test solution. Examples of these fluids are Hank's solution, phosphate buffer solution (PBS) and simulated body fluid (SBF), each with their own particular constituents and ion concentrations. For instance, SBF contains prescribed concentrations of Na^+, K^+, Mg^{2+}, Ca^{2+}, Cl^-, HCO_3^-, HPO_4^{2-} and SO_4^{2-} and is buffered at a pH of 7.4, similar to human blood plasma. Common for corrosion testing is also the weight-loss method in conjunction with one of these test solutions. A further *in vitro* test that is of particular interest for magnesium is based on its dissolution reaction in aqueous solution, where the anodic and cathodic reactions add up to the next overall reaction.

$$Mg + 2H_2O \Rightarrow Mg(OH)_2 + H_2 \uparrow \qquad [13.1]$$

Thus magnesium and water yield magnesium hydroxide and hydrogen gas in fixed ratios. In the hydrogen evolution test, this gas is collected as it emits from the magnesium sample, which is placed in the fluid. The amount of dissolved magnesium can be calculated from the volume of hydrogen gas. An advantage of this test is that it also provides information regarding the time dependence of the corrosion rate.

At present no particular standards exist for the corrosion testing of biomedical magnesium-based materials. As it appears that corrosion and the underlying

Table 13.1 Selection criteria of alloying elements for biomedical magnesium alloys

Element	Technological aspect(s)	Biological aspect(s)
Aluminium	Improves strength and ductility (solid-solution hardening, precipitation, grain refinement), corrosion resistance and castability	Risk factor in generation of Alzheimer's disease; can cause muscle fibre damage; decreases osteoclast activity
Calcium	Improves strength (solid-solution hardening, precipitation, grain refinement) and creep resistance; reduces castability	Most abundant mineral in human body; tightly regulated by homeostasis
Lithium	Reduces strength yet improves ductility/formability (change to bcc lattice structure); reduces corrosion resistance and density	Possible teratogenic effects
Manganese	Improves strength and ductility (grain refinement); improves creep resistance; improves corrosion resistance in combination with aluminium (precipitation that picks up iron)	Essential trace element; important role in metabolic cycle and for the immune system; neurotoxic in higher concentrations
Rare earth elements (including yttrium)	Improve creep resistance and high-temperature strength (solid-solution hardening, precipitation); improve corrosion resistance (surface film); reduce mechanical anisotropy (texture randomisation)	Many rare earth elements exhibit anti-carcinogenic properties
Silicon	Reduces ductility; improves creep resistance and high-temperature strength (precipitation); reduces corrosion resistance and castability	
Strontium	Improves strength and ductility (grain refinement); improves creep resistance and high-temperature strength	
Zinc	Improves strength, yet reduces ductility in high concentrations (solid-solution hardening, precipitation); improves castability	Essential trace element (immune system, co-factor); neurotoxic at higher concentrations
Zirconium	Improves strength, ductility and high-temperature strength (strong grain refinement) in absence of aluminium	

Source: Witte *et al.*, 2008.

mechanisms depend on the test solutions as well as on other experimental basics (such as fluid circulation or not, and stabilisation of the pH value or not), results from different tests can hardly be compared (Xin *et al.*, 2010). Here it should also be noted that the functional conditions for the different applications vary widely and that experimental conditions for materials testing may have to be adapted accordingly. For instance, a stent that is placed in a clogged artery will experience a quite different environmental and mechanical loading than a screw and plate assembly fixing a broken bone. In this respect, a further recurring point of debate is if and how results from such *in vitro* studies relate and can be extrapolated to the actual *in vivo* situation.

As in other sectors, the use of particular alloys/materials for medical applications requires pre-market approval. For devices such as implants this certification is evidently very strict and commonly takes the form of so-called Food and Drug Administration (FDA) approval. The FDA is an agency of the United States Department of Health and Human Services and is responsible for protecting and promoting public health through the regulation and supervision of such products. Due to the many aspects associated with materials for (degradable) implants, pursuing FDA approval for these new products is intense and takes several years. In general the development of implants starts with preclinical testing (*in vitro* testing and *in vivo* animal testing) and, if successful, is followed by early clinical human trials. To the knowledge of the authors, the single magnesium alloy that has received earlier clearance for biomedical use is WE43.

13.2.2 Degradation characteristics and tuning

The degradation of current magnesium alloys in the aggressive physiological environment of the body is generally considered to be too fast to cover the functional phase of most targeted implant applications. Moreover, the corrosion of magnesium is associated with the formation of hydrogen, as mentioned above. Thus the corrosion rate is directly linked to the amount of hydrogen gas that is formed in a given period of time: 1 g Mg translates to 1.08 l H_2 (at atmospheric pressure). This is another reason why the corrosion rate should preferably be low, although the hydrogen dissolution tends to be fast and gas pockets that may occur in the surroundings of the implant may disappear after some time.

Figure 13.4 shows the corrosion rate for a variety of commercial magnesium alloys as calculated from potentiodynamic polarisation testing results (reworked from Erinc *et al.*, 2009). WE43 serves as a kind of baseline alloy as it has been used in many early investigations in the field; ZM21, ZE10, ZK60, AZ31 and AZ80 are other typical wrought alloys. Noting that processing affects the resulting microstructure and hence their mechanical and corrosion behaviour, these six alloys were tested in a common extruded condition (extrusion ratio of 25, temperature of 375 °C, exit speed of 1.5 m/min). AZ91 and AM50 are typical casting alloys that were tested in the as-received condition (ingots). The bar chart reveals that the corrosion rates in SBF vary over a wide range, with the highest

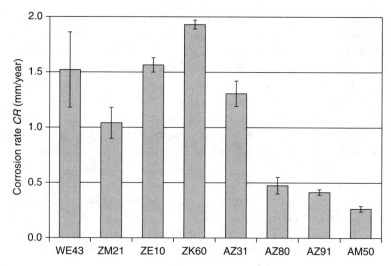

13.4 Corrosion resistance of magnesium in physiological environment. All alloys were tested on Ø5 mm bar in as-extruded condition, except for AZ91 and AM50 that were tested as-cast. Derived from potentiodynamic polarisation tests in SBF at 37 °C for 4 h, according to ASTM standard G59-97.

and lowest values differing by a factor of 7. WE43 and the alloys with zinc as the main alloying element are among the alloys with the highest corrosion rates. The alloys with aluminium as the main alloying element show lower corrosion rates, notably those with high aluminium content. This is generally associated with the presence of the $Mg_{17}Al_{12}$ intermetallic phase. While for the alloys with low aluminium content these particles form galvanic couples with the primary magnesium phase, however for the alloys, with high aluminium content the secondary phase appears rather as a network which shields the primary phase.

As an extension to this, Fig. 13.5 visualises the influence of the processing/temper condition for the alloy ZK60. Samples were tested in the as-received pre-extruded condition (F1), the extruded condition as discussed above (F2), the extruded and subsequently quenched condition (T4), and the extruded, quenched and artificially aged condition (T6). The two last-mentioned conditions followed a recommended aging treatment (Avedesian and Baker, 1999). In all light-optical microscopy (LOM) images, the horizontal direction represents the extrusion direction. For F1, the microstructure has a typical inhomogeneous appearance (with light 'islands' that are solid-solution zones deficient in zinc and zirconium due to alloy segregation), which becomes further refined and distorted for F2. For T4 the microstructure is recrystallised and hence contains primarily equi-axed grains with some remaining banding from the processing history. For T6 the image is quite similar, with the precipitates formed (consisting of a MgZn

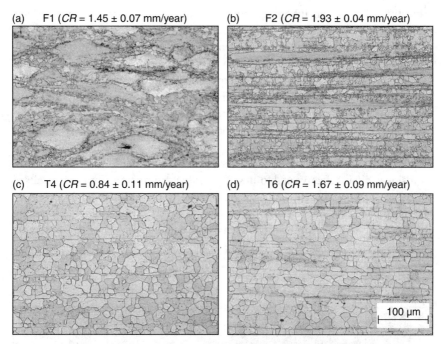

13.5 Micrographs and the associated corrosion resistance in physiological environment of ZK60 in different processing/temper conditions. (a) F1 = as-received; (b) F2 = F1 + extrusion; (c) T4 = F2 + heating from 260 °C to 500 °C in 2 h, solutionising at 500 °C for 2 h, quenching in water to room temperature; (d) T6 = T4 + heating at 130 °C for 48 h. Corrosion rate CR derived from potentiodynamic polarisation tests in SBF at 37 °C for 4 h, longitudinal cross-sections, etchant picric acid, scale bar as indicated for all images.

compound) being too small to be identified by LOM. Nevertheless, Vickers hardness values are 74.2, 74.0, 60.6 and 79.6 GPa for the F1, F2, T4 and T6, respectively, which basically confirms the effectiveness of the aging treatment. As for the associated corrosion rates, the improvement in T4 (with a factor 2.3 over F2 and 2.0 over T6) is notable and presumably associated with the dissolution of secondary phases due to the annealing. This example also illustrates that demands on mechanical and corrosion properties can be contradictory and thus challenging to meet.

Of relevance is not only the corrosion rate, but also the corrosion mechanism. With general (uniform) corrosion of the surface being the preferred mode, localised corrosion including pitting are also common types of corrosion for magnesium alloys in, for instance, simulated physiological fluids. Localised corrosion is characterised by selective attack of weak spots in the (partially) protective film and an inhomogeneous appearance of the magnesium surface. Pitting corrosion shows as spreading porous pits.

Figure 13.6 displays examples of corrosion as encountered in the already introduced potentiodynamic polarisation tests (Erinc et al., 2009). The upper half of the figure displays LOM images at comparably low magnification. For WE43 in particular, corrosion is heavily localised and follows the extrusion direction with associated banding (vertical direction). AZ80 shows pits in an otherwise irregularly corroded surface. The bottom half of the figure displays scanning electron microscopy (SEM) images at comparably high magnification. For ZK60, in which corrosion products are seen on top with the metal below, corrosion again is selective and follows the extrusion direction. For these localised corrosion types, microgalvanic couples between the primary and secondary phases (such as the intermetallic compound $Mg_{17}Al_{12}$ in aluminium-containing alloys) or due to alloy segregation (such as the depleted areas in ZK60) promote

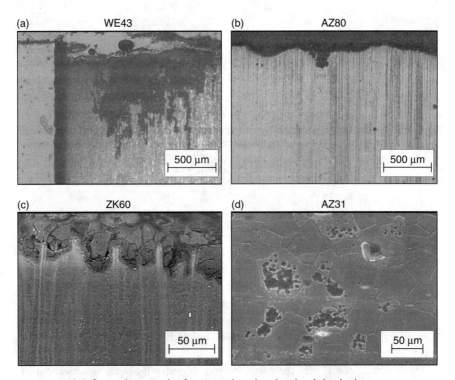

13.6 Corrosive attack of magnesium bar in physiological environment. (a, b) LOM images of corroded surface; (c) SEM image of corroded surface; (d) SEM image of the bulk material. Outcome of potentiodynamic polarisation tests in SBF at 37 °C for 4 h, longitudinal cross-sections, size bar as indicated for each image.

such selective attack, with corrosion of the less noble constituent being accelerated. For AZ31, a peculiar type of corrosion within the grains is visible in which the affected grains are likely to be interconnected. Obviously the progressing corrosion – and in particular the more localised types of attack – will also affect the mechanical integrity of the remaining structure.

As an illustration of how the degradation appears *in vivo*, Fig. 13.7 presents post mortem results of some magnesium samples that were implanted in mice for an extended period of time. In these experiments, electrochemically polished and

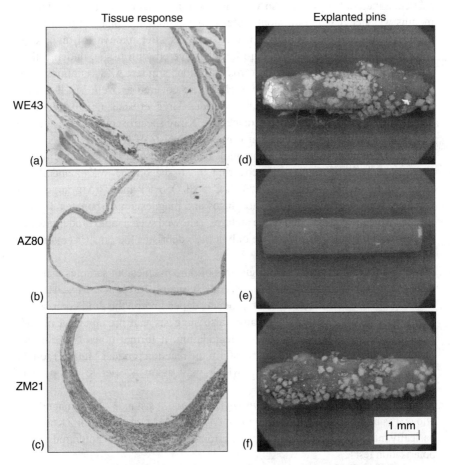

13.7 In vivo degradation of magnesium pins. (a, b, c) Inflammatory tissue response, haematoxylin phloxine saffron (HPS) stain; (d, e, f) SEM images of explanted magnesium alloy samples. Mouse model, intraperitoneal positioning, two months of implantation. (Source: courtesy of Boston Scientific.)

sterilised pins (Ø1 mm × 5 mm) of different alloys were used that were fixed with suture ligations in the body cavity (intraperitoneally) of the mice, which mimics to a certain extent intravascular conditions. The pictures on the left display transverse cross-sections of the tissue from which the pins were removed with an ethylene-diamine-tetra-acetic acid (EDTA) mediated treatment. The tissues are marked with a general histological stain from which inflammatory cell and fibrotic tissue accumulation can be studied. The three represented cases all show a strong tissue embedding with inflammatory response differing from fairly strong for WE43, moderate for AZ80, to mild for ZM21. With tissue embedding being a generally favourable situation, it was also noticed that the inflammatory response of the best-performing magnesium alloys was similar to that of a stainless steel reference material. Notable also was that the inflammatory behaviour after two months was better than one month after implantation. The pictures on the right display the appearance of the explanted samples after removal of any excess substances. In all cases the pins are loosely covered with tissue, underneath of which a thick (cracked) layer of corrosion products hides. AZ80 is almost fully covered with this tissue. WE43 and ZM21 are covered as well, but the tissue appears thicker at some regions and besides large crystals (>100 μm) have precipitated on this tissue. Energy dispersive X-ray spectroscopy (EDS) analysis of the covering tissue revealed that it mainly consists of C and O (H being too light to be detected with this technique), being consistent with organic matter. Further, analysis of the crystals covering the tissue on the WE43 and ZM21 samples reproducibly yielded about 35% C, 55% O, 5% P and 6% Mg, suggesting that they consist of magnesium phosphates and magnesium carbonates. Finally, ZK60 – which is not included in the pictures – showed such a rapid biodegradation that in most of the concerned mice only a little debris of the pins (if anything at all) could be recovered.

The literature indicates that the surfaces of pure magnesium samples that were exposed to such physiological fluids consist mainly of MgO, $Mg(OH)_2$ and $MgCO_3$, the ratio of these corrosion products depending on the fluid composition (Tie et al., 2010). While the literature also suggests that precipitation of such insoluble corrosion products retards degradation, it further notes that chloride ions in the physiological fluid can transform the reaction product $Mg(OH)_2$ on the surface into more soluble $MgCl_2$, which may again accelerate the pace of dissolution (Xin et al., 2010).

The previous tests showed that in vivo corrosion resistance of the magnesium alloys was best for AZ80, moderate for WE43 and ZM21, and poor for ZK60. This basically matches the ranking as obtained from the earlier in vitro potentiodynamic polarisation tests.

As is obvious from the previous results, alloy chemical composition and microstructure have an important influence on the biodegradation of magnesium alloys. In pursuit of a material that meets the various requirements, a wide range of magnesium alloys has been developed and/or explored. Table 13.2 gives a

Table 13.2 Magnesium alloys considered for biomedical applications (ASTM designations or designations with alloying contents in wt%)

Alloy family	Representative alloys	Main phase(s)
Pure Mg	Mg	Mg
Mg-Al-Zn	AZ31	Mg, $Mg_{17}Al_{12}$
	AZ91	Mg, $Mg_{17}Al_{12}$
Mg-Ca	Mg-xCa (x = 1, 2, 3, 4, 5 ...)	Mg, Mg_2Ca
Mg-Zn-Ca	Mg-1Zn-1Ca	Mg, Mg_2Ca, $Ca_2Mg_6Zn_3$
Mg-Zn-Mn-Ca	Mg-2.0Zn-1.2Mn-1Ca	Mg, Mg_2Ca, $Ca_2Mg_6Zn_3$, $Ca_2Mg_5Zn_{13}$
Mg-Si-Ca	Mg-1Si-1Ca	Mg, Mg_2Si, SiMgCa
Mg-Zn	Mg-xZn (x = 1, 3, 10)	Mg, MgZn, Mg_2Zn_3, Mg_7Zn_3
Mg-Mn-Zn	Mg-1Mn-1Zn	Mg, MgZn, Mg_2Zn_3, Mg_7Zn_3
Mg-Mn	Mg-1Mn	Mg, Mn
Rare-earth containing alloys	LAE442	Mg, $Al_{11}RE_3$
	WE43	Mg, $Mg_{12}YNd$, $Mg_{14}YNd_2$
	ZE41	Mg, MgZn(RE)
	AE44	Mg, $Mg_{17}Al_{12}$, $Al_{11}RE_3$, $Al_{12}RE$
	Mg-xGd (x = 5, 10, 15 ...)	Mg, Mg_5Gd
	WZ21	Mg, $MgYZn_3$, Mg_7Zn_3, Mg_3YZn_6
	Mg-8Y	Mg, $Mg_{24}Y_5$, Mg_2Y

Source: Xin *et al.*, 2010.

non-exhaustive overview, including the main phases that affect corrosion and mechanical behaviour. The listed alloys vary widely in corrosion performance and in mechanical properties, with no demonstrated best overall candidate as yet. For instance, pure magnesium has by its very nature excellent biocompatibility, but lacks corrosion resistance as well as strength. Further, it is to be noted that actual biocompatibility (including toxicity) depends not only on the alloying elements/ contents and corrosion rate, but also on the implant volume to be dissolved, the chemical form in which the alloy constituents are released and/or react with the surroundings, and the physiological environment by which the reaction products are absorbed and/or diffused.

A further particular avenue of development that needs to be mentioned here is that of bulk metallic glasses (BMGs). Here, non-equilibrium alloy compositions are synthesised in an amorphous solid state by means of rapid solidification from the melt. As an example, MgZnCa glasses with up to 40% alloying elements exhibit no hydrogen evolution during degradation (Zberg *et al.*, 2009). Due to the required high cooling rates during their production, dimensions of these materials are generally limited.

With corrosion resistance of current magnesium alloys being generally considered inadequate for biomedical applications, surface modification is of great interest. Appropriate coatings or surface treatments should retard corrosion of the implant while maintaining its mechanical integrity during the functional phase, cover complex surfaces completely and evenly, and have good adhesion to the metal substrate and wear resistance to protect the implant during its implantation and throughout the load cycles of its life. It is also of concern that coatings are non-toxic and fully degradable, as well as have good cellular response and attachment to allow the implant to become fully integrated with the biological system. Several surface modification systems are presently being explored for magnesium-based biomaterials (Waterman and Staiger, 2011). Coatings and coating methods include anodisation, metal coatings, calcium phosphates, plasma spray, chemical vapour deposition, pulsed laser deposition, ion beam assisted deposition, solution coatings and electro-deposition. Surface treatments include techniques to change the surface into an amorphous layer and (plasma immersion) ion implantation of elements that create a more protective passivation layer upon corroding. A specific aspect to be considered in the use of coatings is that when the system fails at particular sites, such as by cracks or voids, serious local corrosion attack – which is obviously unfavourable – may result.

13.3 Cardiovascular implants

The previous section has introduced some general aspects relating to the use of magnesium-based materials for biomedical applications. This and the following section will focus on some particular implant categories that are currently anticipated as degradable devices.

In general, a stent is an artificial tubular device that is surgically inserted into a natural passage in the body to counteract or prevent it from clogging. The most common type – which will be dealt with in this text – is the cardiovascular or coronary stent. Such a stent essentially consists of a thin-walled mesh tube with typical overall dimensions being 3 mm in diameter and 20 mm in length. Wall thickness and strut size can be as low as 50–150 µm. The current generation of stents are made of stainless steel, Nitinol or similar non-degradable metallic materials. A stent is implanted by shrinking it onto a balloon catheter and subsequently expanding it upon proper positioning in the artery. Thus, the stent material must comply with such mechanical requirements as strength and ductility in addition to biological demands, as is indicated in Table 13.3 for the envisaged magnesium stent. Further, there are supplementary requirements that are not listed but nevertheless relevant, such as manufacturability, geometrical accuracy and fatigue resistance. Apart from that, stent design (that is, the actual mesh pattern) is an important aspect associated with these requirements; for instance, in relation to the plastic strains during shrinking and expanding and in relation to elastic spring-back (the so-called recoil) upon placement.

The development of magnesium (cardio)-vascular stents is currently being undertaken with initial successes reported (Müller, 2008; Deng *et al.*, 2011). Magnesium stents – with an indicative weight of 5 mg – are among the least

Table 13.3 Primary requirements for biodegradable stent tube

Aspect	Description
Absorption	Scaffolding integrity 3–6 months
	Full dissolution within 1–2 years
Biocompatibility	Non-toxic, no inflammatory tissue response
	No harmful release and/or residence of particles
Mechanical properties	Tensile yield stress *TYS* >200 MPa
	Ultimate tensile strength *UTS* >300 MPa
	Tensile elongation >15–18%
	Elastic recoil <4% (in conjunction with stent design)

mass-critical as compared to other implants. This section outlines the state of the art, and exemplifies some ongoing research efforts for these biomedical devices.

13.3.1 State of the art

During early investigations the commercial-grade alloy WE43 was considered; more recently a variety of other alloy systems have been proposed. These include alloy modifications with aluminium (Erinc *et al.*, 2010), calcium (Hassel *et al.*, 2006) or lithium (Leeflang *et al.*, 2010) as the primary alloying element. The main objective of these developments is to slow down the corrosion rate while maintaining or improving mechanical properties.

Efforts on manufacturing magnesium stent tube are reportedly based on the route used for conventional metal stents. This consists of:

1. Hot working of cast feedstock to obtain a solid rod.
2. Deep drilling of a hole in the rod to obtain a tube.
3. Multiple cold drawing and intermediate annealing of the tube to reduce diameter and wall thickness (Gerold and Müller, 2006).

The mesh is subsequently made by laser cutting, and the cutting edges are smoothed by electro-chemical polishing. An alternative route departs from this in that the hot-working process is conducted by direct extrusion with a mandrel and tubular billets, yielding seamless tube, and the drawing is done at elevated temperature, reducing the number of drawing steps (Hassel *et al.*, 2006). Yet another route employs direct hot extrusion with a porthole die, which yields tubes directly from solid billets, followed by cold drawing (Werkhoven *et al.*, 2011). Each of these methods has its particular assets and drawbacks.

Figure 13.8 shows some cut-off samples of the feedstock as well as of the semi-finished tubes, together with a finished stent as manufactured according to this last-mentioned route. The basic reasoning behind combining an initial extrusion process with subsequent drawing to produce a stent tube is as follows. In hot extrusion, shape alterations imposed in a single processing step are high, but geometrical

13.8 Magnesium stent: from feedstock to finished product. From left to right: cut-offs of a billet, extruded tube, extruded and drawn tube, and a finished stent.

tolerances of the obtained tubes are limited and not high enough to reach the required accuracy of a finished stent tube. Thus hot extrusion is used to provide an intermediary shape. In drawing at room temperature (cold drawing), dimensional accuracy of the product is high, but the deformation that can be imposed in a single drawing step is limited so that a substantial reduction in tube diameter and wall thickness can only be achieved by repetitive drawing with progressively smaller tooling and by implementing intermediary heat treatments to restore the ductility of the material. Thus it is used to provide the final shape of the stent tube.

13.3.2 Ongoing research and development

For magnesium alloy stent tube, the strategy for development is to realise a possibly fine microstructure, driven by the following considerations.

- Small grains enhance yield stress according to the Hall-Petch relationship with a sensitivity that is typically greater for metals such as magnesium that yield by mechanical twinning rather than by dislocation glide (Barnett, 2008). Also, grain refinement reduces (tension-compression) yield anisotropy to become marginal for grains in the micrometer range (Bohlen *et al.*, 2006).
- The literature consistently suggests that, as grain size decreases, the corrosion resistance of magnesium alloys is improved in neutral and in alkaline electrolytes (Ralston and Birbilis, 2010), such as blood.

- A fine microstructure enhances homogeneity across the thin walls and struts of the stent. This will reduce local property variations (mechanical and otherwise, such as in corrosion attack) as well as the risk of releasing large particles during the absorption of the implant.

From the above it follows that the quality of the stent tube in terms of its mechanical and corrosion properties will depend (through its microstructure) on the sequence of forming operations and their interaction. Some illustrations of these (inter) relations will be outlined below. The first relates to the working order of hot extrusion followed by cold drawing, the second to equal channel angular pressing (ECAP) followed by hot extrusion (Werkhoven *et al.*, 2011). The latter is considered here as a development sidetrack to potentially improve the quality of the feedstock prior to extrusion.

Figure 13.9 displays microstructures of magnesium stent tube during the various stages of manufacturing. The as-received feedstock material here corresponds to a pre-extruded condition. Extrusion of the feedstock into

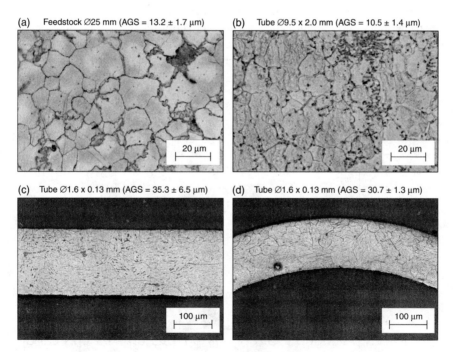

13.9 Microstructural evolution of AZ80 during the extrusion and drawing of stent tube. (a) Billet; (b) extruded; (c, d) extruded and drawn/heat treated. Headings include average grain size AGS (linear intercept values). Longitudinal cross-sections except (d) which is transverse, etchant picric acid, size bar as indicated for each image.

Ø9.5 mm × 2.0 mm tubes at the specified settings (extrusion ratio of 10, temperature of 375 °C, exit speed of 0.3 m/min) refines the microstructure somewhat, as is obvious from the included average grain-size values. The LOM images reveal equi-axed, recrystallised microstructures with precipitates at the grain boundaries, typically being the result of dynamic recrystallisation during the hot-working process. Subsequent drawing of the extruded tubes into Ø1.6 mm × 0.13 mm finished stent tube induces a substantial coarsening of the microstructure. Here it is to be mentioned that the multiple cold-drawing steps were alternated with annealing treatments at a recommended 385 °C (Avedesian and Baker, 1999) and a concluding heat treatment at 340 °C after the final drawing step. The LOM images reveal equi-axed, recrystallised microstructures. Similar microstructures appear after each intermediate annealing step, supporting the view that static recrystallisation is the underlying mechanism.

In ECAP, heated material is repeatedly pushed through an angular die without altering the cross-section; due to the severe plastic deformation, microstructures can be refined. In this case a die is used which redirects the material flow by 90°; channel size is Ø25 mm. Temperature of billet and tooling were set at 275 °C; samples were reversed and turned 90° between individual passes. For AZ80 as well as for two experimental alloys indicated as AZM and AZNd (with modified manganese and neodymium contents, respectively), this effectively refines the microstructure to an average grain size of 3 µm after three passes, which does not further decrease within the explored range of up to nine passes. Figure 13.10

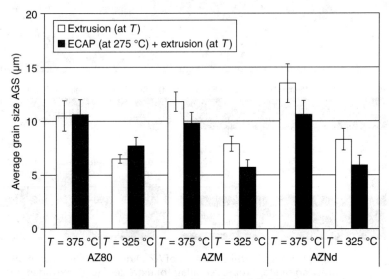

13.10 Microstructural characterisation of semi-finished magnesium stent tube (Ø9.5 mm × 2.0 mm) in dependence on the processing route (average grain size AGS as linear intercept values).

summarises microstructural results for some series of extrusion trials without and with prior ECAPing. From this it appears that for AZ80 the influence of ECAP on the average grain size of the semi-finished tube is insignificant or even detrimental, while for AZM and AZNd it refines the eventually obtained microstructure to some extent. The extrusion temperature, however, is a more effective parameter for controlling eventual grain size with a lower temperature leading to a finer microstructure. Here, press capacity commonly dictates the lowest temperature to be practically feasible.

13.4 Orthopaedic and other implants

The early efforts to introduce magnesium-based materials for implants were directed towards orthopaedic and, in particular, towards traumatological applications such as fixation plates, pegs and screws to secure bone fractures. This section outlines the state of the art, and exemplifies some ongoing research efforts for these biomedical devices.

13.4.1 State of the art

In addition to its biodegradability, there are some further specific arguments in favour of magnesium as an implant material for bone surgery.

Table 13.4 compares some relevant properties of natural bone with those of magnesium (alloys), some conventional metallic implant materials, and the ceramic hydroxyapatite $Ca_{10}(PO_4)_6(OH)_2$ that is commonly used as a bone graft. From this it appears that – among these materials – magnesium does most closely resemble the natural tissue that is to be (temporarily) supported and hence has the best structural biocompatibility. A specific role for the modulus of elasticity of the implant material is that a mismatch with the host material is known to introduce a

Table 13.4 Basic design properties of various implant materials in comparison to natural bone

Material	Density ρ (kg/m³)	Elastic modulus e (GPA)	Compressive yield stress CYS (MPa)	Fracture toughness k_{1c} (MPa.m$^{1/2}$)
Natural bone	1800–2100	3–20	130–180	3–6
Magnesium alloys	1740–2000	41–45	65–100	15–40
Titanium alloys	4400–4500	110–117	758–1117	55–115
Cobalt-chrome alloys	8300–9200	230	450–1000	N/A
Stainless steel	7900–8100	189–205	170–310	50–200
Synthetic hydroxyapatite	3100	73–117	600	0.7

Source: Staiger *et al.*, 2006.

so-called stress-shielding effect, being a common cause of loosening of (permanent) implants as the natural tissue retracts in a situation where it is under-loaded.

Further, *in vivo* (animal) trials have led to the notion that magnesium degradation enhances osteoblastic activity that promotes bone growth and thus healing, based on the observation of high mineral apposition rates and increased bone mass around magnesium-alloy bone implants as compared to a degradable polymer reference (Witte *et al.*, 2005).

Magnesium thus appears to be an obvious candidate for such hard-tissue scaffolds. Table 13.5 lists conceivable applications for orthopaedic and related implants. Notably, the level of ambition for these examples is quite different, ranging from basic mono-material implants to multi-component devices in which a biodegradable part could, for instance, serve for anchoring and drug release. With the variety of implants being quite wide in terms of size, shape and functionality, their manufacturing routes as well as their lists of demands will be equally diverse. Mechanical requirements, however, will in most cases be related to static and dynamic strength (in conjunction, of course, with degradation behaviour).

Cellular/porous structures or foams (sponges) are of particular interest in this respect. They can be used to further adapt stiffness of the implant to that of the bone, provide a suitable architecture for in-growth of (bone) tissue and/or serve as a carrier for drugs. For instance, powder metallurgy is being explored to produce open porous magnesium scaffolds. These are produced by sintering a mixture of magnesium powder and space-holding particles (spacers), in which the spacers evaporate upon heating. Such sponges are reported to be effective in physically releasing gentamicin (a common antibiotic to treat bacterial infections) at rates that can be controlled through different pore morphologies that were realised with spacer contents ranging between 0–25% (Aghion *et al.*, 2010). A concern with these open porous structures is the accelerated degradation of the material due to the high exposed magnesium surface area.

Alternatively, metal matrix composites (MMCs) are being investigated to tailor corrosion rates and mechanical attributes. Here, an example is the combination of magnesium with hydroxyapatite (HA) particles as reinforcements. The literature

Table 13.5 Candidates for magnesium-based or magnesium-containing hard-tissue implants

Area	Products
Joints	Knee implants, hip implants, shoulder implants, ankle implants, elbow implants, bone grafts
Spine	Internal spinal fixation devices, fusion cages, bone allografts, disc replacements, spacers
Dental/oral	Alveolar bone grafts
Other	Fixtures (plates, screws, pins, staples), tamponades, craniomaxillofacial patches, gauzes, sutures and suture anchors

indicates that the mechanical properties of the MMC can be adjusted by the choice of the HA particle size and distribution, and that the HA particles stabilise the corrosion rate with the MMC exhibiting more uniform corrosion attack (Witte et al., 2007).

13.4.2 Ongoing research and development

As stated above, requirements on orthopaedic and other hard-tissue implants are diverse, with no single development approach being applicable for each case. As mere illustrations, some examples of current *in vivo* studies are outlined below. The first relates to the influence of implant surface condition on corrosion, the second to mechanical implant degradation, and the third to the degradation of a coated sponge. The first two examples are concerned with such applications as stabilisation of bone fractures and osteotomy (surgery for bone adjustments), while the last-mentioned example is more directed towards trauma surgery to fill critical bone-size defects.

Figure 13.11 charts the observed occurrences of gas formation in the vicinity of MgCa0.8 cylinders (Ø3 mm × 5 mm) implanted in the so-called cortico-spongy passage of the medial femoral epicondyle of the hind legs of rabbits (adapted from Höh et al., 2009). The implants were prepared from extruded bar by machining, machining and sand blasting, and machining and threading. This yielded smooth (mean roughness value R_a = 3.7 µm), rough (R_a = 32.7 µm), and threaded (screw)

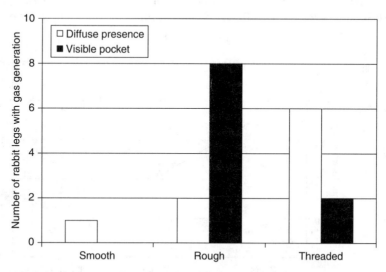

13.11 In vivo generation of gas for MgCa0.8 implants with different surface topographies (out of a total of 48 implants). Rabbit model, hind leg positioning, 3 and 6 months of implantation (pooled representation).

surfaces, respectively. A total of 48 implants remained *in vivo* for three or six months and the legs were subsequently examined by various means with respect to implant condition, new bone formation and other tissue response, as well as with respect to gas formation. The bar chart shows that a fair number of these implants led either to the diffuse presence of gas or clearly visible gas pockets underneath the skin near the implant. Differences between the three implant types are notable and obviously related to the pace of decomposition. Tomography indeed confirmed that the rough implants showed the fastest degradation. Overall it is concluded that all three implant types are well-tolerated for the investigated implantation periods.

Figure 13.12 shows the pace of degradation of cylindrical MgCa0.8 samples (Ø2.5 mm × 25 mm) that were implanted in the marrow cavities of rabbit tibiae (adapted from Thomann *et al.*, 2010). The implants were manufactured by extrusion and provided with a magnesium-fluoride (MgF_2) coating. A total of ten implants remained *in vivo* for three or six months and the tibiae were subsequently examined by various means with respect to implant condition and new bone formation, as well as with respect to residual volume and mechanical properties of the explanted pins. Mechanical testing consisted of three-point bending. The graph illustrates the loss in maximum bending force, amounting to 41% after three months and to 58% after six months of implantation. Notably, the corresponding bending displacement at fracture remained practically the same. Volume loss appeared to be less pronounced than strength loss. Overall it is concluded that fluoride-coated MgCa0.8 implants show only a slight reduction in degradation rate over uncoated implants, but also that they possess good clinical tolerance.

Figure 13.13 presents the appearance of coated AX30 cylindrical sponges (Ø3 mm × 4 mm) after different periods of implantation in the femur of rabbits

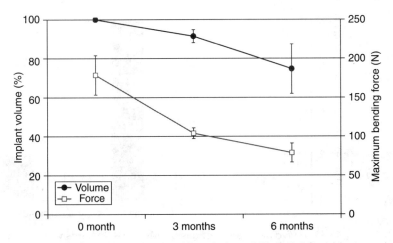

13.12 Relation between *in vivo* degradation of MgCa0.8 fluoride-coated pins (Ø2.5 mm × 25 mm) and implantation time. Mechanical test: three-point bending, supported length 15 mm, cross-head speed 1 mm/min.

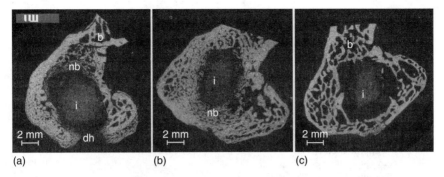

13.13 In vivo degradation of bioglass-coated AX30 sponges after: (a) 6 weeks of implantation (b) 12 weeks of implantation (c) 24 weeks of implantation. Key: b = cancellous bone, i = implant, nb = newly formed bone, dh = drill hole. Obtained with μ-computed tomography.

(reproduced from Lalk *et al.*, 2010). The sponges were produced via gravity casting in which sodium chloride particles were added to the melt; these particles were subsequently washed out (amongst others, with a sodium hydroxide solution), giving a sponge porosity of 250–500 μm. The sponges were coated with bioglass ($CaO\text{-}SiO_2\text{-}P_2O_5$), which is a commercially available series of bio-active glasses. The pictures show a gradual degradation of the implants with increasing implantation time along with a simultaneous remodelling of the bone (to be seen from the closing of the drill hole and changing bone structure). The overall conclusion is that the implants are clinically tolerated well; however, it is also stated that the bioglass coating does not sufficiently retard degradation, with further improvements to be pursued in this respect.

13.5 Future trends

The previous sections have highlighted some recent developments on magnesium implants. Actual introduction of these devices, however, is still pending. This section identifies some directions for further advancement, distinguishing between materials, processes and products but noting that overlaps between these areas occur. Figure 13.14 summarises these issues.

Regarding materials, already known and (probably) new alloy series and chemical compositions will be further studied and developed. As before, improvement of biodegradation and biocompatibility along with mechanical properties (strength, ductility, fatigue) will be the main concerns. Except for reducing corrosion rates in the physiological environment, the target is to achieve general (uniform) corrosion in favour of local corrosion. These developments will likely lead to preferred alloys for distinct product groups (with different property profiles) rather than to a universal biomedical alloy. It is unclear at this stage what the favoured alloying elements and contents will be. In addition, non-conventional

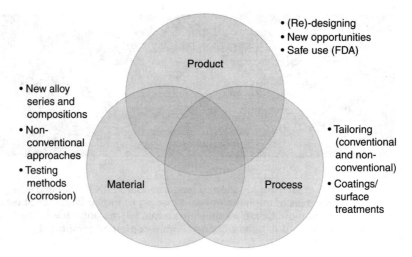

13.14 Research and development issues for biomedical application of magnesium.

material approaches will be further explored and will probably emerge to circumvent restrictions of melt-metallurgical production, including bulk metallic glasses and powder metallurgy. Expected also are the further tailoring and probable standardisation of testing methods (notably for corrosion), as well as an increased emphasis on studies that more closely simulate actual functional conditions such as combined corrosion and mechanical tests and *in vivo* studies.

Regarding processes, developments are due to tailor them for the particular characteristics of magnesium-based materials and establish manufacturing routes fitting product demands in terms of geometrical accuracy, shapes/sizes, production volumes, and so on. This concerns conventional manufacturing processes such as extrusion, drawing, laser cutting and polishing of stents, as well as casting, extrusion, forging and machining of orthopaedic devices. Further, specific applications will call for specific non-conventional manufacturing processes such as melt infiltration and metal injection moulding for sponges, rapid-manufacturing techniques for customised implants (such as skull and other craniomaxillofacial patches to bridge large bone defects), and severe plastic deformation processes for enhancing performance. Key is also the further development of coatings and surface modification treatments. Here the palette of processes is very wide, holding the potential of coating systems with a wide range of functionalities and characteristics including that of drug release in addition to corrosion control.

Regarding products, redesign is in order with due consideration for the engineering properties and manufacturability of magnesium-based materials. For instance, appropriate stent designs must take the generally lower ductility and elastic modulus as compared to the currently used metals into account (the former affects expandability, the latter recoil). A particular concern with magnesium wrought alloys is the directionality of mechanical properties due to the microstructural textures, calling for appropriate material characterisation and product engineering tools. Due to the particular attributes of magnesium, new product opportunities arise as well. The versatility of sponges for stiffness adaptation, in-growth of tissue and drug delivery was already mentioned. Combinations of magnesium with non-degradable and/or other degradable biomaterials may provide hybrid implant solutions with dedicated functionality. Further, new means of (implant) monitoring and new sensors and actuators (e.g. for triggering the release of bioactive substances) may be developed that, for instance, use the electromagnetic properties of magnesium including conductivity and magnetic susceptibility in conjunction with its absorption. Finally, a crucial aspect in view of FDA approval for new products is the demonstration of safe use in the human body, including any conceivable long-term effects.

The obvious scenario for magnesium implants is that small and basic devices consisting of mono-materials will be introduced first (pins, staples, etc.), to be followed by larger, more complicated and hybrid products (bone grafts, fusion cages, etc.). The development of cardio-vascular stents currently seems to be somewhat ahead. Although difficult to estimate, the scope for maturing of this field is typically medium- to long-term (5–10 years) at the time of writing.

13.6 Sources of further information

As outlined in this chapter, magnesium implants are an emerging application field with a strong interdisciplinary character. Textbooks and handbooks are not yet available for this dynamically developing field, but the number of publications in a variety of medical and engineering journals and conference proceedings has steeply increased over the last few years. The reference list for this chapter, albeit a limited representation of what is actually available in the public domain, illustrates that as well. Thus, up-to-date information on research results can best be gathered from on-line literature retrieval systems. With the medical sector being highly perceptive to intellectual property, patents are another good source of information on new alloys, manufacturing processes and coatings, as well as on specific tailored designs of medical devices.

For further general orientation on the subject, review articles are an appropriate source. A well-documented introduction from the orthopaedic perspective is given by Staiger and co-workers (Staiger *et al.*, 2006). A more recent contribution by the same group summarises the various inorganic coating systems and surface treatments that are currently being considered and developed in order to better tune implant degradation (Waterman and Staiger, 2011). Xin and co-workers compare *in vitro* testing methods for magnesium alloys in a simulated physiological environment, amongst others in view of the selection of suitable test solutions (Xin *et al.*, 2010). Finally, the historical perspective with documented *in vitro* and *in vivo* information on previous experience from its beginnings and onwards is addressed by Witte (Witte, 2010).

Although there are currently several research groups from around the globe engaged in the field (with an emphasis on China, Germany, the USA and Australia), some prominent public-funded research programmes are as follows. The NSF Engineering Research Center for Revolutionizing Metallic Biomaterials is a co-operation of mainly US-based universities that is dedicated to research and training on biodegradable metals, biofunctional surface modification, sensors and controlled degradation as well as on controlled release. The so-called DFG CRC 599 ('collaborative research centre') is a co-operation of the medical and technical universities of Hanover, Germany, that targets bio-absorbable and permanent implants of metallic and ceramic materials. Magnesium is a substantial part of these programmes although their actual scope is wider.

13.7 Conclusions

Biomedical implants are an area of growing interest for magnesium. Research efforts have steadily increased over the last five years, with several research groups working in this multidisciplinary field of study. In this period the possibilities and limitations have been explored and the challenges and needs for development identified. The drivers for the development of degradable implants are obvious. With growing expectations on the quality of life, the associated healthcare costs and need to control these are a pressing societal issue. Meanwhile the market for biomedical implants – which is catered for by an international and innovation-driven industrial sector – is booming, especially in regions with an aging population.

The current state of the art can concisely be assessed in terms of its so-called technology readiness level (TRL). Within this methodology, the maturity of evolving technologies prior to incorporating them into (sub)-systems or regular practice is rated from TRL1 (basic principles observed and reported) to TRL9 (system adequacy proven in practice). In this view, magnesium-based materials for biomedical applications would have to be rated TRL3–5, meaning that research and development are directed to the demonstration of technical feasibility (proof of concept) and validation in a laboratory setting and relevant simulated or somewhat realistic environments. Cardiovascular implants are in this context somewhat further along than orthopaedic implants.

If and to what extent magnesium implants will become common practice still remains to be seen. At the time of writing this book chapter, an early generation of magnesium drug-eluting coronary stents entered a clinical study in Europe. A favourable outcome of this milestone study will be a positive signal to the stakeholders that this is a viable route for the development of degradable implants, which will, in turn, trigger further research and development. Efforts to bring such products to the market are vast but can be leveraged by the high societal and commercial stakes. Eventual success will also depend on any breakthroughs with competing biodegradable materials, notably with polymers in view of mechanical performance and with iron in view of the tuning of corrosion rate and release of corrosion products.

13.8 References

Aghion E, Yered T, Perez Y and Gueta Y (2010) 'The prospects of carrying and releasing drugs via biodegradable magnesium foam', *Adv Eng Mat*, **12**/8, B374–B379.

Apelian D (2007) 'Looking beyond the last 50 years: The future of materials science and engineering', *JOM*, **59**/2, 65–73.

Avedesian M M and Baker H (1999) *Magnesium and Magnesium Alloys – ASM Specialty Handbook*, Materials Park, OH, ASM International.

Barnett M R (2008) 'A rationale for the strong dependence of mechanical twinning on grain size', *Scripta Mat*, **59**, 696–698.

Bohlen J, Dobron P, Garcia E M, Chmelík F, Lukáč P, Letzig D and Kainer K U (2006) 'The effect of grain size on the deformation behaviour of magnesium alloys investigated by the acoustic emission technique', *Adv Eng Mat*, **8**/5, 422–427.

Deng C Z, Radhakrishnan R, Larsen S R, Boismer D A, Stinson J S *et al.* (2011) 'Magnesium alloys for bioabsorbable stents: A feasibility assessment', *Magnesium Technology 2011*, pp. 413–418, London, Wiley.

Erinc M, Sillekens W H, Mannens R G T M and Werkhoven R J (2009) 'Applicability of existing magnesium alloys as biomedical implant materials', *Magnesium Technology 2009*, pp. 209–214, London, Wiley,

Erinc M, Zhang X and Sillekens W H (2010) 'Modified AZ80 magnesium alloys for biomedical applications', *Magnesium Technology 2010*, pp. 641–646, London, Wiley.

Feyerabend F, Fischer J, Holtz J, Witte F, Willumeit R *et al.* (2009) 'Evaluation of short-term effects of rare earth and other elements used in magnesium alloys on primary cells and cell lines', *Acta Biomater*, doi: 10.1016/j.actbio.2009.09.024.

Gerold B and Müller H (2006) 'Konzept für biologisch abbaubare Implantate aus Magnesium', *4. Ranshofener Leichtmetalltage 2006*, pp. 173–183, Maschine un Werkzeug, Gilching, Germany.

Hassel Th, Bach F-W and Golovko A N (2006) 'Production and properties of small tubes made from MgCa0,8 for application as stent in biomedical science', *Proc 7th Int Conf Magnesium Alloys and their Applications*, pp. 432–437, London, Wiley.

Höh N von der, Bormann D, Lucas A, Denkena B, Hackenbroich C *et al.* (2009) 'Influence of different surface machining treatments of magnesium-based resorbable implants on the degradation behavior in rabbits', *Adv Eng Mat*, **11**/5, B47–B54.

Lalk M, Reifenrath J, Rittershaus D, Bormann D and Meyer-Lindenberg A (2010) 'Biocompatibility and degradation behaviour of degradable magnesium sponges coated with bioglass: Method establishment within the framework of a pilot study', *Mat-wiss u Werkstofftech*, **41**/12, 1025–1034.

Leeflang M A, Zhou J and Duszczyk J (2010) 'Deformability and extrusion behaviour of magnesium-lithium binary alloys for biomedical applications', *Proc 8th Int Conf Magnesium Alloys and their Applications*, pp. 1182–1188, London, Wiley.

Müller H (2008) 'Biodegradation von Magnesiumlegierungen – Ein neues Konzept für temporäre Gefäßimplantate', *5. Ranshofener Leichtmetalltage 2008*, pp. 111–123, Ranshofen, LKR.

Ralston K D and Birbilis N (2010) 'Effect of grain size on corrosion: A review', *Corrosion*, **66**/7, 075005/1–13.

Staiger M P, Pietak A M, Huadmai J and Dias G (2006) 'Magnesium and its alloys as orthopaedic biomaterials: A review', *Biomat*, **27**, 1728–1734.

Thomann M, Krause C, Angrisani N, Bormann D, Hassel T et al. (2010) 'Influence of a magnesium-fluoride coating of magnesium-based implants (MgCa0.8) on degradation in a rabbit model', *J Biomed Mater Res A*, doi: 10.1002/jbm.a.32639.

Tie D, Feyerabend F, Hort N, Willumeit R and Hoeche D (2010) 'XPS studies of magnesium surfaces after exposure to Dulbecco's modified eagle medium, Hank's buffered salt solution, and simulated body fluid', *Adv Eng Mat*, **12**/12, B699–B704.

Waterman J and Staiger M P (2011) 'Coating systems for magnesium-based biomaterials: – State of the art', *Magnesium Technology 2011*, pp. 403–408, London, Wiley.

Werkhoven R J, Sillekens W H and Lieshout J B J M van (2011) 'Processing aspects of magnesium alloy stent tube', *Magnesium Technology 2011*, pp. 419–424, London, Wiley.

Witte F, Kaese V, Haferkamp H, Switzer E, Meyer-Lindenberg A et al. (2005), '*In vivo* corrosion of four magnesium alloys and the associated bone response', *Biomat*, **26**, 3557–3563.

Witte F, Feyerabend F, Maier P, Fischer J, Störmer M et al. (2007) 'Biodegradable magnesium-hydroxyapatite metal matrix composites', *Biomat*, **28**, 2163–2174.

Witte F, Hort N, Vogt C, Cohen S, Kainer K U, Willumeit R and Feyerabend F (2008) 'Degradable biomaterials based on magnesium corrosion', *Curr Op Sol St Mater Sci*, **12**, 63–72.

Witte F (2010) 'The history of biodegradable magnesium implants: A review', *Acta Biomat*, **6**, 1680–1692.

Xin Y, Hu T and Chu P K (2010) '*In vitro* studies of biomedical magnesium alloys in simulated physiological environment: A review', *Acta Biomat*, doi: 10.1016/j.actbio.2010.12.004.

Zberg B, Uggowitzer P J and Löffler J F (2009) 'MgZnCa glasses without clinically observable hydrogen evolution for biodegradable implants', *Nat Mat*, **8**/11, 887–891.

Zheng Y and Gu X (2011) 'Research activities of biomedical Mg alloys in China', *JOM*, **63**/4, 105–108.

Index

AA6063 alloy, 306
active sheet hydroforming, 411
alloy development, 398, 400–8
   commercial extrusion alloys, 398, 400–8
      nominal composition and tensile properties, 401
   commercial sheet alloys, 405–6
      nominal composition and typical tensile properties, 406
   new extrusion alloys, 401–5
      AM30 alloy, 401–3
      Mg-Zn-Ce (ZE) alloys, 403–5
   new sheet alloys, 406–8
      Mg-Al-based alloys microalloying, 408
      Mg-Li-based alloys, 406–7
      Mg-Zn-Ce (ZE) alloys, 408
alloying, 133–6
ALSIM software, 299
AM30 alloy, 11–13, 401–3
   AM30 and AZ31 alloys extrusion limit diagram, 402
   extruded tubes tension-compression yield ratios, 13
   extrusion limit diagram, strain hardening exponent and rate, 12–13
   temperature effect on tensile properties, 403
AM50+Ce alloy, 11–13
ANSYS Fluent, 295
ASTM B93-94a, 402
asymmetrical sheet processing, 367–8
automotive body, 415–17
   Buick LeSabre, Chevrolet Corvette and Chevrolet Corvette SS Race Car, 416
   Mg sheet panels, 416
automotive engineering
   future trends, 417–21
      aluminium isolator locations, 420
      corrosion and surface finishing, 419–21
      crashworthiness, 418–19
      fatigue and durability, 419
      material challenges, 418
      metals standard reduction potential, 420
      noise, vibration and harshness, 419
      performance challenges, 418–21
      process challenges, 418
      production sedan front end structure, 422
   magnesium alloys applications, 393–422
      alloy development, 398, 400–8
      manufacturing process development, 408–13
   materials properties, 394–8
      mechanical vs physical properties of automotive materials, 395–6
      Mg alloy vs other automotive materials, 394–7
      structural performance and mass-saving potential, 397–8
      thickness and mass ratios, 399
      various materials vs structural beam steel mass saving percentage, 400
      various materials vs structural panel steel mass saving percentage, 400
automotive interior, 413
Avitzur's solution, 332
AZ31 alloy, 286, 306–7, 336, 338–9, 351–2, 354, 364, 436–7
AZ61 alloy, 318, 338
AZ80 alloy, 338–9, 386–7, 436, 438
AZ31 alloys, 19, 108, 112
AZ31-3Li alloys, 19–20
AZ61alloy, 364

B-36 bomber, 413
basal slip, 65–72
   grain size effects, 70–1
   order strengthening, 68–9
   precipitate strengthening, 69–70
   solute strengthening, 67–8
   stacking fault energy, 66
   texture effects, 71–2
   thermal activation, 66–7
Bessimer, 274
bi-modal grain
   structure and twinning, 151–7
      AZ31 microstructure after six passes of warm ECAP, 157

large grains in the microstructure of ZK60, 153
quantitative characterisation of ZK60 grain structure, 154
ZK60 microstructure after eight passes of warm ECAP, 156
ZK60 microstructure after homogenisation, 152
ZK60 microstructure after warm ECAP, 151
bioglass, 449
biomedical applications, 427–52
  biomedical implants trends, 429
  cardiovascular implants, 440–5
  future trends, 449–51
    R & D issues for Mg, 450
  magnesium implants, 429–40
    degradation characteristics and tuning, 433–40
    requirements and testing, 429–33
  orthopaedic and other implants, 445–9
  published works from SciVerse Scopus literature database, 428
Buick LeSabre, 415
bulk metallic glasses, 439
Burgers vector, 72, 78, 92, 189, 205, 207, 208, 215

C-5 Corvette, 415
Cadillac XLR roadster, 415
CANMET configuration, 278–9
cardiovascular implants, 440–5
  biodegradable stent tube requirements, 441
  ongoing research and development, 442–5
    AZ80 alloy microstructural evolution, 443
    Mg stent tube microstructural characterisation, 444
  state of the art, 441–2
    Mg stent photo, 442
cardiovascular stent, 440
casting parameters, 281–2
  casting speed, 281
  liquid metal temperature, 281
  Mg alloys cast using TRC, 281–2
Castrip, 274
chassis, 417
Chevrolet Corvette, 415
Chevrolet Corvette SS Race Car, 415
Chevrolet Corvette Z06, 417
chill zone, 283
Chrysler LLC, 415
coincident site lattice (CSL), 205
columnar dendrites, 283
  microstructures, 284
composite, 81–2
compressive stress, 126
continuous DRX (CDRX), 188
  magnesium and magnesium alloys, 189–99
    continuous reaction resulting in scale grains formation, 198–9

mechanism with balanced contribution of basal and none-basal slips, 193–4
mechanism with basal slip predomination, 191–2
microhardness strain dependence of deformed ZK60 alloy, 197
ZK60 alloy flow curves, 196
conventional extrusion processes, 408–9
conventional rolling see sheet rolling
core structure, 75, 79–80
coronary stent see cardiovascular stent
cover gas systems, 245
cracking, 256–61
  inflatable wipes, 260
  ingot surface reheating, 260
  type of cold crack, 258
  wipers application, 259
critical grain size
  twinning, 148–51
    XK60 and AZ31 stress dependence and dislocation glide stress, 150
critical resolved shear stresses (CRSS), 189
crystallographic texture, 71
crystallography, 112–16
  rolled magnesium twinning, 116
  twinning elements relation to plane shear, 115
  twinning modes and interface dislocation, 114
  twinning showing the magnesium unit cell, 115

Darcy's law of momentum, 294
Deform, 382
deformation
  basal slip, 65–72
  dynamic, 90–2
  magnesium alloys, 63–93
  non-basal slip of <a> type dislocations, 72–7
  non-basal slip of <c+a> type dislocations, 78–83
deformation temperature, 314
deformation twinning, 83–90, 378
  fracture, 89
  grain size effects, 88–9
  kink banding, 89–90
  latent hardening, 85–6
  polycrystal deformation, 83–5
  precipitation effects, 87–8
  solid solution alloy effects, 86–7
differential speed rolling (DSR), 368
direct chill casting, 410
  heat and fluid flow, 235–45
  magnesium extrusion billet and rolling slab, 229–61
  overview, 229–34
    early open mould VDC casting for aluminium or magnesium, 230
    HDC magnesium caster, 234
    magnesium fully continuous DC casting process, 233

Index

magnesium with steel trough and tilting furnace, 231
1950's magnesium DC casting technology, 232
solidified structures and defect formation, 249–61
  alloy type effect on heat flow, stresses and microstructure, 261
  cracking, 256–61
  grain structure, 249
  surface defects, 249–56
technology and engineering, 245–9
  cover gas systems, 245
  electromagnetic DC casting, 249
  gas-pressurised hot top mould, 247–8
  hot top mould development, 247
  melt delivery systems, 245
  open top mould casting, 245–7
  rolling slab casting, 248–9
direct chill ingot casting route, 272
direct extrusion, 304, 323
discontinuous DRX (DDRX), 188
  magnesium and magnesium alloys, 199–202
  flow curve for occurrence and deformed microstructure, 203
  mechanism, 200
  misorientation map indicating DDRX nuclei formation, 201
discontinuous grain boundary nucleation mechanism, 358
dislocation, 129–30
door stop fitting, 386–8
  forging tool, 387
  specimen from AZ80 alloy, 388
double twin (Type I) boundaries, 116
Dow Chemical, 274
drop forging, 377–8
ductility, 124
dynamic deformation, 90–2
dynamic recovery, 309
dynamic recrystallisation, 309, 350–1
  initial structure effect, 207–11
    crystallographic texture on dislocation percentage and dependence, 210
    cutting rod specimen, 209
    Zener-Hollomon parameter effect, 211
    ZK60A alloy true-stress strain curves, 210
  magnesium alloys, 186–219, 211–15
    Mg-Al-Zn alloys, 212–14
    Mg-Zn-Zr alloys, 214–15
    pure magnesium, 211–12
  operating mechanism, 187–207
    CDRX in magnesium and magnesium alloys, 189–99
    DDRX in magnesium and magnesium alloys, 199–202
    TDRX in magnesium and magnesium alloys, 202–7
  plastic deformation, 215–18

electromagnetic DC casting, 249
electron backscattering diffraction (EBSD), 364
energy dispersive X-ray spectroscopy (EDS) analysis, 438
equal channel angular pressing (ECAP), 368, 443–4
  DRX, 216–17
equal channel angular rolling (ECAR), 368
European Commission, 386
exterior composites, 397
extrudability, 6
extrusion billet
  direct chill casting of magnesium and rolling slab, 229–61
  heat and fluid flow, 235–45
  overview, 229–34
  solidified structures and defect formation, 249–61
  technology and engineering, 245–9
extrusion limit diagrams, 305
Extrusion Research and Development Centre, 333
extrusion speed, 336–8
  Mg alloy bar mechanical properties upon different extrusion methods, 338
  Mg alloy bar micrographs and billets macrographs, 337
  Mg alloy test sections data, 337
extrusion temperature, 337–42
  Mg alloy bar inverse pole figures, 340
  Mg alloy bar mechanical properties at different billet temperatures, 341
  Mg alloy bar micrographs upon different extrusion methods and temperatures, 339

FactSage thermochemical software, 285
fatigue, 124–5
fine-structure superplasticity (FSS), 146
finite element method (FEM), 380–2
flash gap, 378
fluid flow, 235–45
  mould cooling, 240–2
  pool depth development measured by dip rod, 238
  solidification front shape in hot top VDC, 235
  sump and shell shape, 236
  water cooling, 242–5
Ford F-150, 415
forging processes, 409
forging technology
  cast, machined and drop forged parts schematic, 377
  drop forging schematic, 377
  magnesium alloys, 376–88
    deformation twins principle, 378
    principles and forming behaviour, 378–9
    process design, 379–84
    magnesium wrought alloys for extrusion and forgings, 379
    material characterisation, 380

process chain, 382–4
  servo-hydraulic forming simulator, 381
  simulation-based forging process by FEA, 380–2
forming limit curves, 362
forming limit diagram, 362
fracture, 89
  superplastic deformation, 168–75
    fibres fibrous structure and ductile rupture, 173
    secondary fibres spacing with Mn_5Al_8 precipitate size, 172
    superplastically deformed AZ31, 171
    terrace-like morphology of fracture surface, 174
    texture in ECAP processed specimen, 174
fracture toughness, 124
friction load, 255
friction stir welding, 217–18
fusion welding process, 365

galvanic corrosion, 365
gas metal arc welding, 412
gas-pressurised hot top mould, 247–8
gas tungsten arc welding, 412
geometrically necessity boundaries (GNB), 190
grain boundary strengthening, 70
grain size, 70–1, 77, 88–9, 131–3
  influence on yield and flow stress, 132
grain structure, 249
graphite, 254

Hall–Petch relationship, 315–16, 335, 442
hammer forging see open die forming
heat flow, 235–45
  mould cooling, 240–2
  pool depth development measured by dip rod, 238
  solidification front shape in hot top VDC, 235
  sump and shell shape, 236
  water cooling, 242–5
heat transfer coefficient, 296–8
  expected change schematic, 297
  metal/roll interface diagram, 298
hexagonal close-packed (hcp) structure, 272
high-angle grain boundaries (HABs), 188, 190, 192, 194, 200–2
high strain rate superplasticity (HSRS), 146
horizontal direct chill (HDC) casting, 232
horizontal TRC configuration, 277
hot cracking, 323–4
hot shortness see hot cracking
hot top mould, 247
Hunter, 274
Hydrex Materials, 336
hydro-mechanical drawing, 411
hydrogen evolution test, 431
hydrostatic extrusion, 305, 323–44, 409
  future trends, 342–3

Mg alloys characteristic features, 344
Mg alloys direct extrusion workability, 324
principle of, 325–30
  hollow sections schematic, 326
  hydrostatic extrusion press, 328
  hydrostatically extruded Mg alloy sections, 327
  Mg alloys surface imperfections, 329
  solid sections schematic, 325
process analysis, 330–3
  estimated thermal effect of Mg alloy bar, 332
  extrusion process limits schematic, 331
  steady-state extrusion upper-bound model, 330
process basics, 324–33
research and development issues, 333–42
  extrusion speed, 336–8
  extrusion temperature, 338–42
  indirect vs hydrostatic extrusion, 333–6

indirect extrusion, 305
  vs hydrostatic extrusion, 333–6
  Mg alloy bar micrographs upon different extrusion methods, 334
  Mg alloy bar properties upon different extrusion methods, 335
Institute of Forming Technology and Machines, 380, 386
integrated computational materials engineering (ICME), 421
internal-stress superplasticity (ISS), 146
inverse segregation, 284
IsoCAST, 246

joining process, 412–13

kink banding, 89–90

latent hardening, 82–3, 85–6
localised corrosion, 435
low-angle boundaries (LABs), 188, 190, 192, 194, 200–2
low temperature DRX (LTDRX), 196–7

MagForming project, 386
magnesium
  continuous DRX (CDRX), 189–99
  direct chill casting of extrusion billet and rolling slab, 229–61
    heat and fluid flow, 235–45
    overview, 229–34
    solidified structures and defect formation, 249–61
    technology and engineering, 245–9
  discontinuous DRX (DDRX), 199–202
  twin roll casting (TRC), 272–300
    industrial perspective, 274–6
    process, 276–82
    process modelling and simulation, 292–9

solidification and strip microstructure, 282–4
thermodynamic calculations, 284–92
twinning DRX (TDRX), 202–7
magnesium alloys
alloy development, 398–408, 400–8
commercial extrusion alloys, 398, 400–408
commercial sheet alloys, 405–6
new extrusion alloys, 401–5
new sheet alloys, 406–8
alloying effects on sheet properties, 354–62
as rolled binary Mg sheets pole figures and max intensities, 359
conventional Mg sheets mechanical properties, 355
H24-temper AZ31 sheet pole figure intensity, 358
mechanical properties orientation dependence, 356
Mg sheets pole figures from texture measurements, 357
ZEK100 texture development, 361
automotive engineering, 393–422
current and future applications, 413–17
historical applications, 413
materials properties, 394–8
Porsche Carrera GT console cover, 414
wrought Mg components applications, 414
deformation mechanism, 63–93
basal slip, 65–72
deformation twinning, 83–90
dynamic deformation, 90–2
non-basal slip of <c+a> type dislocations, 78–83
non-basal slip of <a> type dislocations, 72–7
dynamic recrystallisation (DRX), 186–219, 211–15
initial structure effect, 207–11
operating mechanism, 187–207
plastic deformation, 215–18
forging technology, 376–88
important specimens, 379
manufacturing process development, 408–13
extrusion and forging processes, 408–9
sheet production and forming processes, 410–12
tube bending and hydroforming processes, 409–10
welding and joining techniques, 412–13
near net shape forming, 384–8
examples, 385–8
rolling of, 346–70
sheet formability, 362–4
Mg sheet forming limit curves at various temperatures, 363
sheets potential, 347–8
superplasticity by plastic deformation, 144–79

fracture during superplastic deformation, 168–75
mechanisms and models, 175–8
microstructure evolution during thermomechanical processing, 147–57
overview, 144–7
superplastic behaviour, 157–68
magnesium crystals, 112
Magnesium Front End Research & Development project, 418, 421
magnesium implants, 429–40
degradation characteristics and tuning, 433–40
biomedical Mg alloys, 439
magnesium pins *in vivo* degradation, 437
Mg bar corrosive attack in physiological environment, 436
Mg corrosion resistance in physiological environment, 434
ZK60 alloy corrosion resistance in physiological environment, 435
requirements and testing, 429–33
alloying elements selection for biomedical Mg alloys, 432
degradable implants targeted mechanical behaviour, 430
M1 alloy, 336, 364
'materials performance index' concept, 397
maximum major strain, 362
melt delivery systems, 245
Mercedes S-Class Coupe, 415
metadynamic recrystallisation, 312
metal matrix composites, 446–7
metro-lite trucks, 413
Mg-Al alloys, 213–14
Mg-Al-Zn alloys
DRX, 212–14
Mg-Li alloys, 406–7
Al and/or Zn, 18–23
alloys rolled 15 passes at 150°C, 23
axial ratio, grain structure/size, edge cracking index and texture intensity, 25
AZ31, Mg-2Li-1Zn microstructure, 25–6
AZ31 and Mg-2Li-1Zn tension twin, 24
edge cracking index *vs.* c/a, 23
effect of lithium on Mg, 21
grain size effect, 27
Mg-1Li, Mg-3Li, AZ31 and AZ31-3Ll alloys textures, 20
recrystallisation of AZ31, AZ31-3Li and Mg-1Li alloys, 19
microstructure and tensile properties, 407
phase diagram, 407
rare-earth additions, 23, 28–31
composition and tensile properties of Mg-Li-Zn-Y alloys, 29
extruded alloy microstructure, 31
mechanical properties of Mg-Li-Al-Zn-RE *vs.* AZ31, 28

Mg-Li-Al-Zn-RE alloys compositions, 28
Mg-Li-Zn-Y alloys microstructure, 30
Mg-1Li alloys, 18–20
Mg-2Li alloys, 18, 22
Mg-3Li alloys, 18–19
Mg-Mn alloys, 31–8
  4 min. of annealing of MJ11, 37
  EBSD maps of deformed M16 alloy, 33
  illustration, 32
  inverse pole figures from extruded bars, 35
  longitudinal sections of extruded round bars, 34
  Mg-Mn-Ce alloy composition, 34
  Mg-Mn-Re alloy compositions, 36
  MJ alloy compositions, 36
  surface texture pole figures of alloys, 37
  true strain deformation of M03 alloy, 34
Mg-Mn-Ce alloys, 32
Mg-1Mn (M1) alloy, 35
Mg-1Mn-xSr (MJ) alloys, 35
Mg-rare-earth alloys, 13–18
  EBSD map of Mg-Gd and recrystallised region, 14
  Mg alloys with various rare-earth elements, 15–16
  IPF of GW123, 17
  Mg-Re wrought alloys mechanical properties, 16
Mg-RE-Sn alloys, 16–18
  Al2MM phase in ETA221 alloy, 18
Mg-Zn-alkaline earth alloys, 49–55
  3 min. annealing of ZJ11, 55
  density of basal planes *vs.* deviation angle, 55
  extruded and T6 treated Mg-6Zn-0.4Ag-0.2Ca alloys, 51
  extruded Mg-Zn-Ca alloys, 51
  hardening response of extruded alloys, 52
  rolled/annealed condition alloys microstructure, 54
  rolled condition alloys microstructure, 53
Mg-Zn-Ce (ZE) alloys, 403–5, 408
  pure Mg and Mg-0.2%Ce alloy tensile curves after extrusion, 404
  vs AZ31 and Mg-0.2%Ce alloy tensile properties, 405
Mg-Zn-RE alloys, 38–49
  alloy composition and grain size, 41
  As-cast Mg-4Zn-1Y and Mg-4Zn-1Y-0.2Al, 49
  basal pole figures intensity *vs.* tilt, 40
  cast ZEK100 alloys, 43
  extruded ZK60 alloys, 48
  Mg-Gd-Zn sheets mechanical response, 47
  micro and macro features of alloys, 42
  microstructures after warm rolling and annealing, 44
  particle-stimulated nucleation (PSN), 46
  RE elements on texture evolution during warm rolling, 45
  rolled alloy texture measurements, 39
  rolled and annealed ZE10, 41
  rolled Mg-4Zn-1Y and Mg-4Zn-1Y-0.2Al alloys microstructures, 50
  sheet anisotropy, 46
  TEM images of Mg-4Zn-1Zr-0.2Al, 50
  ZEK100 alloys chemical compositions, 42
Mg-Zn-Zr alloys
  DRX, 214–15
micro-alloying, 319
microstructure
  evolution during thermomechanical processing, 147–57
    published data on superplasticity of ZK60 and AZ31 alloys, 148
    twinning and bi-modal grain structure, 151–7
    twinning and critical grain size, 148–51
minimum edge cracking, 5–6
  twinning shear *vs.* axial ratio, 5
minimum major strain, 362
Mitsubishi Heavy industries, 274
12 MN ASEA hydrostatic extrusion press, 328, 333, 338
modified AZ31 alloy, 6–11
  EBSD of recrystallised grains, 10
  grain structure, 8
  micro-mechanism map during extrusion, 10
  prism planes and texture measurement locations, 9
  yield stress asymmetry, 7
modified AZ61 alloy, 11
mould chill zone (MCZ), 241
mould cooling, 240–2
  shell region measured temperatures, 242
MSC.MARC, 382

near net shape forming, 384–8
  examples, 385–8
    door stop fitting, 386–8
    pulley wheels, 385–6
Nippon steel, 274
non-basal slip <c+a> type dislocations, 78–83
  core structure, 79–80
  latent hardening, 82–3
  polycrystal deformation, 78–9
  precipitation and composite effects, 81–2
  solute effects, 81
  thermal activation, 80–1
non-basal slip <a> type dislocations, 72–7
  core structure, 75
  grain size effects, 77
  polycrystal deformation, 72–4
  precipitation strengthening, 77
  solute strengthening and softening, 75–7
  thermal activation, 74–5
nucleation, 117–18
nucleation stress, 128

open die forming, 377
open top mould casting, 245–7

Index

hot top casting configuration for magnesium VDC casting, 246
open top casting set up for magnesium, 246
order strengthening, 68–9
orifice plate, 252
orthopaedic implants, 445–9
　research and development, 447–9
　　bioglass-coated AX30 sponges *in vivo* degradation, 449
　　MgCa0.8 fluoride-coated pins *in vivo* degradation, 448
　　MgCa0.8 implants *in vivo* gas generation, 447
　state of the art, 445–7
　　implant materials vs natural bone design properties, 445
　　Mg-based or hard-tissue implants candidates, 446

Pandat software, 285
particle stimulated nucleation, 360
Pechiney, 274
phenomenology, 107–12
　extruded alloy AZ31 applied stress normalisation by Schmid factors, 108–9
　finite element simulation, 111
pinch roll, 281
pitting corrosion, 435
planar anisotropy, 355
plastic deformation, 215–18
　DRX under equal channel angular pressing, 216–17
　DRX under friction stir welding, 217–18
　fracture during superplastic deformation, 168–75
　mechanisms and models, 175–8
　　Arrhenius plot for AZ31 activation energy determination, 177
　　AZ31 microstructure after tensile test, 178
　　AZ60 microstructure after tensile test, 178
　　parameters characterising deformation in superplasticity, 175
　　scribed line on ZK60 surface, 176
　microstructure evolution during thermomechanical processing, 147–57
　overview, 144–7
　superplastic behaviour, 157–68
　superplasticity in magnesium alloys, 144–79
polycrystal deformation, 72–4, 78–9, 83–5
POSCO, 275–6
potentiodynamic polarisation test, 431
precipitation, 81–2, 87–8
precipitation strengthening, 69–70, 77
precision forging, 384
ProCAST, 295
pulley wheels, 385–6
　production tool concept, 385
　specimens from AZ31 and AZ61 alloys, 385
pure magnesium, 211–12

quick plastic forming, 412

rare earth additions, 319, 356, 358
'rare earth texture,' 319
recrystallisation, 125
roll setback, 280
rolling
　magnesium alloys, 346–70
　　asymmetrical sheet processing, 367–8
　　enabling technologies and future trends, 365–8
　　magnesium sheets and parts post-processing, 365–6
　　twin roll casting, 366–7
　sheet rolling process and influence on sheet properties, 348–54
　　average grain sizes at different temperatures, 351
　　laboratory rolled ZM21 sheets pole figures, 352
　　process parameters and sheet properties, 350–4
　　sheet rolling, 348–50
rolling slab
　casting, 248–9
　direct chill casting of magnesium extrusion billet, 229–61
　　heat and fluid flow, 235–45
　　overview, 229–34
　　solidified structures and defect formation, 249–61
　　technology and engineering, 245–9
rotationally symmetrical process model, 330

Scheil cooling calculation, 286–7, 290
shear stress, 126
sheet forming processes, 410–12
　forming trials on 125 mm deep pan, 411
sheet production processes, 410
sheet rolling, 348–50
　cast Mg alloys heat treatment, 349
　typical rolling temperatures, 350
Simufact.forming, 382
solid solution alloying, 86–7
solidification microstructure, 282–4
　mid-width solidification microstructure, 283
solute
　effects, 81
　softening, 75–7
　strengthening, 67–8, 75–7
stacking fault energy, 66
static recrystallisation, 186, 353
stent, 440
'Stow-n-Go' system, 415
strain, 129–30
strain rate, 130
stress analysis, 298–9
　stress and fraction solid model, 299
stress–strain curve, 126
structural biocompatibility, 430

structural composites, 394
superheat, 281
superplastic behaviour, 157–68
   AZ31 microstructure processed by six ECAP passes, 164
   AZ31 tensile test processed by one- and two-step deformation, 167
   bi-modal structure formed after six passes of ECAP, 159
   strain rate sensitivity parameter, 165
   strain vs. strain curve, 160–1
   tensile samples and geometry after superplastic tensile deformation, 158
   ZK60 microstructure after rolling and six passes of ECAP, 165
   ZK60 microstructure processed by six ECAP passes, 162–3
   ZK60 tensile test processed by one-step deformation, 166
superplastic deformation
   fracture, 168–75
     fibres fibrous structure and ductile rupture, 173
     secondary fibres spacing with Mn_5Al_8 precipitate size, 172
     superplastically deformed AZ31, 171
     terrace-like morphology of fracture surface, 174
     texture in ECAP processed specimen, 174
superplastic ductility, 157
superplastic forming, 412
superplasticity
   fracture during superplastic deformation, 168–75
   magnesium alloys by plastic deformation, 144–79
   mechanisms and models, 175–8
     Arrhenius plot for AZ31 activation energy determination, 177
     AZ31 microstructure after tensile test, 178
     AZ60 microstructure after tensile test, 178
     parameters characterising deformation in superplasticity, 175
     scribed line on ZK60 surface, 176
   microstructure evolution during thermomechanical processing, 147–57
   overview, 144–7
     published data of magnesium wrought alloys, 145
   superplastic behaviour, 157–68
surface defects, 249–56
   orifice plate geometry arrangements, 253
   steps in cold fold formation during HDC casting, 251
   steps in cold fold formation during HDC casting with OP overhang, 253

Taylor criterion, 65
technology readiness level, 344, 452
temperature, 130–1
temperature control, 383
texture, 71–2, 125–9
thermal activation, 66–7, 74–5, 80–1
Transvalor Forge, 382
tube bending, 409–10
   bent Mg AM30 and AZ31tubes, 410
tube hydroforming, 409
twin axis symmetry, 378
twin roll casting, 366–7
   AZ31 alloy rolling materials microstructures, 367
   industrial perspective, 274–6
     aluminium strip casting technology historical evolution, 274
     as-cast strip AZ31 alloy coil, 276
     research and development activities, 275
     strip width evolution, 275
   magnesium, 272–300
   process, 276–82
     casting parameters, 281–2
     components, 278–81
     equipment layout, 279
     solidification microstructure, 282–4
     strip defects, 284
     twin roll caster photo, 277
   process modelling and simulation, 292–9
     basic phenomena and modelling, 293
     boundary conditions, 295–6
     fluid flow, 293–5
     stress analysis, 298–9
     strip and roll heat transfer coefficient, 296–8
     validation, 299
   solidification and strip microstructure, 282–4
     solidification microstructure, 282–4
     strip defects, 284
   thermodynamic calculations, 284–92
     as-TRC microstructure calculations, 286–91
     AZ31 alloy phase distribution, 287
     AZ31 alloy solute and solidification temperature profile, 288–9
     Mg-Al-1%Zn-0.3%Mn alloy phase diagram, 285
     new wrought alloys development, 291–2
     Scheil cooling scheme phase distribution, 290
     solidification schematic, 288
   traditional route vs TRC process, 273
   vs conventional rolling diagram, 366
twin-roll continuous casting process, 410
twinning
   bi-modal grain structure, 151–7
   critical grain size, 148–51
     XK60 and AZ31stress dependence and dislocation glide stress, 150
   fundamentals, 107–18
     crystallography, 112–16
     nucleation and growth mechanism, 117–18
     phenomenology, 107–12

mechanical response impact, 118–25
 ductility, 124
 fatigue, 124–5
 fracture toughness, 124
 recrystallisation, 125
 work hardening, 120–4
 yield phenomena, 118–20
overview, 105–7
 AZ31 alloy plot transition between slip and dominated deformation, 107
 AZ31 alloy true stress-strain curves, 106
structure and processing impact, 125–36
 alloying, 133–6
 grain size, 131–3
 strain and dislocation density, 129–30
 strain rate, 130
 temperature, 130–1
 texture and stress, 125–9
wrought magnesium alloys, 105–36
twinning DRX (TDRX), 188
magnesium and magnesium alloys, 202–7
 AZ91 alloy nucleation operating mechanism, 206
 mechanism, 204–5
two-step heat treatment, 291

ultra-fine grained (UFG), 187
upper-bound model, 330
US Food and Drug Administration (FDA), 433

vertical direct chill (VDC) casting, 232
Von Mises criterion, 65

water cooling, 242–5
 boiling regimes in DC casting, 243
 water spray configuration and key features, 244
WE43 alloy, 433–4, 438
weakened texture, 4
weight-loss method, 431
welding process, 412
work hardening, 120–4
 stress-strain curve for AZ31 alloy during cycle of deformation, 123
 stress-strain curve for extruded AZ31 alloy, 122
 stress-strain curve from pure magnesium crystals, 121
wrought magnesium alloys, 3–59
 current developments, 6–31
  lithium-containing alloys, 18–31
  Mg-rare-earth alloys, 13–18
  modified Mg-Al alloys, 6–13
 extrudability, 305–9
  as-cast AZ31 extrusion limits, 305
  commercial AA6063 and magnesium alloys extrusion limits, 307
  commercial magnesium alloys compression yield strength, 308
  extrusion limits, 305–6
  mechanical property trade off, 308–9
  role of composition, 306–8
 extrudability enhancement, 304–21
 extrusion alloy development, 319–20
 future trends, 55–9
  current development and programs, 56–8
 microstructural development during extrusion, 309–19
  AZ31, M1 and ME10 alloys texture, 320
  development after deformation, 311–13
  development during hot deformation, 309–11
  extruded M1 billets microstructures and flow curves, 316
  grain size prediction of extrudates, 313–17
  hot deformed Mg alloys grain size range, 318
  partially extruded AZ31 billet microstructures and FEM simulation, 310
  partially extruded AZ31 orientation map, 311
  process predictions incorporation over extrusion limit diagram, 315
  role of composition, 317–19
  Zener-Hollomon parameter influence, 313
 overview, 3–6
  extrudability, 6
  minimum edge cracking, 5–6
  requirements, 4
  yield symmetry and weakened texture, 4
 progress, 31–55
  Mg-Mn alloys, 31–8
  Mg-Zn alloys, 38–55
 twinning, 105–36
  fundamentals, 107–18
  mechanical response impact, 118–25
  overview, 105–7
  structure and processing impact, 125–36

yield phenomena, 118–20
 asymmetric experimental yield surface for ZM61 alloy, 119
yield stress, 127
yield symmetry, 4

ZE10 alloy, 364
ZE20 alloy, 404
Zener-Hollomon parameter, 312
ZK60 alloy, 400–1, 434, 436
ZM20 alloy, 315
ZM21 alloy, 336, 438

图书在版编目（ＣＩＰ）数据

变形镁合金研究进展：加工原理、性能和应用：英文／（澳）科琳·贝特尔（Colleen Bettles），（澳）马修·巴尼特（Matthew Barnett）编著. --长沙：中南大学出版社，2017.10

ISBN 978-7-5487-2991-4

Ⅰ.①变… Ⅱ.①科… ②马… Ⅲ.①镁合金－研究－英文 Ⅳ.①TG146.22

中国版本图书馆 CIP 数据核字（2017）第 230377 号

变形镁合金研究进展：加工原理、性能和应用
BIANXING MEIHEJIN YANJIU JINZHAN：JIAGONGYUANLI、XINGNENG HE YINGYONG

Colleen Bettles　　Matthew Barnett　　编著

□责任编辑	史海燕	
□责任印制	易红卫	
□出版发行	中南大学出版社	
	社址：长沙市麓山南路	邮编：410083
	发行科电话：0731-88876770	传真：0731-88710482
□印　　装	长沙超峰印刷有限公司	
□开　　本	720×1000　1/16　□印张 30.25　□字数 766 千字	
□版　　次	2017 年 10 月第 1 版　□2017 年 10 月第 1 次印刷	
□书　　号	ISBN 978-7-5487-2991-4	
□定　　价	150.00 元	

图书出现印装问题，请与经销商调换